Middleware for Communications

Middleware for Communications

Edited by

Qusay H. Mahmoud
University of Guelph, Canada

John Wiley & Sons, Ltd

Telephone (+44) 1243 779777

Email (for orders and customer service enquiries): cs-books@wiley.co.uk
Visit our Home Page on www.wileyeurope.com or www.wiley.com

Other Wiley Editorial Offices

John Wiley & Sons Inc., 111 River Street, Hoboken, NJ 07030, USA

Jossey-Bass, 989 Market Street, San Francisco, CA 94103-1741, USA

Wiley-VCH Verlag GmbH, Boschstr. 12, D-69469 Weinheim, Germany

John Wiley & Sons Australia Ltd, 33 Park Road, Milton, Queensland 4064, Australia

John Wiley & Sons (Asia) Pte Ltd, 2 Clementi Loop #02-01, Jin Xing Distripark, Singapore 129809

John Wiley & Sons Canada Ltd, 22 Worcester Road, Etobicoke, Ontario, Canada M9W 1L1

Wiley also publishes its books in a variety of electronic formats. Some content that appears in print may not be available in electronic books.

British Library Cataloguing in Publication Data

A catalogue record for this book is available from the British Library

ISBN 0-470-86206-8

Typeset in 10.25/12pt Times by Laserwords Private Limited, Chennai, India

This book is printed on acid-free paper responsibly manufactured from sustainable forestry in which at least two trees are planted for each one used for paper production.

To all those who helped in the creation of this book in one way or another

Contents

Preface

Middleware has emerged as a critical part of the information technology infrastructure. The need for it stems from the increasing growth in network-based applications as well as the heterogeneity of the network computing environment. Communications systems such as computer and telecommunications networks are composed of a collection of heterogeneous devices whose applications need to interact. Middleware systems are used to mask heterogeneity of applications that are made up of several distributed parts running on different computer networks and telecommunications systems. In addition, middleware systems provide value-added services such as naming and transaction services, as well as tools and APIs that offer uniform high-level interfaces to application developers so that applications can be easily constructed. The importance of middleware will continue to grow as long as computing and communications systems continue to be heterogeneous.

All existing books on middleware concentrate on a specific area: fundamentals of middleware, comparing some middleware technologies such as CORBA, RMI, and DCOM, or surveying middleware products that exist on the market. The aim of this book is to fill the gap by providing a state-of-the-art guide to middleware. The book covers all aspects of middleware by including chapters on concepts and fundamentals for beginners to get started, advanced topics, research-oriented chapters, and case studies.

This book provides convenient access to a collection of exemplars illustrating the diversity of communications problems being addressed by middleware technology today and offers an unfolding perspective on current trends in middleware. The lessons learned and issues raised in this book pave the way toward the exciting developments of next generation middleware.

Audience

This book is aimed at students, researchers, and practitioners. It may be used in undergraduate and graduate courses on middleware. Researchers will find the book useful as it provides a state-of-the-art guide to middleware technology, and offers an unfolding perspective on current and future trends in middleware. Practitioners will find the book useful as a means of updating their knowledge on particular topics such as Message-Oriented Middleware (MOM), Adaptive and Reflective Middleware, Transaction Middleware, Middleware for Mobile Computing, Middleware for Peer-to-Peer Systems, QoS-enabled Middleware,

Grid Middleware, Model Driven Middleware, Real-time Middleware, Middleware for Smart Cards, Middleware Performance, and Middleware Security.

Acknowledgments

A large number of people have contributed to this book and I would like to thank them all. First of all, I am deeply grateful to my editors Birgit Gruber and Sally Mortimore for providing me with the opportunity to edit this book but more importantly for providing me with comments, suggestions, and guidelines over the course of its production.

This book would not exist without the authors of the individual chapters that make up this book. I would like to thank all the authors of the individual chapters without whom this book would not have been possible. Also, I am grateful to the anonymous reviewers for the terrific job they did in evaluating and recommending chapters.

Finally, I would like to thank my wife, Reema, and son Yusef for putting up with my strenuous schedule over the past several months.

Qusay H. Mahmoud
Toronto, Canada
June 2004

Contributors

Qusay H. Mahmoud
Department of Computing & Information Science
University of Guelph, Guelph, ON, N1G 2W1
Canada
qmahmoud@cis.uoguelph.ca

Edward Curry
Department of Information Technology
National University of Ireland, Galway
Ireland
edward.curry@nuigalway.ie

Stefan Tai
IBM T.J. Watson Research Center
P.O. Box 704
Yorktown Heights, NY 10598
USA
stai@us.ibm.com

Thomas Mikalsen
IBM T.J. Watson Research Center
P.O. Box 704
Yorktown Heights, NY 10598
USA
tommi@us.ibm.com

Isabelle Rouvellou
IBM T.J. Watson Research Center
P.O. Box 704
Yorktown Heights, NY 10598
USA
rouvellou@us.ibm.com

Markus Oliver Junginger
Germersheimer Str. 33
81541 München
Germany
markus@junginger.biz

Yugyung Lee
School of Computing & Engineering
University of Missouri, Kansas City
5100 Rockhill Rd.
Kansas City, MO 64110
USA
leeyu@umkc.edu

Gregor von Laszewski
Argonne National Laboratories & University of Chicago
9700 S. Cass Ave.
Argonne, IL 60439
USA
gregor@mcs.anl.gov

Kaizar Amin
Argonne National Laboratories & University of North Texas
9700 S. Cass Ave.
Argonne, IL 60439
USA
amin@mcs.anl.gov

Nanbor Wang
5541 Central Ave., Suite 135
Boulder, CO 80301
USA
nanbor@cs.wustl.edu

Christopher D. Gill
Dept. of Computer Science & Engineering
Washington University
One Brookings Drive
St. Louis, MO 63130
USA
cdgill@cse.wustl.edu

Douglas C. Schmidt
Institute for Software Integrated Systems
Vanderbilt University
Box 1829, Station B
Nashville, TN 37235
USA
schmidt@dre.vanderbilt.edu

Aniruddha Gokhale
Institute for Software Integrated Systems
Vanderbilt University
Box 1829, Station B
Nashville, TN 37235
USA
a.gokhale@vanderbilt.edu

Balachandran Natarajan
Institute for Software Integrated Systems
Vanderbilt University
Box 1829, Station B
Nashville, TN 37235
USA
bala@dre.vanderbilt.edu

Joseph P. Loyall
BBN Technologies
10 Moulton Street
Cambridge, MA 02138
USA
jloyall@bbn.com

Richard E. Schantz
BBN Technologies
10 Moulton Street
Cambridge, MA 02138
USA
schantz@bbn.com

Craig Rodrigues
BBN Technologies
10 Moulton Street
Cambridge, MA 02138
USA
crodrigu@bbn.com

Jeff Gray
Department of Computer & Information Science
University of Alabama
1300 University Blvd.
Birmingham, AL 35294
USA
gray@cis.uab.edu

Shikharesh Majumdar
Department of Systems & Computer Engineering
Carleton University
1125 Colonel By Drive, Ottawa, ON, K1S 5B6
Canada
majumdar@sce.carleton.ca

Steven A. Demurjian, Sr.
Computer Science & Engineering Department
371 Fairfield Road, Unit 1155
The University of Connecticut
Storrs, Connecticut 06269-1155
USA
steve@engr.uconn.edu

Keith Bessette
Computer Science & Engineering Department
The University of Connecticut
Storrs, CT 06269-3155
USA
keithbessette@hotmail.com

Thuong Doan
Computer Science & Engineering Department
The University of Connecticut

Storrs, CT 06269-3155
USA
thuongdoan@yahoo.com

Charles Phillips
Department of Electrical Engineering & Computer Science
United Stated Military Academy
West Point, NY 10996
USA
charles.phillips@usma.edu

Panos K. Chrysanthis
Department of Computer Science
University of Pittsburgh
Pittsburgh, PA 15260
USA
panos@cs.pitt.edu

Vincenzo Liberatore
Electrical Engineering & Computer Science Department
Case Western Reserve University
Cleveland, Ohio 44106-7071
USA
vincenzo.liberatore@cwru.edu

Kirk Pruhs
Department of Computer Science
University of Pittsburgh
Pittsburgh, PA 15260
USA
kirk@cs.pitt.edu

Cecilia Mascolo
Department of Computer Science
University College London
Gower Street
London WC1E 6BT
UK
c.mascolo@cs.ucl.ac.uk

Licia Capra
Department of Computer Science
University College London
Gower Street
London WC1E 6BT
UK
l.capra@cs.ucl.ac.uk

Wolfgang Emmerich
Department of Computer Science
University College London
Gower Street
London WC1E 6BT
UK
w.emmerich@cs.ucl.ac.uk

Dr. Guijun Wang
Mathematics & Computing Technology
The Boeing Company
P.O. Box 3707, MC 7L-20
Seattle, WA 98124
USA
guijun.wang@boeing.com

Alice Chen
Mathematics & Computing Technology
The Boeing Company
P.O. Box 3707, MC 7L-20
Seattle, WA 98124
USA
alice.chen@boeing.com

Surya Sripada
Mathematics & Computing Technology
The Boeing Company
P.O. Box 3707, MC 7L-20
Seattle, WA 98124
USA
surya.sripada@boeing.com

Changzhou Wang
Mathematics & Computing Technology
The Boeing Company
P.O. Box 3707, MC 7L-20
Seattle, WA 98124
USA
changzhou.wang@boeing.com

Peter Langendörfer
IHP, Im Technologiepark 25

D-15236 Frankfurt (Oder)
Germany
langendoerfer@ihp-microelectronics.com

Oliver Maye
IHP, Im Technologiepark 25
D-15236 Frankfurt (Oder)
Germany
maye@ihp-microelectronics.com

Zoya Dyka
IHP, Im Technologiepark 25
D-15236 Frankfurt (Oder)
Germany
dyka@ihp-microelectronics.com

Roland Sorge
IHP, Im Technologiepark 25
D-15236 Frankfurt (Oder)
Germany
sorge@ihp-microelectronics.com

Rita Winkler
IHP, Im Technologiepark 25
D-15236 Frankfurt (Oder), Germany
rwinkler@ihp-microelectronics.com

Rolp Kraemer
IHP, Im Technologiepark 25
D-15236 Frankfurt (Oder), Germany
kraemer@ihp-microelectronics.com

Karl R.P.H. Leung
Department of Information & Communications Technology
Hong Kong Institute of Vocational Education (Tsing Yi)
Tsing Yi Island
Hong Kong
kleung@computer.org

Joseph Kee-Yin Ng
Department of Computer Science
Hong Kong Baptist University
Kowloon
Hong Kong
jng@comp.hkbu.edu.hk

Calvin Kin-Cheung Hui
Department of Computer Science
Hong Kong Baptist University
Kowloon
Hong Kong
kchui@comp.hkbu.edu.hk

Harald Vogt
Institute for Pervasive Computing
Haldeneggsteig 4
ETH Zentrum
8092 Zurich
Switzerland
vogt@inf.ethz.ch

Michael Rohs
Institute for Pervasive Computing
Haldeneggsteig 4
ETH Zentrum
8092 Zurich
Switzerland
rohs@inf.ethz.ch

Roger Kilian-Kehr
SAP Corporate Research
Vincenz-Priessnitz-Str. 1, 76131, Karlsruhe
Germany
roger.kilian-kehr@sap.com

Jesus Martinez
Computer Science Department
University of Malaga
Campus de Teatinos s/n. 29071, Malaga
Spain
jmcruz@lcc.uma.es

Luis R. López
Computer Science Department
University of Malaga
Campus de Teatinos s/n. 29071, Malaga
Spain
lramonl@terra.es

Pedro Merino
Computer Science Department
University of Malaga

Campus de Teatinos s/n. 29071, Malaga
Spain
pedro@lcc.uma.es

Arvind S. Krishna
Institute of Software Integrated Systems
P.O.Box 1829, Station B
Nashville, TN 37235
USA
arvindk@dre.vanderbilt.edu

Raymond Klefstad
Dept. of Electrical Engineering & Computer Science
University of California, Irvine
Irvine, CA 92697
USA
klefstad@uci.edu

Angelo Corsaro, Phone: 314-935-6160
Dept. of Computer Science & Engineering
Washington University
St. Louis, MO 63130
USA
corsaro@cs.wustl.edu

Diana Szentivanyi
Department of Computer & Information Science
Linköping University
S-581 83 Linköping
Sweden
diasz@ida.liu.se

Simin Nadjm-Tehrani
Department of Computer & Information Science
Linköping University
S-581 83 Linköping
Sweden
simin@ida.liu.se

Introduction

Traditionally, distributed application developers had to rely on low-level TCP/IP communication mechanisms, such as sockets, in order to establish communication sessions between a client and a server. The use of such low-level mechanisms, involves the design of a protocol or a set of rules that a client and a server must follow in order to communicate. The design of protocols, however, is error-prone.

Remote Procedure Calls (RPCs), invented by Sun Microsystems in the early 1980s, provide a higher-level abstraction where procedures, or operations, can be invoked remotely across different operating system platforms. RPC can be considered as a basis for Object-Oriented middleware platforms such as the Distributed Computing Environment (DCE), the Common Object Request Broker Architecture (CORBA), the Distributed Component Object Model (DCOM), and Java Remote Method Invocation (RMI). In such systems, communication interfaces are abstracted to the level of a local procedure call, or a method invocation. RPC systems are usually synchronous – the caller must block and wait until the called method completes execution [1]–and thus offer no potential for parallelism without using multiple threads. In other words, RPC systems require the client and the server to be available at the same time; such tight coupling may not be possible or even desired in some applications. Message-Oriented Middleware (MOM) systems provide solutions to these problems: they are based on the asynchronous interaction model, and provide the abstraction of a message queue that can be accessed across a network. This generalization of the well-known operating system construct 'the mailbox' is flexible as it allows applications to communicate with each other without requiring same-time availability or causing anyone to block.

What is Middleware?

Middleware is a distributed software layer that sits above the network operating system and below the application layer and abstracts the heterogeneity of the underlying environment. It provides an integrated distributed environment whose objective is to simplify the task of programming and managing distributed applications [2] and also to provide value-added services such as naming and transactions to enable distributed application development. Middleware is about integration and interoperability of applications and services running on heterogeneous computing and communications devices.

Middleware platforms such as CORBA, DCOM, and RMI offer higher-level distributed programming models that extend the native OS network environment. Such platforms allow application developers to develop distributed applications much like stand-alone applications. In addition, the Simple Object Access Protocol (SOAP) can be considered as another emerging distribution middleware technology that is based on a lightweight and simple XML-based protocol allowing applications to exchange structured and typed information on the Web. SOAP is designed to enable automated Web services based on open Web infrastructure, and more importantly, SOAP applications can be written in a wide range of programming languages. In addition, middleware approaches such as Jini, JavaSpaces, Enterprise JavaBeans (EJBs), and Message Queuing Series (MQSeries) have received widespread attention.

All such middleware platforms are designed to mask the heterogeneity that developers of distributed applications must deal with. They mask the heterogeneity of networks and hardware as well as operating systems. Some middleware platforms go further and mask the heterogeneity of programming languages, and CORBA goes even further by masking the heterogeneity among vendor implementations of the same standard. In addition, middleware platforms provide higher-level, domain-independent value-added services that free application developers from worrying about the lower-level details and thus enable them to concentrate on the business logic [3]. For example, application developers do not need to write code that handles database connection pooling, threading, or security because such tasks are bundled as reusable components into middleware platforms. Examples of middleware value-added services include the following: CORBA services provide a wide variety of services including event notification, logging, persistence, global time, multimedia streaming, fault tolerance, concurrency control, and transactions; Sun Microsystems' EJBs allow developers to create multi-tier distributed applications simply by linking prebuilt software component services, or beans, without having to write code from scratch.

The role of middleware in communications systems will become increasingly important, especially in emerging technology areas such as mobile computing where the integration of different applications and services from different wired and wireless networks becomes increasingly important. As an example, the wireless mobile Internet will be different from simply accessing the Internet wirelessly. Users with handheld wireless devices – being mostly mobile – have different needs, motivations, and capabilities from wired users. Users will employ mobile services that will act as brokers and communicate with other applications and services to address their needs. In order to achieve such seamless interaction, applications and services must be aware of and able to adapt to the user's environment. One approach is to rewrite all existing applications and services to the wireless mobile Internet. A more intelligent solution would be to implement generic functionality in a middleware layer of the infrastructure.

Middleware Challenges

There are several challenges that need to be addressed in order to design and optimize middleware for communications systems. The most significant challenge facing middleware is facilitating the wireless mobile Internet application integration.

Conventional middleware platforms designed for wired networks expect static connectivity, reliable channels, and high bandwidth, which are limited in wireless networks.

Therefore, new middleware platforms are needed to support devices and applications running across mixed wireless networks, and more importantly, to provide reliable and secure wireless access to content.

Integrated networks (or the convergence of computer and telecommunications networks) enable a new kind of application to emerge where the network is populated with services and devices with different levels of granularity and thus have different requirements for Quality of Service (QoS). Such integrated networks increase the level of system heterogeneity, but hiding such heterogeneity becomes harder or even undesirable since middleware should adapt to changes in location and connectivity. Therefore, there is a need for more flexible middleware platforms where flexibility should not be at the expense of dependability or security.

Reflective and adaptive techniques are important as they allow for the development of reflective and adaptive middleware platforms that can adapt to new environments and react to changes in contexts. Middleware systems, however, rely heavily on indirect communication mechanisms that lead to some performance penalties; adaptive middleware makes the situation even worse simply because there is additional indirect communication.

The chapters in this book address these and several other challenges.

Overview of the Book

The book contains a collection of related chapters, both introductory and advanced, that are authored by leading experts in the field. The chapters cover topics that are related to communications and the future of communications. The first few chapters cover different categories of middleware including MOM, adaptive, reflective, and transaction middleware, followed by chapters that discuss middleware technologies that target specific application domains, including mobile computing, grid computing, peer-to-peer systems video streaming, real-time systems, smart cards, and e-commerce. In addition, several chapters address emerging research issues such as Quality of Service (QoS) and Model-Driven Architecture (MDA). Finally, there are chapters covering security, performance, and faculty-tolerance issues. Here is an overview of the individual chapters:

Chapter 1: Message-Oriented Middleware. In this chapter, Edward Curry provides a detailed introduction to Message-Oriented Middleware (MOM). Unlike RPC mechanisms, MOM provides distributed communication based on the asynchronous interaction model where the sender does not block waiting for a reply. Curry offers guidelines for using MOM as well as a detailed discussion of messaging models and their performance. While several MOM implementations are available, Curry selects the Java Message Service (JMS), which is a standard Java-based interface to the message services of MOM, for discussion.

Chapter 2: Reflective and Adaptive Middleware. In this chapter, Edward Curry explores adaptive and reflective middleware. Reflection refers to the capability of a system to reason about and act upon itself. A reflective middleware platform is more adaptable to its environment and capable of coping with change, and this is important for middleware-based applications that are deployed in a dynamic environment, such as mobile computing. Curry provides a state-of-the-art survey as well as directions for future research in the field of reflective and adaptive middleware.

Chapter 3: Transaction Middleware. This chapter by Stefan Tai, Thomas Mikalsen, and Isabelle Rouvellou provides a complete middleware view of transaction processing. The aim of transaction middleware is to enable application developers to write and deploy reliable, scalable transactional applications. Tai et al discuss how transactions are supported in OO middleware, MOM middleware, and Web service-oriented middleware, emphasizing the two main transaction models supported by CORBA and J2EE/EJB: direct transaction processing (DTP) and queued transaction processing (QTP).

Chapter 4: Peer-to-Peer Middleware. Markus Oliver Junginger and Yugyung Lee discuss the emerging field of Peer-to-Peer (P2P) Middleware. File-sharing applications such as Napster and Gnutella are, to some extent, examples of P2P applications. Junginger and Lee present challenges and solutions for P2P middleware platforms such as JXTA and their own P2P Messaging System that provides a scalable and robust middleware suited for P2P applications that require group communication facilities. Although decentralized networks have their own advantages, Junginger and Lee argue the need for such networks to collaborate with centralized networks, and therefore the need for hybrid solutions.

Chapter 5: Grid Middleware. In this chapter, Gregor von Laszewski and Kaizer Amin present research activities to develop grid middleware. Such middleware is a natural extension of established parallel and distributed computing research. Laszewski and Amin discuss issues related to grid middleware that must be resolved before the Grid can be truly universal and highlight a number of projects, such as the widely used Globus middleware toolkit, that address some of these challenging issues.

Chapter 6: QoS-enabled Middleware. Distributed real-time and embedded (DRE) systems that control processes and devices in industrial mission-critical applications have distinguished characteristics than conventional desktop applications in the sense that if the right answer is delivered late, it can become the wrong answer that is, failure to meet QoS requirements can lead to catastrophic consequences. In this chapter, Nanbor Wang, Christopher D. Gill, Douglas C. Schmidt, Aniruddha Gokhale, Balachandran Natarajan, Joseph P. Loyall, Richard E. Schantz, and Craig Rodrigues describe how the Component Integrated ACE ORB (CIAO) middleware can be used to support static QoS provisioning by preallocating resources for DRE applications. In addition, they describe how BBN's QuO Qosket middleware framework provides powerful abstractions that help define and implement reusable dynamic QoS provisioning behaviors and how CIAO and QuO Qosket can be combined to provide an integrated end-to-end QoS provisioning solution for DRE applications.

Chapter 7: Model-Driven Middleware. The Model-Driven Architecture (MDA) is a declarative approach for building distributed applications; it is well-suited to address the highly distributed and constantly changing environments such as those found in telecommunications networks. In this chapter, Aniruddha Gokhale, Douglas C. Schmidt, Balachandran Natarajan, Jeff Gray, and Nanbor Wang continue the theme of QoS by discussing how QoS component middleware can be combined with MDA tools to rapidly develop, generate, assemble, and deploy flexible DRE applications that can be tailored to meet the needs of multiple simultaneous QoS requirements.

Chapter 8: High-Performance Middleware-Based Systems. In this chapter, Shikharesh Majumdar is concerned with the performance of middleware systems. Majumdar presents three different approaches to engineering performance into middleware-based systems. The first two approaches deal with the engineering of performance into the middleware platform itself and therefore would be of interest to middleware platform builders, and the third approach covers performance optimization at the application level and therefore would be of interest to application developers.

Chapter 9: Concepts and Capabilities of Middleware Security. In this chapter, S. Demurjian, K. Bessette, T. Doan, and C. Phillips discuss the state of the art in support of middleware security by focusing on the security abilities of CORBA, J2EE, and .NET. Demurjian et al also demonstrate the utilization of middleware to realize complex security capabilities such as role-based access control (RBAC) and mandatory access control (MAC).

Chapter 10: Middleware for Scalable Data Dissemination. Panos K. Chrysanthis, Vincenzo Liberatore, and Kirk Pruhs discuss the problem of scalable dissemination over the Internet and provide a middleware solution that unifies and extends support for data management in multicast data dissemination. The authors provide a detailed discussion of their middleware and describe two applications (a scalable Web server, and a scalable healthcare alert system that disseminates information about prodromes and diseases over a wide geographical area) of middleware that ensure the scalability of bulk data dissemination.

Chapter 11: Principles of Mobile Computing Middleware. Middleware will play a significant role in mobile computing as existing and future mobile networks are and will continue to be heterogeneous as the market is not controlled by a single vendor. In this section Cecilia Mascolo, Licia Capra, and Wolfgang Emmerich begin their chapter by discussing how the requirements (such as heterogeneity and fault tolerance) associated with distributed systems are affected by physical mobility issues and how traditional middleware can fulfill such requirements. Finally, Mascolo et al provide a framework and classification showing recent advances and future directions in mobile computing middleware.

Chapter 12: Application of Middleware Technologies to Mobile Enterprise Information Services. In this chapter, Guijun Wang, Alice Chen, Surya Sripada, and Changzhou Wang continue the theme of mobile computing by focusing on middleware technologies for Mobile Enterprise Information Systems (EIS). Extending EIS with mobility (mobile EIS) requires solutions to challenges such as dealing with diverse EIS, unpredictable connectivity, and heterogeneous data contents. In addition, mobile EIS must be integrated with diverse EIS such as email and directory services. The authors present a middleware solution for mobile EIS that is based on J2EE, JMS, and XML.

Chapter 13: Middleware for Location-based Services: Design and Implementation Issues. P. Langendörfer, O. Maye, Z. Dyka, R. Sorge, R. Winkler, and R. Krämer continue the theme of mobile computing further by focusing on middleware for location-based services. Such services require sensitive information such as the current location of the user, the interests of the user, and the capability of his or her profile. This really means that application programmers will need to deal with a diverse set of mobile devices, positioning systems, and wireless communication protocols. The authors present a Platform Supporting Mobile Applications or PLASMA, which is a middleware platform that consists of components for location, event, and profile handling. To achieve high performance, replication and filtering mechanisms are used to minimize internal communications.

Chapter 14: QoS-Enabled Middleware for MPEG Video Streaming. QoS is important not only in distributed real-time and embedded (DRE) systems but also in video Internet applications such as an MPEG video–streaming service. Karl R. P. H. Leung, Joseph Kee-Yin Ng, and Calvin Kin-Cheung Hui provide a distributed MPEG video–streaming middleware that enables clients to use their favorite HTTP MPEG video player to enjoy the video–streaming services over the open network. The middleware components work together to tune the QoS for each individual client by means of a QoS tuning scheme.

Chapter 15: Middleware for Smart Cards. As smart card applications continue to gain popularity, so will the need for application developers who are able to deal with the large number of diverse, manufacturer-dependent interfaces offered by smart card services and smart card readers. In this chapter, Harald Vogt, Michael Rohs, and Roger Kilian-Kehr discuss middleware for smart cards that mediate between diverse smart card services and readers so that application developers do not need to deal with different smart card interfaces.

Chapter 16: Application-Oriented Middleware for E-commerce. The theme of smart cards is continued in this chapter by Jesús Martínez, Luis R. López, and Pedro Merino. The authors describe a middleware ticketing solution for the integration of smart cards in a public transport ticketing service. The resulting system allows client applications to be physically decoupled from the reader and the card, and this is valuable when smart cards are used for more complex applications than just access control.

Chapter 17: Real-time CORBA Middleware. In this chapter, Arvind S. Krishna, Douglas C. Schmidt, Raymond Klefstad, and Angelo Corsaro show how well-designed Real-time CORBA middleware architecture can minimize footprint and facilitate customization for Object Request Brokers (ORBs) to support various classes of applications. The authors also describe how highly optimized ORB designs can improve the performance and predictability of DRE systems and show that Java and Real-time Java features can be applied to an ORB to increase ease of use and predictability for Java-based DRE applications. Finally, the authors describe the ZEN ORB which is a next generation of Real-time CORBA middleware that combines Real-time Java with Real-time CORBA resulting in an easy-to-use, extensible, and flexible middleware with an appropriate footprint and QoS to meet the needs of many DRE systems.

Chapter 18: Middleware Support for Fault Tolerance. Telecommunication management systems are large and complex pieces of software that are costly to maintain. One contributing factor to the complexity is that system availability is dealt with at the application level by ad hoc fault tolerance mechanisms. A more intelligent solution is to build fault tolerance capabilities into the middleware itself. In this chapter, Diana Szentiványi and Simin Nadjm-Tehrani report on experiences with building two different platforms supporting fault tolerance in CORBA applications: FT-CORBA that deals only with failures of the application and FA_CORBA (fully available), which uniformly handles failures of the application as well as the infrastructure.

Finally, I hope readers will find the selection of chapters for this book useful. What I have tried to provide is convenient access to exemplars illustrating the diversity of the communications problems being addressed by middleware technology. I believe that the topics covered in this book pave the way toward the exciting developments of next generation middleware platforms.

Bibliography

[1] Bernstein, P. A. (1996) Middleware: A Model for Distributed System Services. *Communications of the ACM*, **39**(2) 86–98.

[2] Campbell, A., Coulson, G., and Kounavis, M. (1999) *Managing Complexity: Middleware Explained*. IT Professional, IEEE Computer Society, **1**(5), 22–28.

[3] Schantz, R. E. and Schmidt, D. C. (2002) Research Advanced in Middleware for Distributed Systems: Start of the Art. *Proceedings of the IFIP World Computer Congress*, Montreal, Canada, August 2002 (also in Kluwer Academic Publishers "Communications Systems: The State of the Art").

1

Message-Oriented Middleware

Edward Curry
National University of Ireland

1.1 Introduction

As software systems continue to be distributed deployments over ever-increasing scales, transcending geographical, organizational, and traditional commercial boundaries, the demands placed upon their communication infrastructures will increase exponentially. Modern systems operate in complex environments with multiple programming languages, hardware platforms, operating systems and the requirement for dynamic flexible deployments with 24/7 reliability, high throughput performance and security while maintaining a high Quality-of-Service (QoS). In these environments, the traditional direct Remote Procedure Call (RPC) mechanisms quickly fail to meet the challenges present.

In order to cope with the demands of such systems, an alternative to the RPC distribution mechanism has emerged. This mechanism called Message-Oriented Middleware or MOM provides a clean method of communication between disparate software entities. MOM is one of the cornerstone foundations that distributed enterprise systems are built upon. MOM can be defined as any middleware infrastructure that provides messaging capabilities.

A client of a MOM system can send messages to, and receive messages from, other clients of the messaging system. Each client connects to one or more servers that act as an intermediary in the sending and receiving of messages. MOM uses a model with a peer-to-peer relationship between individual clients; in this model, each peer can send and receive messages to and from other client peers. MOM platforms allow flexible cohesive systems to be created; a cohesive system is one that allows changes in one part of a system to occur without the need for changes in other parts of the system.

1.1.1 Interaction Models

Two interaction models dominate distributed computing environments, synchronous and asynchronous communication. This section introduces both interaction models; a solid

Middleware for Communications. Edited by Qusay H. Mahmoud
 ISBN 0-470-86206-8

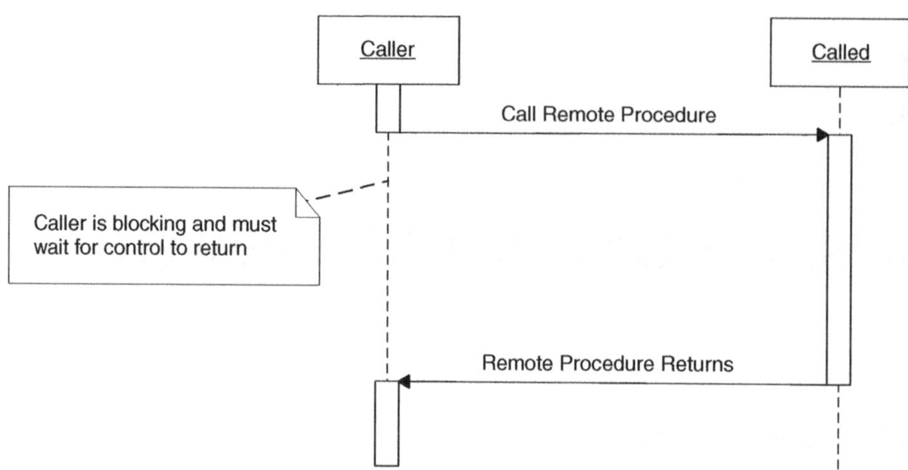

Figure 1.1 Synchronous interaction model

knowledge of these models and the differences between them is key to understanding the benefits and differences between MOM and other forms of distribution available.

1.1.2 Synchronous Communication

When a procedure/function/method is called using the synchronous interaction model, the caller code must block and wait (suspend processing) until the called code completes execution and returns control to it; the caller code can now continue processing. When using the synchronous interaction model, as illustrated in Figure 1.1, systems do not have processing control independence; they rely on the return of control from the called systems.

1.1.3 Asynchronous Communication

The asynchronous interaction model, illustrated in Figure 1.2, allows the caller to retain processing control. The caller code does not need to block and wait for the called code to return. This model allows the caller to continue processing regardless of the processing state of the called procedure/function/method. With asynchronous interaction, the called code may not execute straight away. This interaction model requires an intermediary to handle the exchange of requests; normally this intermediary is a message queue.

While more complex than the synchronous model, the asynchronous model allows all participants to retain processing independence. Participants can continue processing, regardless of the state of the other participants.

1.1.4 Introduction to the Remote Procedure Call (RPC)

The traditional RPC model is a fundamental concept of distributed computing. It is utilized in middleware platforms including CORBA, Java RMI, Microsoft DCOM, and XML-RPC. The objective of RPC is to allow two processes to interact. RPC creates the façade of making both processes believe they are in the same process space (i.e., are the one

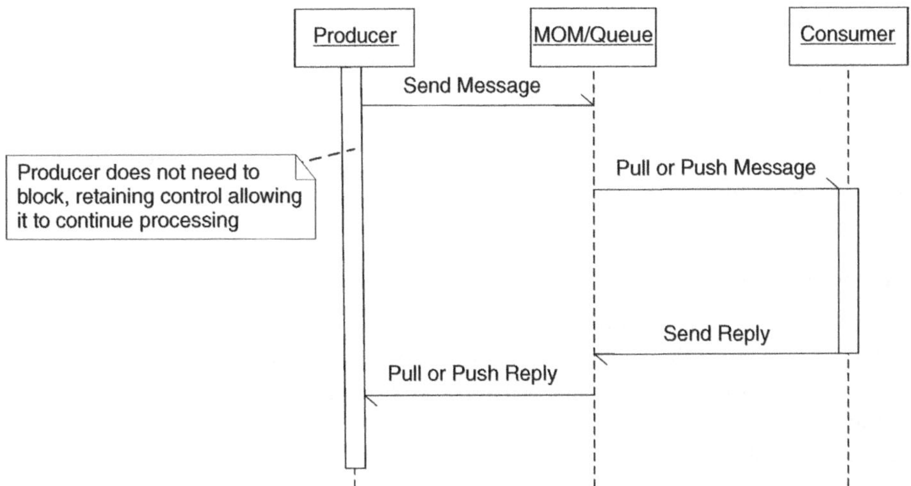

Figure 1.2 Asynchronous interaction model

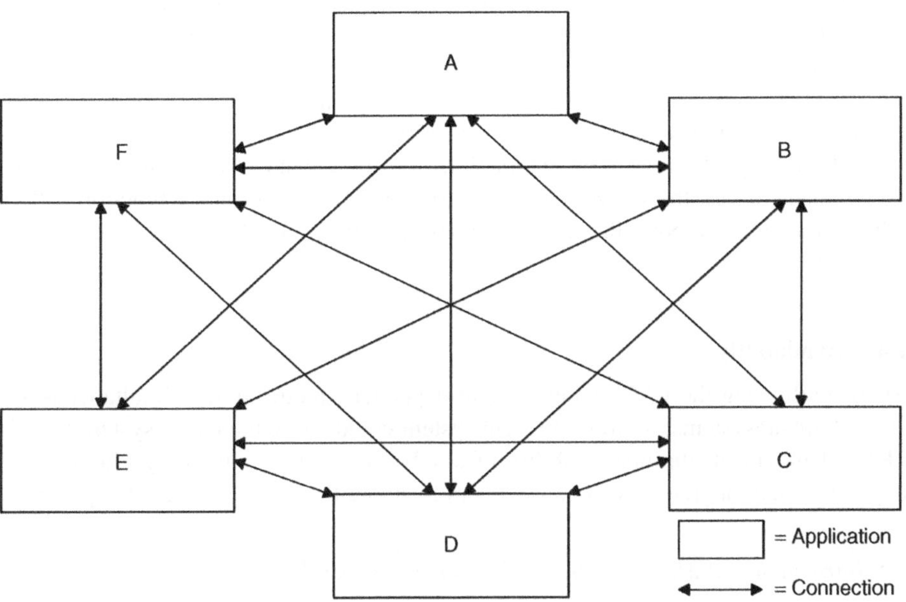

Figure 1.3 An example remote procedure call deployment

process). Based on the synchronous interaction model, RPC is similar to a local procedure call whereby control is passed to the called procedure in a sequential synchronous manner while the calling procedure blocks waiting for a reply to its call. RPC can be seen as a direct conversation between two parties (similar to a person-to-person telephone conversation). An example of an RPC-based distributed system deployment is detailed in Figure 1.3.

1.1.4.1 Coupling

RPC is designed to work on object or function interfaces, resulting in the model producing tightly coupled systems as any changes to the interfaces will need to be propagated through the codebase of both systems. This makes RPC a very invasive mechanism of distribution. As the number of changes to source or target systems increase, the cost will increase too. RPC provides an inflexible method of integrating multiple systems.

1.1.4.2 Reliability

Reliable communications can be the most important concern for distributed applications. Any failure outside of the application – code, network, hardware, service, other software or service outages of various kinds (network provider, power, etc) – can affect the reliable transport of data between systems. Most RPC implementations provide little or no guaranteed reliable communication capability; they are very vulnerable to service outages.

1.1.4.3 Scalability

In a distributed system constructed with RPC, the blocking nature of RPC can adversely affect performance in systems where the participating subsystems do not scale equally. This effectively slows the whole system down to the maximum speed of its slowest participant. In such conditions, synchronous-based communication techniques such as RPC may have trouble coping when elements of the system are subjected to a high-volume burst in traffic. Synchronous RPC interactions use more bandwidth because several calls must be made across the network in order to support a synchronous function call. The implication of this supports the use of the asynchronous model as a scalable method of interaction.

1.1.4.4 Availability

Systems built using the RPC model are interdependent, requiring the simultaneous availability of all subsystems; a failure in a subsystem could cause the entire system to fail. In an RPC deployment, the unavailability of a subsystem, even temporally, due to service outage or system upgrading can cause errors to ripple throughout the entire system.

1.1.5 Introduction to Message-Oriented Middleware (MOM)

MOM systems provide distributed communication on the basis of the asynchronous interaction model; this nonblocking model allows MOM to solve many of the limitations found in RPC. Participants in a MOM-based system are not required to block and wait on a message send, they are allowed to continue processing once a message has been sent. This allows the delivery of messages when the sender or receiver is not active or available to respond at the time of execution.

MOM supports message delivery for messages that may take minutes to deliver, as opposed to mechanisms such as RPC (i.e., Java RMI) that deliver in milliseconds or seconds. When using MOM, a sending application has no guarantee that its message

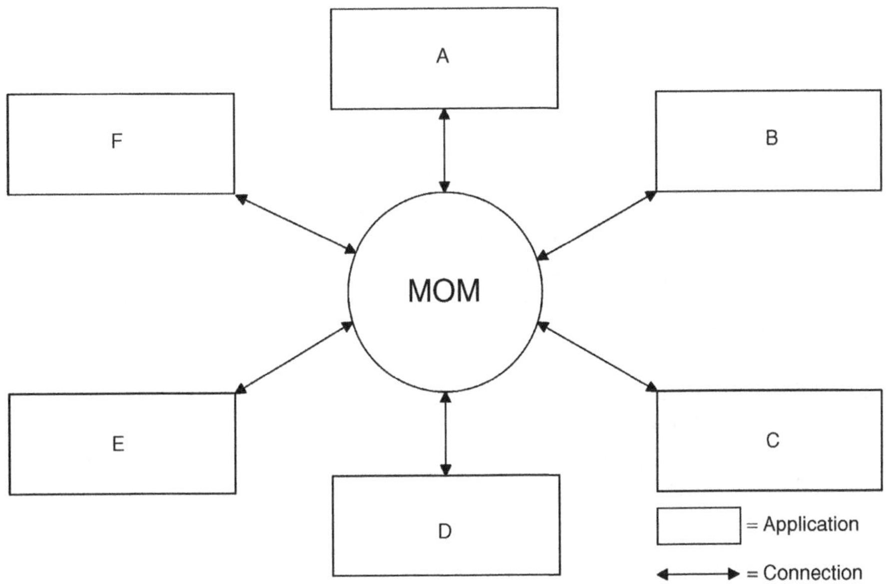

Figure 1.4 An example message-oriented middleware deployment

will be read by another application nor is it given a guarantee about the time it will take the message to be delivered. These aspects are mainly determined by the receiving application.

MOM-based distributed system deployments, as shown in Figure 1.4, offer a service-based approach to interprocess communication. MOM messaging is similar to the postal service. Messages are delivered to the post office; the postal service then takes responsibility for safe delivery of the message [1].

1.1.5.1 Coupling

MOM injects a layer between senders and receivers. This independent layer acts as an intermediary to exchange messages between message senders and receivers, see Figure 1.2 for an illustration of this concept. A primary benefit of MOM is the loose coupling between participants in a system – the ability to link applications without having to adapt the source and target systems to each other, resulting in a highly cohesive, decoupled system deployment [2].

1.1.5.2 Reliability

With MOM, message loss through network or system failure is prevented by using a store and forward mechanism for message persistence. This capability of MOM introduces a high level of reliability into the distribution mechanism, store and forward prevents loss of messages when parts of the system are unavailable or busy. The specific level-of-reliability is typically configurable, but MOM messaging systems are able to guarantee that a message will be delivered, and that it will be delivered to each intended recipient exactly once.

1.1.5.3 Scalability

In addition to decoupling the interaction of subsystems, MOM also decouples the performance characteristics of the subsystems from each other. Subsystems can be scaled independently, with little or no disruption to other subsystems. MOM also allows the system to cope with unpredictable spikes in activity in one subsystem without affecting other areas of the system. MOM messaging models contain a number of natural traits that allow for simple and effective load balancing, by allowing a subsystem to choose to accept a message when it is ready to do so rather than being forced to accept it. This load balancing technique will be covered in more detail later in the chapter. State-of-the-art enterprise-level MOM platforms have been used as the backbone to create massively scalable systems with support for handling 16.2 million concurrent queries per hour and over 270,000 new order requests per hour [3].

1.1.5.4 Availability

MOM introduces high availability capabilities into systems allowing for continuous operation and smoother handling of system outages. MOM does not require simultaneous or "same-time" availability of all subsystems. Failure in one of the subsystems will not cause failures to ripple throughout the entire system. MOM can also improve the response time of the system because of the loose coupling between MOM participants. This can reduce the process completion time and improve overall system responsiveness and availability.

1.1.6 When to use MOM or RPC

Depending on the scenario they are deployed in, both MOM and RPC have their advantages and disadvantages. RPC provides a more straightforward approach to messaging using the familiar and straightforward synchronous interaction model. However, the RPC mechanism suffers from inflexibility and tight coupling (potential geometric growth of interfaces) between the communicating systems; it is also problematic to scale parts of the system and deal with service outages. RPC assumes that all parts of the system will be simultaneously available; if one part of the system was to fail or even become temporarily unavailable (network outage, system upgrade), the entire system could stall as a result.

There is a large overhead associated with an RPC interaction; RPC calls require more bandwidth than a similar MOM interaction. Bandwidth is an expensive performance overhead and is the main obstacle to scalability of the RPC mechanism [4]. The RPC model is designed on the notion of a single client talking to a single server; traditional RPC has no built-in support for one-to-many communications. The advantage of an RPC system is the simplicity of the mechanism and straightforward implementation.

An advantage that RPC has over MOM is the guarantee of sequential processing. With the synchronous RPC model, you can control the order in which processing occurs in the system. For example, in an RPC system you can be sure that at any one time all the new orders received by the system have been added to the database and that they have been added in the order of which they were received. However, with an asynchronous MOM approach this cannot be guaranteed, as new orders could exist in queues waiting to be added to the database. This could result in a temporal inaccuracy of the data in the database. We are not concerned that these updates will not be applied to the database, but that a snapshot of the current database would not accurately reflect the actual state of orders placed. RPC is slow but consistent; work is always carried out in the correct order. These are important considerations for sections in a system that requires data to have 100% temporal integrity. If this type of integrity is more important than performance, you will need to use the RPC model or else design your system to check for these potential temporal inaccuracies.

MOM allows a system to evolve with its operational environment without dramatic changes to the application assets. It provides an integration infrastructure that accommodates functionality changes over time without disruption or compromising performance and scalability. The decoupled approach of MOM allows for flexible integration of clients, support for large numbers of clients, and client anonymity. Commercial MOM implementations provide high scalability with support for tens of thousands of clients, advanced filtering, integration into heterogeneous networks, and clustering reliability [3].

The RPC model is ideal if you want a strongly typed/Object-Oriented (OO) system with tight coupling, compile-time semantic checking and an overall more straightforward system implementation.

If the distributed systems will be geographically dispersed deployments with poor network connectivity and stringent demands in reliability, flexibility, and scalability, then MOM is the ideal solution.

1.2 Message Queues

The message queue is a fundamental concept within MOM. Queues provide the ability to store messages on a MOM platform. MOM clients are able to send and receive messages to and from a queue. Queues are central to the implementation of the asynchronous interaction model within MOM. A queue, as shown in Figure 1.5, is a destination where messages may be sent to and received from; usually the messages contained within a queue are sorted in a particular order. The standard queue found in a messaging system is the First-In First-Out (FIFO) queue; as the name suggests, the first message sent to the queue is the first message to be retrieved from the queue.

Many attributes of a queue may be configured. These include the queue's name, queue's size, the save threshold of the queue, message-sorting algorithm, and so on. Queuing is of particular benefit to mobile clients without constant network connectivity, for example, sales personnel on the road using mobile network (GSM, GRPS, etc) equipment to remotely send orders to head office or for remote sites with poor communication infrastructures. These clients can use a queue as a makeshift inbox, periodically checking the queue for new messages. Potentially each application may have its own queue, or applications may share a queue, there is no restriction on the setup. Typically, MOM platforms support multiple queue types, each with a different purpose. Table 1.1 provides a brief description of the more common queues found in MOM implementations.

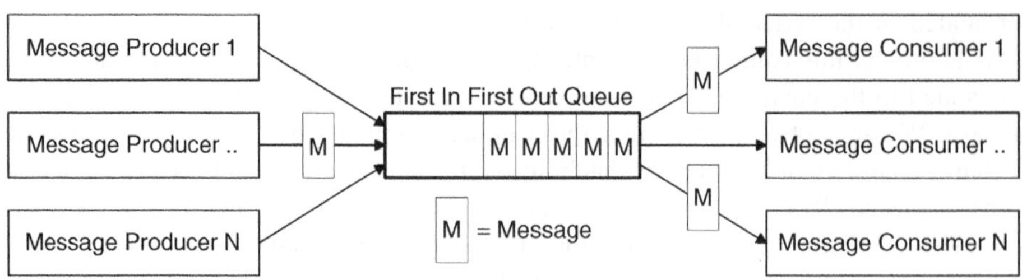

Figure 1.5 Message queue

Table 1.1 Queue formats

Queue type	Purpose
Public Queue	Public open access queue
Private Queue	Require clients to provide a valid username and password for authentication and authorization
Temporary Queue	Queue created for a finite period, this type of queue will last only for the duration of a particular condition or a set time period
Journal Queues	Designed to keep a record of messages or events. These queues maintain a copy of every message placed within them, effectively creating a journal of messages
Connector/Bridge Queue	Enables proprietary MOM implementation to interoperate by mimicking the role of a proxy to an external MOM provider. A bridge handles the translation of message formats between different MOM providers, allowing a client of one provider to access the queues/messages of another
Dead-Letter/Dead-Message Queue	Messages that have expired or are undeliverable (i.e., invalid queue name or undeliverable addresses) are placed in this queue

1.3 Messaging Models

A solid understanding of the available messaging models within MOM is key to appreciate the unique capabilities it provides. Two main message models are commonly available, the point-to-point and publish/subscribe models. Both of these models are based on the exchange of messages through a channel (queue). A typical system will utilize a mix of these models to achieve different messaging objectives.

1.3.1 Point-to-Point

The point-to-point messaging model provides a straightforward asynchronous exchange of messages between software entities. In this model, shown in Figure 1.6, messages from producing clients are routed to consuming clients via a queue. As discussed earlier, the most common queue used is a FIFO queue, in which messages are sorted in the order

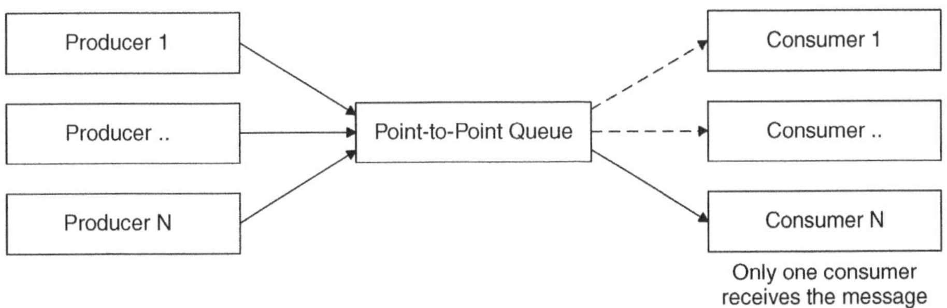

Figure 1.6 Point-to-point messaging model

in which they were received by the message system and as they are consumed, they are removed from the head of the queue.

While there is no restriction on the number of clients who can publish to a queue, there is usually only a single consuming client, although this is not a strict requirement. Each message is delivered only once to only one receiver. The model allows multiple receivers to connect to the queue, but only one of the receivers will consume the message. The technique of using multiple consuming clients to read from a queue can be used to easily introduce smooth, efficient load balancing into a system. In the point-to-point model, messages are always delivered and will be stored in the queue until a consumer is ready to retrieve them.

Request-Reply Messaging Model

This model is designed around the concept of a request with a related response. This model is used for the World Wide Web (WWW), a client requests a page from a server, and the server replies with the requested web page. The model requires that any producer who sends a message must be ready to receive a reply from consumers at some stage in the future. The model is easily implemented with the use of the point-to-point or publish/subscribe models and may be used in tandem to complement.

1.3.2 Publish/Subscribe

The publish/subscribe messaging model, Figure 1.7, is a very powerful mechanism used to disseminate information between anonymous message consumers and producers. This one-to-many and many-to-many distribution mechanism allows a single producer to send a message to one user or potentially hundreds of thousands of consumers.

In the publish/subscribe (pub/sub) model, the sending and receiving application is free from the need to understand anything about the target application. It only needs to send the information to a destination within the publish/subscribe engine. The engine will then send it to the consumer. Clients producing messages 'publish' to a specific topic or channel, these channels are then 'subscribed' to by clients wishing to consume messages. The service routes the messages to consumers on the basis of the topics to which they have subscribed as being interested in. Within the publish/subscribe model, there is no restriction on the role of a client; a client may be both a producer and consumer of a topic/channel.

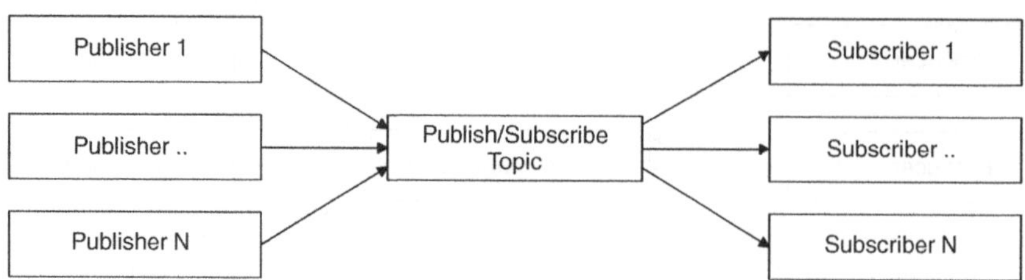

Figure 1.7 Publish/subscribe messaging model

A number of methods for publish/subscribe messaging have been developed, which support different features, techniques, and algorithms for message filtering [5], publication, subscription, and subscription management distribution [6]. For a comprehensive list of current research visit the IEEE Distributed Systems Online Resource on Event-Based Computing – http://dsonline.computer.org/middleware/em/emlinks.htm.

PUSH and PULL

When using these messaging models, a consuming client has two methods of receiving messages from the MOM provider.

Pull

A consumer can poll the provider to check for any messages, effectively **pulling** them from the provider.

Push

Alternatively, a consumer can request the provider to send on relevant messages as soon as the provider receives them; they instruct the provider to **push** messages to them.

1.3.2.1 Hierarchical Channels

Hierarchical channels or topics are a destination grouping mechanism in pub/sub messaging model. This type of structure allows channels to be defined in a hierarchical fashion, so that they may be nested under other channels. Each subchannel offers a more granular selection of the messages contained in its parent. Clients of hierarchical channels subscribe to the most appropriate level of channel in order to receive the most relevant messages. In large-scale systems, the grouping of messages into related types (i.e., into channels) helps to manage large volumes of different messages [7].

The relationship between a channel and subchannels allows for super-type subscriptions, where subscriptions that operate on a parent channel/type will also match all subscriptions of descendant channels/types. A channel hierarchy for an automotive trading service may be structured by categorizing messaging into buys or sells, then further subcategorization breaking down for commercial and private vehicle types. An example hierarchy illustrating this categorizing structure is presented in Figure 1.8. A subscription to the "`Sell.Private_Vehicles`" channel would receive all messages classified as a private vehicle sale, whereas subscribing to "`Sell.Private_Vehicles.Cars`" would result in only receiving messages classified as a car sale.

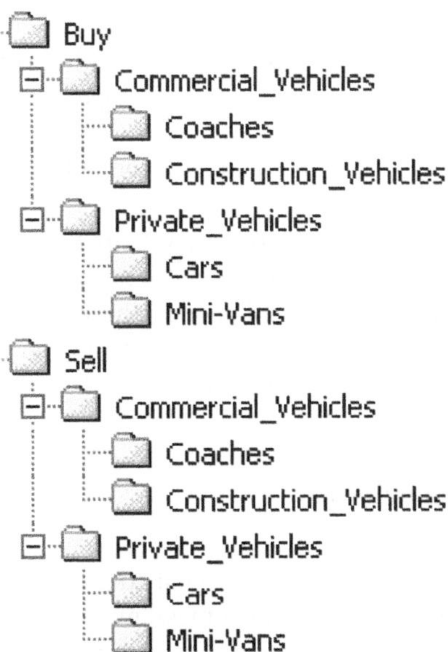

Figure 1.8 An automotive hierarchical channel structure

Hierarchical channels require the channel namespace schema be both well defined and universally understood by the participating parties. Responsibility for choosing a channel in which to publish messages is left to the publishing client. Hierarchical channels are used in routing situations that are more or less static; however, research is underway on defining reflective hierarchies, with adaptive capabilities, for dynamic environments [8].

Consumers of the hierarchy are able to browse the hierarchy and subscribe to channels. Frequently used in conjunction with the publish/subscribe messaging model, hierarchical channels allow for the dissemination of information to a large number of unknown consumers. Hierarchical channels can compliment filtering as a mechanism of routing relevant messages to consumers; they provide a more granular approach to consumer subscription that reduces the number of filters needed to exclude unwanted messages, while supporting highly flexible easy access subject-based routing.

1.3.3 Comparison of Messaging Models

The two models have very different capabilities and most messaging objectives can be achieved using either model or a combination of both. The fundamental difference between the models boils down to the fact that within the publish/subscribe model every consumer to a topic/channel will receive a message published to it, whereas in point-to-point model only one consumer will receive it. Publish/subscribe is normally used in a broadcast scenario where a publisher wishes to send a message to 1-N clients. The publisher has no real control over the number of clients who receive the message, nor have they a guarantee any will receive it. Even in a one-to-one messaging scenario, topics can be useful to categorize different types of messages. The publish/subscribe model is the more powerful messaging model for flexibility; the disadvantage is its complexity.

In the point-to-point model, multiple consumers may listen to a queue, although only one consumer will receive each message. However, point-to-point will guarantee that a consumer will receive the message, storing the messages in a queue until a consumer is ready to receive the message; this is known as '*Once-and-once-only*' messaging. While the point-to-point model may not be as flexible as the publish/subscribe model, its power is in its simplicity.

A common application of the point-to-point model is for load balancing. With multiple consumers receiving from a queue, the workload for processing the messages is distributed between the queue consumers. In the pull model, the exact order of how messages are assigned to consumers is specific to the MOM implementation, but if you utilize a pull model, a consumer will receive a message only when they are ready to process it.

1.4 Common MOM Services

When constructing large-scale systems, it is vital to utilize a state-of-the-art enterprise-level MOM implementation. Enterprise-level messaging platforms will usually come with a number of built-in services for transactional messaging, reliable message delivery, load balancing, and clustering; this section provides an overview of these services.

1.4.1 Message Filtering

Message filtering allows a message consumer/receiver to be selective about the messages it receives from a channel. Filtering can operate on a number of different levels. Filters use Boolean logic expressions to declare messages of interest to the client, the exact format of the expression depends on the implementation but the WHERE clauses of SQL-92 (or a subset of) is commonly used as the syntax. Filtering models commonly operate on the properties (name/value pairs) of a message; however, a number of projects have extended filtering to message payloads [9]. Message filtering is covered further in the Java Message Service section of this chapter.

Listed in Table 1.2 are a some common filtering capabilities found in messaging systems, it is useful to note that as the filtering techniques get more advanced, they are able to replicate the techniques that proceed them. For example, subject-based filtering is able to replicate channel-based filtering, just like content-based filtering is able to replicate both subject and channel-based filtering.

1.4.2 Transactions

Transactions provide the ability to group tasks together into a single unit of work. The most basic straightforward definition of a transaction is as follows:

All tasks must be completed or all will fail together

In order for transactions to be effective, they must conform to the following properties in Table 1.3, commonly referred to as the **ACID** transaction properties.

In the context of transactions, any asset that will be updated by a task within the transaction is referred to as a resource. A resource is a persistent store of data that is

Table 1.2 Message filters

Filter type	Description
Channel-based	Channel-based systems categorize events into predefined groups. Consumers subscribe to the groups of interest and receive all messages sent to the groups
Subject-based	Messages are enhanced with a tag describing their subject. Subscribers can declare their interests in these subjects flexibly by using a string pattern match on the subject, for example, all messages with a subject starting of "Car for Sale"
Content-based	As an attempt to overcome the limitations on subscription declarations, content-based filtering allows subscribers to use flexible querying languages in order to declare their interests with respect to the contents of the messages. For example, such a query could be giving the price of stock 'SUN' when the volume is over 10,000. Such a query in SQL-92 would be "stock_symbol = 'SUN' AND stock_volume >10,000"
Content-based with Patterns (Composite Events)	Content-based filtering with patterns, also known as *composite events* [10], enhances content-based filtering with additional functionality for expressing user interests across multiple messages. Such a query could be giving the price of stock 'SUN' when the price of stock 'Microsoft' is less than $50

Table 1.3 The properties of a transaction

Property	Description
Atomic	All tasks must complete, or no tasks must complete
Consistent	Given an initial consistent state, a final consistent state will be reached regardless of the result of the transaction (success/fail)
Isolated	Transactions must be executed in isolation and cannot interfere with other concurrent transactions
Durable	The effect of a committed transaction will not be lost subsequent to a provider/broker failure

participating within a transaction that will be altered; a message broker's persistent message store is a resource. A resource manager controls the resource(s); they are responsible for managing the resource state.

MOM has the ability to include a message being sent or received within a transaction. This section examines the main types of transaction commonly found in MOM. When examining the transactional aspects of messaging systems, it is important to remember that MOM messages are autonomous self-contained entities. Within the messaging domain, there are two common types of transactions, Local Transactions and Global Transactions. Local transactions take place within a single resource manager such as a single messaging broker. Global transactions involve multiple, potentially distributed heterogeneous resource managers with an external transaction manager coordinating the transaction.

1.4.2.1 Transactional Messaging

When a client wants to send or retrieve messages within a transaction, this is referred to as *Transactional Messaging*. Transactional messaging is used when you want to perform several messaging tasks (send 1-N messages) in a way that all tasks will succeed or all will fail. When transactional messaging is used, the sending or receiving application has the opportunity to commit the transaction (all the operations have succeeded), or to abort the transaction (one of the operations failed) so all changes are rolled back. If a transaction is aborted, all operations are rolled back to the state when the transaction was invoked.

Messages delivered to the server in a transaction are not forwarded on to the receiving client until the sending client commits the transaction. Transactions may contain multiple messages. Message transactions may also take place in transactional queues. This type of queue is created for the specific purpose of receiving and processing messages that are sent as part of a transaction. Nontransactional queues are unable to process messages that have been included in a transaction.

1.4.2.2 Transaction Roles

The roles played by the message producer, message consumer, and message broker are illustrated in Figure 1.9 and Figure 1.10.

Producer

The producer sends a message or set of messages to the broker.
On Commit, the broker stores then sends the message(s)
On Rollback, the broker disposes of the message(s).

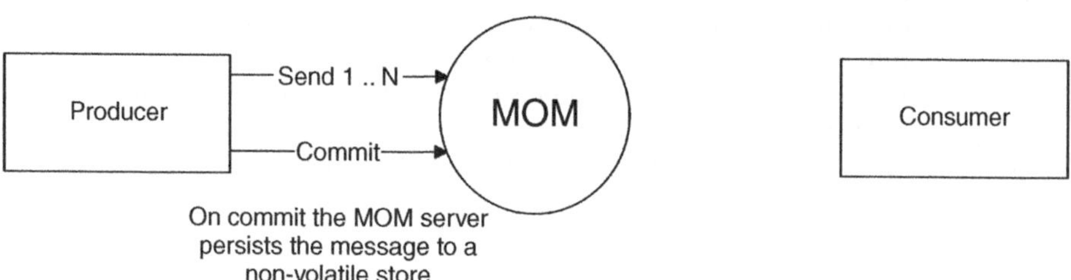

Figure 1.9 Role of a producer in a transaction

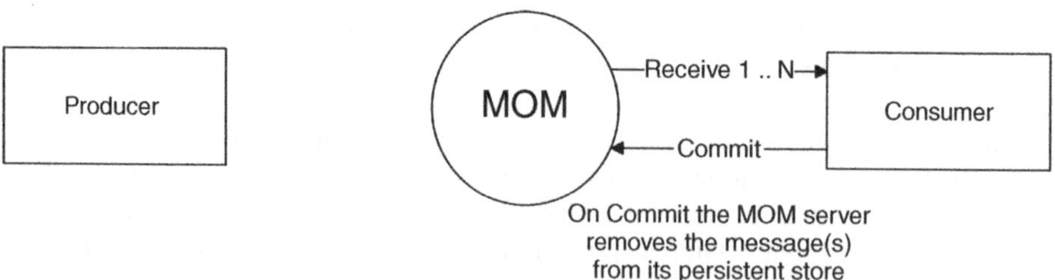

Figure 1.10 Role of a consumer in a transaction

Consumer

The consumer receives a message/set of messages from the broker.
On Commit, the broker disposes of the set of messages
On Rollback, the broker resends the set of messages.

To summarize the roles of each party in a message transaction, the message producer has a contract with the message server; the message server has a contract with the message consumer. Further information on integrating messaging with transactions is available in [11, 12] and Chapter 3 (Transaction Middleware).

1.4.2.3 Reliable Message Delivery

A MOM service will normally allow the configuration of the Quality-of-Service (QoS) delivery semantics for a message. Typically, it is possible to configure a message delivery to be of *at-most once*, *at-least-once*, or *once-and-once-only*. Message acknowledgment can be configured, in addition to the number of retry attempted on a delivery failure. With persistent asynchronous communication, the message is sent to the messaging service that stores it for as long as it takes to deliver the message, unless the *Time-to-Live* (TTL) of the message expires.

1.4.3 Guaranteed Message Delivery

In order for MOM platforms to guarantee message delivery, the platform must save all messages in a nonvolatile store such as a hard disk. The platform then sends the message to the consumer and waits for the consumer to confirm the delivery. If the consumer does not acknowledge the message within a reasonable amount of time, the server will resend the message. This allows for the message sender to '*fire-and-forget*' messages, trusting the MOM to handle the delivery. Certified message delivery is an extension of the guaranteed message delivery method. Once a consumer has received the message, a consumption report (receipt) is generated and sent to the message sender to confirm the consumption of the message.

1.4.4 Message Formats

Depending on the MOM implementation, a number of message formats may be available to the user. Some of the more common message types include Text (including XML), Object, Stream, HashMaps, Streaming Multimedia [13], and so on. MOM providers can offer mechanisms for transforming one message format into another. They can also provide facilities to transform/alter the format of the message contents, some implementations allow eXtensible Stylesheet Language: Transformations (XSLT) on messages containing XML payloads. Such MOM providers are often referred to as Message brokers and are used to "broker" the difference between diverse systems [14].

1.4.5 Load Balancing

Load balancing is the process of spreading the workload of the system over a number of servers (in this scenario, a server can be defined as a physical hardware machine or software server instance or both). A correctly load balanced system should distribute work between servers, dynamically allocating work to the server with the lightest load.

Two main approaches of load balancing exist, '*push*' and '*pull*'. In the push model, an algorithm is used to balance the load over multiple servers. Several algorithms exist, which attempt to guess the least-burdened and push the request to that server. The algorithm, in conjunction with load forecasting, may base its decision on the performance record of each of the participating servers or may guesstimate the least-burdened server. The push approach is an imperfect, but acceptable, solution to load balancing a system.

In the pull model, the load is balanced by placing incoming messages into a point-to-point queue, consuming servers can then pull messages from this queue at their own pace. This allows for true load balancing, as a server will only pull a message from the queue once they are capable of processing it. This provides the ideal mechanism as it more smoothly distributes the loads over the servers.

1.4.6 Clustering

In order to recover from a runtime server failure, the server's state needs to be replicated across multiple servers. This allows a client to be transparently migrated to an alternative server, if the server it is interacting with fails. Clustering is the distribution of an application over multiple servers to scale beyond the limits, both performance and reliability, of a single server. When the limits of the server software or the physical limits of the hardware have been reached, the load must be spread over multiple servers or machines to scale the system further. Clustering allows us to seamlessly distribute over multiple servers/machines and still maintain a single logical entity and a single virtual interface to respond to client requests. The grouping of clusters creates highly scalable and reliable deployments while minimizing the number of servers needed to cope with large workloads.

1.5 Java Message Service

A large number of MOM implementations exist, including WebSphere MQ (formerly MQSeries) [15], TIBCO [16] SonicMQ [17], Herald [18], Hermes [7], [19], SIENA [20], Gryphon [21], JEDI [22], REBECA [23] and OpenJMS [24]. In order to simplify the development of systems utilizing MOMs, a standard was needed to provide a universal interface to MOM interactions. To date, a number of MOM standardization have emerged such as the CORBA Event Service [25], CORBA Notification Service [26] and most notably the Java Message Service (JMS).

The Java Message Service (JMS) provides a common way for Java programs to create, send, receive, and read, an enterprise messaging system's messages [27]. The JMS provides a solid foundation for the construction of a messaging infrastructure that can be applied to a wide range of applications. The JMS specification defines a general purpose Application Programming Interface (API) to an enterprise messaging service and a set of semantics that describe the interface and general behavior of a messaging service. The goal of the JMS specification is to provide a universal way to interact with multiple heterogeneous messaging systems in a consistent manner. The learning curve associated with many proprietary-messaging systems can be steep, thus the powerful yet simple API defined in the JMS specification can save a substantial amount of time for developers in a pure Java environment.

This API is designed to allow application programmers to write code to interact with a MOM. The specification also defines a Service Provider Interface (SPI). The role of the SPI is to allow MOM developers to hook up their proprietary MOM implementation to the API. This allows you to write code once using the API and plug-in the desired MOM provider, making client-messaging code portable between MOM providers that implement the JMS specification, reducing vendor lock-in and offering you a choice. It should be noted that JMS is an API specification and does not define the implementation of a messaging service. The semantics of message reliability, performance, scalability, and so on, are not fully defined. JMS does not define an *'on-the-wire'* format for messages. Effectively, two JMS compatible MOM implementations cannot talk to each other directly and will need to use a tool such as a connector/bridge queue to enable interoperability.

1.5.1 Programming using the JMS API

The format of a general message interface to a MOM would need to have the following minimum functionality detailed in Table 1.4. The JMS API provides this basic functionality through its programming model, illustrated in Figure 1.11, allowing it to be compatible with most MOM implementations. This section gives a brief overview of the programming model and presents an example of its usage. The code presented in the section is pseudocode to illustrate the main points of the JMS API, in order to conserve space; error/exception handling code has been omitted.

1.5.1.1 Connections and Sessions

When a client wants to interact with a JMS-compatible MOM platform, it must first make a connection to the message broker. Using this connection, the client may create one or more sessions; a JMS session is a single-threaded context used to send and receive messages to and from queues and topics. Each session can be configured with individual transactional and acknowledgment modes.

Table 1.4 General MOM API interface

Action	Description
SEND	Send a message to a specific queue
RECEIVE (BLOCKING)	Read a message from a queue. If the queue is empty, the call will block until it is nonempty
RECEIVE (NONBLOCKING POLL)	Read a message from the queue. If the queue is empty, do not block
LISTENER (NOTIFY)	Allows the message service to inform the client of the arrival of a message using a callback function on the client. The callback function is executed when a new message arrives in the queue

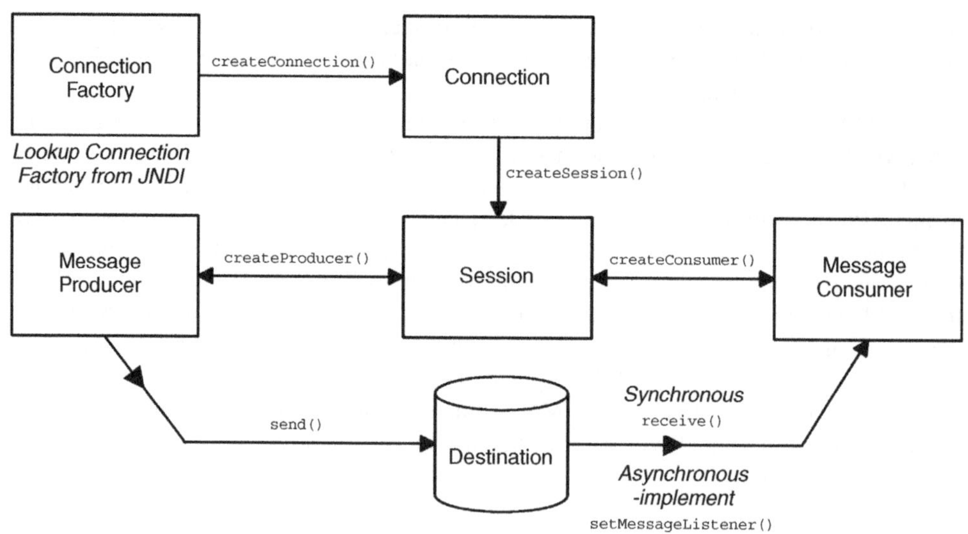

Figure 1.11 The JMS API programming model

```
try {
  // Create a connection
  javax.jms.QueueConnectionFactory queueConnectionFactory
           = (QueueConnectionFactory) ctx.lookup("QueueConnectionFactory");
  javax.jms.QueueConnection queueConnection
                          = queueConnectionFactory.createQueueConnection();

  // Create a Session
  javax.jms.QueueSession queueSession
                             = queueConnection.createQueueSession(false,
                             Session.AUTO_ACKNOWLEDGE);
  // Create Queue Sender and Receiver
  javax.jms.Queue myQueue = queueSession.createQueue("MyQueue");
  javax.jms.QueueSender queueSender = queueSession.createSender(myQueue);
  javax.jms.QueueReceiver queueReceiver
                                 = queueSession.createReceiver(myQueue);

  // Start the Connection
  queueConnection.start();

  // Send Message
  javax.jms.TextMessage message = queueSession.createTextMessage();
  message.setText(" Hello World ! ");
  queueSender.send(message);

  // Synchronous Receive
  javax.jms.Message msg = queueReceiver.receive();
  if (msg instanceof TextMessage) {
    javax.jms.TextMessage txtMsg = (TextMessage) msg;
    System.out.println("Reading message: " + txtMsg.getText());
```

```
} else {
  // Handle other message formats
}

// Asynchronous Receive
MessageListener msgListener = new MessageListener() {
 public void onMessage(javax.jms.Message msg) {
  if (msg instanceof TextMessage) {  // Only supports text messages
   javax.jms.TextMessage txtMsg = (TextMessage) msg;
   System.out.println("Reading message: " + txtMsg.getText());
  }
 }
};
queueReceiver.setMessageListener(msgListener);

} catch (javax.jms.JMSException jmse) {
  // Handle error
}
```

1.5.1.2 Message Producers and Consumers

In order for a client to send a message to or receive a message from a JMS provider, it must first create a message producer or message consumer from the JMS session. For the publish/subscribe model, a `javax.jms.TopicPublisher` is needed to send messages to a topic and a `javax.jms.TopicSubscriber` to receive. The above pseudocode is an example of the process of connecting to a JMS provider, establishing a session to a queue (point-to-point model) and using a `javax.jms.QueueSender` and `javax.jms.QueueReceiver` to send and receive messages. The steps involved in connecting to a topic are similar.

Receive Synchronously and Asynchronously

The JMS API supports both synchronous and asynchronous message delivery. To synchronously receive a message, the `receive()` method of the message consumer is used. The default behavior of this method is to block until a message has been received; however, this method may be passed a *time-out* value to limit the blocking period. To receive a message asynchronously, an application must register a `Message Listener` with the message consumer. Message listeners are registered with a message consumer object by using the `setMessageListener(javax.jms.MessageListener msgL)` method. A message listener must implement the `javax.jms.MessageListener` interface. Further detailed discussion and explanations of the JMS API are available in [27–29].

1.5.1.3 Setting Message Properties

Message properties are optional fields contained in a message. These user-defined fields can be used to contain information relevant to the application or to identify messages. Message properties are used in conjunction with message selectors to filter messages.

1.5.1.4 Message Selectors

Message selectors are used to filter the messages received by a message consumer, and they assign the task of filtering messages to the JMS provider rather than to the application. The message consumer will only receive messages whose headers and properties match the selector. A message selector cannot select messages on the basis of the content of the message body. Message selectors consist of a string expression based on a subset of the SQL-92 conditional expression syntax.

"`Property_Vehicle_Type` = 'SUV' AND `Property_Mileage` =< 60000"

1.5.1.5 Acknowledgment Modes

The JMS API supports the acknowledgement of the receipt of a message. Acknowledgement modes are controlled at the sessions level, supported modes are listed in Table 1.5.

1.5.1.6 Delivery Modes

The JMS API supports two delivery modes for a message. The default `PERSISTENT` delivery mode instructs the service to ensure that a message is not lost because of system failure. A message sent with this delivery mode is placed in a nonvolatile memory store. The second option available is the `NON_PERSISTENT` delivery mode; this mode does not require the service to store the message or guarantee that it will not be lost because of system failure. This is a more efficient delivery mode because it does not require the message to be saved to nonvolatile storage.

1.5.1.7 Priority

The priority setting of a message can be adjusted to indicate to the message service urgent an message that should be delivered first. There are ten levels of priority ranging from 0 (lowest priority) to 9 (highest priority).

Table 1.5 JMS acknowledgement modes

Acknowledgment modes	Purpose
`AUTO_ACKNOWLEDGE`	Automatically acknowledges receipt of a message. In asynchronous mode, the handler acknowledges a successful return. In synchronous mode, the client has successfully returned from a call to receive()
`CLIENT_ACKNOWLEDGE`	Allow a client to acknowledge the successful delivery of a message by calling its acknowledge() method
`DUPS_OK_ACKNOWLEDGE`	A lazy acknowledgment mechanism that is likely to result in the delivery of message duplicates. Only consumers that can tolerate duplicate messages should use this mode. This option can reduce overhead by minimizing the work to prevent duplicates

1.5.1.8 Time-to-Live

JMS messages contain a use-by or expiry time known as the Time-to-Live (TTL). By default, a message never expires; however, you may want to set an expiration time. When the message is published, the specified TTL is added to the current time to give the expiration time. Any messages not delivered before the specified expiration times are destroyed.

1.5.1.9 Message Types

The JMS API defines five message types, listed in Table 1.6, that allow you to send and receive data in multiple formats. The JMS API provides methods for creating messages of each type and for filling in their contents.

1.5.1.10 Transactional Messaging

JMS clients can include message operations (sending and receiving) in a transaction. The JMS API session object provides commit and rollback methods that are used to control the transaction from a JMS client. A detailed discussion on transactions and transactional messaging is available in Chapter 3.

1.5.1.11 Message Driven Enterprise Java Beans

The J2EE includes a Message Driven Bean (MDB) as a component that consumes messages from a JMS topic or queue, introduced in the Enterprise Java Beans (EJB) 2.0 specification they are designed to address the integration of JMS with EJBs. An MDB is a stateless, server-side, transaction-aware component that allows J2EE applications to process JMS and other messages such as HTTP, ebXML, SMTP, and so on, asynchronously. Traditionally a proxy was needed to allow EJBs to process an asynchronous method invocation. This approach used an external Java program that acted as the listener, and on receiving a message, invoked a session bean or entity bean method synchronously using RMI/JRMP or RMI/IIOP. With this approach, the message was received outside the EJB container. MDB solves this problem by allowing the message-processing code access to the infrastructure services available from an EJB container such as transactions, fault-tolerance, security, instances pooling, and so on. The EJB 2.0 specification also provides concurrent processing for MDBs with pooling of bean instances. This allows for the

Table 1.6 JMS message types

Message type	Message contains
`javax.jms.TextMessage`	A java.lang.String object
`javax.jms.MapMessage`	A set of name/value pairs, with names as strings and values as java primitive types. The entries can be accessed by name
`javax.jms.BytesMessage`	A stream of uninterrupted bytes
`javax.jms.StreamMessage`	A stream of Java primitive values, filled and read sequentially
`javax.jms.ObjectMessage`	A Serializable Java object

simultaneous processing of messages received, allowing MDBs a much higher throughput with superior scalability than traditional JMS clients.

An MDB is a message listener that can reliably consume messages from a single JMS destination (queue or topic). Similar to a message listener in a standalone JMS client, an MDB contains an `onMessage(javax.jms.Message msg)`. The EJB container invokes this method when it intercepts an incoming JMS message, allowing the bean to process the message. A detailed discussion on implementing MDBs is presented in [29].

1.6 Service-Oriented Architectures

The problems and obstacles encountered during system integration pose major challenges for an organizations IT department:

> "70% of the average IT department budget is devoted to data integration projects"–IDC
>
> "PowerPoint engineers make integration look easy with lovely cones and colorful boxes"–Sean McGrath, CTO, Propylon
>
> "A typical enterprise will devote 35%–40% of its programming budget to programs whose purpose is solely to transfer information between different databases and legacy systems."–Gartner Group

Increasing pressure to cut the cost of software development is driving the emergence of open nonproprietary architectures to utilize the benefits of reusable software components. MOM has been used to create highly open and flexible systems that allow the seamless integration of subsystems. MOM solves many of the transport issues with integration. However, major problems still exist with the representation of data, its format, and structure. To develop a truly open system, MOM requires the assistance of additional technologies such as XML and Web Services. Both of these technologies provide a vital component in building an open cohesive system. Each of these technologies will be examined to highlight the capabilities they provide in the construction of open system architectures.

1.6.1 XML

The eXtensible Mark-up Language (XML) provides a programming language and platform-independent format for representing data. When used to express the payload of a message, this format eliminates any networking, operating system, or platform binding that a binary proprietary protocol would use. XML provides a natural independent way of representing data. Once data is expressed in XML, it is trivial to change the format of the XML using techniques such as the eXtensible Stylesheet Language: Transformations

(XSLT). In order for XML to be used as a message exchange medium, standard formats need to be defined to structure the XML messages. There are a number of bodies working on creating these standards such as ebXML and the OASIS Universal Business Language (UBL). These standards define a document layout format to provide a standard communicative medium for applications. With UBL, you convert your internal message formats to the standard UBL format and export to the external environment. To import messages, you mirror this process. An extensive examination of this process and relevant standards is outside the scope of this chapter, for further information see [30, 31].

1.6.2 Web Services

Web Services are platform- and language-independent standards defining protocols for heterogeneous system integration. In their most basic format, web services are interfaces that allow programs to run over public or private networks using standard protocols such as the Simple Object Access Protocol (SOAP). They allow links to be created between systems without the need for massive reengineering. They can interact with and/or invoke one another, fulfilling tasks and requests that in turn carry out specific parts of complex transactions or workflows. Web services can be seen in a number of ways, such as a business-to-business/enterprise application integration tool or as a natural evolution of the basic RPC mechanism. The key benefit of a web services deployment is that they act as a façade to the underlying language or platform, a web service written in C and running on Microsoft's Internet Information Server can access a web service written in Java running on BEA's Weblogic server.

1.6.2.1 SOAP

The Simple Object Access Protocol (SOAP) provides a simple and lightweight mechanism for exchanging structured and typed information between peers in a decentralized, distributed environment using XML [32]. SOAP messages contain an envelope, message headers, and a message body. The protocol allows you to bind it to a transport mechanism such as SMTP, HTTP, and JMS. SOAP can be used to implement both the synchronous and asynchronous messaging models. SOAP has a number of uses; it can be used as a document exchange protocol, a heterogeneous interoperability standard, a wire protocol standard (something not defined in JMS), and an RPC mechanism. A detailed discussion on SOAP is available in [32]. For the purposes of this section, SOAP is seen as a document-exchange protocol between heterogeneous systems.

1.6.3 MOM

Message-Oriented Middleware provides an asynchronous, loosely coupled, flexible communication backbone. The benefits of utilizing a MOM in distributed systems have been examined in this chapter. When the benefits of a neutral, independent message format and the ease of web service integration are combined with MOM, highly flexible open systems may be constructed. Such an approach is more likely to be robust with respect to change over a systems lifecycle.

1.6.4 Developing Service-Oriented Architectures

A service is a set of input messages sent to a single object or a composition of objects, with the return of causally related output messages

Through the combination of these technologies, we are able to create Service-Oriented Architectures (SOA). The fundamental design concept behind these architectures is to reduce application processing to logic black boxes. Interaction with these black boxes is achieved with the use of a standard XML message format; by defining the required XML input and output formats, we are able to create black box services. The service can now be accessed by transmitting the message over an asynchronous messaging channel.

With traditional API-based integration, the need to create adaptors and data format converters for each proprietary API is not a scalable solution. As the number of APIs increases, the adaptors and data converters required will scale geometrically. The important aspect of SOAs is their message-centric structure. Where message formats differ, XML-based integration can convert the message to and from the format required using an XML transformation pipeline, as is shown in Figure 1.12.

In this approach, data transformation can be seen as just another assembly line problem, allowing services to be true black box components with the use of an XML-in and XML-out contract. Users of the services simply need to transform their data to the services contract XML format. Integration via SOA is significantly cheaper than integration via APIs; with transformations taking place outside of the applications, it is a very noninvasive method of integration. Service-Oriented Architecture with transforming XML is a new and fresh way of looking at Enterprise Application Integration (EAI) and distributed systems. They also provide a new way of looking at web services that are often touted as a replacement for traditional RPC. Viewed in this light, they are an evolution of the RPC mechanism but still suffer from many of its shortcomings.

With Service-Oriented Architectures, creating connections to trading partners and legacy systems should be as easy as connecting to an interdepartmental system. Once the initial infrastructure has been created for the architecture, the amount of effort to connect to further systems is minimal. This allows systems created with this framework to be highly dynamic, allowing new participants to easily join and leave the system.

Figure 1.13 illustrates an example SOA using the techniques advocated. In this deployment, the system has been created by interconnecting six subsystems and integrating

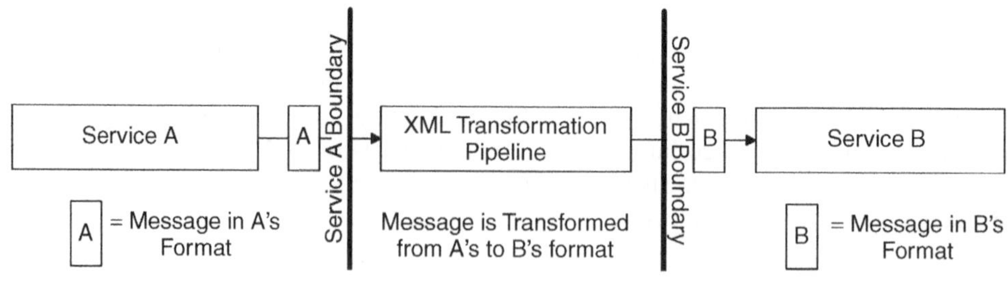

Figure 1.12 XML transformation pipeline

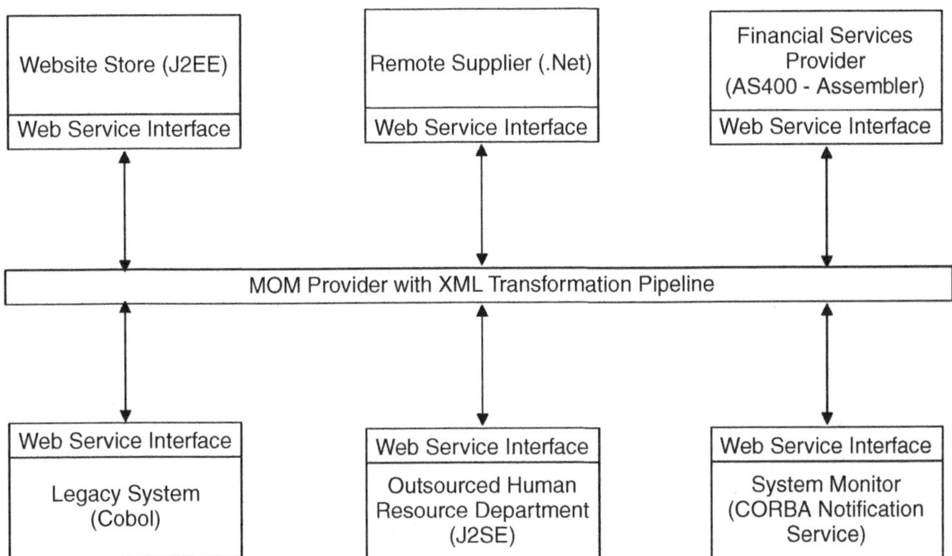

Figure 1.13 System deployed using web service, XML messages and MOM to create a SOA

them, each of the subsystems is built using a different technology for their primary implementation.

The challenges faced in this deployment are common difficulties faced daily by system developers. When developing a new system it is rare for a development team not to have some form of legacy system to interact with. Legacy systems may contain irreplaceable business records, and losing this data could be catastrophic for an organization. It is also very common for legacy systems to contain invaluable business logic that is vital for an organization's day-to-day operation. It would be preferable to reuse this production code, potentially millions of lines, in any new system. Transforming a legacy system into an XML service provides an easy and flexible solution for legacy interaction. The same principle applies to all the other subsystems in the deployment, from the newly created J2EE *Web Store*, or the *Financial Services Provider* running on an AS400 to the *Remote Supplier* running the latest Microsoft .NET systems. Each of these proprietary solutions can be transformed into a service and join the SOA. Once a subsystem has been changed into a service, it can easily be added and removed from the architecture. SOAs facilitate the construction of highly dynamic systems, allowing functionality (services) such as payroll, accounting, sales, system monitors, and so on, to be easily added and removed at run time without interruptions to the overall system. The key to developing a first-rate SOA is to interconnect services with a MOM-based communication; MOM utilization will promote loose coupling, flexibility, reliability, scalability, and high-performance characteristics in the overall system architecture.

$$\frac{\text{XML + Web Services + MOM}}{\text{Service Oriented Architecture}} = \text{Open System}$$

1.7 Summary

Distribution middleware characterizes a high-level remote-programming mechanism designed to automate and extend the native operating system's network programming facilities. This form of middleware streamlines the development of distributed systems by simplifying the mechanics of the distribution process.

Traditionally, one of the predominate forms of distribution mechanisms used is Remote Procedure Calls (RPC). This mechanism, while powerful in small- to medium-scale systems, has a number of shortcomings when used in large-scale multiparticipant systems. An alternative mechanism to RPC has emerged to meet the challenges presented in the mass distribution of large-scale enterprise-level systems.

Message-Oriented Middleware or MOM is a revolutionary concept in distribution allowing for communications between disparate software entities to be encapsulated into messages. MOM solves a number of the inadequacies inherent in the RPC mechanism. MOM can simplify the process of building dynamic, highly flexible enterprise-class distributed systems.

MOM can be defined as any middleware infrastructure that provides messaging capabilities. They provide the backbone infrastructure to create cohesive, dynamic, and highly flexible, distributed applications and are one of the cornerstone foundations that distributed enterprise-class systems are built upon.

The main benefits of MOM come from the asynchronous interaction model and the use of message queues. These queues allow each participating system to proceed at its own pace without interruption. MOM introduces transaction capability and a high Quality of Service (QoS). It also provides a number of communication messaging models to solve a variety of different messaging challenges.

MOM-based systems are proficient in coping with traffic bursts while offering a flexible and robust solution for disperse deployments. Remote systems do not need to be available for the calling program to send a message. Loose coupling exists between the consumers and producers, allowing flexible systems to grow and change on demand. MOM also provides an abstract interface for communications. When MOM is used in conjunction with XML messages and web services, we are able to create highly flexible service-oriented architectures. This form of architecture allows for the flexible integration of multiple systems.

Bibliography

[1] Tanenbaum, A. S. and Steen, M. V. (2002) *Distributed Systems: Principles and Paradigms*, 1st ed., Prentice Hall, New York, USA.

[2] Banavar, G., Chandra, T., Strom, R. E., and Sturman, D. C. (1999) A Case for Message Oriented Middleware. *Proceedings of the 13th International Symposium on Distributed Computing*, Bratislava, Slovak Republic.

[3] Naughton, B. (2003) *Deployment Strategies Focusing on Massive Scalability*, Massive Scalability Focus Group–Operations Support Systems (OSS) through Java Initiative, pp. 1–46. http://java.sun.com/products/oss/pdf/MassiveScalability-1.0.pdf.

[4] Curry, E., Chambers, D., and Lyons, G. (2003) A JMS Message Transport Protocol for the JADE Platform. *Proceedings of the IEEE/WIC International Conference on Intelligent Agent Technology*, Halifax, Canada; IEEE Press, Los Alamitos, CA.

[5] Hinze, A. and Bittner, S. (2002) Efficient Distribution-Based Event Filtering. *Proceedings of the 1st International Workshop on Distributed Event-Based Systems (DEBS'02)*, Vienna, Austria; IEEE Press.

[6] Carzaniga, A., Rosenblum, D.S., and Wolf, A. L. (2001) Design and Evaluation of a Wide-Area Event Notification Service. *ACM Transactions on Computer Systems*, **19**(3), 332–383.

[7] Pietzuch, P. R. and Bacon, J. M. (2002). Hermes: a distributed event-based middleware architecture, *1st International Workshop on Distributed Event-Based Systems (DEBS'02)*, Vienna, Austria; IEEE Computer Society, Los Alamitos, CA.

[8] Curry, E., Chambers, D., and Lyons, G. (2003) Reflective Channel Hierarchies. *Proceedings of the 2nd Workshop on Reflective and Adaptive Middleware, Middleware 2003*, Rio de Janeiro, Brazil; Springer-Verlag, Heidelberg, Germany.

[9] Mühl, G. and Fiege, L. (2001) Supporting Covering and Merging in Content-Based Publish/Subscribe Systems: Beyond Name/Value Pairs. *IEEE Distributed Systems Online*, **2**(7).

[10] Pietzuch, P. R., Shand, B., and Bacon, J. (2003) A Framework for Event Composition in Distributed Systems. *Proceedings of the ACM/IFIP/USENIX International Middleware Conference (Middleware 2003)*, Rio de Janeiro, Brazil; Springer-Verlag, Heidelberg, Germany.

[11] Tai, S. and Rouvellou, I. (2000) Strategies for Integrating Messaging and Distributed Object Transactions. *Proceedings of the Middleware 2000*, New York, USA; Springer-Verlag, Berlin.

[12] Tai, S., Totok, A., Mikalsen, T., and Rouvellou, I. (2003) Message Queuing Patterns for Middleware-Mediated Transactions. *Proceedings of the SEM 2002*, Orlando, FL; Springer-Verlag.

[13] Chambers, D., Lyons, G., and Duggan, J. (2002) A Multimedia Enhanced Distributed Object Event Service. *IEEE Multimedia*, **9**(3), 56–71.

[14] Linthicum, D. (1999) *Enterprise Application Integration*, Addison-Wesley, New York, USA.

[15] Gilman, L. and Schreiber, R. (1996) *Distributed Computing with IBM MQSeries*, John Wiley, New York.

[16] Skeen, D. (1992) An Information Bus Architecture for Large-Scale, Decision-Support Environments. *Proceedings of the USENIX Winter Conference*, San Francisco, CA.

[17] Sonic Software. *Sonic MQ*, http://www.sonicmq.com, April 14, 2004.

[18] Cabrera, L. F., Jones, M. B., and Theimer, M. (2001) Herald: Achieving a Global Event Notification Service. *8th Workshop on Hot Topics in Operating Systems*, Elmau, Germany.

[19] Pietzuch, P. R. (2002) Event-based middleware: a new paradigm for wide-area distributed systems? *6th CaberNet Radicals Workshop Funchal, Madeira Island, Portugal*, http://www.newcastle.research.ec.org/cabernet/workshops/radicals/2002/index.htm.

[20] Carzaniga, A., Rosenblum, D. S., and Wolf, A. L. (2000) Achieving Expressiveness and Scalability in an Internet-Scale Event Notification Service. *Proceedings of the Nineteenth ACM Symposium on Principles of Distributed Computing (PODC2000)*, Portland, OR.

[21] Strom, R., Banavar, G., Chandra, T., Kaplan, M., Miller, K., Mukherjee, B., Sturman, D., and Ward, M. (1998) Gryphon: An Information Flow Based Approach to Message Brokering. *Proceedings of the International Symposium on Software Reliability Engineering*, Paderborn, Germany.

[22] Cugola, G., Nitto, E. D., and Fuggetta, A. (2001) The JEDI Event-Based Infrastructure and its Application to the Development of the OPSS WFMS. *IEEE Transactions on Software Engineering*, **27**(9), 827–850.

[23] Fiege, L. and Mühl, G. *REBECA*, http://www.gkec.informatik.tu-darmstadt.de/rebeca/, April 14, 2004.

[24] ExoLab Group. *OpenJMS*, http://openjms.sourceforge.net/, April 14, 2004.

[25] Object Management Group (2001) *Event Service Specification.*

[26] Object Management Group (2000) *Notification Service Specification.*

[27] Sun Microsystems (2001) *Java Message Service: Specification.*

[28] Haase, K. and Sun Microsystems. *The Java Message Service (JMS) Tutorial*, http://java.sun.com/products/jms/tutorial/, April 14, 2004.

[29] Monson-Haefel, R. and Chappell, D. A. (2001) *Java Message Service*, Sebastopol, CA; O'Reilly & Associates.

[30] Bosak, J. and Crawford, M. (2004) *Universal Business Language (UBL) 1.0 Specification*, OASIS Open, Inc. pp. 1–44, http://www.oasis-open.org/committees/ubl/.

[31] Lyons, T. and Molloy, O. (2003) Development of an e-Business Skillset Enhancement Tool (eSET) for B2B Integration Scenarios. *Proceedings of the IEEE Conference on Industrial Informatics (INDIN 2003)*, Banff, Canada.

[32] W3C (2001) *SOAP Version 1.2.*

2

Adaptive and Reflective Middleware

Edward Curry
National University of Ireland

2.1 Introduction

Middleware platforms and related services form a vital cog in the construction of robust distributed systems. Middleware facilitates the development of large software systems by relieving the burden on the applications developer of writing a number of complex infrastructure services needed by the system; these services include persistence, distribution, transactions, load balancing, clustering, and so on.

The demands of future computing environments will require a more flexible system infrastructure that can adapt to dynamic changes in application requirements and environmental conditions. Next-generation systems will require predictable behavior in areas such as throughput, scalability, dependability, and security. This increase in complexity of an already complex software development process will only add to the already high rates of project failure.

Middleware platforms have traditionally been designed as monolithic static systems. The vigorous dynamic demands of future environments such as large-scale distribution or ubiquitous and pervasive computing will require extreme scaling into large, small, and mobile environments. In order to meet the challenges presented in such environments, next-generation middleware researchers are developing techniques to enable middleware platforms to obtain information concerning environmental conditions and adapt their behavior to better serve their current deployment. Such capability will be a prerequisite for any next-generation middleware; research to date has exposed a number of promising techniques that give middleware the ability to meet these challenges head on.

Adaptive and reflective techniques have been noted as a key emerging paradigm for the development of dynamic next-generation middleware platforms [1, 2]. These

Middleware for Communications. Edited by Qusay H. Mahmoud
 ISBN 0-470-86206-8

techniques empower a system to automatically self-alter (adapt) to meet its environment and user needs. Adaptive and reflective systems support advanced adaptive behavior. Adaptation can take place autonomously or semiautonomously, on the basis of the systems deployment environment, or within the defined policies of users or administrators [3].

The objective of this chapter is to explore adaptive and reflective techniques, their motivation for use, and introduce their fundamental concepts. The application of these techniques will be examined, and a summary of a selection of middleware platforms that utilize these techniques will be conducted. The tools and techniques that allow a system to alter its behavior will be examined; these methods are vital to implementing adaptive and reflective systems. Potential future directions for research will be highlighted; these include advances in programming techniques, open research issues, and the relationship to autonomic computing systems.

2.1.1 Adaptive Middleware

Traditionally, middleware platforms are designed for a particular application domain or deployment scenario. In reality, multiple domains overlap and deployment environments are dynamic, not static; current middleware technology does not provide support for coping with such conditions. Present research has focused on investigating the possibility of enabling middleware to serve multiple domains and deployment environments. In recent years, platforms have emerged that support reconfigurability, allowing platforms to be customized for a specific task; this work has led to the development of adaptive multipurpose middleware platforms.

Adapt–a. To alter or modify so as to fit for a new use

An adaptive system has the ability to change its behavior and functionality. Adaptive middleware is software whose functional behavior can be modified dynamically to optimize for a change in environmental conditions or requirements [4]. These adaptations can be triggered by changes made to a configuration file by an administrator, by instructions from another program, or by requests from its users. The primary requirements of a runtime adaptive system are *measurement*, *reporting*, *control*, *feedback*, and *stability* [1].

2.1.2 Reflective Middleware

The groundbreaking work on reflective programming was carried out by Brian Smith at MIT [5]. Reflective middleware is the next logical step once an adaptive middleware has been achieved. A reflective system is one that can examine and reason about its capabilities and operating environment, allowing it to self-adapt at runtime. Reflective middleware builds on adaptive middleware by providing the means to allow the internals of a system to be manipulated and adapted at runtime; this approach allows for the automated self-examination of system capabilities and the automated adjustment and optimization of those capabilities. The process of self-adaptation allows a system to provide an improved service for its environment or user's needs. Reflective platforms support advanced adaptive behavior; adaptation can take place autonomously on the basis of the status of the system's, environment, or in the defined policies of its users or administrators [3].

Reflect–v. To turn (back), cast (the eye or thought) on or upon something

Reflection is currently a hot research topic within software engineering and development. A common definition of reflection is a system that provides a representation of its own behavior that is amenable to inspection and adaptation and is causally connected to the underlying behavior it describes [6]. Reflective research is also gaining speed within the middleware research community. The use of reflection within middleware for advanced adaptive behavior gives middleware developers the tools to meet the challenges of next-generation middleware, and its use in this capacity has been advocated by a number of leading middleware researchers [1, 7].

Reflective middleware is self-aware middleware [8]

The reflective middleware model is a principled and efficient way of dealing with highly dynamic environments yet supports the development of flexible and adaptive systems and applications [8]. This reflective flexibility diminishes the importance of many initial design decisions by offering late-binding and runtime-binding options to accommodate actual operating environments at the time of deployment, instead of only anticipated operating environments at design time [1].

A common definition of a reflective system [6] is a system that has the following:
Self-Representation: A description of its own behavior
Causally Connected: Alterations made to the self-representation are mirrored in the system's actual state and behavior
Causally Connected Self Representation (CCSR)

Few aspects of a middleware platform would not benefit from the use of reflective techniques. Research is ongoing into the application of these techniques in a number of areas within middleware platforms. While still relatively new, reflective techniques have already been applied to a number of nonfunctional areas of middleware. One of the main reasons nonfunctional system properties are popular candidates for reflection is the ease and flexibility of their configuration and reconfiguration during runtime, and changes to a nonfunctional system property will not directly interfere with a systems user interaction protocols. Nonfunctional system properties that have been enhanced with adaptive and reflective techniques include distribution, responsiveness, availability, reliability, fault-tolerance, scalability, transactions, and security.

Two main forms of reflection exist, behavioral and structural reflection. Behavioral reflection is the ability to intercept an operation and alter its behavior. Structural reflection is the ability to alter the programmatic definition of a programs structure. Low-level structural reflection is most commonly found in programming languages, that is, to change the definition of a class, a function, or a data structure on demand. This form of reflection is outside the scope of this chapter. In this chapter, the focus is on behavioral reflection, specifically altering the behavior of middleware platforms at runtime, and structural reflection concerned with the high-level system architecture and selection of plugable service implementations used in a middleware platform.

2.1.3 Are Adaptive and Reflective Techniques the Same?

Adaptive and reflective techniques are intimately related, but have distinct differences and individual characteristics:

—An adaptive system is capable of changing its behavior.
—A reflective system can inspect/examine its internal state and environment.

Systems can be both adaptive and reflective, can be adaptive but not reflective, as well as reflective but not adaptive. On their own, both of these techniques are useful, but when used collectively, they provide a very powerful paradigm that allows for system inspection with an appropriate behavior adaptation if needed. When discussing reflective systems, it is often assumed that the system has adaptive capabilities.

Common Terms

Reification
The process of providing an external representation of the internals of a system. This representation allows for the internals of the system to be manipulated at runtime.

Absorption
This is the process of enacting the changes made to the external representation of the system back into the internal system. Absorbing these changes into the system realizes the casual connection between the model and system.

Structural Reflection
Structural reflection provides the ability to alter the statically fixed internal data/functional structures and architecture used in a program. A structural reflective system would provide a complete reification of its internal methods and state, allowing them to be inspected and changed. For example, the definition of a class, a method, a function, and so on, may be altered on demand.
Structural reflection changes the internal makeup of a program

Behavioral Reflection
Behavioral reflection is the ability to intercept an operation such as a method invocation and alter the behavior of that operation. This allows a program, or another program, to change the way it functions and behaves.
Behavioral reflection alters the actions of a program

Nonfunctional Properties
The nonfunctional properties of a system are the behaviors of the system that are not obvious or visible from interaction with the system. Nonfunctional properties include distribution, responsiveness, availability, reliability, scalability, transactions, and security.

2.1.4 Triggers of Adaptive and Reflective Behavior

In essence, the reflective capabilities of a system should trigger the adaptive capabilities of a system. However, what exactly can be inspected in order to trigger an appropriate adaptive behavior? Typically, a number of areas within a middleware platform, its functionality, and its environment are amenable to inspection, measurement, and reasoning as to the optimum or desired performance/functionality. Software components known as *interceptors* can be inserted into the execution path of a system to monitor its actions. Using interceptors and similar techniques, reflective systems can extract useful information from the current execution environment and perform an analysis on this information.

Usually, a reflective system will have a number of interceptors and system monitors that can be used to examine the state of a system, reporting system information such as its performance, workload, or current resource usage. On the basis of an analysis of this information, appropriate alterations may be made to the system behavior. Potential monitoring tools and feedback mechanisms include performance graphs, benchmarking, user usage patterns, and changes to the physical deployments infrastructure of a platform (network bandwidth, hardware systems, etc).

2.2 Implementation Techniques

Software development has evolved from the 'on-the-metal' programming of assembly and machine codes to higher-level paradigms such as procedural, structured, functional, logic, and Object-Orientation. Each of these paradigms has provided new tools and techniques to facilitate the creation of complex software systems with speed, ease, and at lower development costs.

In addition to advancements in programming languages and paradigms, a number of techniques have been developed that allow flexible dynamic systems to be created. These techniques are used in adaptive systems to enable their behavior and functionality changes. This section provides an overview of such techniques, including meta-level programming, components and component framework, and generative programming.

2.2.1 Meta-Level Programming

In 1991, Gregor Kiczale's work on combining the concept of computational reflection and object-oriented programming techniques lead to the definition of a meta-object protocol [9]. One of the key aspects of this groundbreaking work was in the separation of a system into two levels. The base-level provides system functionality, and the meta-level contains the policies and strategies for the behavior of the system. The inspection and alteration of this meta-level allows for changes in the system's behavior.

The base-level provides the implementation of the system and exposes a meta-interface that can be accessed at the meta-level. This meta-interface exposes the internals of the base-level components/objects, allowing it to be examined and its behavior to be altered and reconfigured. The base-level can now be reconfigured to maximize and fine-tune the systems characteristics and behavior to improve performance in different contexts and operational environments. This is often referred to as the *Meta-Object Protocol* or MOP. The design of a meta-interface/MOP is central to studies of reflection, and the interface should be sufficiently general to permit unanticipated changes to the platform but should also be restricted to prevent the integrity of the system from being destroyed [10].

Meta Terms Explained

Meta-

Prefixed to technical terms to denote software, data, and so on, which operate at a higher level of abstraction – Oxford English Dictionary

Meta-Level

The level of software that abstracts the functional and structural level of a system. Meta-level architectures are systems designed with a base-level (implementation level) that handles the execution of services and operations, and a meta-level that provides an abstraction of the base-level.

Meta-Object

The participants in an object-oriented meta-level are known as meta-objects

Meta-Object Protocol

The protocol used to communicate with the meta-object is known as the Meta-Object Protocol (MOP)

2.2.2 *Software Components and Frameworks*

With increasing complexity in system requirements and tight development budget constraints, the process of programming applications from scratch is becoming less feasible. Constructing applications from a collection of reusable components and frameworks is emerging as a popular approach to software development.

A software component is a functional discrete block of logic. Components can be full applications or encapsulated functionality that can be used as part of a larger application, enabling the construction of applications using components as building blocks. Components have a number of benefits as they simplify application development and maintenance, allowing systems to be more adaptive and respond rapidly to changing requirements. Reusable components are designed to encompass a reusable block of software, logic, or functionality. In recent years, there is increased interest in the use of components as a mechanism of building middleware platforms; this approach has enabled middleware platforms to be highly flexible to changing requirements.

Component frameworks are a collection of interfaces and interaction protocols that define how components interact with each other and the framework itself, in essence frameworks allow components to be plugged into them. Examples of component frameworks include Enterprise Java Beans (EJB) [11] developed by Sun Microsystems, Microsoft's .NET [12] and the CORBA Component Model (CMM) [13]. Components frameworks have also been used as a medium for components to access middleware services, for example, the EJB component model simplifies the development of middleware applications by providing automatic support for services such as transactions, security, clustering, database connectivity, life-cycle management, instance pooling, and so on. If components are analogous to building blocks, frameworks can be seen as the cement that holds them together.

The component-oriented development paradigm is seen as a major milestone in software construction techniques. The process of creating applications by composing preconstructed program 'blocks' can drastically reduce the cost of software development. Components and component frameworks leverage previous development efforts by capturing key implementation patterns, allowing their reuse in future systems. In addition, the use of replaceable software components can improve reliability, simplify the implementation, and reduce the maintenance of complex applications [14].

2.2.3 Generative Programming

Generative programming [15] is the process of creating programs that construct other programs. The basic objective of a generative program, also known as a program generator [16], is to automate the tedious and error-prone tasks of programming. Given a requirements specification, a highly customized and optimized application can be automatically manufactured on demand. Program generators manufacture source code in a target language from a program specification expressed in a higher-level Domain Specific Language (DSL). Once the requirements of the system are defined in the higher-level DSL, the target language used to implement the system may be changed. For example, given the specification of a text file format, a program generator could be used to create a driver program to edit files in this specified format. The program generator could use Java, C, Visual Basic (VB) or any other language as the target language for implementation; two program generators could be created, a Java version and a C version. This would allow the user a choice for the implementation of the driver program.

Generative programming allows for high levels of code reuse in systems that share common concepts and tasks, providing an effective method of supporting multiple variants of a program; this collection of variants is known as a *program family*. Program generation techniques may also be used to create systems capable of adaptive behavior via program recompilation.

2.3 Overview of Current Research

Adaptive and reflective capabilities will be commonplace in future next-generation middleware platforms. There is consensus [1, 8] that middleware technologies will continue to incorporate this new functionality. At present, these techniques have been applied to a number of middleware areas. There is growing interest in developing reflective middleware with a large number of researchers and research groups carrying out investigations in this area. A number of systems have been developed that employ adaptive and reflective techniques, this section provides an overview of some of the more popular systems to have emerged. For a comprehensive list of current research visit the IEEE Distributed Systems Online Reflective and Adaptive Middleware Resource Collection – http://dsonline.computer.org/middleware/rmlinks.html.

2.3.1 Reflective and Adaptive Middleware Workshops

Reflective and adaptive middleware is a very active research field with the completion of a successful workshop on the subject at the IFIP/ACM Middleware 2000 conference [17].

Papers presented at this workshop cover a number of topics including reflective and adaptive architectures and systems, mathematical models and performance measurements of reflective platforms. Building on the success of this event a second workshop took place at Middleware 2003, The 2nd Workshop on Reflective and Adaptive Middleware [18] covered a number of topics including nonfunctional properties, distribution, components, and future research trends. A third Workshop, RM2004, will take place at Middleware 2004.

2.3.2 *Nonfunctional Properties*

Nonfunctional properties of middleware platforms have proved to be very popular candidates for enhancement with adaptive and reflective techniques. These system properties are the behaviors of the system that are not obvious or visible from interaction with the system. This is one of the primary reasons they have proved popular with researchers, because they are not visible in user/system interactions any changes made to these properties will not affect the user/system interaction protocol. Nonfunctional properties that have been enhanced with adaptive and reflective techniques include distribution, responsiveness, availability, reliability, scalability, transactions, and security.

2.3.2.1 Security

The specialized Obol [19] programming language provides flexible security mechanisms for the Open ORB Python Prototype (OOPP). In OOPP, the flexible security mechanisms based on Obol is a subset of the reflective features of the middleware platform that enables programmable security via Obol. Reflective techniques within OOPP provide the mechanisms needed to access and modify the environment; Obol is able to access the environment meta-model making it possible to change and replace security protocols without changing the implementation of the components or middleware platform.

2.3.2.2 Quality-of-Service

A system with a Quality-of-Service (QoS) demand is one that will perform unacceptably if it is not carefully configured and tuned for the anticipated environment and deployment infrastructure. Systems may provide different levels of service to the end-user, depending on the deployment environment and operational conditions. An application that is targeted to perform well in a specific deployment environment will most likely have trouble if the environment changes. As an illustration of this concept, imagine a system designed to support 100 simultaneous concurrent users, if the system was deployed in an environment with 1000 or 10,000 users it will most likely struggle to provide the same level of service or QoS when faced with demands that are 10 or 100 times greater than what it is designed to handle.

Another example is a mobile distributed multimedia application. This type of application may experience drastic changes in the amount of bandwidth provided by the underlying network infrastructure from the broadband connections offered by residential or office networks to the 9600 bps GSM connection used during transit. An application designed to operate on a broadband network will encounter serious difficulties when deployed over the substantially slower GSM-based connection. Researchers at Lancaster University have developed a reflective middleware platform [10] that adapts to the underlying network infrastructure in order to improve the QoS provided by the application. This research alters the methods used to deliver the content to the mobile client, achieved by using an appropriate video and audio compression component for the network bandwidth available or the addition of a jitter-smoothing buffer to a network with erratic delay characteristics.

2.3.2.3 Fault-Tolerant Components

Adaptive Fault-Tolerance in the CORBA Component Model (AFT-CCM) [20] is designed for building component-based applications with QoS requirements related to fault-tolerance. AFT-CCM is based on the CORBA Component Model (CCM) [13] and allows an application user to specify QoS requirements such as levels of dependability or availability for a component. On the basis of these requirements, an appropriate replication technique and the quantity of component replicas will be set to achieve the target. These techniques allow a component-based distributed application to be tolerant of possible component and machine faults. The AFT-CCM model enables fault-tolerance in a component with complete transparency for the application without requiring changes to its implementation.

2.3.3 Distribution Mechanism

A number of reflective research projects focus on improving the flexibility of application distribution. This section examines the use of adaptive and reflective techniques in enhancing application distribution mechanisms.

2.3.3.1 GARF and CodA

Projects such as GARF and CodA are seen as a milestone in reflective research. GARF [21] (automatic generation of reliable applications) is an object-oriented tool that supports the design and programming of reliable distributed applications. GARF wraps the distribution primitives of a system to create a uniform abstract interface that allows the basic behavior of the system to be enhanced. One technique to improve application reliability is achieved by replicating the application's critical components over several machines. Group-communication schemes are used to implement these techniques by providing multicasting to deliver messages to groups of replicas. In order to implement this group-communication, multicasting functionality needs to be mixed with application functionally. GARF acts as an intermediate between group-communication functionality and applications; this promotes software modularity by clearly separating the implementation of concurrency, distribution, and replication from functional features of the application.

The CodA [22] project is a pioneering landmark in reflective research. Designed as an object meta-level architecture, its primary design goal was to allow for decomposition by logical behavior. Through the application of the decomposition OO technique, CodA eliminated the problems existing in 'monolithic' meta-architectures. CodA achieves this by using multiple meta-objects, with each one describing a single small behavioral aspect of an object, instead of using one large meta-object that describes all aspects of an objects behavior. Once the distribution concern has been wrapped in meta-objects, aspects of the systems distribution such as message queues, message sending, and receiving can be controlled. This approach offers a fine-grained approach to decomposition.

2.3.3.2 Reflective Architecture Framework for Distributed Applications

The Reflective Architecture Framework for Distributed Applications (RAFDA) [23] is a reflective framework enabling the transformation of a nondistributed application into a flexibly distributed equivalent one. RAFDA allows an application to adapt to its environment by dynamically altering its distribution boundaries. RAFDA can transform a local object into a remote object, and vice versa, allowing local and remote objects to be interchangeable.

As illustrated in Figure 2.1, RAFDA achieves flexible distribution boundaries by substituting an object with a proxy to a remote instance. In the example in Figure 2.1, objects A and B, both hold references to a shared instance of object C, all objects exist in a single address space (nondistributed). The objective is to move object C to a new address space. RAFDA transforms the application so that the instance of C is remote to its reference holders; the instance of C in address space A is replaced with a proxy, Cp, to the remote implementation of C in address space B.

The process of transformation is performed at the bytecode level. RAFDA identifies points of substitutability and extracts an interface for each substitutable class; every reference to a substitutable class must then be transformed to use the extracted interface. The proxy implementations provide a number of transport options including Simple Object Access Protocol (SOAP), Remote Method Invocation (RMI), and Internet Inter-ORB Protocol (IIOP). The use of interfaces makes nonremote and remote versions of a class interchangeable, thus allowing for flexible distribution boundaries. Policies determine substitutable classes and the transportation mechanisms used for the distribution.

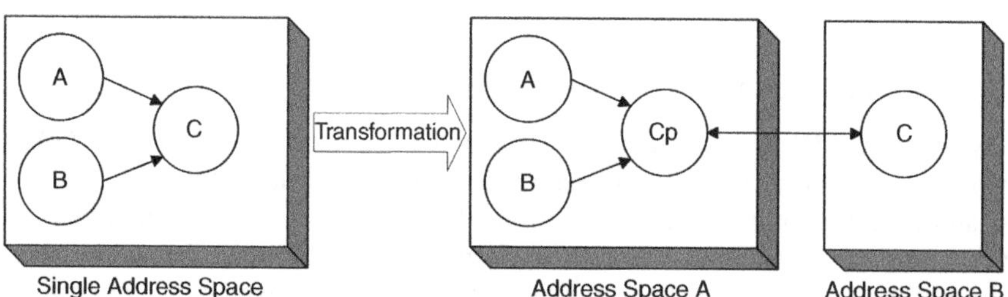

Figure 2.1 RAFDA redistribution transformation. Reproduced by permission of Alan Dearle, in Portillo, A. R., Walker, S., Kirby, G., *et al.* (2003) A Reflective Approach to Providing Flexibility in Application Distribution. *Proceedings of the 2nd Workshop on Reflective and Adaptive Middleware, Middleware 2003*, Rio de Janeiro, Brazil

2.3.3.3 mChaRM

The Multi-Channel Reification Model (mChaRM) [24] is a reflective approach that reifies and reflects directly on communications. The mChaRM model does not operate on base-objects but on the communications between base-objects, resulting in a communication-oriented model of reflection. This approach abstracts and encapsulates interobject communications and enables the meta-programmer to enrich and or replace the predefined communication semantics. mChaRM handles a method call as a message sent through a logical channel between a set of senders and a set of receivers. The model supports the reification of such logical channels into logical objects called *multi-channels*. A multi-channel can enrich the messages (method calls) with new functionality. This technique allows for a finer reification reflection granularity than those used in previous approaches, and for a simplified approach to the development of communication-oriented software. mChaRM is specifically targeted for designing and developing complex communication mechanism from the ground up, or for extending the behavior of a current communication mechanism; it has been used to extend the standard Java RMI framework to one supporting multicast RMI.

2.3.3.4 Open ORB

The Common Object Request Broker Architecture (CORBA) [25] is a popular choice for research projects applying adaptive and reflective techniques. A number of projects have incorporated these techniques into CORBA Object Request Brokers or ORBs.

The Open ORB 2 [10] is an adaptive and dynamically reconfigurable ORB supporting applications with dynamic requirements. Open ORB has been designed from the ground up to be consistent with the principles of reflection. Open ORB exposes an interface (framework) that allows components to be plugable; these components control several aspects of the ORBs behavior including thread, buffer, and protocol, management. Open ORB is implemented as a collection of configurable components that can be selected at build-time and reconfigured at runtime; this process of component selection and configurability enables the ORB to be adaptive.

Open ORB is implemented with a clear separation between base-level and meta-level operations. The ORBs meta-level is a causally connected self-representation of the ORBs base-level (implementation) [10]. Each base-level component may have its own private set of meta-level components that are collectively referred to as the components meta-space. Open ORB has broken down its meta-space into several distinct models. The benefit of this approach is to simplify the interface to the meta-space by separating concerns between different system aspects, allowing each distinct meta-space model to give a different view of the platform implementation that can be independently reified. As shown in Figure 2.2, models cover the interfaces, architecture, interceptor, and resource

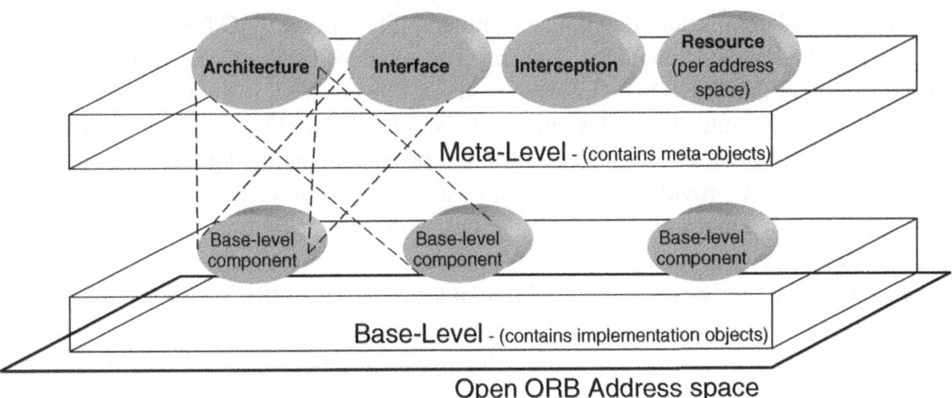

Figure 2.2 Open ORB architecture. Reproduced by permission of IEEE, in Blair, G. S., Coulson, G., Andersen, A., *et al.* (2001) The Design and Implementation of Open ORB 2, *IEEE Distributed Systems Online*, **2**(6)

meta-spaces. These models provide access to the underlying platform and component structure through reflection; every application-level component offers a meta-interface that provides access to an underlying meta-space, which is the support environment for the component.

Structural Reflection

Open ORB version 2 uses two meta-models to deal with structural reflection, one for its external interfaces and one for its internal architecture. The interface meta-model acts similar to the Java reflection API allowing for the dynamic discovery of a component's interfaces at runtime. The architecture meta-model details the implementation of a component broken down into two lines, a component graph (a local-binding of components) and an associated set of architectural constraints to prevent system instability [10]. Such an approach makes it possible to place strict controls on access rights for the ORBs adaptation. This allows all users the right to access the interface meta-model while restricting access rights to the architecture meta-model permitting only trusted third parties to modify the system architecture.

Behavioral Reflection

Two further meta-models exist for behavioral reflection, the interception, and resource models. The interception model enables the dynamic insertion of interceptors on a specific interface allowing for the addition of prebehavior and postbehavior. This technique may be used to introduce nonfunctional behavior into the ORB. Unique to Open ORB is its use of a resource meta-model allowing for access to the underlying system resources, including memory and threads, via resource abstraction, resource factories, and resource managers [10].

2.3.3.5 DynamicTAO – Real-Time CORBA

Another CORBA-based reflective middleware project is DynamicTAO [26]. DynamicTAO is designed to introduce dynamic reconfigurability into the TAO ORB [27] by adding reflective and adaptive capabilities. DynamicTAO enables on-the-fly reconfiguration and customization of the TAO ORBs internal engine, while ensuring it is maintained in a consistent state. The architecture of DynamicTAO is illustrated in Figure 2.3; in this architecture, reification is achieved through a collection of component configurators. Component implementations are provided via libraries. DynamicTAO allows these components to be dynamically loaded and unloaded from the ORBs process at runtime, enabling the ORB to be inspected and for its configuration to be adapted. Component implementations are organized into categories representing different aspects of the ORB's internal engine such as concurrency, security, monitoring, scheduling, and so on. Inspection in DynamicTAO is achieved through the use of interceptors that may be used to add support for monitoring, these interceptors may also be used to introduce behaviors for cryptography, compression, access control, and so on. DynamicTAO is designed to add reflective features to the TAO ORB, reusing the codebase of the existing TAO ORB results in a very flexible, dynamic, and customizable system implementation.

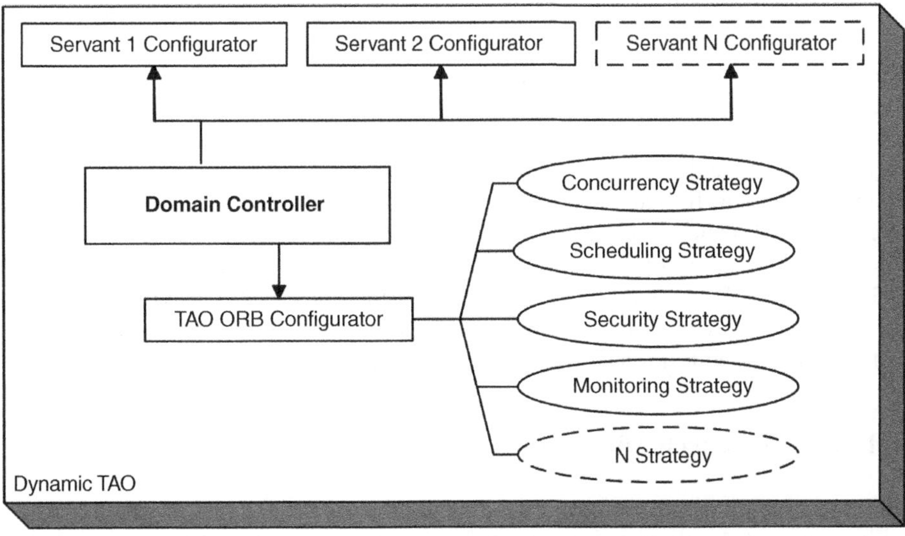

Figure 2.3 Architecture of dynamicTAO. Reproduced by permission of Springer, in Kon, F., Román, M., Liu, P., *et al.* (2002) Monitoring, Security, and Dynamic Configuration with the dynamicTAO Reflective ORB. *Proceedings of the IFIP/ACM International Conference on Distributed Systems Platforms and Open Distributed Processing (Middleware'2000)*, New York

2.3.3.6 Reflective Message-Oriented Middleware with Chameleon

The Chameleon message framework [28, 29] provides a generic technique for extending Message-Oriented Middleware (MOM) services using interception, specifically it has been used to apply reflective techniques to MOM services. Reflective capabilities focus on the application of such techniques within message queues and channel/topic hierarchies used as part of the publish/subscribe messaging model.

The publish/subscribe messaging model is a very powerful mechanism used to disseminate information between anonymous message consumers and producers. In the publish/subscribe model, clients producing messages 'publish' these to a specific topic or channel. These channels are 'subscribed' to by clients wishing to consume messages of interest to them. Hierarchical channel structures allows channels to be defined in a hierarchical fashion so that channels may be nested under other channels. Each subchannel offers a more granular selection of the messages contained in its parent channel. Clients of hierarchical channels subscribe to the most appropriate level of channel in order to receive the most relevant messages. For further information on the publish/subscribe model and hierarchical channels, please refer to Chapter 1 (Message-Oriented Middleware).

Current MOM platforms do not define the structure of channel hierarchies. Application developers must therefore manually define the structure of the hierarchy at design time. This process can be tedious and error-prone. To solve this problem, the Chameleon messaging architecture implements reflective channel hierarchies [28, 29, 41] with the ability to autonomously self-adapt to their deployment environment. The Chameleon architecture exposes a causally connected meta-model to express the set-up and configuration of the queues and the structure of the channel hierarchy, which enables the runtime inspection and adoption of the hierarchy and queues within MOM systems.

Chameleon's adaptive behavior originates from its reflection engine whose actions are governed by plugable reflective policies (Interceptors); these intelligent policies contain the rules and strategies used in the adaptation of the service. Potential policies could be designed to optimize the distribution mechanism for group messaging using IP multicasts or to provide support for federated messaging services using techniques from Hermes [30] or Herald [31]. Policies could also be designed to work with different levels of filtering (subject, content, composite) or to support different formats of message payloads (XML, JPEG, PDF, etc). Policies could also be used to introduce new behaviors into the service; potential behaviors include collaborative filtering/recommender systems, message transformations, monitoring and accounting functionality.

2.4 Future Research Directions

While it is difficult to forecast the future direction of any discipline, it is possible to highlight a number of developments on the immediate horizon that could affect the direction taken by reflective research over the coming years. This section will look at the implications of new software engineering techniques and highlight a number of open research issues and potential drawbacks that effect adaptive and reflective middleware platforms. The section will conclude with an introduction to autonomic computing and the potential synergy between autonomic and reflective systems.

2.4.1 Advances in Programming Techniques

The emergence of multifaceted software paradigms such as Aspect-Oriented Programming (AOP) and Multi-Dimensional Separation of Concerns (MDSOC) will have a profound effect on software construction. These new paradigms have a number of benefits for the application of adaptive and reflective techniques in middleware systems. This section provides a brief overview of these new programming techniques.

2.4.1.1 Aspect-Oriented Programming (AOP)

We can view a complex software system as a combined implementation of multiple concerns, including business-logic, performance, logging, data and state persistence, debugging and unit tests, error checking, multithreaded safety, security, and various other concerns. Most of these are system-wide concerns and are implemented throughout the entire codebase of the system; these system-wide concerns are known as crosscutting concerns.

The most popular practice for implementing adaptive and reflective systems is the Object-Oriented (OO) paradigm. Excluding the many benefits and advantages object-oriented programming has over other programming paradigms, object-oriented and reflective techniques have a natural fit. The OO paradigm is a major advancement in the way we think of and build software, but it is not a silver bullet and has a number of limitations. One of these limitations is the inadequate support for crosscutting concerns. The Aspect-Oriented-Programming (AOP) [32] methodology helps overcome this limitation,

AOP complements OO by creating another form of separation that allows the implementation of a crosscutting concern as a single unit. With this new method of concern separation, known as an *aspect*, crosscutting concerns are more straightforward to implement. Aspects can be changed, removed, or inserted into a systems codebase enabling the reusability of crosscutting code.

A brief illustration would be useful to explain the concept. The most commonly used example of a crosscutting concern is that of logging or execution tracking; this type of functionality is implemented throughout the entire codebase of an application making it difficult to change and maintain. AOP [32] allows this functionality to be implemented in a single aspect; this aspect can now be applied/weaved throughout the entire codebase to achieve the required functionality.

Dynamic AOP for Reflective Middleware

The Object-Oriented paradigm is widely used within reflective platforms. However, a clearer separation of crosscutting concerns would be of benefit to meta-level architectures. This provides the incentive to utilize AOP within reflective middleware projects.

A major impediment to the use of AOP techniques within reflective systems has been the implementation techniques used by the initial incarnations of AOP [33]. Traditionally, when an aspect is inserted into an object, the compiler weaves the aspect into the objects code; this results in the absorption of the aspect into the object's runtime code. The lack of preservation of the aspect as an identifiable runtime entity is a hindrance to the dynamic adaptive capabilities of systems created with aspects. Workarounds to this problem exist in the form of dynamic system recompilation at runtime; however, this is not an ideal solution and a number of issues, such as the transference of the system state, pose problems.

Alternative implementations of AOP have emerged that do not have this limitation. These approaches propose a method of middleware construction using composition [34] to preserve aspect as runtime entities, this method of creation facilitates the application of AOP for the construction of reflective middleware platforms. Another approach involving Java bytecode manipulation libraries such as Javassist [35] provide a promising method of implementing AOP frameworks (i.e. AspectJ 1.2+, AspectWerkz and JBossAOP) with dynamic runtime aspect weaving.

One of the founding works on AOP highlighted the process of performance optimization that bloated a 768-line program to 35,213 lines. Rewriting the program with the use of AOP techniques reduced the code back to 1039 lines while retaining most of the performance benefits. Grady Booch, while discussing the future of software engineering techniques, predicts the rise of multifaceted software, that is, software that can be composed in multiple ways at once, he cites AOP as one of the first techniques to facilitate a multifaceted capability [36].

2.4.1.2 Multi-Dimensional Separation of Concerns

The key difference between AOP and Multi-Dimensional Separation of Concerns [37] (MDSOC) is the scale of multifaceted capabilities. AOP will allow multiple crosscutting aspects to be weaved into a program, thus changing its composition through the addition of these aspects. Unlike AOP, MDSOC multifaceted capabilities are not limited to the use of aspects; MDSOC allows for the entire codebase to be multifaceted, enabling the software to be constructed in multiple dimensions.

MDSOC also supports the separation of concerns for a single model [38], when using AOP you start with a base and use individually coded aspects to augment this base. Working from a specific base makes the development of the aspects more straightforward but also introduces limitations such as restrictions on aspect composition [38]; you cannot have an aspect of an aspect. In addition, aspects can be tightly coupled to the codebase for which they are designed; this limits their reusability.

MDSOC enables software engineers to construct a collection of separate models, each encapsulating a concern within a class hierarchy specifically designed for that concern [38]. Each model can be understood in isolation, any model can be augmented in isolation, and any model can be augmented with another model. These techniques streamline the division of goals and tasks for developers. Even with these advances, the primary benefit of MDSOC comes from its ability to handle multiple decompositions of the same software simultaneously, some developers can work with classes, others with features, others with business rules, other with services, and so on, even though they model the system in substantially different ways [38].

To further illustrate these concepts an example is needed, a common scenario by Ossher [38] is of a software company developing personnel management systems for large international organizations. For the sake of simplicity, assume that their software has two areas of functionality, personal tracking that records employees' personal details such as name, address, age, phone number, and so on, and payroll management that handles salary and tax information.

Different clients seeking similar software approach the fictitious company, they like the software but have specific requirements, some clients want the full system while others do not want the payroll functionality and refuse to put up with the extra overhead within their system implementation.

On the basis of market demands, the software house needs to be able to mix and match the payroll feature. It is extremely difficult to accomplish this sort of dynamic feature selection using standard object-oriented technology. MDSOC allows this flexibility to be achieved within the system with on-demand remodularization capabilities; it also allows the personnel and payroll functionality to be developed almost entirely separate using different class models that best suit the functionality they are implementing.

2.4.2 Open Research Issues

There are a number of open research issues with adaptive and reflective middleware systems. The resolution of these issues is critical to the wide-scale deployment of adaptive and reflective techniques in production and mission critical environments. This section highlights a number of the more common issues.

2.4.2.1 Open Standards

The most important issue currently faced by adaptive and reflective middleware researchers is the development of an open standard for the interaction of their middleware platforms. An international consensus is needed on the interfaces and protocols used to interact with these platforms. The emergence of such standards is important to support the development of next-generation middleware platforms that are configurable and reconfigurable and also to offer applications portability and interoperability across proprietary implementation of such platforms [10]. Service specifications and standards are needed to provide a stable base upon which to create services for adaptive and reflective middleware platforms. Because of the large number of application domains that may use these techniques, one generic standard may not be enough; a number of standards may be needed.

As adaptive and reflective platforms mature, the ability of such system to dynamically discover components with corresponding configuration information at runtime would be desirable. Challenges exist with this proposition, while it is currently possible to examine a components interface at runtime; no clear method exists for documenting the functionality of neither a component nor its performance or behavioral characteristics. A standard specification is needed to specify what is offered by a component.

2.4.2.2 System Cooperation

One of the most interesting research challenges in future middleware platforms is the area of cooperation and coordination between middleware services to achieve a mutual beneficial outcome. Middleware platforms may provide different levels of services, depending on environmental conditions and resource availability and costs. John Donne said 'No man is an island'; likewise, no adaptive or reflective middleware platform, service, or component is an island, and each must be aware of both the individual consequences and group consequences of its actions. Next-generation middleware systems must coordinate/trade with each other in order to maximize the available resources to meet the system requirements.

To achieve this objective, a number of topics need to be investigated. The concept of negotiation-based adaptation will require mechanisms for the trading of resources and resource usage. A method of defining a resource, its capabilities, and an assurance of the QoS offered needs to be developed. Trading partners need to understand the commodities they are trading in. Resources may be traded in a number of ways from simple barter between two services to complex auctions with multiple participants, each with their own tradable resource budget, competing for the available resource. Once a trade has been finalized, enforceable contracts will be needed to ensure compliance with the trade agreement. This concept of resource trading could be used across organizational boundaries with the trading of unused or surplus resources in exchange for monetary reimbursement.

2.4.2.3 Resource Management

In order for next-generation middleware to maximize system resource usage and improve quality-of-service, it must have a greater knowledge with regard to the available resources and their current and projected status. Potentially, middleware platforms may wish to participate in system resource management. A number of resource support services need to be developed, such as mechanisms to interact with a resource, obtain a resource's status, and coordination techniques to allow a resource to be reserved for future usage at a specified time. Techniques are required to allow middleware to provide resource management policies to the underlying system-level resource managers or at the minimum to influence these policies by indicating the resources it will need to meet its requirements.

2.4.2.4 Performance

Adaptive and reflective systems may suffer in performance because of additional infrastructure required to facilitate adaptations and the extra self-inspection workload required by self-adaptation; such systems contain an additional performance overhead when compared to a traditional implementation of a similar system. However, under certain circumstances, the changes made to the platform through adaptations can improve performance and reduce the overall workload placed on the system. This saving achieved by adaptations may offset the performance overhead or even write it off completely.

System speed may not always be the most important measurement of performance for a given system, for example, the Java programming language is one of the most popular programming languages even though it is not the fastest language; other features such as its cross-platform compatibility make it a desirable option. With an adaptive and reflective platform, a performance decrease may be expected from the introduction of new features, what limits in performance are acceptable to pay for a new feature? What metrics may be used to measure such a trade-off? How can a real measurement of benefit be achieved?

Reflective systems will usually have a much larger codebase compared to a nonreflective one, such code-bloat is due to the extra code needed to allow for the system to be inspected and adapted as well as the logic needed to evaluate and reason about the systems adaptation. This larger codebase results in the platform having a larger memory footprint. What techniques could be used to reduce this extra code? Could this code be made easily reusable within application domains?

2.4.2.5 Safe Adaptation

Reflection focuses on increasing flexibility and the level of openness. The lack of safe-bounds for preventing unconstrained system adaptation resulting in system malfunctions is a major concern for reflective middleware developers. This has been seen as an 'Achilles heel' of reflective systems [39]. It is important for system engineers to consider the impact that reflection may have on system integrity and to include relevant checks to ensure that integrity is maintained. Techniques such as architectural constraints are a step in the right direction to allowing safe adaptations. However, more research is needed in this area, particularly where dynamically discovered components are introduced into a platform. How do we ensure that such components will not corrupt the platform? How do we discover the existence of such problems? Again, standards will be needed to document component behavior with constant checking of its operations to ensure it does not stray from its contracted behavior.

2.4.2.6 Clearer Separation of Concerns

The clearer separation of concerns within code is an important issue for middleware platforms. A clear separation of concerns would reduce the work required to apply adaptive and reflective techniques to a larger number of areas within middleware systems. The use of dynamic AOP and MDSOC techniques to implement nonfunctional and crosscutting concerns eases the burden of introducing adaptive and reflective techniques within these areas.

The separation of concerns with respect to responsibility for adaptation is also an important research area, multiple subsystems within a platform may be competing for specific adaptations within the platform, and these adaptations may not be compatible with one another. With self-configuring systems and specifically when these systems evolve to become self-organizing groups, who is in charge of the group's behavior? Who performs the mediations between the conflicting systems? Who chooses what adaptations should take place? These issues are very important to the acceptance of a self-configuration system within production environments.

The Object Management Group (OMG) Model Driven Architecture (MDA) [40] defines an approach for developing systems that separates the specification of system functionality from the specification of the implementation of that functionality with a specific technology. MDA can be seen as an advance on the concept of generative programming. The MDA approach uses a Platform Independent Model (PIM) to express an abstract system design that can be implemented by mapping or transforming to one or more Platform Specific Models (PSMs). The major benefit of this approach is that you define a system model over a constantly changing implementation technology allowing your system to be easily updated to the latest technologies by simple switching to the PSMs for the new technology. The use of MDA in conjunction with reflective component-based middleware platforms could be a promising approach for developing future systems.

2.4.2.7 Deployment into Production Environments

Deployment of adaptive and reflective systems into production mission critical environments will require these systems to reach a level of maturity where system administrators feel comfortable with such a platform in their environment. Of utmost importance to reaching this goal is the safe adaptation of the system with predictable results in the systems behavior. The current practices used to test systems are inadequate for adaptive and reflective systems. In order to be accepted as a deployable technology, it is important for the research community to develop the necessary practices and procedures to test adaptive and reflective systems to ensure they perform predictably. Such mechanisms will promote confidence in the technology. Adaptive and reflective techniques must also mature enough to require only the minimum amount of system adoption necessary to achieve the desired goal. Once these procedures are in place, an incremental approach to the deployment of these systems is needed; the safe coexistence of both technologies will be critical to acceptance, and it will be the responsibility of adaptive and reflective systems to ensure that their adaptations do not interfere with other systems that rely on them or systems they interact with.

2.4.3 Autonomic Computing

As system workloads and environments become more unpredictable and complex, they will require skilled administration personnel to install, configure, maintain, and provide 24/7 support. In order to solve this problem, IBM has announced an autonomic computing initiative. IBM's vision of autonomic computing [41] is an analogy with the human autonomic nervous system; this biological system relieves the conscious brain of the burden of having to deal with low-level routine bodily functions such as muscle use, cardiac muscle use (respiration), and glands. An autonomic computing system would relieve the burden of low-level functions such as installation, configuration, dependency

Table 2.1 Fundamental characteristics of autonomic systems

Characteristic	Description
Self-Configuring	The system must adapt automatically to its operating environment. Hardware and software platforms must possess a self-representation of their abilities and self-configure to their environment
Self-Healing	Systems must be able to diagnose and solve service interruptions. For a system to be self-healing, it must be able to recognize a failure and isolate it, thus shielding the rest of the system from its erroneous activity. It then must be capable of recovering transparently from failure by fixing or replacing the section of the system that is responsible for the error
Self-Optimizing	On a constant basis, the system must be evaluating potential optimizations. Through self-monitoring and resource tuning, and through self-configuration, the system should self-optimize to efficiently maximize resources to best meet the needs of its environment and end-user needs
Self-Protecting	Perhaps the most interesting of all the characteristics needed by an autonomic system is self-protection, systems that protect themselves from attack. These systems must anticipate a potential attack, detect when an attack is under way, identify the type of attack, and use appropriate countermeasures to defeat or at least nullify the attack. Attacks on a system can be classified as Denial-of-Service (DoS) or the infiltration of an unauthorized user to sensitive information or system functionality

management, performance optimization management, and routine maintenance from the conscious brain, the system administrators.

The basic goal of autonomic computing is to simplify and automate the management of computing systems, both hardware and software, allowing them to self-manage, without the need for human intervention. Four fundamental characteristics are needed by an autonomic system to be self-managing; these are described in Table 2.1. The common theme shared by all of these characteristics is that each of them requires the system to handle functionality that has been traditionally the responsibility of a human system administrator.

Within the software domain, adaptive and reflective techniques will play a key role in the construction of autonomic systems. Adaptive and reflective techniques already exhibit a number of the fundamental characteristics that are needed by autonomic systems. Thus, reflective and adaptive middleware provide the ideal foundations for the construction of autonomic middleware platforms. The merger of these two strands of research is a realistic prospect. The goals of autonomic computing highlight areas for the application of reflective and adaptive techniques, these areas include self-protection and self-healing, with some work already initiated in the area of fault-tolerance [20].

2.5 Summary

Middleware platforms are exposed to environments demanding the interoperability of heterogeneous systems, 24/7 reliability, high performance, scalability and security while maintaining a high QoS. Traditional monolithic middleware platforms are capable of coping with such demands as they have been designed and fine-tuned in advance to meet these specific requirements. However, next-generation computing environments such as large-scale distribution, mobile, ubiquitous, and pervasive computing will present middleware with dynamic environments with constantly changing operating conditions, requirements, and underlying deployment infrastructures. Traditional static middleware platforms will struggle when exposed to these environments, thus providing the motivation to develop next-generation middleware systems to adequately service such environments.

To prepare next-generation middleware to cope with these scenarios, middleware researchers are developing techniques to allow middleware platforms to examine and reason about their environment. Middleware platforms can then self-adapt to suit the current operating conditions based on this analysis; such capability will be a prerequisite for next-generation middleware.

Two techniques have emerged that enable middleware to meet these challenges. Adaptive and reflective techniques allow applications to examine their environment and self-alter in response to dynamically changing environmental conditions, altering their behavior to service the current requirements. Adaptive and reflective middleware is a key emerging paradigm that will help simplify the development of dynamic next-generation middleware platforms [1, 2].

There is a growing interest in developing reflective middleware with a large number of researchers and research group's active in the area. Numerous architectures have been developed that employ adaptive and reflective techniques to allow for adaptive and self-adaptive capabilities; these techniques have been applied in a number of areas within middleware platforms including distribution, responsiveness, availability, reliability, concurrency, scalability, transactions, fault-tolerance, and security.

IBM's autonomic computing envisions a world of self-managing computer systems, such autonomic systems will be capable of self-configuration, self-healing, self-optimization, and self-protection against attack, all without the need for human intervention. Adaptive and reflective middleware platforms share a number of common characteristics with autonomic systems and will play a key role in their construction.

Bibliography

[1] Schantz, R. E. and Schmidt, D. C. (2001) Middleware for Distributed Systems: Evolving the Common Structure for Network-centric Applications, *Encyclopedia of Software Engineering*, New York, Wiley & Sons pp. 801–813.

[2] Geihs, K. (2001) Middleware Challenges Ahead. *IEEE Computer*, **34**(6), 24–31.

[3] Blair, G. S., Costa, F. M., Coulson, G., Duran, H. A., Parlavantzas, N., Delpiano, F., Dumant, B., Horn, F., and Stefani, J.-B. (1998) The Design of a Resource-Aware Reflective Middleware Architecture, *Proceedings of the Second International Conference on Meta-Level Architectures and Reflection (Reflection'99)*, Springer, St. Malo, France, pp. 115–134.

[4] Loyall, J., Schantz, R., Zinky, J., Schantz, R., Zinky, J., Pal, P., Shapiro, R., Rodrigues, C., Atighetchi, M., Karr, D., Gossett, J. M., and Gill, C. D. (2001) Comparing and Contrasting Adaptive Middleware Support in Wide-Area and Embedded

Distributed Object Applications. *Proceedings of the 21st International Conference on Distributed Computing Systems*, Mesa, AZ.

[5] Smith, B. C. (1982) *Procedural Reflection in Programming Languages*, PhD Thesis, MIT Laboratory of Computer Science.

[6] Coulson, G. (2002) What is Reflective Middleware? *IEEE Distributed Systems Online*, http://dsonline.computer.org/middleware/RMarticle1.htm.

[7] Geihs, K. (2001) Middleware Challenges Ahead. *IEEE Computer*, **34**(6), 24–31.

[8] Kon, F., Costa, F., Blair, G., and Campbell, R. H. (2002) The Case for Reflective Middleware. *Communications of the ACM*, **45**(6), 33–38.

[9] Kiczales, G., Rivieres, J. d., and Bobrow, D. G. (1992) *The Art of the Metaobject Protocol*, MIT Press, Cambridge, MA.

[10] Blair, G. S., Coulson, G., Andersen, A., Blair, L., Clarke, M., Costa, F., Duran-Limon, H., Fitzpatrick, T., Johnston, L., Moreira, R., Parlavantzas, N., and Saikoski, K. (2001) The Design and Implementation of Open ORB 2. *IEEE Distributed Systems Online*, **2**(6).

[11] DeMichiel, L.G. (2002) Enterprise JavaBeansTM Specification, Version 2.1. 2002, Sun Microsystems, Inc.

[12] Overview of the .NET Framework White Paper. 2001, Microsoft.

[13] Object Management Group (2002) *CORBA Components OMG Document formal/ 02-06-65*.

[14] Szyperski, C. (1997) *Component Software: Beyond Object-Oriented Programming*, Addison-Wesley, New York, USA.

[15] Czarnecki, K. and Eisenecker, U. (2000) *Generative Programming: Methods, Tools, and Applications*, Addison-Wesley, New York, USA.

[16] Cleaveland, C. (2001) *Program Generators with XML and Java*, Prentice Hall, Upper Saddle River, NJ.

[17] Kon, F., Blair, G. S., and Campbell, R. H. (2000) Workshop on Reflective Middleware. *Proceedings of the IFIP/ACM Middleware 2000*, New York, USA.

[18] Corsaro, A., Wang, N., Venkatasubramanian, N., Coulson, G., and Costa, F. M. (2003) The 2nd Workshop on Reflective and Adaptive Middleware. *Proceedings of the Middleware 2003*, Rio de Janeiro, Brazil.

[19] Andersen, A., Blair, G. S., Stabell-Kulo, T., Myrvang, P. H., and Rost, T.-A. N. (2003) Reflective Middleware and Security: OOPP meets Obol. *Proceedings of the Workshop on Reflective Middleware, Middleware 2003*, Rio de Janeiro, Brazil; Springer-Verlag, Heidelberg, Germany.

[20] Favarim, F., Siqueira, F., and Fraga, J. (2003) Adaptive Fault-Tolerant CORBA Components. *Proceedings of the 2nd Workshop on Reflective and Adaptive Middleware, Middleware 2003*, Rio de Janeiro, Brazil.

[21] Garbinato, B., Guerraoui, R., and Mazouni, K.R. (1993) Distributed Programming in GARF, *Proceedings of the ECOOP Workshop on Object-Based Distributed Programming*, Springer-Verlag, Kaiserslautern, Germany, pp. 225–239.

[22] McAffer, J. (1995) Meta-level Programming with CodA. *Proceedings of the European Conference on Object-Oriented Programming (ECOOP)*, Åarhus, Denmark.

[23] Portillo, A. R., Walker, S., Kirby, G., and Dearle, A. (2003) A Reflective Approach to Providing Flexibility in Application Distribution. *Proceedings of the 2nd Workshop on Reflective and Adaptive Middleware, Middleware 2003*, Rio de Janeiro, Brazil; Springer-Verlag, Heidelberg, Germany.

[24] Cazzola, W. and Ancona, M. (2002) mChaRM: a Reflective Middleware for Communication-Based Reflection. *IEEE Distributed System On-Line*, **3**(2).

[25] Object Management Group (1998) *The Common Object Request Broker: Architecture and Specification.*

[26] Kon, F., Román, M., Liu, P., Mao, J., Yamane, T., and Magalhães, L. C., and Campbell, R. H. (2002) Monitoring, Security, and Dynamic Configuration with the dynamicTAO Reflective ORB. *Proceedings of the IFIP/ACM International Conference on Distributed Systems Platforms and Open Distributed Processing (Middleware'2000)*, New York.

[27] Schmidt, D. C. and Cleeland, C. (1999) Applying Patterns to Develop Extensible ORB Middleware. *IEEE Communications Special Issue on Design Patterns*, **37**(4), 54–63.

[28] Curry, E., Chambers, D., and Lyons, G. (2003) Reflective Channel Hierarchies. *Proceedings of the 2nd Workshop on Reflective and Adaptive Middleware, Middleware 2003*, Rio de Janeiro, Brazil; Springer-Verlag, Heidelberg, Germany.

[29] Curry, E., Chambers, D., and Lyons, G. (2004) Extending message-oriented middleware using interception. Proceedings of the *Third International Workshop on Distributed Event-Based Systems (DEBS '04)*, Edinburgh, Scotland, UK.

[30] Pietzuch, P. R. and Bacon, J. M. (2002) *Hermes: A Distributed Event-Based Middleware Architecture.*

[31] Cabrera, L. F., Jones, M. B., and Theimer, M. (2001) Herald: Achieving a Global Event Notification Service. *Proceedings of the 8th Workshop on Hot Topics in Operating Systems*, Elmau, Germany.

[32] Kiczales, G., Lamping, J., Mendhekar, A., Maeda, C., Lopes, C. V., Loingtier, J. M., and Irwin, J. (1997) Aspect-Oriented Programming. *Proceedings of the European Conference on Object-Oriented Programming*, Jyväskylä, Finland.

[33] Kiczales, G., Hilsdale, E., Hugunin, J., Kersten, M., rey Palm, J., and Griswold, W. G. (2001) An Overview of AspectJ. *Proceedings of the European Conference on Object-Oriented Programming (ECOOP)*, Budapest, Hungary.

[34] Bergmans, L. and Aksit, M. (2000) Aspects and Crosscutting in Layered Middleware Systems. *Proceedings of the IFIP/ACM (Middleware2000) Workshop on Reflective Middleware*, Palisades, New York.

[35] Chiba., S. (1998) Javassist – A Reflection-based Programming Wizard for Java. *Proceedings of the Workshop on Reflective Programming in C++ and Java at OOPSLA'98*, Vancouver, Canada.

[36] Booch, G. (2001) *Through the Looking Glass*, Software Development.

[37] Tarr, P., Ossher, H., Harrison, W., and Sutton Jr, S. M. (1999) N Degrees of Separation: Multi-Dimensional Separation of Concerns. *Proceedings of the International Conference on Software Engineering ICSE'99*, Los Angeles, CA.

[38] Ossher, H. and Tarr, P. (2001) Using Multidimensional Separation of Concerns to (re)shape Evolving Software. *Communications of the ACM*, **44**(10), 43–50.

[39] Moreira, R. S., Blair, G. S., and Garrapatoso, E. (2003) Constraining Architectural Reflection for Safely Managing Adaptation. *Proceedings of the 2nd Workshop on Reflective and Adaptive Middleware, Middleware 2003*, Rio de Janeiro, Brazil; Springer-Verlag, Heidelberg, Germany.

[40] OMG (2001) *Model Driven Architecture - A Technical Perspective. OMG Document: ormsc/01-07-01.*

[41] Ganek, A. and Corbi, T. (2003) The Dawning of the Autonomic Computing Era. *IBM Systems Journal*, **42**(1), 5–18.

3

Transaction Middleware

Stefan Tai, Thomas Mikalsen, Isabelle Rouvellou
IBM T.J. Watson Research Center

3.1 Introduction

Transactions are a commonly employed approach to model and build reliable, fault-tolerant systems. The "system" might be as simple as a standalone database application running on a single machine, or as complex as a distributed application involving a network of shared resources executing over the Internet. The basic idea is that the transaction transforms the system from one consistent state to another consistent state, despite any failures that may occur along the way. Starting from a consistent initial state, the transaction applies one or more operations (producing a set of potentially inconsistent intermediate states) such that a consistent final state is reached. Even when things go wrong, a consistent final state (perhaps the initial state) is achieved. When supported directly by the programming environment and underlying system, this simple idea provides a powerful abstraction that cleanly separates an application's normal program execution logic from system failure/recovery logic.

A good example of a transaction is a funds transfer from an account in one bank to an account in another bank. The funds transfer involves executing a debit operation on one account, and executing a credit operation on the other account. Implemented as a transaction, the idea is to ensure that no inconsistent state with funds being lost or generated due to one of the operations failing and the other one succeeding can occur.

Supporting transactions, however, is not trivial. In database and other persistent data management systems, durable-media failures must be considered. In multiuser/process environments, the effect of concurrently executing transactions must be considered. In distributed systems, communication failures and remote system availability must be considered. Supporting transactions over the Internet and in other open, heterogeneous environments poses additional challenges.

Transactions also introduce programming paradigms and execution styles that must be integrated with programming languages, APIs, and business process/workflow modeling

Middleware for Communications. Edited by Qusay H. Mahmoud
 ISBN 0-470-86206-8

languages. Further, communication middleware and protocols need to be extended to support transactional semantics. Architectures such as *transaction processing monitors* (a.k.a. TP Monitors) have emerged as *transaction processing middleware* that fill the gap between operating system services and the requirements of reliable, scalable, distributed computing.

The goal of a *transaction processing middleware*, in general, is to make it easier for the programmer to write and deploy reliable, scalable transactional applications. Rather than interacting directly with transactional resources and transaction management functions, application programs are built around a middleware that mediates access to these shared resources. Such middleware provides abstractions that are more closely aligned with business objectives, allowing the programmer to focus on business logic instead of low-level transaction management.

This chapter takes a middleware view of transaction processing. In particular, we look at how transactions are supported in *object-oriented middleware*, *message-oriented middleware*, and *(Web) service-oriented middleware.*

We begin in Section 3.2 by reviewing some fundamentals on which middleware for transactions have been built. Our coverage starts by defining ACID transactions, their properties, and some of the techniques used to support them. We then show how *distributed transactions* extend this model to support the ACID properties in a distributed setting. Following this, we describe some commonly employed variations of the ACID model that overcome some of its shortcomings. And finally, we elaborate on some of the programming models used for transaction processing.

Building on our discussion of the distributed transaction model, we present *distributed object transactions* as supported by object-oriented middleware, *transactional messaging* as supported by message-oriented middleware, and *Web transactions* as supported by the Web services platform, in Sections 3.4, 3.5, and 3.6, respectively. We first examine transactions in the context of CORBA OTS and Enterprise JavaBeans. We then describe messaging concepts and queued transaction processing as an important architectural pattern. Finally, we present emerging open standards addressing distributed transactions and reliable messaging for Web services.

In Section 3.6, we introduce three research projects that describe advanced transaction middleware. The *Long Running Unit of Work (LRUOW)* framework defines a transaction model to support long-running business transactions for J2EE/EJB environments. *Dependency Spheres (D-Spheres)* is a middleware to support transactional applications that are built across both traditional object-oriented middleware and message-oriented middleware. *Transactional Attitudes* is a new approach to automate the transactional composition of Web services while maintaining autonomy of the individual services.

We conclude the chapter in Section 3.7 with a summary.

3.2 Transaction Processing Fundamentals

Many transaction-processing concepts and techniques have been developed. These include transaction protocols (such as two-phase locking and two-phase commit), transactional communication mechanisms (such as transactional remote procedure calls), and transaction-processing architectures (such as X/Open Distributed Transaction Processing). In the following, we look briefly at some of these concepts as they shape middleware support for transactions.

3.2.1 ACID Transactions

We begin with a simple and elegant model for transactions, sometimes called *classical transactions* or *flat transactions*. This model is often characterized by the so-called *ACID* properties of *atomicity*, *consistency*, *isolation/independence*, and *durability* [14]. A transaction is an *atomic* unit-of-work that transforms the system from one *consistent* state to another consistent state, executing without interference from other concurrently executing transactions (*isolation/independence*), and, once committed, cannot be undone by system failures (*durability*).

Following this, we examine the ACID properties in more detail, highlighting some of the protocols and techniques employed to support them.

3.2.1.1 Atomicity

ACID transactions are an all or nothing affair: a transaction either commits, and the system state reflects the effect of the transaction, or it aborts, and the transaction's initial state is restored. In many systems, atomicity is achieved by way of the two-phase commit (2PC) and DO-UNDO-REDO protocols [14].

The two-phase commit (2PC) protocol achieves atomicity in environments where operations are applied to independent/portable resource managers (as in the case of distributed transactions discussed below) and where certain work is deferred until commit time (e.g., checking database integrity constraints and writing back cached data). The 2PC protocol achieves atomicity by requiring that all participating resource managers agree to commit the transaction (phase 1) before actually committing (phase 2). A *transaction manager*, or *coordinator*, solicits the votes of the resource managers: if all vote yes, the transaction is committed and the coordinator tells the resource managers to commit; otherwise, the transaction is aborted, and the coordinator tells the resource managers to abort.

The DO-UNDO-REDO protocol is used to achieve atomicity in centralized systems. It says that operations on transactional resources should be designed around *do*, *undo*, and *redo* functions, where the *do* function produces a log record that can later be used to *undo* or *redo* the operation. *Do* functions are typically called as part of normal transaction execution, while *undo* and *redo* functions are called during transaction rollback and recovery. If the transaction aborts, the transaction manager can scan the log and execute the undo records associated with the aborting transaction, thereby achieving atomicity in the event of a rollback.

3.2.1.2 Consistency

While consistency is often specified and enforced by the application, the ACID model does guarantee that, given a consistent initial state and an *individually* correct transaction (more on this below), a consistent final state will be reached. While the transaction-processing system does not necessarily help the application define and enforce individually correct transaction (though in some cases it does, as with SQL integrity constraints [11]), the model allows the programmer to focus on writing individually correct programs, and not worry about, for example, interference between concurrently executing transaction programs or complex transaction recovery logic [10].

3.2.1.3 Isolation

In multiuser/process and distributed environments, it is desirable to execute multiple transactions concurrently. Yet, this can lead to certain problems, the most famous of which is the *lost update* anomaly. The problem is that if two transactions read and then update the same object, the second update can potentially overwrite the first; that is, the first update can be lost. There are other problems as well, such as the *uncommitted dependencies problem* and *inconsistent analysis problem* [11]. These problems are sometimes expressed in terms of the following interference anomalies:

- *Dirty reads*, where one transaction sees the uncommitted changes of another transaction.
- *Nonrepeatable reads*, where a transaction reads the same object twice and sees different values each time because of the updates of another transaction.
- *Phantoms*, where new objects appear in a transaction's query results because of the updates of another transaction.

The isolation property of ACID transactions guarantees that such interference problems do not occur. This is achieved by controlling concurrent access to shared resources. One approach might be to simply restrict execution to one transaction at a time; however, this would be highly inefficient and difficult to achieve in distributed systems. Rather, the approach typically taken is for the system to *appear* as if transactions are executing one at a time, one after the other. If a set of concurrently executing, interleaved transactions produce a result equivalent to some serially ordered execution of those same transactions, then that interleaved execution is said to be *serializable* and considered to be correct (i.e., no interference has occurred) [10].

A common approach to ensuring serializable (i.e., correct) execution of interleaved transactions is the *two-phase locking protocol*. The protocol says that a transaction must acquire a *shared lock* (S lock) on any object it reads, and an *exclusive lock* (X lock) on any object that it writes, and that all locks must be acquired (phase 1) prior to releasing any lock (phase 2); locks granted to one transaction can delay another transaction attempting to acquire a *conflicting* lock:

- S locks conflict with X locks
- X locks conflict with X locks.

If all transactions obey this protocol, then all possible interleaved execution of concurrent transactions will be serializable and therefore correct. This protocol is typically implemented by the transaction manager, lock manager, and resource manager components of the system, and is thus not directly a concern of the programmer.

3.2.1.4 Durability

ACID transactions are durable; that is, the effects of committed transactions are not lost because of failures. In particular, the only way to change the durable state of the system is to commit a transaction.

The *DO-UNDO-REDO* protocol (discussed above), *write-ahead log* (WAL) protocol, and *force-at-commit rule* together allow for system state recovery after a failure [14]. The idea is that prior to writing uncommitted changes to persistent storage, a corresponding REDO-UNDO record must be written to the persistent log. The force-at-commit rule ensures that a transaction's log records are written to persistent storage prior to committing the transaction. When restarting after a failure, a consistent system state can be restored by redoing the log records for all committed transactions, undoing the log records for all aborted transactions, and then completing any active transactions.

3.2.2 Distributed Transactions

The *distributed transaction model* extends the classical transaction model to provide ACID properties for operations on independent transactional resources, potentially executing on different nodes in a network. Unlike in centralized transactions, where transactional resource management can be tightly coupled with transaction management (e.g., using a common log, shared recovery mechanisms, proprietary APIs), distributed transactions must accommodate a certain degree of autonomy in the participating transactional resources.

The challenge introduced by this model is that more things can go wrong: distributed transactions introduce the possibility of communication failures and independent node failures that can affect both normal processing and recovery. To tolerate these additional failure scenarios, the mechanisms described above must be extended to support the commitment and recovery of distributed resources.

3.2.2.1 Transactional Remote Procedure Calls

Transactional remote procedure calls (TRPC) are an extension to basic remote procedure calls (RPC) that allow remote resources to participate in global transactions. Often, communication middleware is extended to support transactional RPC, as illustrated in Figure 3.1.

In the figure, a client interacts with a local transaction manager to begin and complete transactions. A transactional RPC mechanism allows the client to operate on a remote server within the context of its transaction. The communication middleware interacts with a local transaction manager to export and import transactions when remote procedure calls occur, and to register remote transaction managers as participants in the transaction.

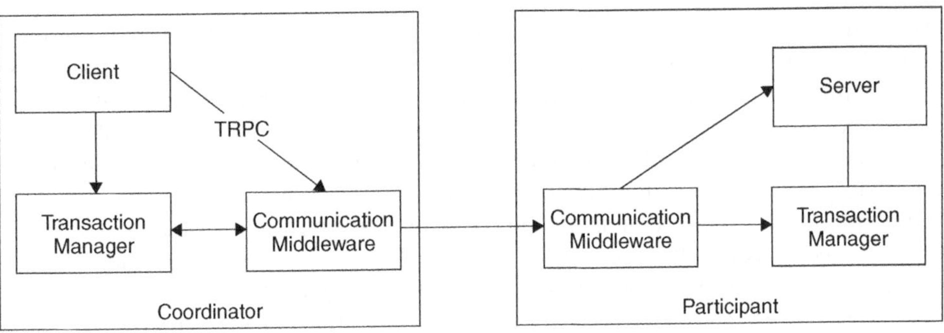

Figure 3.1 TRPC and communication middleware

3.2.2.2 Distributed Two-Phase Commit

For operations of a distributed transaction to behave as an atomic unit, the participating distributed resources must either all commit or all abort in accordance with the outcome of the transaction. For this, the 2PC protocol is extended to include remote transaction managers as resource managers in a transaction. While this may seem straightforward, there are a few complications.

A resource manager that agrees to commit a transaction must wait for a decision from the coordinator. In the case of a remote participant, communication failures or a coordinator crash could potentially force the participant to wait forever. The problem is amplified in the Web services environments, where autonomous services are coordinated over the Internet: failures are more likely and services are less willing to be held hostage to the coordinator.

The practical application of the 2PC protocol for distributed transactions typically requires additional mechanisms that avoid forcing participants to wait endlessly for a decision. Often, *heuristic decisions* are made, where a participant unilaterally decides the outcome of its local transaction (e.g., an educated guess based on the outcome of previous transactions or a policy based on the "type" of transaction) before the coordinator's decision is known; if the participant later receives (from the coordinator) a decision that contradicts its heuristic decision, it responds with a heuristic mismatch message (e.g., X/Open XA specification defines XA_HEURCOM, XA_HEURRB, and XA_HEURMIX), which tells the coordinator that atomicity has been violated. Resolving heuristic mismatches may require ad-hoc recovery processes, involving manual steps, such as a human analyzing transaction log records.

3.2.2.3 X/Open DTP

The X/Open Distributed Transaction Processing (DTP) [38] model is a standard architecture for coordinating shared, transactional resources in a distributed environment. The model is illustrated in Figure 3.2.

In the figure, TX is the transaction demarcation API; XA is a specification for a portable resource manager; XA+ extends XA to support portable communication resource managers (CRMs); and OSI-TP specifies an interoperable, distributed 2PC protocol.

3.2.3 Common Extensions

The ACID transaction model discussed thus far offers an elegant solution to many problems. However, the mechanisms used to support this model impose certain restrictions and constraints that make strict ACID semantics impractical in many applications. The lock-based approach typically used is practical only when transactions are short-lived; long-lived transactions would force the system to hold locks on resources for unacceptably long periods of time. Further complicating matters is that a unit-of-work is also a *unit-of-failure*; if a transaction aborts, all the work performed by that transaction is lost. Also of concern is that flat transactions limit modularity as the model cannot be easily applied to the composition constructs found in many programming languages.

To address these and other issues, a number of variations of the classical transaction model have been introduced [17]. In the following, we look at a few common extensions.

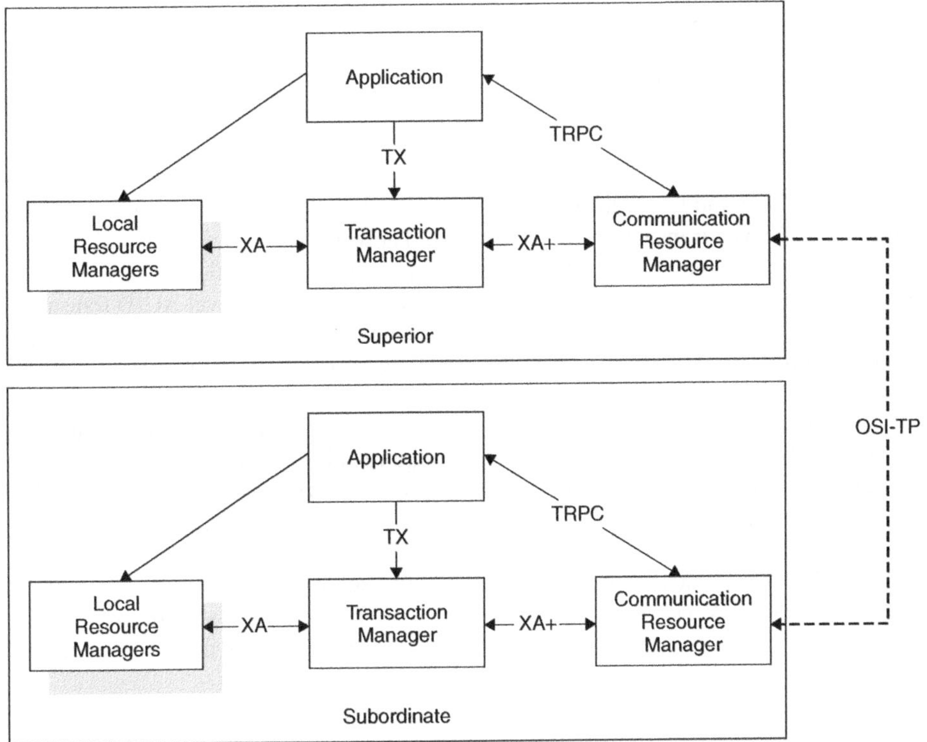

Figure 3.2 X/Open DTP

3.2.3.1 Isolation Levels

Our discussion above on isolation assumed that serializability was the only criteria for correctness. Yet, many applications, such as read-only applications, can tolerate some degree of interference from concurrently executing transactions and still produce correct results.

Isolation levels are a mechanism for weakening the serializability requirement of a transaction and can be expressed in terms of the anomalies that they permit. For example, SQL defines the following isolation levels, listed in order of increasing interference [11]:

- *Serializable.* No interference is allowed.
- *Repeatable Read.* Phantoms are allowed.
- *Read Committed.* Phantoms and nonrepeatable reads are allowed.
- *Read Uncommitted.* Phantoms, nonrepeatable reads, and dirty reads are allowed.

3.2.3.2 Optimistic Concurrency Control

Lock-based concurrency control mechanisms, such as the two-phase locking, are often said to be *pessimistic*, as they leave little to chance: transactions are delayed to avoid potential conflicts. *Optimistic* concurrency control mechanisms [19] offer an alternative. In these models, concurrently executing transactions operate on *copies* of shared objects in hope that no conflicts exist when it is time to commit; at commit time, if a conflict is detected, the conflicting transaction is aborted and redone. This approach can lead to greater concurrency, when compared to pessimistic approaches, though transaction throughput degrades when conflicts occur and transactions must be redone.

3.2.3.3 Nested Transactions

Nested transaction models improve transaction modularity/composition and offer finer grained control over transaction recovery [26]. These models, sometimes called *closed nested transactions*, use lock-based concurrency controls to regulate access to shared objects. The idea is to nest nondurable *subtransactions* within top-level ACID transactions (as well as within other subtransactions). To ensure the serializability of concurrently executing (sub) transactions, nested transaction-locking rules are defined[1]:

- A subtransaction is granted a lock if the lock is free or held by its parent (or ancestor).
- When a subtransaction commits, its parent acquires all of its locks.
- When a subtransaction aborts, its locks are released (without affecting locks held by its parent or ancestors).

This model, like most lock-based models, is most appropriate for short-running transactions. The LRUOW [1] model, discussed in Section 3.6, is a novel approach to *long-running nested transaction*; the model employs a *relaxed-optimistic* approach using *multi-version* objects.

3.2.3.4 Compensation-based Recovery

Compensation-based recovery models, as exemplified by Sagas [13], are often employed to address the problems introduced by long-running transactions. Similar to the isolation levels discussed above, these models loosen the strict serializability requirement of ACID transactions.The idea is that an otherwise long-running ACID transaction is broken down into a series of shorter running ACID transactions. The result is a *logical transaction* with certain transactional properties, such as *eventual atomicity*, but that is not itself an ACID transaction; in particular, the effects of intermediate ACID transactions are visible to other logical transactions. To achieve atomicity in the event of a rollback (of the logical transaction), a series of *compensating* ACID transactions, corresponding to the already completed ACID transactions, are executed.

Compensation-based recovery is also important in loosely coupled systems. Therefore, transactional messaging applications and Web services transactions commonly employ compensation to undo previously completed activities. Middleware can support managing compensating activities; an example of an advanced transaction middleware that has built-in support for compensation is the *Dependency-Spheres* middleware [32] presented in Section 3.6.

3.2.4 Programming Models for Transactions

So far, we have focused on transaction models and techniques. We now turn briefly to the programmer's view of transactions and discuss some of the common programming

[1] To simplify the presentation, the locking rules presented here only consider X locks.

models available for writing transactional programs. Further details on these programming models are provided in later sections when we introduce specific transaction middleware.

3.2.4.1 Transaction Demarcation APIs

Often, the transaction-processing environment provides an explicit *transaction demarcation API* for starting and completing (committing and aborting) transactions. The API typically includes verbs such as BEGIN, COMMIT, and ABORT. The programmer brackets the transactional part of the program using the BEGIN and COMMIT/ABORT verbs.

While explicit transaction demarcation offers flexibility, it also leaves some of the transaction-processing responsibility with the programmer. For example, the program must know when to start new transactions, remember to commit/rollback transactions, and understand how the completion of one transaction might effect or trigger the start of another transaction. Declarative transactions and transactional process modeling can shift some of this burden to the transaction-processing system.

3.2.4.2 Declarative Transactions

Some transaction-processing environments, such as Component Broker [7] and Enterprise JavaBeans [31], support *declarative transactions*. In these environments, the programmer does not have to explicitly begin and complete transactions. Rather, transaction policies are associated with component methods, and specified as part of the component's deployment descriptor.

3.2.4.3 Transactional Business Process Modeling

Transaction and reliability semantics can also be supported directly as part of a business process/workflow modeling language [22, 23]. In doing so, process execution guarantees and transactional semantics can be tightly integrated and supported through consistent process modeling abstractions. Further, top-level process models can be validated to ensure compatibility with subordinate activities and resource managers.

3.3 Distributed Object Transactions

In this section, we present the *distributed object transaction* processing model, and describe how the model is supported by distributed object middleware. We explain the explicit, API-based (client-side) programming model of the CORBA Object Transaction Service and review the implicit, declarative, container-managed (server-side) programming model of transactional Enterprise JavaBeans.

3.3.1 Transaction Model

Distributed object middleware evolved from the idea of remote procedure calls (RPCs), introducing object-oriented principles to the development of distributed systems. The

Common Object Request Broker Architecture (CORBA) and the Java 2 Enterprise Edition (J2EE) are examples of distributed object middleware standards.

In order for an object to initiate, invoke, and participate in a distributed transaction, additional APIs and/or deployments steps are needed. Distributed object transactions essentially follow the X/Open DTP model previously introduced. The *CORBA Object Transaction Service (OTS)* [27] can be used for implementing distributed transactions in a CORBA-based system. The *Java Transaction Service (JTS)* defines a distributed transaction architecture for Java-based systems. JTS supports the *Java Transaction API* (JTA) [30] as its high-level programming model; this API includes transaction demarcation operations and interfaces for coordinating XA resource managers. JTS also implements the Java mapping of the OTS interfaces; this allows a JTS implementation to be portable over different ORB implementations and provides for interoperability between different JTS/OTS implementations.

OTS standardizes a sct of interfaces for transactional CORBA clients and transactional CORBA servers. If implicit transaction context propagation is used, the transaction operations are provided by the `Current` object; if explicit transaction contexts are used, the `TransactionFactory`, `Control`, `Terminator`, and `Coordinator` objects are used. In both cases, the client demarcates the transaction explicitly. The `Coordinator` object is also used by recoverable CORBA servers to register `Resource` objects. Resources support the two-phase commit protocol, which is driven by the object transaction middleware service that implements the OTS. Figure 3.3 illustrates the participants of an OTS transaction.

As noted earlier, a variety of extended transaction models has been proposed to overcome the limitations of ACID transactions. The implementation of such extended transaction models requires specialized middleware, which typically can make use of services that are already available with transaction-processing middleware such as OTS implementations. The *Additional Structuring Mechanisms for the OTS* specification has recognized this concern, and proposed the *Activity Service Framework* [16]. The framework provides general-purpose mechanisms for implementing diverse extended transaction models. For

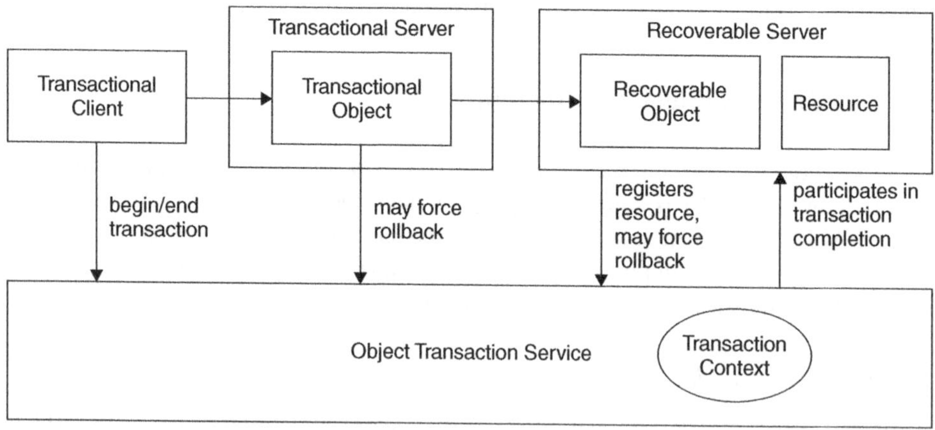

Figure 3.3 CORBA OTS

example, event signaling can be programmed to coordinate transaction activities according to the protocols of a specific transaction model under consideration.

3.3.2 Transaction APIs

One concern in programming distributed object transactions is how to initiate and complete the transaction. The OTS requires explicit, API-based (client-side) transaction demarcation. A client can begin and commit (or abort) a distributed transaction using the standard OTS interfaces defined. The J2EE platform similarly supports explicit transaction demarcation through the `javax.transaction.UserTransaction` interface; the X/Open DTP's TX interface is yet another API for programming ACID transactions.

The following pseudo-code example illustrates the client use of the `UserTransaction` interface to define a Java transaction. First, a `UserTransaction` object is created (alternatively, the object can be looked up using a JNDI name). Next, the `begin()` method to start a transaction is called on the `UserTransaction` object. The `commit()` method is used to complete the transaction; `rollback()` is called if an exceptional state is encountered.

```
UserTransaction ut = new javax.transaction.UserTransaction();
try
  {
     ut.begin();
     // Perform operations...
     ut.commit();
  }
  catch(Exception e)
  {
     ut.rollback();
  }
```

Also of concern is how to implement and deploy transactional servers/resources. The J2EE architecture supports multiple models for implementing and deploying transactional objects. They include *Java Database Connectivity (JDBC)*, *J2EE Connector Architecture (J2CA)* [28], and *EJB*. With the exception of EJB, such resources are (in general) expected to be provided by independent software vendors. EJB, however, are a model for writing custom transactional servers/resources. The transaction middleware provides an interface for registering resources to participate in transaction protocols like the two-phase commit protocol.

Further, there is an issue of binding to a transaction server/resource such that invocations on objects occur within the correct transactional context. In J2EE, a combination of *Java Naming and Directory Interface (JNDI)* and the transactional resource models are used to locate and connect to transactional objects.

3.3.3 Container-Managed Transactions

The J2EE architecture defines EJB as a component model for distributed systems [31]. EJB support the transaction programming model of *container-managed transactions*, which is described in this section.

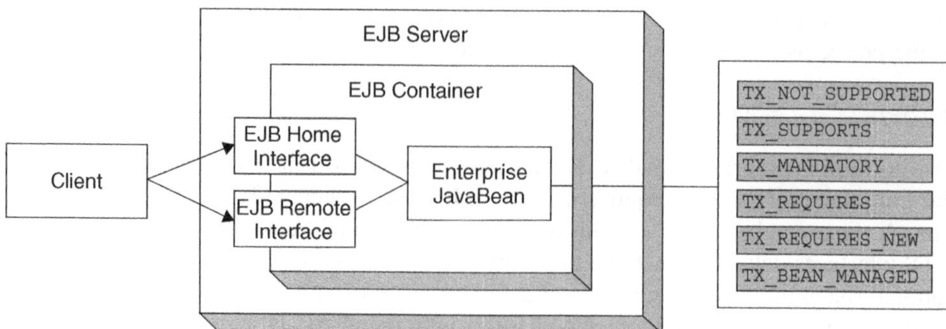

Figure 3.4 Enterprise JavaBeans

The EJB *container* is a runtime system for one or more enterprise beans that also provides tools for deploying beans. Deployment content includes, among access control policies, JNDI names and type information, *transaction policies (attributes)* that govern the transactional behavior of (entity and session) EJB and its operations. Transaction policies cover the transaction scope and the isolation level.

The EJB specification defines different values for the transactional scope. The values describe whether an enterprise bean does not support a global transaction (TX_NOT_SUPPORTED) or supports a global transaction in that it can execute an invocation in a transaction context (TX_SUPPORTS), requires an existing transaction context (TX_MANDATORY), will always execute the invocation in an existing or in a new transaction context (TX_REQUIRED), will always create a new transaction context (TX_REQUIRES_NEW), or that the transaction behavior is managed by the enterprise bean implementation (TX_BEAN_MANAGED).

The EJB transaction isolation levels describe the degree to which the serializability of a transaction may be violated (as described in Section 3.2). Nonrepeatable reads and phantoms may be allowed, but dirty reads are disallowed (TRANSACTION_READ_COMMITTED), all three violations may be allowed (TRANSACTION_READ_UNCOMMITTED), only phantoms may be allowed (TRANSACTION_REPEATABLE READ), or no violation may be allowed (TRANSACTION_SERIALIZABLE).

The model of container-managed transactions refers to the container being responsible for implementing the transaction scope and isolation level specified. Instead of the bean provider implementing transaction demarcation and management, the container is responsible for generating and executing appropriate code. The operations of an entity bean to which the transactional behavior applies are specified in the deployment descriptor of the EJB server. The transaction policies (of scope and isolation level) are set declaratively; container-managed transactions therefore are also referred to as *declarative transactions.*

The EJB 2.0 specification introduced a new type of enterprise bean, the *Message-driven Bean (MDB).* An MDB provides a mechanism implemented through the EJB container for dealing with asynchronous message delivery (using the *Java Message Service, JMS,* see below) to an EJB. An MDB is stateless and is accessed only by the container in response to JMS messages (therefore, an MDB has no home and no remote interface); an MDB correspondingly has different transactional behavior than entity and session EJB, as will be described in the following section on messaging transactions. Nevertheless, the transactional behavior of an MDB is also specified using declarative transaction policies.

3.4 Messaging Transactions

Transactional messaging refers to transaction processing in messaging environments. In this section, we present transactional messaging as supported by message-oriented middleware (MOM). We describe MOM concepts, and examine queued transaction processing as an important architectural pattern.

In the RPC model, the client of a request waits synchronously for the server to execute the request and return a reply. The RPC model is used in distributed transaction processing, as previously described. The model requires client and server to be available at the same time, and the client to block while waiting for the server to return the reply. Such tight coupling of client and server may not be possible or desirable in some environments. For example, a client may wish to issue a request even if the server is not available, and may also wish to continue its processing and not to be blocked while waiting for a response from the server. The inverse case may also be true for a server sending a reply to a client. Unlike in the RPC model, a client may also wish not to address a single, specific server, but to send a message to multiple recipients, or to have a message processed by any server of a group of servers that is able to process the request.

Message-oriented middleware is a solution to these problems. The middleware acts as a mediator between client and server, buffering or persistently storing requests and replies. Applications interact with the MOM, but not directly with each other. MOM ensures guaranteed message delivery, so that applications can "fire-and-forget" a message. MOM naturally implements an asynchronous communication model, enabling applications to communicate with each other without requiring same-time availability or causing anyone to block.

3.4.1 Messaging Models

Two principle messaging models exist: point-to-point messaging and publish/subscribe messaging. With *point-to-point messaging*, or message queuing, message *queues* are used as the mediators between communicating applications. A queue is a persistent store for messages. A queue is referred to using a logical name, and it is managed by a queue manager (a messaging daemon). A messaging architecture will typically comprise one or more (distributed) queue managers, each maintaining multiple queues. A queue also is a transactional resource that can be integrated in direct transaction processing. That is, writing a message to a queue or reading a message from a queue is permanent or can be undone, depending on whether the transaction that created the message commits or aborts. This describes a very powerful and important concept, and we will come back to transactional queuing later in this section.

Publish/subscribe messaging describes another messaging model, which employs the concept of a topic (in a namespace) to publish messages to, or to subscribe to in order to receive published messages. Unlike the transactional message publication and consumption model of message queues, a message posted to a topic may be read ("consumed") by multiple recipients. Many forms of publish/subscribe messaging have been proposed, supporting different features and algorithms for message publication, subscription, subscription distribution, message dispatch, and so on. A substantial body of literature and ongoing research exists.[2]

For both point-to-point messaging and publish/subscribe messaging, different message delivery semantics (Quality-of-Service, QoS) can be defined. For example, a messaging client can specify the desired message delivery to be of at-most-once, at-least-once, or exactly-once semantics. Messages can also be acknowledged differently, and retry modes should any delivery attempts fail can be configured. Other QoS properties include the specification of message priorities and of message expiration based on timestamps.

3.4.2 Programming Models

A number of commercial MOM products exist. Two prominent examples are Websphere MQ (MQSeries), mostly known for its point-to-point messaging model, and TIBCO Rendezvous, a publish/subscribe middleware. Various messaging standardization efforts have also been carried out, including the CORBA Notification Service and, most notably, the J2EE Java Message Service (JMS) [29]. The J2EE framework also introduces Message-driven beans (MDB). More recently, messaging standards for Web services (SOAP [2]) and protocols for reliable Web services messaging (WS-Reliable Messaging [12]) have been developed.

MOM products typically support proprietary APIs, messaging formats, and protocols. The JMS specification can be regarded as a result of unifying the most common programming models. JMS is a standard API, defining a set of interfaces for both point-to-point and publish/subscribe messaging. JMS does not prescribe a particular implementation, and therefore, the semantics of messaging reliability, performance, scalability, and so on, are not fully specified.

JMS requires a messaging client to first create a connection to a JMS middleware, from which messaging sessions can then be created. A JMS session is a single-threaded context that is used to send and receive messages from and to queues and topics. Regarding the transactional aspects of JMS, the notion of a *transacted session* is introduced. A JMS session that is declared to be transactional supports commit and rollback semantics, so that the publication and consumption of a set of messages are atomic. Intermediate results are not visible unless the transacted session commits.

The following pseudo-code illustrates the use of a JMS-transacted session. A transacted session is created by setting the first parameter passed into the method that creates the session (here, a `QueueSession`) to "true":

```
QueueSession qs = connection.createQueueSession(true,
                                Session.AUTO_ACKNOWLEDGE);
```

All messages in the transacted session will become part of the transaction. Invoking the `send()` method on a sender object buffers the message in an internal storage of the client runtime; invoking `receive()` on a receiver object also buffers the message so that the message can be thrown back into the queue should the transaction fail. The messages are only actually sent out or retrieved from the queue should the transaction commit.

2 http://dsonline.computer.org/middleware/em/EMLinks.htm

```
QueueReceiver receiver = qs.createReceiver(queue);
TextMessage t1 = (TextMessage) receiver.receive();

QueueSender sender = qs.createSender(queue);
TextMessage t2 = qs.createTextMessage();
t2.setText(someString);
sender.send(t2);

session.commit();
```

A Message-driven Bean (MDB), as mentioned earlier, is a type of EJB that addresses the integration of JMS and EJB. A given MDB is associated with a single JMS destination (a queue or a topic) and a subscription durability (in case of a topic), which are set using the deployment descriptor. The EJB container intercepts incoming JMS messages, and locates and invokes the MDB's `onMessage()` method, which is defined in the JMS `MessageListener` interface. The bean can then extract the message body content.

Notice that an MDB has different transactional semantics than a transactional enterprise bean. JMS messages do not carry client transaction contexts, so that an MDB is never invoked and executed in an existing client transaction context. Valid transaction scope values for the `onMessage()` for container-managed transactions are `TX_NOT_SUPPORTED` and `TX_REQUIRED`; the `TX_REQUIRED` policy will always create a new transaction context.

In addition to transactional messaging using JMS and/or MDBs, the general architectural pattern of integrating message queues in direct transaction processing (as described in Section 3.3) is of fundamental importance (see below).

Attention must also be paid to the topology and configuration of the messaging architecture, transactional or nontransactional, in order for the architecture to be scalable. In the queuing model, for example, a simple load-balancing strategy can be implemented by configuring a set of replicated receivers to access the same queue; requests are distributed to each receiver consuming incoming requests from the queue. MOM products often provide other features addressing scalability and performance, including support for clustering and WAN (versus LAN) connectivity.

3.4.3 Queued Transaction Processing

Queued Transaction Processing (QTP) is an architectural pattern common and critical to many messaging applications. In QTP, clients and servers interact with each other using MOM and a series of distributed, direct transactions. Each transaction integrates message queues as transactional resources with other resources used by the applications (e.g., databases).

The series of these transactions can be considered to constitute a messaging workflow, or business transaction. The execution of the transactions may be managed and supported by workflow technology, business rule technology, or directly by the applications.

Figure 3.5 illustrates the QTP pattern. In the figure, three separate transactions are used to ensure that the interaction between application A (client) and application B (server) is reliable. In transaction 1, application A accesses local resources and sends a request message to application B; however, the message is not visible to application B until transaction 1 commits. In transaction 2, application B consumes the message, accesses

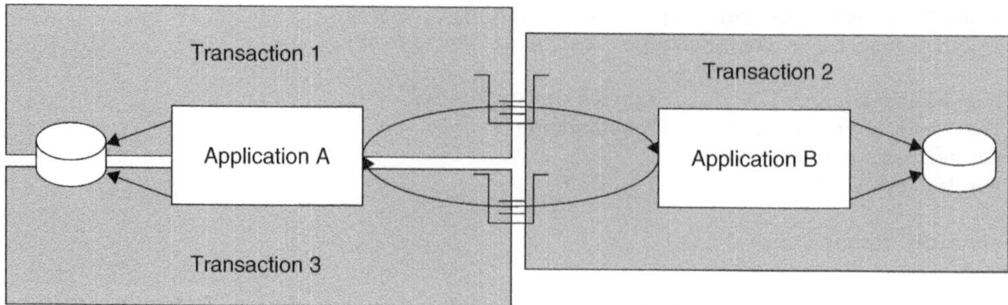

Figure 3.5 Queued transaction processing

local resources, and sends a response message to application A; the response is not visible to application A until transaction 2 commits. In transaction 3, application A consumes the response message from application B and accesses local resources. A failure during any single transaction returns the interaction to a well-defined, and locally consistent, state.

QTP allows the client and server to maintain control of their processing in that they each own a separate transaction. Should any of the applications fail and become unavailable, the other one is not affected, but can continue its processing. For example, client requests can still be issued and be stored persistently by the MOM should a server become unavailable; once the server is available again, it can process the stored requests on the basis of the message (back)log created.

The loose coupling between applications in QTP, however, can force the applications to implement potentially complex logic for coordinating the effects and outcomes of message delivery and processing with the effects and outcomes of other activities (e.g., database updates); if, for example, the response message in Figure 3.5 indicates that the outcome of application B's processing is inconsistent with application A's processing, application A cannot simply abort its transaction and expect to return to a consistent state (as it would be the case with direct transaction processing); rather, application A will have to actively correct for the inconsistency, perhaps initiating additional transactions and interactions with application B (for example, a compensating transaction that reverses the work performed in transaction 1).

In Section 3.6, we present research on advanced transaction middleware that proposes techniques for application-to-application reliability that overcome some of these disadvantages.

3.5 Web Transactions

Web services are a distributed programming model promoting service-oriented architectures [18]. With the integration of diverse applications as Web services in distributed, intra- and interenterprise systems, there is a strong need to ensure the (transactional and messaging) reliability of service interactions. Specifications addressing Web services reliability have consequently been proposed; these include WS-Coordination, WS-Transaction, and WS-ReliableMessaging. These specifications complement the basic Web services standards of WSDL (Web Services Description Language), UDDI (Universal Description, Discovery, and Integration), and SOAP to support distributed transactions and queued transactions in a Web environment.

These specifications are the first, important step to supporting transactional Web services. In addition, service-oriented middleware that intelligently organizes and distributes the components of Web transactions within a "Web" of services is needed. Yet, designing and developing Web middleware architectures is not a simple matter of "porting" existing middleware concepts to the Web. For example, the various dimensions of transparency (location transparency, concurrency transparency, etc.) typically provided by middleware in closed-distributed environments are challenged by the decentralized, autonomous nature of the Web [24].

3.5.1 Web Services Coordination and Transactions

WS-Coordination [4] describes a framework for use by transaction-processing systems, workflow systems, or other systems that wish to coordinate multiple Web services. The framework enables the creation of (transaction) contexts, the propagation of contexts among services, and the registration of services for specific coordination protocols. WS-Coordination is related to the work on extensible frameworks for extended transactions (such as the Activity service described earlier).

Using WS-Coordination, a specific coordination model is represented as a coordination type supporting a set of coordination protocols; a coordination protocol (such as a completion protocol or a synchronization protocol) is the set of well-defined messages that are exchanged between coordination participants. WS-Coordination defines an extensible coordination context and two coordination services for context creation and participant registration.

WS-Transaction [3] defines two particular coordination types leveraging WS-Coordination, called "Atomic Transaction (AT)" and "Business Activity (BA)". ATs principally compare to distributed transactions as described in Section 3.3. They support the property of atomicity on the basis of the premise that data and physical resources can be held, and are suggested for the coordination of short-lived activities within a trust domain. BAs address the coordination of short- or long-lived activities of potentially different trust domains. They do not require resources to be held, but suggest compensation-based recovery and business logic to be applied to handle exceptions.

WS-Coordination and WS-Transaction AT and BA are specifications developed by IBM, Microsoft, and BEA. Other specifications for transactional Web services, developed by other groups, have also been proposed. These include the Business Transaction Protocol (BTP) [5], and the Web Services Composite Application Framework (WS-CAF) [21].

3.5.2 Programming model

The Business Process Execution Language for Web Services (BPEL) [8] addresses the specification and execution of business processes and business interactions (message exchange protocols) that span multiple Web services. BPEL introduces a number of language constructs, including exception and compensation handlers, to define the various elements of a "business transaction". Using BPEL, both the processes and the process partners are modeled as WSDL services. BPEL is being standardized by the Organization for the Advancement of Structured Information Standards (OASIS).

In combination with BPEL, the WS-Coordination and WS-Transaction specifications allow the BPEL composition model to be extended with distributed coordination capabilities [9, 36]. The fault- and compensation-handling relationship between BPEL scopes can be expressed as a WS-Coordination coordination type, and distributed BPEL implementations can register for fault handling and compensation notifications using the coordination framework. WS-Coordination defines the coordination context for use in environments where BPEL scopes are distributed or span different vendor implementations; a context that is understood across the participants (BPEL implementations) is required in such environments.

Declarative policy assertions indicating required or supported transactional semantics can also be attached to BPEL processes and WSDL definitions. Languages such as WS-Policy [15] can be used to express transaction policies, and middleware can be used to match, interpret, and execute transactional business processes. This model, built around the standard Web services specifications, is described in detail in [36]. In Section 3.6, we describe another example of a policy-driven Web services transaction environment that allows for the composition of diverse, seemingly incompatible transactional services.

3.5.3 Web Services Messaging

The messaging protocol most commonly used for Web services is the Simple Object Access Protocol (SOAP) [2]. SOAP is an XML message format and encoding mechanism that allows messages to be sent over a variety of transports, including HTTP and JMS.

SOAP messaging can take different forms of reliability, depending on the underlying transport chosen. While SOAP-over-HTTP is not reliable, SOAP-over-HTTPR [37] ensures that messages are delivered to their specified destination. Similar reliability guarantees can be made for SOAP-over-JMS and SOAP-over-Websphere MQ. On the other hand, a SOAP message itself can be extended to include reliability properties, for example, using the recently proposed WS-ReliableMessaging specification [12]. These "extended SOAP" messages then carry relevant reliability information in the SOAP header, which must be understood and supported by a messaging infrastructure (that may or may not employ other reliable MOM).

The transactional features for Web services messaging vary accordingly. SOAP libraries offer no or very limited transactional support, whereas a MOM typically supports some notion of a transacted session, as described in Section 3.4, and, in the case of message queues, supports the integration of messaging endpoints into distributed transactions. A QTP model for Web services can be implemented by either using a particular MOM to exchange SOAP-formatted messages or by using SOAP messaging with a reliability protocol such as WS-ReliableMessaging, which in turn requires the use of some persistent storage and a MOM implementing the reliability protocol. For a more detailed discussion, see [34].

3.6 Advanced Transactions

Existing middleware supports the traditional transaction models, as described in Section 3.3 and Section 3.4. However, modern enterprise systems beg for more flexible transactional support. This section gives an overview of three projects at IBM Research that extend the current state of transactional middleware.

3.6.1 Long Running Unit of Work (LRUOW)

As previously mentioned, traditional TP monitors and lock-based models are mostly appropriate for short-running transactions. However, many business processes are long running, multiuser processes. Those long-running business processes (LRPB) cannot be naively implemented on a traditional TP system because of the interaction between their long duration and required concurrent access. As a result, developers must develop custom-built infrastructure – on an application-by-application basis – to support transactional LRBP (Long-Running Business Processes).

The LRUOW framework [1] is based on an advanced nested transaction model that enables concurrent access to shared data without locking resources. The model supports the concurrent execution of long-running units of work with a relaxed optimistic concurrency control and nested units of work. It relies on traditional ACID transactions for reliable execution. The LRUOW model as such does not provide an execution model; that is, the scheduling of actions is not part of the model. How extended transaction models, such as LRUOW, can be supported by business process servers (such as workflow management systems and other related technologies) is explored in [23].

The LRUOW framework provides three major pieces of functionality to the LRBP application developer.

- Packaging control of business activities into a unit of work so that the set of activities can be committed or rolled back as a unit
- Visibility control so that the objects created or updated are only visible within well-defined scopes rather than visible to everyone
- Concurrency control that manages the possibility for multiple users to add or change the same data.

The framework regards a LRBP as a directed, acyclic graph, whose nodes consist of units of work, each of which is a nestable long-running transaction [26]. Each UOW has one parent UOW (except for the root UOW) and may have multiple children UOWs. The root UOW owns all objects in the system and is never committed. In general, each subtask of the LRBP is mapped to a node in the UOW tree. All activities done within the course of an LRBP are done within the context of some UOW. The UOW context is established when the client either obtains or creates a new UOW, and joins the UOW using join() on the UOW object. Subsequent method invocations are performed within the scope of that UOW. A transaction can be committed or rolled back by invoking the respective method on the UOW object.

In the client view of the LRUOW programming model, user interactions occur only with UOW objects (provided by the framework) and base objects (arbitrary, non UOW-aware, objects). The framework enforces the protection implied by the visibility rules and, when a participant commits, propagates the objects' state changes to the parent UOW. To control visibility, the framework takes the base objects and creates versions that are associated with the UOWs. It transparently maps method invocations under a given UOW context onto the set of objects associated with the UOW. The client accesses a facade object that, in turn, delegates the client's method invocations to version objects that are associated with individual UOWs. The transaction context is implicitly propagated between the distributed EJB components that participate in the transaction, using request interceptors.

A UOW executes in two phases: a long-running phase (rehearsal phase), and a short-running phase (performance phase). During UOW rehearsal, no work is actually committed (in a transactional sense). More precisely, although user work can be made persistent (so that if the system crashes, user activity will resume from the last synchronization point), the UOW does not commit and make its work visible to a parent UOW context until the user invokes UOW.commit(). If a participant instead invokes UOW.rollback(), the work will be rolled back in traditional fashion. Because each UOW operates on a private set of data (the versions discussed above), protection from concurrent activity is automatically provided, making lock constraints unnecessary. Following the rehearsal phase, the performance phase is, in effect, a short, traditional transaction (with ACID properties) that modifies the versions of the objects in the parent UOW. During the performance phase, the framework deals with the concurrency issues that were ignored during the rehearsal phase. It supports two different relaxed-optimistic concurrency control mechanisms: conflict detection/resolution and predicate/transform.

The first concurrency mechanism relies on the framework to keep snapshots of the objects as they are first versioned in the child UOW. Upon commit, the framework checks to see if any data in the parent UOW has changed since the object was copied to the child UOW. If no changes are found, the parent versions are updated to reflect the data in the leaf UOW, else application-specific resolution managers are invoked, and they apply the business logic necessary to resolve conflicts and arrive at the desired parent data state (in a similar way to what the manual procedures accomplish in a single-threaded multiuser system). Resolution managers are provided by the application developer and can be plugged in the framework. The second concurrency mechanism uses the concept of predicates and transforms. The application programmer has to code invocations in terms of predicates and transforms (e.g., a debit transform is predicated by a check for positive balance). The predicates and transforms are implicitly logged during the UOW rehearsal phase as invocations are made on the objects versions it owns. Both the transforms and the predicates are replayed during the performance phase against the versions of the objects owned by the parent UOW. If a transform cannot be replayed (against its parent's state) because an associated predicate is no longer true, the UOW is rolled back. Concurrency will not be a problem, that is, the replay will succeed, as long as the predicates associated with the child's transforms are not violated by the current state of the parent's version. More details can be found in [1].

The LRUOW system prototype is illustrated in Figure 3.6.

3.6.2 Conditional Messaging and D-Spheres

As described in Section 3.4, messaging and messaging middleware are often used for purposes of enterprise application integration. Applications create, manipulate, store, and (typically asynchronously) communicate messages using MOM. The middleware guarantees reliable and eventual message delivery to intermediary destinations such as message queues, but does not readily support the management of application-defined conditions

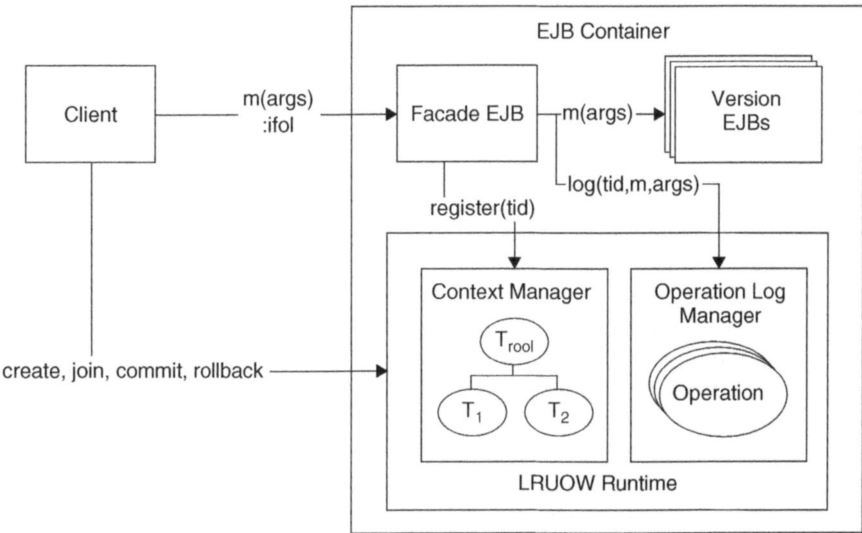

Figure 3.6 LRUOW system architecture

on messages, for example, to determine the timely receipt or the timely processing of a message by a set of final, transactional recipients.

Conditional messaging [33] is an extension to standard MQ/JMS middleware

- to define diverse conditions on message delivery and message processing in a structured and flexible way, and to represent these conditions independently of messages as distinct condition objects
- to send messages associated with conditions using a simple indirection API to standard messaging middleware
- to monitor the delivery to and the processing by final recipients using automatically generated internal acknowledgment messages of receipt and of (transactional) processing
- to evaluate conditions to determine a message outcome of success or failure
- to perform actions on the basis of the outcome of a message, including the sending of success notifications to all destinations in case of a message success, or the sending of compensation messages in case of a message failure.

Conditional messaging uniquely shifts the responsibilities for implementing the management of conditions on messages from an application to the middleware. Details can be found in [33].

The *Dependency Spheres (D-Spheres)* [32] project builds on the concept of conditional messaging and addresses the problem of integrating transactions and messaging for combined OOM and MOM system environments. Most enterprise applications need to combine various existing legacy components along with new ones. Since legacy applications generally operate on an asynchronous messaging scheme (which, for example, may involve one-way messages, on-commit message publication, and persistent message stores that are transaction resource managers), integrating asynchronous messaging with the synchronous, tightly coupled notion of a distributed (object) transaction is of high practical interest and importance. Existing middleware solutions to integrating object transactions and messaging are very limited, either restricting the programming flexibility drastically or allowing arbitrary, unstructured integration without any Quality-of-Service guarantees.

The D-Spheres service introduces a layer of abstraction above standard transactional OOM and MOM, presenting an integrated view of the base transaction and messaging services used for transactional application development. The service advances the current state of the art and state of the practice in transactional messaging in that

- conditional messages, as defined above, can be sent to remote queues at any point in time during the ongoing object transaction, and
- the outcome of conditional messages and the outcome of the object transaction are interdependent.

The D-Spheres approach defines a new type of a global transaction context inside of which conventional object transactions and distributed messages may occur. A D-Sphere makes transactional synchronous object requests dependent on asynchronous messages, and vice versa.

A D-Sphere is a kind of *middleware-mediated transaction* [20]. The D-Spheres middleware provides a simple transaction API for explicit transaction demarcation, which is used in place of the OOM transaction API. Beneath its transaction API, the D-Spheres middleware relies on the underlying OOM to implement the management of object transactions. D-Spheres can accommodate a variety of kinds of transaction models for short- or long-running transactions, including the CORBA OTS, EJB, or LRUOW.

As far as the messaging model is concerned, D-Spheres require the ability to determine the success of message delivery to final recipients and the success of message processing by a set of final recipients. D-Spheres use the conditional messaging model described previously to that end.

Message outcomes in D-Spheres are treated analogously to the commit or abort outcomes of OOM transactions. This achieves the interdependence of outcomes between messages and transactions. When a D-Sphere succeeds, then the transaction and messaging operations it contains have been successful, and these are simply allowed to stand. When a D-Sphere fails, though, the transaction is aborted and some additional action must be taken to negate the effects of any messages that have been sent. The D-Sphere can handle the transaction abort (if necessary) by instructing the transaction manager to abort the transaction. Since MOM offers no operation analogous to abort for messages, the D-Sphere must execute some compensating actions. These may include the retracting of messages that have not been delivered to their final recipients, and the sending of compensating messages for messages that have been delivered to their final recipients.

Figure 3.7 illustrates the D-Sphere service architecture. More details on D-Sphere can be found in [32] [35].

3.6.3 Transactional Attitudes (TxA)

The *Transactional Attitudes (TxA)* framework [25] addresses the reliable composition of Web services. Unlike the Web services transaction models presented earlier, the TxA

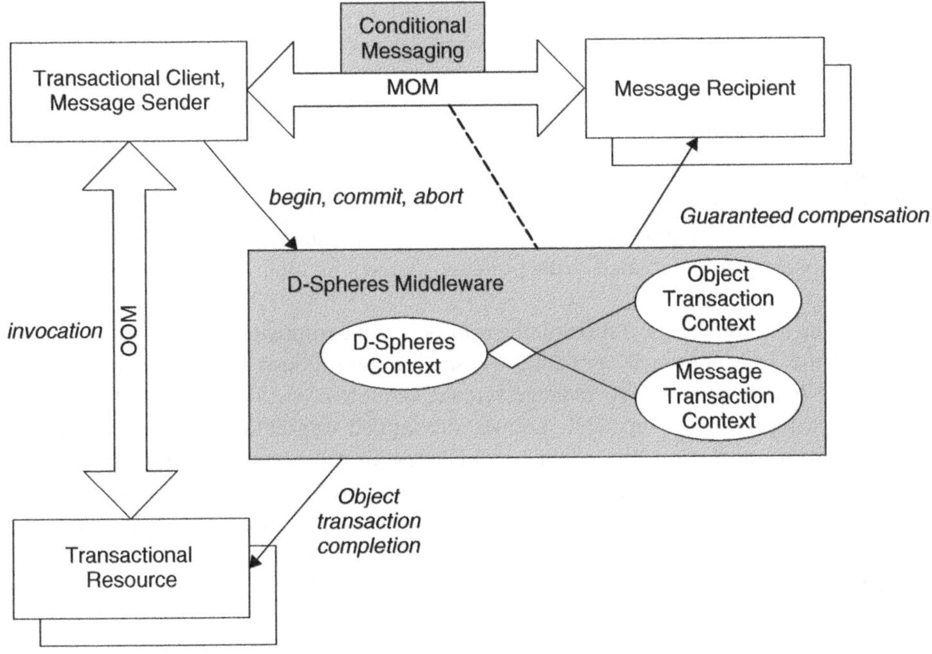

Figure 3.7 D-spheres service architecture

approach does not assume any common transaction semantics, transaction context representation, or coordination protocol to exist among the transaction participants. The participants are assumed to be autonomous (with respect to their implementation and execution environment), and seemingly incompatible transaction models and middleware technologies may be involved in the same Web transaction. Context representation and service coordination and management are achieved in a decoupled, decentralized manner.

The TxA model is based on the notion of *"transactional attitudes"*, where Web service providers declare their individual transactional capabilities and semantics, and Web service clients declare their transactional requirements (of providers). Provider transactional attitudes (PTAs) make a Web service's transactional semantics explicit. Client transactional attitudes (CTAs) make a client's expectation on the transactional composition of a group of services explicit. Finally, the TxA meta-coordination middleware uses PTAs and CTAs to automate the reliable execution of Web transactions, that is, it automates the composition of individual transactional Web services into larger transactional patterns, while maintaining the autonomy of the individual services.

PTAs are a mechanism for Web service providers to explicitly describe their specific transactional behavior. A PTA includes the name of an abstract transactional pattern, plus any additional port-specific information needed to make the pattern concrete. The abstract pattern implies a well-defined state machine describing the valid transactional states, state transitions, and transition-triggering events. The name and semantic of each state is implied by the name of the transactional pattern. State transitions are triggered by externally observable events, such as operation invocations on ports and time-outs.

TxA uses the WSDL extension mechanism [6] to annotate existing WSDL definitions with PTA descriptions. The TxA assertions vocabulary defines three common provider transactional attitudes: "*pending-commit*" (PC), "*group-pending-commit*" (GPC), and "*commit-compensate*" (CC). For example, the *pending-commit* attitude describes a transactional port where the effect of a single *forward operation* invocation can be held in a pending state; the operation-effect remains pending until the subsequent occurrence of an event (e.g., the invocation of a commit or abort operation) triggers either acceptance or rejection (of the operation-effect). Forward, commit, and abort operations are regular WSDL operations annotated with policies identifying their respective transactional semantic.

CTAs make a client's expectation on the transactional composition of a group of services explicit. In the TxA approach, a client's transactional attitude is established by its use of a particular WSDL port type to manage (create, complete, etc.) web transactions, where the port type represents some well- known, predefined transactional pattern. Within the scope of a web transaction, the client executes one or more named *actions*, where each action represents a provider transaction (associated with some PTA) executing within the context of the larger Web transaction. The client initiates an action by binding to an *action port*, which serves as a proxy to a participating provider's transactional port. Each action port represents a unique provider transaction executing within the context of the client's web transaction.

When using an action port, the client may invoke *only the forward operations* of the provider; that is to say, the client cannot invoke *completion* operations (commit, abort, or compensation operations). Several client attitudes have been identified. For example, the *flexible atom (FA)* attitude describes a client transaction where a set of client actions (i.e., provider transactions) are grouped into an atomic group that can have one out of a set of defined group outcomes; that is to say, some actions are declared to be *critical* to the success of the transaction, whereas other actions are part of the transaction, though not pivotal to its success. The client specifies the acceptable outcomes as an *outcome condition*, described in terms of the success or failure of the individual actions, and when ready (i.e., after executing the forward operations of these actions), requests the completion of the flexible atom according to that condition.

The TxA middleware attempts to reach a defined outcome, according to the client's CTA, by invoking the appropriate completion operations on the providers associated with the client's actions. The *Smart Attitude Monitor*, a.k.a. *SAM*, is a prototype middleware that uses PTAs and CTAs to automate the reliable execution of Web transactions. SAM is itself a Web service, and serves as an intermediary between a transactional client and one or more transactional providers.

Figure 3.8 illustrates SAM, within the context of a Travel Booking scenario. SAM comprises three types of ports and a recovery log. The TxA port provides general web transaction and configuration functions, and is independent of any particular CTA. The CTA ports are used to manage web transactions associated with specific CTAs (in Figure 3.8, a port supporting the *flexible atom* CTA is shown). The action ports are *proxy* ports through which a client interacts with providers. An action port is created dynamically (by SAM) whenever a client initiates a provider transaction within the context of a web transaction. SAM intercepts all messages exchanged through the action ports, and uses

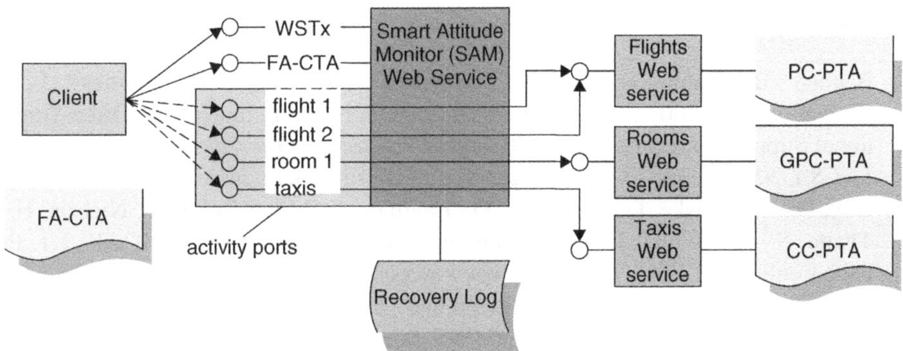

Figure 3.8 TxA system prototype

the provider's PTA to determine the meaning of these messages within the context of the provider's transaction. In doing so, SAM reflects the state of all provider transactions associated with the client's actions. SAM records the events associated with critical state transitions in its recovery log, and uses these records to recover in the event of a failure. When completing a web transaction, SAM drives the associated provider transaction's to completion according to the client's CTA. For example, when completing a flexible atom web transaction, SAM evaluates the actions named in the outcome condition, and attempts to satisfy this condition by invoking the appropriate completion operations (e.g., commit, abort, compensate) on the providers. Details can be found in [25].

3.7 Conclusion

Transaction processing is critical to the development of nearly any enterprise system. Communication middleware in the form of object-oriented middleware, message-oriented middleware, and (Web) service-oriented middleware all consequently support transaction processing. In this chapter, we explored how the different middleware systems address transaction processing.

We focused attention on object-oriented, message-oriented, and service-oriented enterprise middleware, and emphasized the two main transaction models supported by these systems: direct transaction processing (DTP) and queued transaction processing (QTP). We did not cover other transaction middleware including database systems and transaction monitors like IBM's CICS, as these have been well-covered elsewhere in the existing literature. Middleware such as CORBA, J2EE/EJB, and the Web services framework, as selected for presentation in this chapter, also have the additional goal of integrating applications and business processes across heterogeneous implementations.

We described transaction-processing fundamentals and surveyed, from a middleware-viewpoint, specific models for transaction processing. We also described research on advanced transaction middleware that introduces new ideas and overcomes some of the limitations identified with traditional middleware.

Bibliography

[1] Bennett, B., Hahm, B., Leff, A., Mikalsen, T., Rasmus, K., Rayfield, J., and Rouvellou, I. (2000) A Distributed Object-Oriented Framework to Offer Transactional Support for Long Running Business Processes, *Proceedings Middleware 2000, LNCS 1795*, pp. 331–348, Springer-Verlag, Berlin, Germany.

[2] Box, D., Ehnebuske, D., Kakivaya, G., Layman, A., Mendelsohn, N., Nielsen, H. F., Thatte, S., and Winer, D. (2000) Simple Object Access Protocol (SOAP) 1.1. Published online as W3C Note, http://www.w3.org/TR/2000/NOTE-SOAP-20000508

[3] Cabrera, F., Copeland, G., Cox, B., Freund, T., Klein, J., Storey, T., and Thatte, S. (2002) Web Services Transaction (WS-Transaction). Published online by IBM, Microsoft and BEA, http://www-106.ibm.com/developerworks/library/ws-transpec.

[4] Cabrera, F., Copeland, G., Freund, T., Klein, J., Langworthy, D., Orchard, D., Shewchuk, J., and Storey, T. (2002) Web Services Coordination (WS-Coordination). Published online by IBM, Microsoft and BEA, http://www-106.ibm.com/developerworks/library/ws-coor

[5] Ceponkus, A., Dalal, S., Fletcher, T., Furniss, P., Green, A., and Pope, B. (2002) Business Transaction Protocol. Published online, http://www.oasis-open.org/committees/business-transactions/documents/specification/2002-06-03.BTP_cttee_spec_1.0.pdf.

[6] Christensen, E., Curbera, F., Meredith, G., and Weerawarana, S. (2001) Web Services Description Language (WSDL) 1.1. Published online, http://www.w3.org/TR/wsdl.

[7] Codella, C., Dillenberger, D., Ferguson, D., Jackson, R., Mikalsen, T., and Silva-Lepe, I. (1998) Support for Enterprise JavaBeans in Component Broker. *IBM System Journal*, **37**(4), 502–538.

[8] Curbera, F., Goland, Y., Klein, J., Leymann, F., Roller, D., Thatte, S., and Weerawarana, S. (2002) Business Process Execution Language for Web Services 1.0. Published online by IBM, Microsoft and BEA, http://www106.ibm.com/developerworks/library/ws-bpel

[9] Curbera, F., Khalaf, R., Mukhi, N., Tai, S., and Weerawarana, S. (2003) Web Services, The Next Step: Robust Service Composition. *Communications of the ACM: Service-Oriented Computing*, **46**(10), 29–34.

[10] Date, C.J. (1983) *An Introduction to Database Systems*, Volume II, Addison-Wesley Publishing Company Reading, MA.

[11] Date, C.J. (2000) *An Introduction to Database Systems*, 7th ed., Addison-Wesley Longman, Reading, MA.

[12] Ferris, C. and Langworthy, D. (eds.) (2003) Web Services Reliable Messaging Protocol (WS-ReliableMessaging). Published online by BEA, IBM, Microsoft and TIBCO, ftp://www6.software.ibm.com/software/developer/library/ws-reliablemessaging.pdf

[13] Garcia-Molina, H. and Salem, K. (1987) Sagas. *Proceedings ACM SIGMOD*, San Francisco, CA.

[14] Gray, J. and Reuter, A. (1993) *Transaction Processing: Concepts and Techniques*, Morgan Kaufmann, San Francisco, CA.

[15] Hondo, M. and Kaler, C., (eds) (2002) Web Services Policy Framework (WS-Policy). Published online by IBM, Microsoft, and BEA at http://www-106.ibm.com/developerworks/webservices/library/ws-polfram

[16] Houston, I., Little, M., Robinson, I., Shrivastava, S., and Wheater, S. (2001) The CORBA Activity Service Framework for Supporting Extended Transactions, *Proceedings Middleware 2001, LNCS 2218*, Springer-Verlag, Berlin, Germany.

[17] Jajodia, S. and Kerschberg, L., (eds) (1997) *Advanced Transaction Models and Architectures*, Kluwer Academic Publishers, Norwell, MA.

[18] Khalaf, R., Curbera, F., Nagy, W., Tai, S., Mukhi, N., and Duftler, M. (2004) Understanding Web Services, in *Practical Handbook of Internet Computing* (ed. M. Singh), CRC Press (in press).

[19] Kung, H. T. and Robinson, J. T. (1981) On Optimistic Methods for Concurrency Control. *ACM Transactions on Database System*, **6**(2), 213–226.

[20] Liebig, C. and Tai, S. (2001) Middleware-Mediated Transactions, *Proceedings DOA 2001*, IEEE, Los Alamitos, CA.

[21] Little, M. and Newcomer, E., (eds) (2003) Web Services Composite Application Framework (WS-CAF) Version 1.0. Published online by Arjuna, Iona, Sun, Oracle, and Fujitsu at http://www.arjuna.com/standards/ws-caf/

[22] Leymann, F. and Roller, D. (2000) *Production Workflow: Concepts and Techniques*, Prentice Hall, Upper Saddle River, NJ.

[23] Mikalsen, T., Rouvellou, I., Sutton, S., Tai, S., Chessel, M, Griffen, C., and Vines, D. (2000) Transactional Business Process Servers: Definition and Requirements. *Proceedings OOPSLA 2000 Workshop on Business Object Components*, Minneapolis, MN.

[24] Mikalsen, T., Rouvellou, I., and Tai, S. (2001) Reliability of Composed Web Services: From Object Transactions to Web Transactions. *Proceedings OOPSLA 2001 Workshop on Object-Oriented Web Services*, Tampa, FL.

[25] Mikalsen, T., Tai, S., and Rouvellou, I. (2002) Transactional Attitudes: Reliable Composition of Autonomous Web Services, *Proceedings DSN 2002 Workshop on Dependable Middleware Systems*, IEEE, Los Alamitos, CA.

[26] Moss, J. E. B. (1985) *Nested Transactions: An Approach to Reliable Distributed Computing*, MIT Press, Cambridge, MA.

[27] OMG. (2000) Transaction Service v1.1, TR OMG Document formal/2000-06-28.

[28] Sun Microsystems (2001) Java 2 Enterprise Edition: J2EE Connector Architecture Specification 1.0, http://java.sun.com/j2ee/connector/

[29] Sun Microsystems (2002) Java Message Service API Specification v1.1, http://java.sun.com/products/jms/

[30] Sun Microsystems (2002) Java Transaction API (JTA) 1.0.1B, http://java.sun.com/products/jta/

[31] Sun Microsystems (2002) Enterprise JavaBeans Specification Version 2.1, http://java.sun.com/products/ejb/docs.html

[32] Tai, S., Mikalsen, T., Rouvellou, I., and Sutton, S. (2001) Dependency-Spheres: A Global Transaction Context for Distributed Objects and Messages, *Proceedings EDOC 2001*, IEEE, Los Alamitos, CA.

[33] Tai, S., Mikalsen, T., Rouvellou, I., and Sutton, S. (2002) Conditional Messaging: Extending Reliable Messaging with Application Conditions, *Proceedings ICDCS 2002*, IEEE, Los Alamitos, CA.

[34] Tai, S., Mikalsen, T., and Rouvellou, I. (2003) Using Message-oriented Middleware for Reliable Web Services Messaging, *Proceedings CAiSE 2003 Workshop on Web Services, e-Business, and the Semantic Web*, Springer-Verlag, Berlin, Germany (in press).

[35] Tai, S., Totok, A., Mikalsen, T., and Rouvellou, I. (2002) Message Queuing Patterns for Middleware-Mediated Transactions, *Proceedings SEM 2002, LNCS 2596*, Springer-Verlag, Berlin, Germany.

[36] Tai, S., Mikalsen, T., Wohlstadter, E., Desai, N., and Rouvellou, I. (2004) Transaction Policies for Service-Oriented Computing.

[37] Todd, S., Parr, F., and Conner, M. (2001) A Primer for HTTPR: An Overview of the Reliable HTTP Protocol. Published online, http://www-106.ibm.com/developerworks/webservices/library/ws-phtt/

[38] X/Open (1996) Distributed Transaction Processing: Reference Model, Version 3, X/Open Company Limited, Reading, UK.

4

Peer-to-Peer Middleware

Markus Oliver Junginger[1], Yugyung Lee[2]
[1]*IEEE Member*
[2]*University of Missouri*

4.1 Introduction

Peer-to-Peer networking has a great potential to make a vast amount of resources accessible [19]. Several years ago, file-sharing applications such as Napster [15] and Gnutella [8] impressively demonstrated the possibilities for the first time. Because of their success, Peer-to-Peer mistakenly became synonymous for file sharing. However, the fundamental Peer-to-Peer concept is general and not limited to a specific application type. Thus, a broader field of applications can benefit from using Peer-to-Peer technology. Content delivery [11], media streaming [22], games [14], and collaboration tools [9] are examples of applications fields that use Peer-to-Peer networks today.

Although Peer-to-Peer networking is still an emerging area, some Peer-to-Peer concepts are already applied successfully in different contexts. Good examples are Internet routers, which deliver IP packages along paths that are considered efficient. These routers form a decentralized, hierarchical network. They consider each other as peers, which collaborate in the routing process and in updating each other. Unlike centralized networks, they can compensate node failures and remain functional as a network. Another example of decentralized systems with analogies to Peer-to-Peer networks is the Usenet [31]. Considering those examples, many Peer-to-Peer concepts are nothing new. However, Peer-to-Peer takes these concepts from the network to the application layer, where software defines purpose and algorithms of virtual (nonphysical) Peer-to-Peer networks.

Widely used web-based services such as Google, Yahoo, Amazon, and eBay can handle a large number of users while maintaining a good degree of failure tolerance. These centralized systems offer a higher level of control, are easier to develop, and perform more predictably than decentralized systems. Thus, a pure Peer-to-Peer system would be an inappropriate choice for applications demanding a certain degree of control, for example, "who may access what." Although Peer-to-Peer systems cannot replace centralized system, there are areas where they can complement them. For example, Peer-to-Peer

Middleware for Communications. Edited by Qusay H. Mahmoud
 ISBN 0-470-86206-8

systems encourage direct collaboration of users. If a centralized system in between is not required, this approach can be more efficient because the communication path is shorter. In addition, a Peer-to-Peer (sub)system does not require additional server logic and is more resistant to server failures. For similar reasons, because they take workload and traffic off from servers to peers, Peer-to-Peer could reduce the required infrastructure of centralized systems. In this way, Peer-to-Peer networks could cut acquisition and running costs for server hardware. This becomes increasingly relevant when one considers the growing number of end users with powerful computers connected by high bandwidth links.

4.1.1 Peer-to-Peer and Grids

An approach similar to Peer-to-Peer is the Grid [7], which is discussed in detail in Chapter 5 (Grid Middleware). Both are decentralized networks committed to collaborate and to share resources. In addition, both aim at failure tolerance and have to cope with a heterogeneous network. Nevertheless, there are also differences. A Grid consists of mostly high-end computers (supercomputers) with enormous resources, whereas a Peer-to-Peer network consists of less powerful desktop computers. In addition, Peer-to-Peer networks have to deal with a much higher fluctuation as peers join and leave frequently. However, probably the most important difference is the purpose. A Grid consists of machines that collaborate to solve complex computation submitted by users. In contrast, Peer-to-Peer networks are user-centric and thus focus on the collaboration and communication of individuals.

4.1.2 Lack of Peer-to-Peer Middleware

Widely known Peer-to-Peer applications include the office collaboration suite Groove [9] and the IP telephony service Skype [22]. Nevertheless, the number of applications beyond file sharing is kept on a low level, despite the general character of Peer-to-Peer networking. The reasons are not only the popularity and omnipresence of web-based services but also the lack of Peer-to-Peer enabling technologies. Although Groove and Skype do not rely on a standard middleware product but on an individual Peer-to-Peer network, many companies cannot afford the lengthy and costly development of such an infrastructure from scratch. In times of Rapid-Application-Development and short time-to-market requirements, middleware products are needed. CORBA, EJB, COM+, SOAP, and various messaging systems ease the development of distributed applications, but they do not incorporate features needed for decentralized Peer-to-Peer environments [24]. Although there are no counterparts to these established middleware products yet, there exist several efforts of developing Peer-to-Peer middleware. One of the most advanced ones is Project JXTA [25], which we will review in more detail. Note that we intentionally do not review systems dedicated to file sharing, although there are very interesting systems available. Examples are the P-Grid [1], which offers scalable searches, and Freenet [3], which addresses security and anonymity. Nevertheless, their scope is limited and we do not consider them a general-purpose middleware.

4.1.3 Group Communication

We assume group communication especially relevant for Peer-to-Peer applications, because the focus shifts from the individual network node (centralized) towards a group of nodes (decentralized). The potential of Peer-to-Peer technology lies in the collaboration of several peers, allowing each peer to profit from others. As collaboration of a group usually requires communication among its members, we see an increasing need for Peer-to-Peer group communication [2]. Consequently, one of our focuses is group communication.

4.1.4 Challenges

Designing Peer-to-Peer middleware is a complex task. One cannot rely on a static network of dedicated and mostly centralized service providers, which is still a common approach. Instead, Peer-to-Peer networks confront us with several new challenges:

- *Shared environment:* Infrastructures tend to become shared platforms for several independent applications that may have contrary requirements and may interfere with each other.
- *Scalability:* The large number of nodes within a Peer-to-Peer network may affect performance, latency, and reliability.
- *Dynamic network:* Nodes are unreliable because they may join or leave the network unpredictably.
- *Dynamic node characteristics:* Since nodes are autonomous, their characteristics may change. For example, concurrent applications may decrease the available bandwidth.
- *Network heterogeneity:* Each peer is unique according to its networking and computing power, location in the physical network, and provided services.
- *Quality-of-Service:* Given an unpredictable network, the quality of network-dependent services might not be guaranteed but improved.
- *Security:* The middleware should be resistant to malicious peers.

4.1.5 Chapter Outline

Peer-to-Peer middleware is still an open and emerging topic, and thus a final review cannot be given yet. Consequently, we exemplarily present current solutions and outline a possible model for integrating centralized and decentralized systems. The reader will (hopefully) get a notion of the Peer-to-Peer middleware and the general Peer-to-Peer paradigm with its problems and its potential.

This chapter is divided into three sections. First, Sun Microsystem's Project JXTA will be reviewed. JXTA is a Peer-to-Peer middleware offering communication primitives, security features, and higher-level services. Second, we will present the P2P Messaging System, a middleware subsystem specialized in group communication. It bases on a dynamic, hierarchical topology that adjusts automatically to a changing network. Third and last, we propose models for a hybrid middleware combining Peer-to-Peer networking with centralized systems.

4.2 JXTA

> "JXTA is a set of open, generalized Peer-to-Peer (P2P) protocols that allow any connected device on the network - from cell phone to PDA, from PC to server - to communicate and collaborate as peers. The JXTA protocols are independent of any programming language, and multiple implementations (called bindings in Project JXTA) exist for different environments." [27]

JXTA is a Peer-to-Peer platform and middleware. Introduced in April 2001, JXTA is the first significant Peer-to-Peer platform that is not tied to a specific task, such as file sharing or instant messaging. Instead, JXTA provides a general middleware usable for a variety of applications. Almost two years after it was launched, when JXTA was upgraded to version 2.0, it was downloaded over a million times, and over ten thousand developers registered as JXTA community members [28]. Meanwhile, competing Peer-to-Peer platforms are emerging and beginning to challenge JXTA. One of the most promising competitors is Microsoft with its Peer-to-Peer SDK (Software Development Kit), which was released as a beta version in February 2003 [17]. Pastry [20] is another example of a general Peer-to-Peer platform, but it is and has always been research project. Nevertheless, further research projects based on Pastry were initiated. Among these projects, JXTA is the most widely used and most matured Peer-to-Peer platform. It will therefore be further reviewed.

4.2.1 Overview

From a high-level point of view, JXTA could be divided into three layers: core, service, and application layer. The core layer provides fundamental functionalities like transport, discovery, and security primitives. The service layer relies on the core layer and provides services that are typically used in Peer-to-Peer applications like searching and indexing, file sharing, and security services. Applications usually rely on both core and service layers, although the service layer may be more important to application developers. The separation into layers can be demonstrated with a file-sharing application. The application layer provides a GUI to let the user search, download, and upload files. Then, the application layer delegates the actions to JXTA's file-sharing service. The service layer realizes the search, download, and upload functions using the discovery and transport primitives of the core layer.

Figure 4.1 depicts an exemplarily JXTA network with six peers, each involved in a different activity as outlined in the figure. The following sections will explain the concepts behind JXTA in more detail. After discussing resources that are typically advertised within peer groups, the relationship between services, modules, and protocols will be outlined. Thereafter, we will present how peers use pipes to exchange messages and the security mechanisms JXTA supports. Finally, different peer types and group communication issues will be discussed.

4.2.2 Resources and Advertisements

Advertisements, which represent resources, are a fundamental concept. All objects like peers, peer groups, services, and communication facilities are considered a resource and can be described as advertisements based on XML. Resources and advertisements are

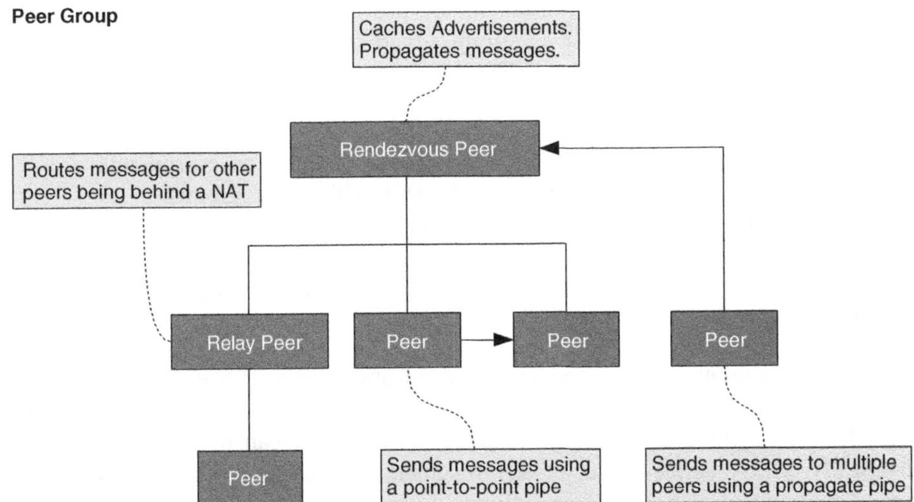

Figure 4.1 Exemplarily JXTA overview

distinguished by unique IDs. All advertisement definitions, such as JXTA's protocols and source code, are openly available [25]. JXTA's programming API has methods to create functional resource objects out of advertisements. Unlike resource objects (e.g., Java objects), advertisements can be sent to other peers regardless of peers' platforms and programming languages. Advertisements are usually published within the network to make them accessible to any peer. JXTA's Discovery Service provides methods for publishing and searching advertisements.

Although caching of advertisements is not essentially required by JXTA bindings, it optimizes the handling of advertisements. Once advertisements are found, peers store them in a local cache to make them immediately available for upcoming queries. To avoid advertisements to prevail forever in the network and caches, advertisements have a defined lifetime.

4.2.3 Peer Groups

A peer consists of software running one or more JXTA protocols, and data including a name and a unique PeerID. Peers that belong to one application may assemble within peer groups, which form subnets within the global JXTA network and therefore contribute to the overall network scalability. All peers are members of the global "Net Peer Group" and may additionally join any number of groups. Most services operate within the environment of a peer group. Particularly, the Membership Service determines which peers may join a group. Individual group access restrictions can be considered by implementing new Membership Services.

4.2.4 Services and Modules

Many of the presented concepts are provided as services, which are a generic concept. JXTA differentiates between services provided by a single peer and services provided by a peer group. While peer services have a single point of failure, peer group services remain functional as long as peers stay present in the peer group. Information on peer group services is usually embedded in peer group advertisements, and can thus be accessed immediately after a group was joined. The following list shows JXTA's core services to provide a general overview.

Discovery Service	Publishes and retrieves resource advertisements
Membership Service	Restricts membership within peer groups
Access Service	Checks whether peers are eligible to access actions
Pipe Service	Establishes the pipe communication infrastructure
Resolver Service	Manages queries and responses among peers
Monitoring Service	Monitors other peers.

Modules implement services and represent an abstraction of a piece of software or code. Modules are not bound to any language and can be Java classes, dynamic libraries, XML files, or anything else. Using modules, new services can be loaded and invoked dynamically. JXTA defines a set of advertisements to clarify a module's purpose and usage. This concept also allows an individual service being implemented for multiple platforms. All of JXTA's built-in services are implemented using modules.

JXTA also provides higher-level services that usually rely on the presented core services. An interesting example is the Content Management Service (CMS) that offers basic file-sharing functionalities. Simple methods exist for making files available to other peers, searching for files, and downloading files. Internally, files are identified with MD5 hashes and are published to others using advertisements. File downloads rely on JXTA's pipe communication facilities, which CMS sets up automatically.

4.2.5 Protocols

Services demand exchanging messages among participants over the network. Protocols standardize the message-exchange by defining the message format and sequence. JXTA's open protocols [26] build on a message layer that abstracts from networking protocols such as TCP and HTTP. Messages are XML documents, and each service has a defined set of XML message schemas. JXTA has six core protocols:

Peer Resolver Protocol	Abstract Query/response protocol
Peer Discovery Protocol	Resource advertisement publication and discovery
Peer Information Protocol	Acquisition of status information from other peers
Pipe Binding Protocol	Establishment of pipes
Endpoint Routing Protocol	Routing to peers that are not directly accessible
Rendezvous Protocol	Message propagation to multiple subscribers.

Detailed information on protocols is out of our scope and can be retrieved from the official protocol specifications [26].

4.2.6 Messages and Pipes

JXTA's fundamental communication facilities are message-oriented and asynchronous. Messages are either binary or XML encoded. JXTA's internal protocols rely on a defined sequence of messages.

Applications usually use pipes as virtual communication channels to send and receive messages. Pipes hide the network's complexity, which may result from indirect delivery that requires routing. They also abstract from underlying protocols such as TCP and HTTP. In JXTA terminology, sending pipe-endpoints are called "output pipes," while receiving pipe-endpoints are called "input pipes". Different pipe types satisfy different application requirements. The most basic type provides a simple point-to-point communication channel, which is insecure and unreliable. A secure pipe variation extends this pipe and establishes a virtual TLS (Transport Layer Security) connection [5] to encrypt data. The third and last type is the propagate pipe, which may consist of multiple input and output endpoints. Thus, messages propagated by one of the pipe's senders will be delivered to all receivers listening to the pipe.

In a strict sense, the message delivery through pipes is unreliable. In practice, however, reliability is typically high because messages are delivered using reliable transport protocols such as TCP and HTTP. Therefore, messages are only lost in stress situations, when message queues are full or peers are disconnected.

4.2.7 Security

Security in decentralized networks is a very challenging topic, because peers may not be trustful, and a trusted central authority lacks. At this point, JXTA makes a compromise because it does not guarantee that peers are the ones that they claim to be. Nevertheless, JXTA provides a security infrastructure that offers established technologies. Messages can be encrypted and signed with several mechanisms. JXTA supports both secret key and public/private key encryptions. By default, JXTA uses RC4 algorithms for secret keys, and RSA algorithms for public/private keys. Signatures and hashes can be created with either SHA1 or MD5 algorithms.

The secure pipe is a higher-level construct, which automatically handles security issues for developers. Using TLS, it uses the RSA public/private keys to establish a secure connection initially. For the actual data flow, it uses Triple-DES and digital signatures. The public key certificates needed for initialization are created by the JXTA platform itself without centralized certificate authorities.

4.2.8 Relay and Rendezvous Peers

JXTA is not a pure Peer-to-Peer network, because some peers provide additional services to others. These peers, namely relay and rendezvous peers, have some characteristics of a server. Despite being called "peers," they are at least partially committed to serve others. Thus, JXTA can be considered a hybrid Peer-to-Peer network (note that the hybrid network presented later differs fundamentally). The additional functionality is assigned by the JXTA configuration or programmatically by developers.

Relay peers take care of routing messages to peers, and also serve as a proxy for peers, which cannot directly access the JXTA network because of a restrictive network such as firewall/NAT. In this case, peers connect to relay peers, which forward messages on behalf of peers. Because relay peers support the HTTP protocol, most firewalls can be passed.

Rendezvous peers provide services that are essential for most peers of the JXTA network. First, rendezvous peers propagate messages to all group members. Higher-level

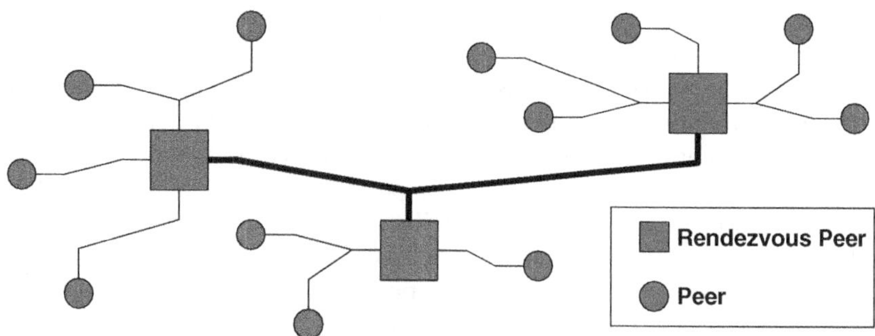

Figure 4.2 Rendezvous network

functionality such as propagate pipes and queries provided by the Resolver Service rely on this propagation. Second, rendezvous peers take care of discovery queries. If they cannot answer queries directly, they forward discovery messages to other rendezvous peers. Third, to support these queries, they cache advertisements requested by connected peers.

After peers join a group, they automatically seek a rendezvous peer and establish a permanent connection to it until they leave the group. If no rendezvous peer exists in a group, peers themselves become rendezvous peers and are thus available to serve for others. Because rendezvous peers within a group have to collaborate, they are connected with each other. An idealized network structure is depicted in Figure 4.2.

4.2.9 Group Communication

A very interesting aspect of Peer-to-Peer computing is group communication. As seen previously, JXTA's group communication relies on a rendezvous network. Messages that are addressed to multiple recipients are initially sent to rendezvous peers, which forward them to other peers. If multiple rendezvous peers are present, the messages will be forwarded in cooperation with them. However, JXTA does not dynamically elect rendezvous peers. Thus, the resulting network topology may not always be ideal. Unless the rendezvous infrastructure is setup manually, it is not ensured that the infrastructure is able to serve all peers reliably. For example, when a peer with low bandwidth joins an initially empty group, it becomes a rendezvous peer automatically when it notices that no rendezvous peers are present. When other peers join the group subsequently and connect to the rendezvous peer, its available bandwidth may be quickly exhausted. Further, because the resources of the rendezvous infrastructure do not grow when the number of peers increases, scalability is limited. For any rendezvous infrastructure, the number of peers that can be served reliably is restricted. Regarding the rendezvous network, JXTA reveals some similarities with centralized networks, and thus the same limitations apply. Figure 4.3 illustrates such a centralized rendezvous network.

Another limitation of JXTA's rendezvous network results from the fluctuation, which typically occurs in Peer-to-Peer networks. When a rendezvous peer gets disconnected for any reason, the dependent peers lose the services that were provided by the rendezvous peer. The services stay unavailable until one of other peers becomes a rendezvous peer.

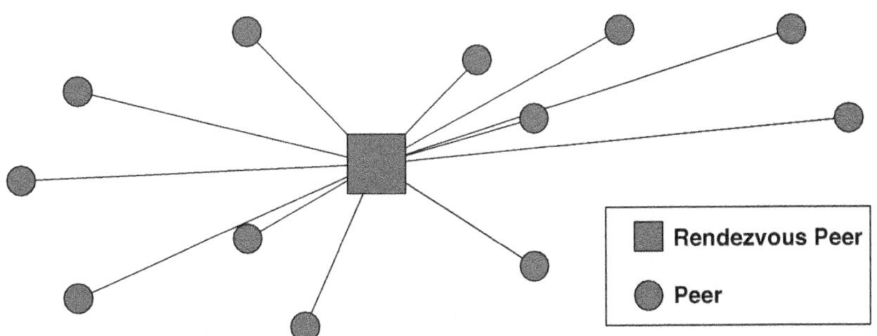

Figure 4.3 Centralized rendezvous network

4.2.10 Applications using JXTA

Besides various open source projects hosted on the Project JXTA homepage itself [25], industry-driven applications [29] were developed using JXTA. One of them, Momentum [10], titles itself a "shared information manager," which combines Peer-to-Peer with a client/server approach. Among Momentum features are collaboration on documents in real time, document versioning, and integration with standard office software. Momentum uses JXTA to discover users and for communication of them.

4.2.11 Challenges

After we reviewed JXTA, we investigate how it performs in regard to the Peer-to-Peer challenges presented in the introduction:

- *Shared environment:* Dynamic creation of peer groups provides subenvironments, while all peers remain in a global peer group.
- *Scalability:* Separation into peer groups enables a large number of groups. However, the number of peers in a group is limited, as the rendezvous peer subnetwork does not scale with the number of peers.
- *Dynamic network:* Dynamics in a network are only rudimentarily considered. Only if no rendezvous peers are present in a group, an arbitrary-chosen peer becomes a rendezvous peer.
- *Dynamic node characteristics:* Changing node characteristics are not considered. For example, there is no handling of congested rendezvous peers.
- *Network heterogeneity:* Individual node characteristics can be measured [30] but are not directly considered yet. For example, a less powerful peer may be selected as a rendezvous although peers that are more powerful are present.
- *Quality-of-Service:* Quality-of-Service is not guaranteed, and applications cannot influence the service.
- *Security:* Although trust is not guaranteed, key-based encryption is supported.

In short, JXTA does not consider many characteristics of a Peer-to-Peer network. It provides fundamental Peer-to-Peer services, but the mechanisms are not optimal for a Peer-to-Peer platform.

4.2.12 Summary

We outlined the fundamental principles of JXTA, which is probably the most widespread general-purpose Peer-to-Peer middleware. Resources are described with XML advertisements, which are published in the JXTA network. Virtual subnetworks based on peer groups assemble peers of specific applications. JXTA's functionality is accessed by services and modules, which implement XML-based protocols. Peers usually use pipes, message-oriented facilities, to communicate with each other. Because JXTA is a decentralized network, it made compromises regarding its security infrastructure. It supports common security techniques such as on keys, signatures, and hashes. Relay and rendezvous peers provide additional services for regular peers and are an integral part. The group communication facilities and their limitations were presented. Although JXTA is one of the most advanced Peer-to-Peer platforms and is used by several applications, it does not meet many of the challenges imposed to Peer-to-Peer networks.

4.3 P2P Messaging System

The goal of the P2P Messaging System [12] is to provide a scalable, robust, and fast middleware suited for Peer-to-Peer applications requiring group communication facilities. It supplies message-oriented group communication based on the Publish/Subscribe model, which is also used by established messaging systems in different domains. This model is usually based on topic-based groups, which consist of Publishers and Subscribers. Group members can be Subscribers, Publishers, or both. Messages are sent by Publishers and are delivered to all Subscribers. In the P2P Messaging System, a peer can become a member of any number of groups in order to send and receive messages. A fundamental concept of the system is its multiring topology, which will be discussed in more detail.

Although the presented group communication concepts can be used by themselves, the intention is to collaborate with existing Peer-to-Peer networks like JXTA. This symbiosis allows usage of existing Peer-to-Peer middleware functionality (e.g., looking up peers), while providing a Publish/Subscribe service to Peer-to-Peer applications. The actual message delivery is realized by an infrastructure, which is set up dynamically and independently from existing middleware.

4.3.1 Self-Organizing Overlay Networks

Limitations to IP multicasting initiated intense research on alternative group communication techniques [6] such as overlay networks. The P2P Messaging System uses virtual overlay networks, which are dynamically created for each group. Figure 4.4 shows a Peer-to-Peer network, in which the highlighted peers have subscribed to a topic. These peers are connected to each other and have thus created an overlay network for the group (highlighted links in the figure). In this way, groups cannot interfere with each other, and every virtual network is organized according to individual group resources and requirements. Another advantage of building virtual networks is that Quality-of-Service parameters can directly be considered. The Peer-to-Peer overlay network is not replaced, but complemented by multiple group-based overlay networks, which provide an efficient group communication infrastructure.

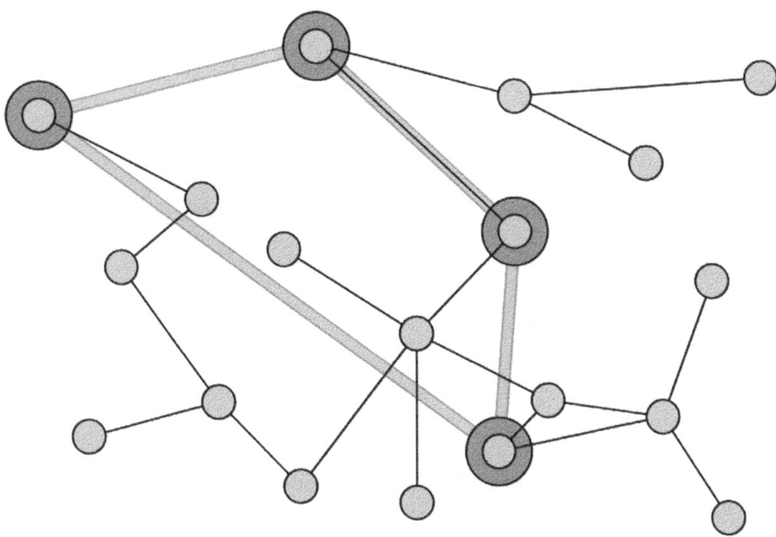

Figure 4.4 Peer-to-peer and peer group layers

The next crucial question regarding overlay networks concerns the topology, which defines how the network nodes are connected. The multiring topology [13] was designed according to requirements of Peer-to-Peer group communication. There are several reasons why the topology is based on rings. First, rings are scalable in means of bandwidth, because the workload of any node is independent of the total number of nodes. Second, they are easy to manage in a decentralized environment, because a node depends only on its neighbors. Third, they can easily be made robust because of their manageability. Each ring node is connected to two neighbor nodes in order to form a ring. When a node wants to broadcast messages to other ring nodes, it sends the messages to its neighbors in both directions. Each node that receives a message from one of its neighbors passes them on to its other neighbor. This forwarding process is continued until the message is delivered to all participants. Internally, arriving messages are stored in FIFO (First-In-First-Out) message queues until a sender process forwards the messages to neighboring nodes. In addition, the node keeps track of received messages, which are associated with their unique ID, and ignores messages that arrived more than once.

To address some of the latency issues that occur in simple rings, the multiring topology forms additional *inner rings*, which provide shortcuts to distant ring sections and therefore decrease latency. Inner ring nodes remain in outer rings (Figure 4.5). The formation of inner rings is recursive, as inner rings can have further inner rings. A system parameter q defines the ratio of inner ring nodes to outer ring nodes. For example, if this parameter is set to 10 and the total number of nodes (and most outer ring nodes) is 900, then the first inner ring has 90 nodes, and the most inner ring has 9 nodes.

Because inner ring nodes forward messages to neighbors in two or more rings, bandwidth consumption of inner ring nodes increases. This does not directly result in negative consequences because the system exploits the heterogeneity of node capabilities and elects preferably nodes with a high bandwidth as inner ring nodes. Consequently,

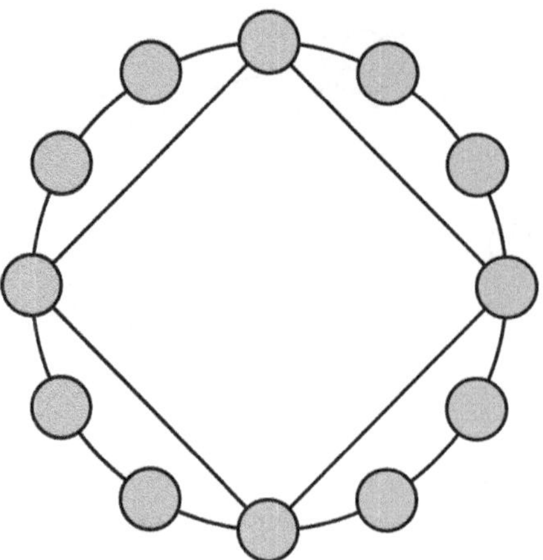

Figure 4.5 Outer and inner ring

the mean bandwidth of inner ring nodes is higher than that of outer rings. Because high bandwidth networking hardware often implies a superior latency, this suggests that the mean latency is lower in inner rings than in outer rings. Another effect of the inner ring formation is that powerful nodes are no longer limited by less powerful ones because they communicate directly.

The election process requires keeping track of nodes' bandwidth potential by measuring their data throughput. Nodes periodically verify their bandwidth as the available bandwidth may vary according to concurrent traffic. After nodes have evaluated their bandwidth, they compare the values with neighboring nodes within a defined range. The information needed for bandwidth comparisons is gathered by exchanging system messages among nodes. If a node has a superior bandwidth among the evaluated nodes, the node is elected as an inner ring node. When nodes that are more powerful arrive, or the available bandwidth changes because of concurrent network traffic, less capable nodes are abandoned from inner rings.

The latency scalability of the multiring topology relies on a balanced occurrence of inner ring nodes. If these nodes get unbalanced, some sections of the ring may suffer from increased latency. To avoid this situation, the system balances inner ring nodes. Each inner ring node determines the distance (in the outer ring) to its inner ring neighbors. If the distances differ by two or more, the node swaps its position with a neighbor node to balance the occurrence of inner ring nodes. This implies that the distances are periodically measured by exchanging system message with neighbor nodes.

4.3.2 Failure Tolerance

Two concepts contributing to failure tolerance were already outlined: sending messages in two directions, and inner rings. If messages get lost at one node, they still can arrive from other paths. Nonetheless, the P2P Messaging System uses backup links to further reduce negative effects resulting from node failures.

A controlled disconnection allows the P2P Messaging System to adjust networks links before nodes leave the network. However, nodes may also leave without any notification

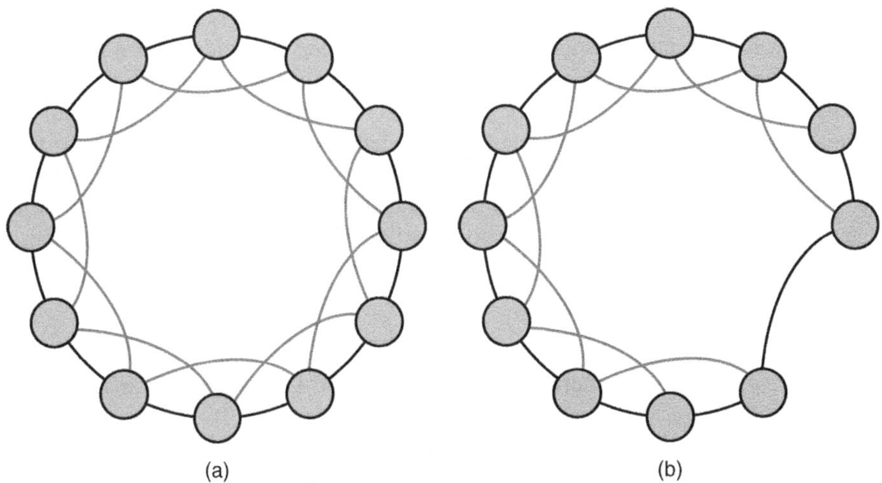

Figure 4.6 Backup links: inactive and activate

for any reason. Because this seems more likely to happen in highly dynamic networks, special precaution must be taken to ensure failure tolerance. A node failure may affect other nodes. However, the simplicity of the ring topology allows establishing backup procedures easily. The proposed mechanism is based on backup links, which remain inactive until needed. Each node has two backup links to the nodes next to its neighbors (left ring in Figure 4.6).

When the P2P Messaging System detects broken links (due to failed nodes), it removes them and activates backup links immediately (right ring in Figure 4.6). The activated backup link is converted to a regular link and new backup links are established to restore the ring completely. Backup links are also used when a neighbor node is congested because it does not receive, process, or forward messages fast enough. These mechanisms increase robustness and overall throughput of the network.

4.3.3 Implicit Dynamic Routing

Although the ring topology predefines routes, messages may arrive from several sources, especially when inner rings are involved. This improves overall stability because node failures may be compensated by alternative routes. However, this may also cause wasting network traffic because messages might be delivered to nodes more than once. Dual-mode links are intended to avoid this scenario while keeping the improved stability. The dual-mode link mechanism decides whether a message is forwarded immediately or not. Decisions are made on the basis of the following policy. The mode of each directed link is considered either primary or secondary. Primary links allow immediate sending of messages, while secondary links must announce the availability of messages before sending. Only if the receiving node needs the announced messages, the messages are actually sent to it. This simple idea has a global impact on the network as the network dynamically adapts to the message flow by adjusting the mode of its links. Thus, the evolved routes automatically consider an efficient message flow.

The initial mode of all links is primary. However, the mode is switched to secondary if a link delivers obsolete messages frequently. The receiver decides when to change the mode by keeping track of the last y messages. To remain primary, a link must deliver at least x new messages out of the last y ones. If the link does not fulfill this condition, its mode is switched to secondary. Similarly, secondary links are switched back to primary when they fulfill the condition again. The switching speed depends on the values of x and y. Since switching requires only little overhead, we propose low values to adjust to changes quickly. Switching the mode is done by the receiver that sends a system messages to the sender.

4.3.4 Quality-of-Service

Dynamic and heterogeneous networks such as Peer-to-Peer networks may have huge impact on the message delivery and thus on the application. To compensate, eventual consequences may be prevented if the application has influence on the message delivery. Therefore, Quality-of-Service (QoS) parameters are especially important in this domain because they let applications specify their primary needs. Despite their importance, Peer-to-Peer middleware like JXTA barely offer QoS parameters. The P2P Messaging System supports the following QoS parameters for messages: priority, preservation priority, expiration time, and delivery mode. Persistence-related QoS parameters like exactly once delivery are currently not.

Message priorities, which are a key QoS parameter, significantly influence the message delivery process. The P2P Messaging System assures that high-priority messages are delivered before messages of lower priority. Internally, the priority handling is implemented by priority-queues, which are used for both sending and receiving processes. *Preservation priority* defines which messages should be preserved over others. In critical situations when message queues are full, the system drops lower preservation priority messages first. *Expiration time* specifies when a message loses its relevance, and can thus be discarded by the system to reduce network traffic. The *delivery mode* indicates the main concern that the delivery tries to satisfy. The current implementation offers modes for latency, throughput, or redundancy optimization. The system decides if a message is forwarded within inner rings depending on the delivery mode.

Message filtering is a common technique among messaging systems to avoid unnecessary network traffic. In this way, the saved bandwidth can be utilized to improve the delivery quality of relevant messages. Peers within a group may have individual preferences and may not be interested in receiving all messages that were published to the group. Each peer with a special preference can specify message filters to sort out irrelevant messages and to assure that the peer obtains only the messages it needs. Ideally, messages are delivered to only those peers, which do not have message filters or whose message filters accepted the messages. This optimal scenario is difficult to implement in decentralized networks because there is no central management of all message filters. Nevertheless, the P2P Messaging System utilizes two concepts to avoid network traffic. Especially when message filtering is either occasionally or frequently used, the system's message filtering is very effective. Backup and inner ring links are used to bypass peers, not interested in the messages. This implies that each peer knows about the message filters of its two neighbors. Using backup links is very effective unless there is a chain of peers, which do not need the message. In the worst case, when none of the peers needs a message, this technique alone avoids half of the network traffic, because the message is delivered only to every second peer. However, using the same concept with the inner ring links compensates such scenarios. Here, inner ring peers know about message filters of the peers on the way to the next inner ring peer. If none of these peers needs the message, the inner ring node bypasses them and delivers the message directly to the next inner ring node.

4.3.5 System Model

The basic system model of the P2P Messaging System is depicted by Figure 4.7. Individual nodes within a ring are represented by RingNode. Messages are sent to and received from the group using the RingNode. Messages may be delivered multiple times, as they may arrive from multiple sources. Although this scenario may rarely occur because of the dual-mode links, it must be assured that the application receives messages only once. To achieve this, a RingNode object has a MessageTracker object, which keeps track of all the received messages. An incoming message is only forwarded to the application if it is new. The RingNode communicates with its neighbors using ModeLinks. Links are virtual, message-oriented communication channels between two endpoints. ModeLinks extend Links with the dual-mode link concept that was introduced earlier.

The ModeLink sends messages directly if its mode is primary. However, if the mode is secondary, the ModeLink takes care of the following steps. Messages are announced by sending system messages containing the message ID. Until the response is received, the ModeLink stores the message in the MessageStore temporary. The ModeLink on the other side receives the system messages, extracts the message ID, and asks the RingNode if the message was delivered before. On the basis of this result, a response message is sent indicating whether the announced message has been rejected or accepted. Finally, the ModeLink, which receives the response message, removes the actual message from the MessageStore and depending on response, sends or drops the message.

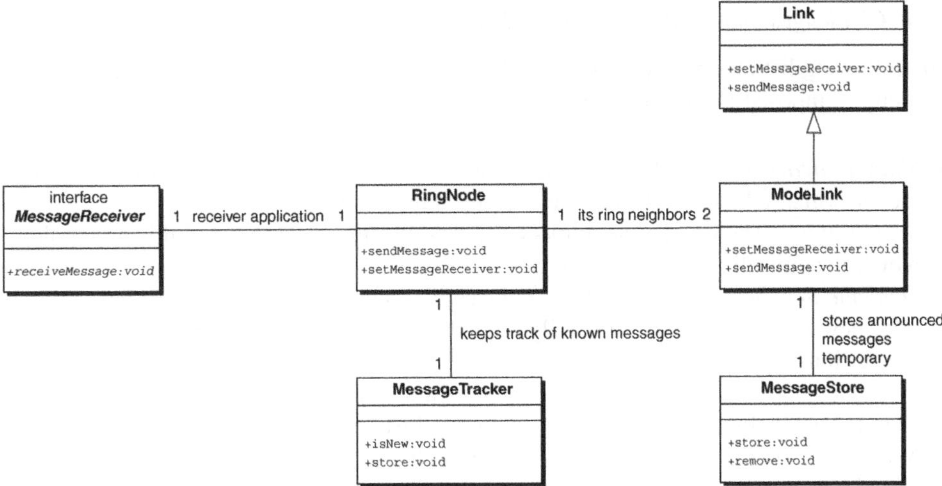

Figure 4.7 Conceptual system model of the P2P messaging system

4.3.6 Network Abstraction Layer

The core system has a network abstraction layer. Thus, it does not depend on a specific network, but on a set of functions that networks must provide. Given that a specific interface is implemented, the core system can collaborate with a variety of networks:

- *Single virtual machine network:* All nodes reside in a single virtual machine. This "network" allows large-scale unit testing.
- *Simulation network:* Similar to the single virtual machine network, but extended functionality to enable large-scale simulations to collect statistics
- *JXTA network:* Uses JXTA services
- *P2PMS Hybrid network:* Uses Centralized Service Providers (see below)
- *Global network:* Meta network that provides access to all distributed networks like the JXTA and the P2PMS Hybrid network.

As an example, it will be outlined how the JXTA network module provides the functionality needed by the P2P Messaging System. First, the module uses JXTA's Discovery Service to find peers in the same group. Second, in case the ring is cut into segments, JXTA's propagate pipes are used to collect disconnected peers in order to reconnect the ring. Third, the module provides group authentication facilities to restrict access to groups.

4.3.7 Implementation

The P2P Messaging System is implemented in Java and thus is platform independent. Further, because its network protocol is language independent, implementations in other programming languages and platforms are possible. Experiments, which directly compared the P2P Messaging System with JXTA, provided evidence for a superior performance, reliability, and scalability compared to JXTA's group communication facilities. Detailed results are available in [12] and [13].

4.3.8 Challenges and Comparison with JXTA

The following list outlines how the P2P Messaging System considers the earlier discussed challenges imposed to Peer-to-Peer networks:

- *Shared environment:* Dynamic creation of overlay networks provides subenvironments, which take care of individual group requirements.
- *Scalability:* The overlay networks are formed using the scalable multiring topology, which dynamically adjusts to changes.
- *Dynamic network:* The topology reduces the dependency on single nodes. In addition, new routes evolve dynamically, and backup links compensate node failures and congestions.
- *Dynamic node characteristics:* Nodes relocate within the topology according to their characteristics. In addition, implicit routing adjusts promptly to a changed network.
- *Network heterogeneity:* Positions within the topology are assigned according to individual peer capabilities. In this way, less powerful peers do not slow down others, while peers that are more powerful communicate directly with each other.

- *Quality-of-Service:* Quality-of-Service is not guaranteed, but applications can influence the delivery process according to individual requirements expressed by Quality-of-Service parameters.
- *Security:* Because decentralized networks still lack trust, security is deferred to centralized service providers. Encryption depends on the underlying network module.

The P2P Messaging System meets more challenges than JXTA. This is advantaged by the narrower application scope of the P2P Messaging System as it concentrates on group communication, while JXTA offers an entire platform with a broad functionality. JXTA also offers group communication and there seem to be parallels as both systems separate groups of common interest. Nonetheless, the notion of a group is different. Within a JXTA peer group, there may exist multiple communication groups (propagate pipes), whereas the P2P Messaging System completely separates each communication groups based on topics. This higher degree of separation reduces interference of groups. Another important difference is that JXTA relies on rendezvous peers to multiplex messages to receiving peers. In contrast to this centralized approach with a single point of failure, the P2P Messaging System is completely decentralized and builds on a hierarchical topology, which adjusts dynamically to the network.

4.3.9 Summary

Unlike JXTA, the P2P Messaging does not try to provide a complete Peer-to-Peer platform. Instead, it specializes in group communication and is intended to collaborate with existing networks, for example, JXTA. We illustrated how the P2P Messaging System creates a self-organizing overlay network for each group on the basis of a hierarchical topology. The formation of the network considers dynamic characters like a nodes' bandwidth. Node failures can be compensated by backup links, which preserve the connection of active nodes. Messages are routed implicitly to along paths by dual-mode links in order to adjust promptly to changes in the network. Although Quality-of-Service cannot be guaranteed, the application can use Quality-of-Service parameters to influence the service. We presented the system model and the network abstraction, which makes the system perform on any network that is able to offer the needed functionality. Finally, we illustrated how the system considers many characteristics of Peer-to-Peer networks and thus meets the challenges we outlined earlier.

4.4 Hybrid Middleware – a Conceptual Proposal

Hybrid Peer-to-Peer middleware [21] may combine advantages of both Peer-to-Peer and centralized networks. Peer-to-Peer is scalable and reduces server workload, but lacks efficient (centralized) control and may process distributed data inefficiently. We do not promote one single network type. Instead, we try to exploit the advantages of the available solutions and propose a hybrid approach.

The characteristics of the two network types are summarized and contrasted in Table 4.1. Peer-to-Peer networks have obvious advantages over centralized networks. Nevertheless, they lack crucial requirements for today's distributed applications, namely control, trustfulness, and integration of legacy systems. Although the lack of control has been a key

Table 4.1 Characteristics of Peer-to-Peer and centralized networks

	Peer-to-Peer network	Centralized network
Scalability	High	Limited
Resource availability	High	Limited
Fault tolerance	High	Limited
Infrastructure	Self-organizing	Needs setup and administration
Infrastructure costs	Low	High
Storage of global data	No	Yes
Control	No	Yes
Trusted	No	Yes
Enterprise/legacy system integration	No	Yes

factor for the success of Peer-to-Peer file-sharing applications, it prohibits enterprise-driven applications, which require a certain level of control. This includes questions such as "who may access what," or "who did what and when." Only this functionality enables user management, restricted access, and paid services. Another issue with pure Peer-to-Peer networks is that there's no guarantee of the peers' trustfulness. Any peer could be malicious, and, to make things worse, it is hardly possible to identify these peers. Finally, integration of legacy systems or extending systems with Peer-to-Peer functionality is difficult because of topological and conceptual differences.

Our definition of a hybrid middleware is general. Hybrid Peer-to-Peer middleware uses different distributed computing concepts and topologies according to individual task requirements. Because the distributed computing concept and topology task uses may vary from one environment to another, it is intentionally left undefined. Considering file sharing, for example, indexing and search functionality may be implemented either by a server, super-peers, or by peers themselves. Each variation has its individual set of advantages and disadvantages, and therefore assigning tasks should not be predetermined. Note that this hybrid model is significantly different from the widely known hybrid Peer-to-Peer model introduced by Napster, which has fixed assignment of tasks.

A highly abstract model of hybrid middleware is depicted in Figure 4.8. The Peer-to-Peer network and Service Providers build the lowest layer, which could also be considered as the network layer. Although it would be possible to access each network directly, we want to take advantage of existing middleware (middle layer in the figure) for each network type. The top layer is the hybrid middleware, which builds upon network-specific middleware and provides the "glue," making them collaborate. Developers, using a hybrid

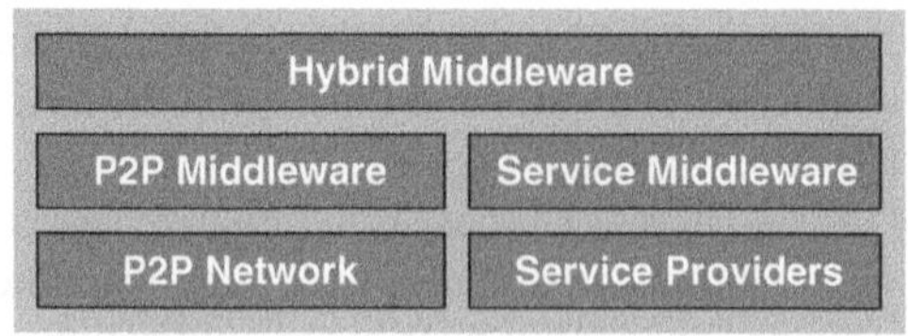

Figure 4.8 Hybrid middleware concept

middleware such as this can rely on a consistent interface and do not have to deal with complexity.

4.4.1 Service Providers

In addition to the Peer-to-Peer concept illustrated above, the notion of Service Providers still needs to be clarified. A Service Provider represents a more or less centralized service. However, the topology for Service Providers is not predetermined and can thus be a single server, a server cluster, Web Services, or even a service grid. In fact, a Service Provider may consist of a decentralized network of serving nodes. Nevertheless, Service Providers differ from typical Peer-to-Peer networks in the following aspects:

- Higher availability and reliability of single nodes
- Easier to locate
- Dedicated server role
- Limited number of nodes.

The higher availability and reliability are due to fluctuation, which is typical for Peer-to-Peer networks, but highly atypical for a server-like infrastructure. For a similar reason, Service Providers are easier to locate, for example, by known IP addresses of access points. Unlike peers, which are both a client and a server, Service Providers are dedicated to serve clients of Peer-to-Peer networks. Finally, the number of nodes is limited in static Service Provider networks as opposed to the highly dynamic Peer-to-Peer networks.

4.4.2 Conceptual Model and Services

A more detailed conceptual model including typical Services is illustrated by Figure 4.9. All Services are accessed by *Service Connectors*. They abstract the concrete implementations and protocols streamlining the interface to the requested Service. The hybrid

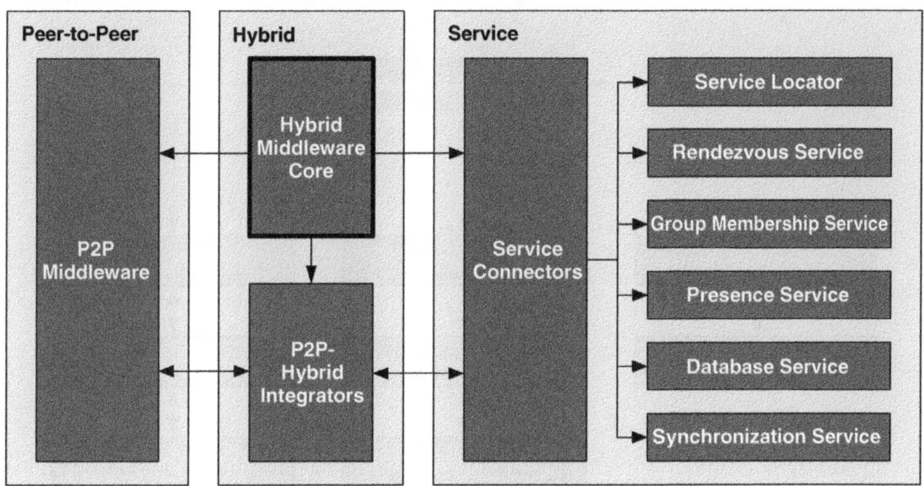

Figure 4.9 Hybrid middleware conceptual model

system may directly access Services, using corresponding Connectors, if the functionality does not involve the P2P Middleware. *P2P-Hybrid Integrators* provide functionality that depends on both P2P Middleware and Service Providers. Last, the Hybrid System may also delegate functionality to the P2P middleware directly if no dependencies to Services exist.

Next, we will outline several Services that are typically considered attractive for hybrid middleware. The *Service Locator* is a special Service, returning a preferred location of a service. This behavior introduces the flexibility needed for features such as public service registries or load balancing. The *Rendezvous Service* provides initial contacts needed to set up groups. The *Group Membership Service* determines who may enter a group, and what rights are retained within the group. The *Presence Service* keeps track of who is online and at which location. The provided functionality of this service is similar to SIP (Session Initiation Protocol) registries, or Instant Messaging server components. The *Database Service* integrates central databases into the hybrid middleware and may serve for other Services as well. The *Synchronization Service* provides data synchronization for groups. This process may include collaboration with the corresponding Peer-to-Peer synchronization module to let members synchronize themselves when they are online at the same time.

4.4.3 Service Connectors

Using Services requires network protocols to invoke service modules and to pass data. However, to avoid dependencies on individual protocols, we utilize Service Connectors as interfaces on a software architecture level. In this way, we have the freedom to choose any distributed computing architecture and protocol including SOAP [23], CORBA, HTTP, RMI, and EJB. In addition, it is open to new protocols and customized solutions.

The principles of Service Connectors including three variations are outlined by Figure 4.10. Common to all variants is a specialized Client Service Connector, which implements a well-known interface. Internally, however, they take care of specific protocols and other individual requirements. At the Service Provider level, a Service Connector represents the counterpart and communication partner to the Client Service Connector. Thus, both

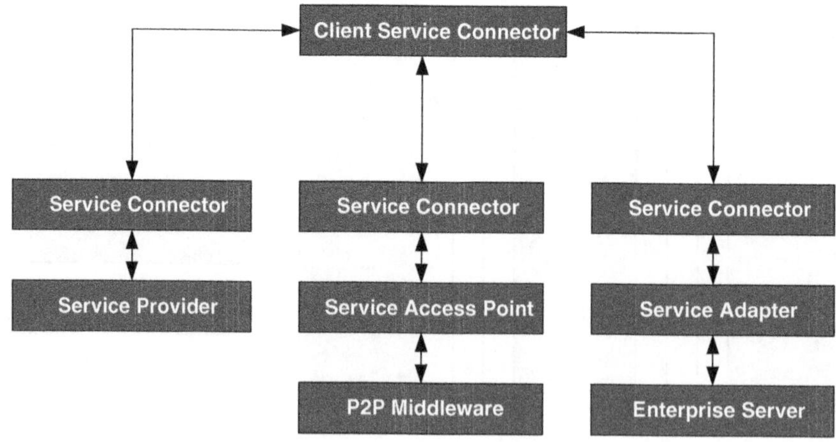

Figure 4.10 Service Connectors

Connectors must have agreed on a common protocol. To make Services accessible by multiple protocols, a Connector pair for each protocol has to be defined.

Service Providers can be implemented in various ways. The first variation, depicted on the left side of Figure 4.10, is the simplest. The Service Connector directly accesses a single Service Provider. Although separation is the recommended approach, it is possible to combine Service Connector and Provider for simple, single protocol services. The second variation, depicted in the middle of the figure, represents a distributed Service accessible by multiple Service Access Points, each providing a partial or full service. Service Access Points are interconnected by, for example, a P2P Middleware because the Service Access Points are peers within the Service network. The third variation, depicted on the right side in the figure, integrates an existing Enterprise Server or legacy system. Here, a Service Adapter may be useful as a mediator between the Service Connector and the Enterprise Server.

4.4.4 Peer Group Membership Service

To exemplify what Services are and how they collaborate with P2P Middleware and particularly with the P2P Messaging System, we will introduce the Peer Group Membership Service and the Synchronization Service. The Peer Group Membership Service may be a more or less central instance, which restricts group membership to certain peers. Further, it may determine the peers restricted to solely receiving messages and the peers additionally allowed to send data to others.

Group access and individual rights are determined when peers try joining a group. The sequence of a new peer entering a group and connecting to two existing peers is outlined in Figure 4.11. The initial step for new peers is to apply for membership at the Group Membership Service. Depending on the access regulation, this step may require transmission of further data like login name and password. Once the Service grants access to the group and determines its individual privileges, the new peer connects to an existing peer in the group. At this point, both peers will query the Service to verify each other

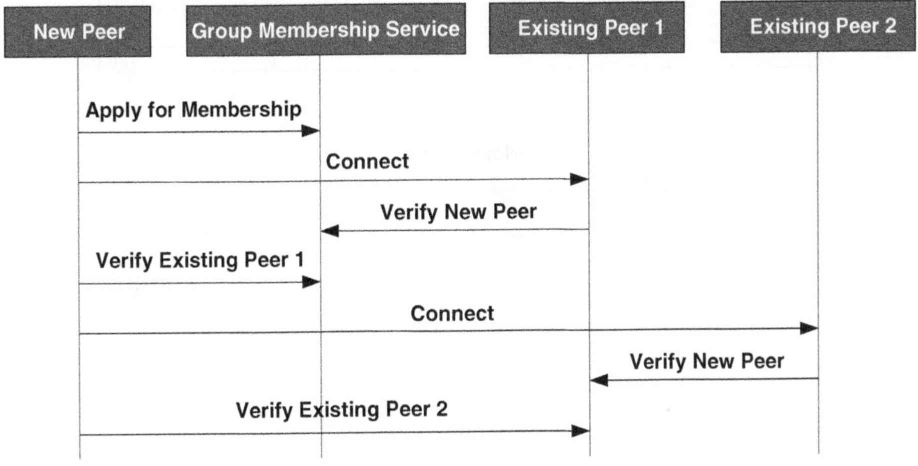

Figure 4.11 Sequence of membership verification

and their individual rights. The connection is established only if both verifications were successful. Besides querying the Service directly, it is possible to query peers, which were already verified and are therefore considered trustworthy. Peers store trust information of already verified peers and may therefore reject messages that were sent without proper rights.

4.4.5 Synchronization Service and P2P Synchronization

The Synchronization Service provides consistency for distributed data within a group. This Service collaborates with the Peer-to-Peer synchronization module and is therefore an interesting study of how Hybrid Middleware overcomes limitations of pure Peer-to-Peer systems. If a piece of data is changed at a peer, the other group members need to reflect the changed data. As there is no global data storage belonging to a group, peers must rely on their local data storage, which has to be synchronized with the other peers. We start with presenting the Peer-to-Peer system and then extend it with the Synchronization Service to a hybrid system.

The Peer-to-Peer-based module ensures synchronization among peers that are currently online. Once a modification takes place, a message containing the updated information is created and sent to other group members. Figure 4.12 shows the involved components and emphasizes the process flow in one particular direction. The distinction in the Figure

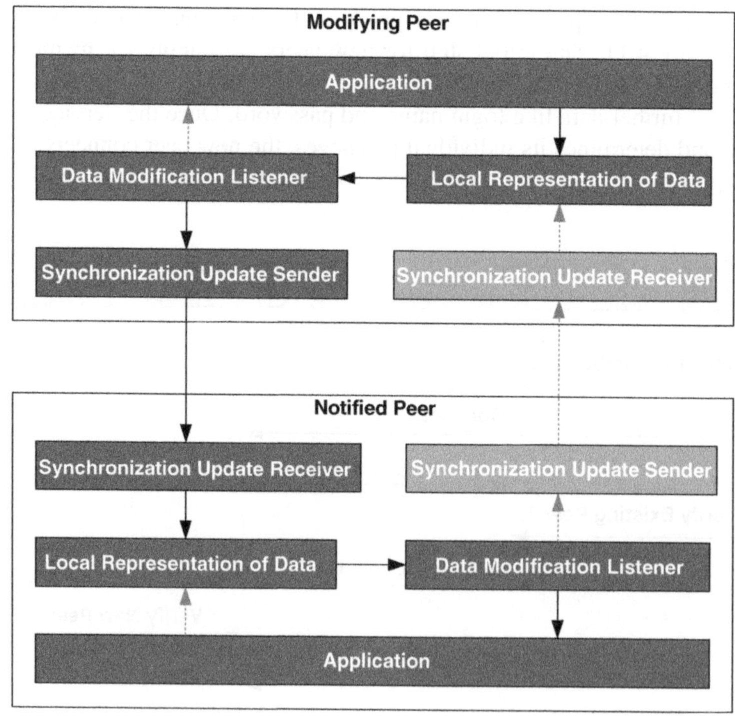

Figure 4.12 Peer-to-peer synchronization model

between "Modifying Peer" and "Notified Peer" is solely to simplify the following discussion. The flow in the other direction, which is depicted with a lighter color and dashed arrows, is equivalent.

Applications directly work on local representations of distributed data. Each peer has its local copy of the data, which, from a conceptual viewpoint, belongs to the group. Another central concept is the usage of Data Modification Listeners. They are invoked when the application modifies the local data. The modification listener concept has proved its practicability in many object-oriented systems, for example, JavaBeans. The concept notifies registered listeners when modifications have been made. Although the synchronization module relies on it, this approach may be a good choice in general because of its decoupling character. Once the Synchronization Update Sender is notified using a Data Modification Listener, it collects information related to the modified data and wraps the data into a update message, which is then sent to the other members of the synchronization group using Peer-to-Peer group communication facilities. Received update messages are processed by the Synchronization Update Receiver, which extracts the information contained in the message and checks if the modification time of the update is newer than the local data representation. In this case, the local data representation is updated to the new value, triggering once again the invocation of the Data Modification Listeners. This time, however, it is the application that may react to the notification.

This approach does not exclude the possibility of two or more peers unknowingly modifying the same piece of data. Although only the modification with the most recent timestamp will prevail in the group, data may be inconsistent within the group for a short period. The benefit of such a relaxed synchronization model is its performance and slimness because no complex operations are involved. If applications have stricter synchronization requirements, this model can be extended. A straightforward approach uses distributed locks, which are similar to semaphores. The application may require obtaining a lock before modification of a certain piece of data is permitted. Since a lock is given to exactly one instance, the data cannot be modified by others until the lock is released. To avoid problems resulting from locks that are not properly released for any reason, locks can only be leased for a short period.

At this point, problems of a pure Peer-to-Peer system are exposed. Locks have a centralized character, which makes them hard to implement in a decentralized Peer-to-Peer network. On the other hand, a centralized solution is simple and efficient, and therefore has advantages over decentralized solutions. Another issue reveals the limitations of pure Peer-to-Peer systems even more dramatically: unavailability of data. When peers, which store recently modified data, are offline, the modifications are not available to peers that were not present at the time of the modification. The unavailability of data is particularly high if the group is small and fluctuation of peers is high. Again, we see the solution in a hybrid approach, because a dedicated Service Provider typically has a high availability. Therefore, peers can query the Service Provider if the current data is not available from other peers.

The Synchronization Service and the Peer-to-Peer synchronization module do not depend on each other. Each can be used standalone if the other is unavailable. Nonetheless, the components are intended to collaborate closely to form a consistent hybrid system that surpasses the individual systems.

When a new peer enters a group, it first tries to synchronize with another group member to obtain the newest data. If no other peer exists or the synchronization fails for any other reason, the peer connects to the Synchronization Service to obtain the latest data updates. To avoid unnecessary synchronizations, each piece of data contains metadata needed to implement a more efficient synchronization process. The metadata consists of a modification time, synchronization point, and modification flag. The modification time specifies the time when the latest modification occurred. Synchronization points serve as starting points for subsequent synchronizations, and are set by Synchronization Service. The modification flag is a Boolean value indicating whether the data was modified since the last synchronization with the Service. With this information, it is possible to implement synchronizations without determining whether each piece of data is up-to-date. Peers remember their last synchronization point and can therefore reduce further synchronizations to only newly modified data. When contacting the Service, the peer transmits its last synchronization point. On the basis of this information, the Service collects the necessary updates and sends it to the peer. However, peers that joined a group do not have to query the Service if the information is available from other peers. Usually, a peer only contacts the service directly if there are no group members available.

Updating the Synchronization Service is a little trickier because the peer group is more complex and unreliable. The questions are who performs the update and when. We do not promote updating after each modification because we can take advantage of the Peer-to-Peer synchronization module while peers are online. In this way, multiple modifications on a single piece of data may be transmitted only once, as each modification obsoletes the previous one. Theoretically, it would be sufficient to let the last peer in a group perform the update when he leaves the group. Practically, however, this may fail, especially when considering the unreliable character of peers. Consequently, this update mechanism is supported by two other mechanisms to improve reliability: users trigger the update process manually or peers can also periodically report modifications to the Service. The latter is performed by the peer who performed the latest modification after a defined period. When more than one peer tries to report modifications simultaneously, the Service recognizes the modifications from the peer first connected.

As mentioned before, the Service can provide a lightweight and efficient lock mechanism for groups. Locks may be associated with a resource consisting of multiple pieces of data, a singe piece of data, or a free textual description. The latter is useful to let application developers operate exclusively on individual defined data sets, such as multiple resource objects. We complete the discussion on the Synchronization Service with the global clock mechanism. Synchronization relies on timing because data modifications are only accepted if they are newer than the existing local piece of data. To avoid side effects resulting from differing clocks of peers, the Service provides a global clock.

4.4.6 Summary

Peer-to-Peer networks have advantages over centralized networks but are also limited. For example, they lack control, a prerequisite for many applications. Hybrid networks can overcome these limitations while preserving the advantages. We outlined the concept of Service Providers, and how they can be accessed by a variety of protocols using Service Connectors. Conceptual models of a hybrid middleware were introduced, and two Services were discussed exemplarily in more detail. Especially the synchronization process demonstrated how the hybrid solution utilizes both the Synchronization Service and the Peer-to-Peer synchronization module to provide a new solution that combines the advantages of each.

4.5 Conclusion

This chapter presented challenges and current solutions for Peer-to-Peer middleware platforms. Peer-to-Peer networks are highly complex and demand a specialized middleware. JXTA is one of the most advanced Peer-to-Peer platforms. It provides various interesting concepts such as advertisements, peer groups, services, and pipes. JXTA's group communication, which is based on a virtual subnetwork of rendezvous peers, has nevertheless some limitations. In contrast, the P2P Messaging System focuses on group communication and can collaborate with existing networks regardless of their organization. By considering various characteristics of Peer-to-Peer networks, this middleware provides an efficient, scalable, and robust infrastructure. Although decentralized networks have several advantages, we see the need to collaborate with centralized networks. Consequently, we pointed out the need for hybrid solutions.

Although Peer-to-Peer advanced in the last years, it remains an emerging technology with open questions. Its future depends on building new infrastructures for general or more specific purposes [16]. Security, trust, searching, and performance are still subject for further investigations [4, 18]. Besides ongoing research, there is the need for Peer-to-Peer middleware to become mature in order to achieve higher levels of functionality, reliability, and performance. Hybrid middleware can become a powerful alternative to pure Peer-to-Peer systems in the future. On the basis of refined models and further research, future systems may be able to combine advantages of the decentralized Peer-to-Peer network with centralized networks.

Bibliography

[1] Aberer, K. (2001) P-Grid: A Self-Organizing Access Structure for P2P Information Systems. *Sixth International Conference on Cooperative Information Systems*, Trento, Italy; Lecture Notes in Computer Science 2172, pp. 179–194; Springer-Verlag, Heidelberg, Germany.

[2] Amir, Y., Nita-Rotaru, C., Stanton, J., and Tsudik, G. (2003) Scaling Secure Group Communication: Beyond Peer-to-Peer. *Proceedings of DISCEX3*, Washington, DC, April 22–24, pp. 226–237.

[3] Clarke, I., Sandberg, O., Wiley, B., and Hong, T. (2000) Freenet: A Distributed Anonymous Information Storage and Retrieval System. *Proceedings of the ICSI Workshop on Design Issues in Anonymity and Unobservability*, Berkeley, CA, pp. 46–66.

[4] Daswani, N., Garcia-Molina, H., and Yang, B. (2003) Open Problems in Data-Sharing Peer-to-Peer Systems. *Proceedings of the 9th International Conference on Database Theory*, pp. 1–15; Springer-Verlag, Heidelberg, Germany.

[5] Dierks, T. and Allen, C. (1999) *The TLS Protocol*, RFC 2246. Available at http://www.ietf.org/rfc/rfc2246.txt

[6] El-Sayed, A., Roca, V., and Mathy, L. (2003) A Survey of Proposals for an Alternative Group Communication Service. *IEEE Network*, **17**(1), 46–51.

[7] Foster, I., Kesselman, C., and Tuecke, S. (2001) The Anatomy of the Grid: Enabling Scalable Virtual Organizations. *International Journal of Supercomputer Applications*, **15**(3), 200–222.

[8] Gnutella (2003) Available at http://gnutella.wego.com/

[9] Groove Networks (2003) Groove Networks Homepage. Available at http://www.groove.net

[10] In ViewSoftware (2003) How Peer-to-Peer Computing Helps Momentum Users Organize and Manage Information, *P2P Journal*, **July**, 10–17. Available at http://www.inviewsoftware.com/products/momentum/momentum_white_paper.pdf

[11] Joltid (2003) Joltid Homepage. Available at: http://www.joltid.com

[12] Junginger, M. O. (2003) *A High Performance Messaging System for Peer-to-Peer Networks*. Master Thesis, University of Missouri-Kansas City. Available at http://www.junginger.biz/thesis

[13] Junginger, M. O. and Lee, Y. (2002) The Multi-Ring Topology – High-Performance Group Communication in Peer-To-Peer Networks. *Proceedings of Second International Conference on Peer-to-Peer Computing*, pp. 49–56.

[14] JXTA Chess (2003) JXTA Chess Homepage. Available at http://chess.jxta.org.

[15] Lam, C. K. M. and Tan, B. C. Y. (2001) The Internet is Changing the Music Industry. *Communications of the ACM*, **44**(8), 62–68.

[16] Loo, A. W. (2003) The Future of Peer-to-Peer Computing. *Communications of the ACM*, **46**(9), 57–61.

[17] Microsoft (2003) *Microsoft Windows XP Peer-to-Peer Software Development Kit (SDK)*. Available at http://msdn.microsoft.com/library/default.asp?url =/downloads/list/winxppeer.asp

[18] Ooi, B. C., Liau, C. Y., and Tan, K. L. (2003) Managing Trust in Peer-to-Peer Systems Using Reputation-Based Techniques. *International Conference on Web Age Information Management (WAIM'03)*, LNCS 2762, pp. 2–12; Springer, Chengdu, China, August; Keynote.

[19] Oram, A., ed. (2001) *Peer-to-Peer: Harnessing the Power of Disruptive Technologies*, O'Reilly & Associates, Sebastopol, CA.

[20] Rowstron, A. and Druschel, P. (2001) Pastry: Scalable, Distributed Object Location and Routing for Large-Scale Peer-to-Peer Systems. *Proceedings of IFIP/ACM International Conference on Distributed Systems Platforms (Middleware)*, Heidelberg, Germany, November, pp. 329–350.

[21] Schollmeier, R. (2001) A Definition of Peer-to-Peer Networking for the Classification of Peer-to-Peer Architectures and Applications. *Proceedings of the First International Conference on Peer-to-Peer Computing*, Linköping, Sweden, pp. 101–102.

[22] Skype (2003) Skype Homepage. Available at http://www.skype.com/

[23] SOAP Specification (2003) *Simple Object Access Protocol (SOAP) 1.1*. Available from http://www.w3.org/TR/SOAP/

[24] Stal, M. (2002) Web Services: Beyond Component-based Computing. *Communications of the ACM*, **45**(10), 71–76.

[25] Sun Microsystems (2003) Project JXTA Homepage. Available at http://www.jxta.org

[26] Sun Microsystems (2003) *JXTA v2.0 Protocols Specification*. Available at http://spec.jxta.org/nonav/v1.0/docbook/JXTAProtocols.pdf

[27] Sun Microsystems (2003) *Project JXTA v2.0: Java Programmer's Guide*. Available at http://www.jxta.org/docs/JxtaProgGuide_v2.pdf

[28] Sun Microsystems (2003) *Sun Continues to Innovate as Project JXTA Speeds Past One Million Downloads*. Sun Microsystems press release, March 4. Available at http://www.sun.com/smi/Press/sunflash/2003-03/sunflash.20030304.2.html

[29] Sun Microsystems (2003) *Project JXTA Solutions Catalog*. Available at http://www.jxta.org/project/www/Catalog/index-catalog.html

[30] Sun Microsystems (2003) *JXTA Metering and Monitoring Project*. Available at http://meter.jxta.org

[31] Sundsted, T. (2001) *The Practice of Peer-to-Peer Computing: Introduction and History*, IBM developerworks. Available at http://www-106.ibm.com/developerworks/java/library/j-p2p/

5

Grid Middleware

Gregor von Laszewski[1,2] and Kaizar Amin[1,3]

[1] *Argonne National Laboratory*
[2] *University of Chicago*
[3] *University of North Texas*

> The reason why we are on a higher imaginative level is not because we have finer imagination, but because we have better instruments.
>
> Alfred North Whitehead [1]

The creation of Grid middleware is an essential aspect of the software engineering effort to create an evolving layer of infrastructure residing between the network and applications [2]. In order to be effective, this Grid middleware manages security, access, and information exchange to

- develop collaborative approaches for a large number of users, including developers, users, and administrators, that can be extended to the larger set of Internet and network users;
- transparently share a wide variety of distributed resources, such as computers, data, networks, and instruments, among the scientific, academic, and business communities; and
- develop advanced tools to enable, manage, and expedite collaboration and communication.

Scenarios that influence the state-of-the-art middleware development include challenging worldwide efforts to deliver scientific experiment support for thousands of physicists at hundreds of laboratories and universities. Such an advanced environment is possible only if the middleware supports high-end and high-performance resource requirements to allow managing and analyzing petabytes of data while using end-to-end services under quality and security provisions.

Middleware for Communications. Edited by Qusay H. Mahmoud
 ISBN 0-470-86206-8

In this chapter, we present information about research activities to develop such computing environments that are supported through Grid middleware. We address the following questions:

- What does the term Grid mean?
- How does the Grid middleware help develop a collaborative science and business environment?
- What trends exist in Grid middleware development?
- What applications exist that use Grid middleware?
- Where can we find more information about Grids?
- What will the future bring?

5.1 The Grid

Because of the recent popularity of the term *Grid*, we revisit its definition. Throughout this chapter, we follow the definition as introduced in [2].

In general, we distinguish between (a) the *Grid approach*, or paradigm, that represents a general concept and idea to promote a vision for sophisticated international scientific and business-oriented collaborations and (b) the physical instantiation of a *production Grid* on the basis of available resources and services to enable the vision for sophisticated international scientific and business-oriented collaborations. Both ideas are often referred to in the literature simply as *Grid.* In most cases, the usage of "Grid" is apparent from its context.

The term "Grid" is chosen in analogy to the electric power grid that allows pervasive access to electric power. Similarly, computational Grids provide access to pervasive collections of compute-related resources and services.

However, we stress that our current understanding of the Grid approach is far beyond simply sharing compute resources [3] in a distributed fashion; indeed, we are dealing with far more than the distribution of a single commodity to its customers. Besides supercomputer and compute pools, Grids include access to information resources (such as large-scale databases) and access to knowledge resources (such as collaborative interactions between colleagues). We believe it is crucial to recognize that Grid resources also include actual scientists or business consultants that may share their knowledge with their colleagues [2]. Middleware must be developed to support the integration of the human in the loop to guide distributed collaborative problem-solving processes. In order to enable the Grid approach, an infrastructure must be provided that allows for flexible, secure, coordinated resource sharing among dynamic collections of individuals, resources, and organizations.

When a *production Grid* is created in order to support a particular user community, we refer to it as *community production Grid.*

Although the resources within such a community may be controlled in different administrative domains, they can be accessed by geographically and organizationally dispersed community members. The management of a community production Grid is handled as part of a *virtual organization* [4]. While considering different organizations, we have to make sure that the policies between the organizations are properly defined. Not only may organizations belong to different virtual organizations but also part of an organization (organizational unit) may be part – but not all – of a virtual organization (Figure 5.1). This is just one aspect addressed within the creation of a Grid; we will address many more challenging issues throughout the chapter.

Although typical Grids contain high-performance resources such as supercomputers, we avoid the term “high performance” and replace it in favor of “high end” in order to stress the fact that performance is just one of many qualities required of a Grid. Fault tolerance and robustness are other qualities that are viewed as important.

In parallel to the electrical power Grid, we term producers and contributors of resources a *Grid plant*. Intermediaries that trade resource commodities are termed *Grid brokers*. Users are able to access these resources through *Grid appliances*, devices, and tools that can be integrated into a Grid while providing the user with a service that enables access to Grid resources. Such appliances can be as simple as computers, handheld devices, or cell phones, but may be as complex as collaborative spaces enabling group-to-group communication or sensor networks as part of sophisticated scientific infrastructure. Many Grid appliances are accessed and controlled through sophisticated portals that enable easy access, utilization, and control of resources available through a Grid by the user.

One important concept that was originally not sufficiently addressed within the Grid community was the acknowledgment of *sporadic and ad hoc Grids* that promote the creation of time-limited services. This concept was first formulated as part of an initial Grid application to conduct structural biology and computed microtomography experiments at Argonne National Laboratory’s Advanced Photon Source (APS) [5]. In these applications, it was not possible to install, on long-term basis, Grid-related middleware on the resources, because of policy and security considerations. Hence, besides the provision for a pervasive infrastructure, we require Grid middleware to enable sporadic and *ad hoc* Grids that provide services with limited lifetime. Furthermore, the administrative overhead of installing such services must be small, to allow the installation and maintenance to be conducted by the nonexpert with few system privileges. This requirement is common in

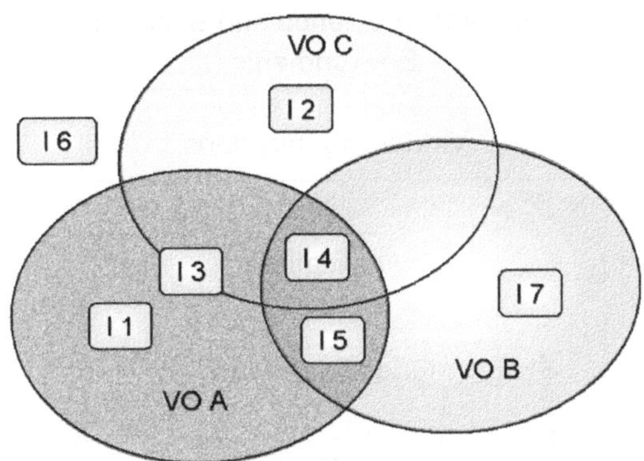

Figure 5.1 Virtual organizations typically contain multiple organizations with complex policy rules. The Grid provides mechanisms to transparently include resources from a number of organizations into a virtual organization

many peer-to-peer networks [6] in which connections between peers are performed on an *ad hoc* or sporadic basis.

5.2 Grid Architecture

In order to develop software supporting the Grid approach, it is important to design an architecture that corresponds with our vision.

A review of the literature [2] shows that a Grid architecture combines traditional architectural views. Such architectural views include layers, roles, and technologies. However, we believe it is important to recognize that the architecture of the Grid is multifaceted and an architectural abstraction should be chosen that best suits the area of the particular Grid research addressed by the architectural view.

The multifaceted aspects of the Grid motivate an architecture that is (a) *layered*, allowing us to bootstrap the infrastructure from a low to a high level, (b) *role-based*, allowing us to abstract the Grid functionality to interface with single resources to a collective multiple resources, (c) *service-oriented*, allowing a convenient abstraction model that others can easily deploy, create, and contribute to.

Let us consider the architectural diagram depicted in Figure 5.2. As mentioned previously, a virtual organization contains resources that are governed by policies and accessed by people that are part of the virtual organization. The *Grid fabric* contains protocols, application interfaces, and toolkits that allow development of services and components to access locally controlled resources, such as computers, storage resources, networks, and sensors. The *application layer* comprises the users' applications that are used within a

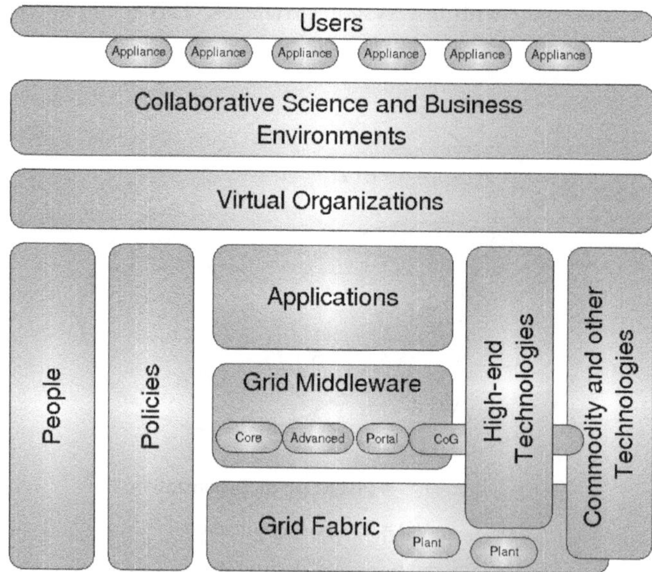

Figure 5.2 Grid middleware is an evolving layer of software infrastructure residing between the Grid Fabric and applications

virtual organization. Additional layers within the *Grid middleware* are defined from top down as follows [2, 7]:

- The *connectivity layer* includes the necessary Grid-specific core communication and authentication support to perform secure network transactions with multiple resources within the Grid fabric. This includes protocols and services allowing secure message exchange, authentication, and authorization.
- The *resource layer* contains protocols that enable secure access and monitoring by collective operations.
- The *collective layer* is concerned with the coordination of multiple resources and defines collections of resources that are part of a virtual organization. Popular examples of such services are directories for resource discovery and brokers for distributed task and job scheduling.

A benefit of this architecture is the ability to bootstrap a sophisticated Grid framework while successively improving it on various levels. Grid middleware is supported with an immensely rich set of application interfaces, protocols, toolkits, and services provided through commodity technologies and developments within high-end computing. This interplay between different trends in computer science is an essential asset in our development for Grids, as pointed out in [8].

Recently, it has become apparent that a future Grid architecture will be based on the Web services model, but will also need additional convenient frameworks of abstraction [9]. From the perspective of Grid computing, we define a service as a platform-independent software component that is self-describing and published within a directory or registry by a service provider (see Figure 5.3). A service requester can locate a set of services with a query to the registry, a process known as *resource discovery*. Binding allows a suitable service to be selected and invoked.

The usefulness of the service-based Grid architecture can be illustrated by scheduling a task in a virtual organization containing multiple supercomputers. First, we locate sets of compute resources that fulfill the requirements of our task under Quality-of-Service

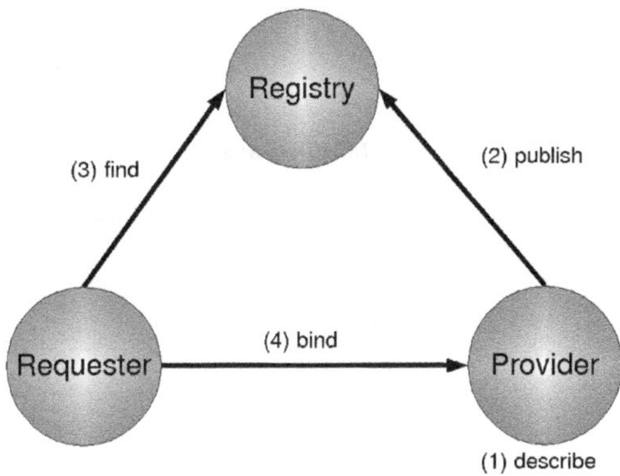

Figure 5.3 The service model allows the description of the provider service by the provider that can be published in a registry and be found and bound by a requester

constraints. Next, we select one of these sets and establish a service binding to reserve the resources. Finally, we execute the task. Clearly, it is desirable to develop complex flows between services. Since this service-based model involves the use of asynchronous services, it will be important to deal appropriately with service guarantees in order to avoid deadlocks and other computational hazards.

The service-based concept has been in wide use, not only by the Grid community but also by the business community. This fact has led to recent collaborative efforts between the Grid and the business community. An example of such an activity is the creation of the Open Grid Service Architecture, which is described in more detail in Section 5.7.3.

5.3 Grid Middleware Software

Developing middleware is not an easy undertaking. It requires the development of software that can be shared, reused, and extended by others in a multitude of higher-level frameworks. Choosing the right paradigm for a software engineering framework in which the middleware is developed provides the means of its usefulness within the community. Hence, developing just an API as part of the middleware development is possible, but does not provide the abstraction level needed in the complex Grid approach. Figure 5.4 depicts a useful subset of technologies that should be provided by state-of-the-art Grid middleware. It includes protocols, data structures, and objects that can be accessed through convenient APIs and classes. Appropriate schemas need to be available to describe services and to deliver components for easy reuse in rapid prototyping of the next generation of Grid software. In many cases, it is not sufficient to concentrate just on one of these aspects. It is important that any middleware that is developed strives to fulfill the needs

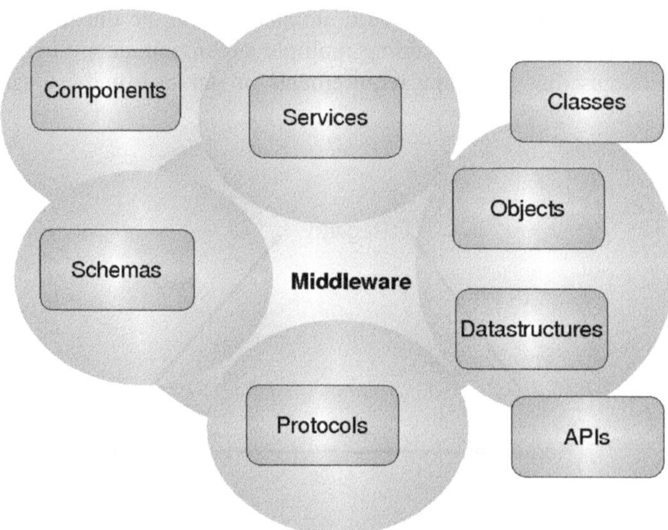

Figure 5.4 Grid middleware needs to support a variety of technologies in order to accommodate a wide variety of uses

of the user community reusing this middleware. In case of the Grid, the user community is so diverse that it is essential to also address the diverse needs of the community while choosing the appropriate frameworks and other tools that must be part of the middleware layer [10].

5.4 Grid Middleware Challenges

The current challenge in developing middleware software for the Grid is to address the complexity of developing and managing Grids. As we address a wide user community ranging from application developers to system administrators, and also a wide range of infrastructure as part of diversified virtual organizations, we must identify a path that leads to an integral software environment developed by the community. Besides emphasizing the interoperability issues, one needs to be concerned about the maintenance, management, and policy issues that go beyond the actual delivery of a technical solution. In many cases, it may be difficult to establish such management collaboration because of the complex and contradictory operating guidelines institutions may have. Hence, building a Grid has not only a technical challenge but also a social challenge.

5.5 Grid Middleware Standardization

In order for any middleware to achieve acceptance with a large community, it must be developed and maintained with the help of a community standards body. In the early days of the Grid development, there was no standard for Grid development; thus, several noninteroperable Grid frameworks resulted [7, 11, 12, 13]. The Global Grid Forum (GGF) [14] has presented itself as the much required body that coordinates the standardization of the Grid development.

The GGF is a community-initiated forum of more than 5000 individual researchers and practitioners working on distributed computing or Grid technologies. The primary objective of GGF is to promote and support the development, deployment, and implementation of Grid technologies and applications through the creation and documentation of "best practices" – technical specifications, user experiences, and implementation guidelines. Additionally, the efforts of the GGF are aimed at the development of a broadly based integrated Grid architecture that can serve to guide the research, development, and deployment activities of emerging Grid communities. Defining such an architecture will advance the Grid approach through the broad deployment and adoption of fundamental basic services and by sharing code among different applications with common requirements.

5.6 Grid Middleware Services

Now that we have an elementary understanding on what the Grid approach constitutes, we ask the question, "Which elementary Grid services must we provide as part of Grid middleware and which services must be enhanced to make the middleware more useful?"

5.6.1 Elementary Grid Middleware Services

The most elementary Grid services include *job execution services, security services, information services*, and *file transfer services*.

Job execution services provide mechanisms to execute jobs on a remote resource, while appropriate security mechanisms are applied to allow only authenticated and authorized users to use the desired compute resources. Job execution services also allow users to interface with batch queuing systems [15]. Hence, a job execution service must be able to submit a job asynchronously and be able to check its status at a later time. As multiple remote resources may be part of the Grid, jobs must have a unique identifier so that they can be properly distinguished.

Security services that are part of the Grid infrastructure should provide the ability to perform authentication and authorization to protect resources and data on the Grid. Encryption and other elementary security mechanism are usually included in Grid security services through the use of commodity technologies such as Public Key Infrastructure (PKI) or Kerberos. Because of the large number of resources in the Grid, it is important to establish policies and technologies that allow the users to perform a *single sign-on* to access the Grid. The Globus Grid security infrastructure provides the framework for such a mechanism [16]. A user can sign on through a single interaction with the Grid. A security credential is created on his behalf that is used through a delegation process to sign onto other resources. Naturally, the user must be authenticated to use the resource. In general, all Grid services must integrate Grid security mechanism that can be established by reusing middleware protocols, APIs, and data structures so they can be reused to develop more sophisticated security services.

Information services provide the ability to query for information about resources and participating services on the Grid. Typically, a virtual organization maintains a Grid Information Service relevant to its user community. Community production Grids may maintain different information services exposed through different protocols. This is in part motivated by the different scale and functionality of the diverse user communities as part of a virtual organization. An application user may be interested in only knowing on which computer he may be able to compile a sophisticated FORTRAN 90 application. Other users may want to know on where to execute this application under certain resource constraints. Hence, an information services framework must be able to deal with the diverse nature of queries directed to it. Many different information services can be integrated and the capabilities be announced and made accessible to the users of the community Grid. Services such as querying the amount of free memory for a compute resource may pose immediate problems if the information service is not properly implemented. Assume a user queries the amount of memory every millisecond, but the memory change is updated only every second. Many resources are wasted for a useless query. Hence, as pointed out in [17], information services must provide protocols and mechanisms that can dynamically change their behavior toward unreasonable requests by the users. We expect that such smart Grid services will become increasingly available and developed by the large Grid community. Agglomeration and filter services as part of the information service framework can be used to join or reduce the amount of information that is available to the user. The capabilities of such services must be self-described, not only through the availability of schemas and objects that can be retrieved but also through the semantics attached with the information through metadata annotations.

File transfer services must be available to move files from one location to the other location. Such file transfer services must be sophisticated enough to allow access to high-performance and high-end mass storage systems. Often, such systems are coupled with each other, and it should be possible to utilize their special capabilities in order to move files between each other under performance considerations. Striping and parallel streams are some obvious strategies to be included in file transfer services between file servers. Ideally, the middleware for file transfers must be exposed to the user community as simply as possible through, for example, a command like "copy file A from server Y to file B on server Z as fast as possible." New protocols and services must be developed that support this notion. An API that provides this functionality is not sufficient.

5.6.2 Advanced Grid Management Services

With the help of the basic Grid middleware services, a variety of more advanced services can be developed. Such services include *file management*, *task management*, and *information management*. Such an advanced service may manage the agglomeration and filtering of information for a community. There exist many more examples of such advanced services. As pointed out in the preceding section, users wish to manage file transfers in an easy and transparent fashion through the invocation of an elementary set of protocols and APIs. These should hide the underlying complexity of the Grid as much as possible from the user but should provide enough flexibility that individual control about a file transfer can be issued. Users may want to use advanced file transfer services that dynamically adapt to changing network conditions and fault situations [18], without user intervention.

Task management of multiple tasks prepared by a user is a common problem. Envision a computational biologist conducting a comparative study in genome analysis. As part of the problem-solving process, multiple, similar experiments must be run. These could involve many thousands of subtasks. Organizing and scheduling these tasks should be performed by a sophisticated service that utilizes the capabilities of the Grid by reusing the current generation of Grid services (see Figure 5.5).

Taking a closer look at our two examples, we observe that in Grid literature a distinction is made between file transfers, task execution, and information queries. However, we believe this distinction is useful only for developers of basic Grid middleware and not for the larger community developing even more advanced Grid services. We have demonstrated in [17] that an abstraction of all of these can be, and should be, performed. Using such an abstraction will allow us to introduce more easily uniform service specifications that integrate higher-level functionality such as fault tolerance and self-configuration. Without this abstraction, the development of future Grid middleware is rather limited and will not address the more complex management issues.

Although middleware typically does not contain user-level applications, one must recognize the fact that middleware must be provided to simplify the development of user applications that reuse Grid middleware [19, 20, 21]. As part of this effort, tools are created within the NSF Middleware Initiative (NMI) to provide portal middleware for Grids [19]. Additional efforts are already under way to provide similar functionality as part of JavaBeans by, for example, the Java CoG Kit.

Significant effort also is currently being spent in developing tools that make the task of the Grid system administrator easier. Administrative and monitoring services are under

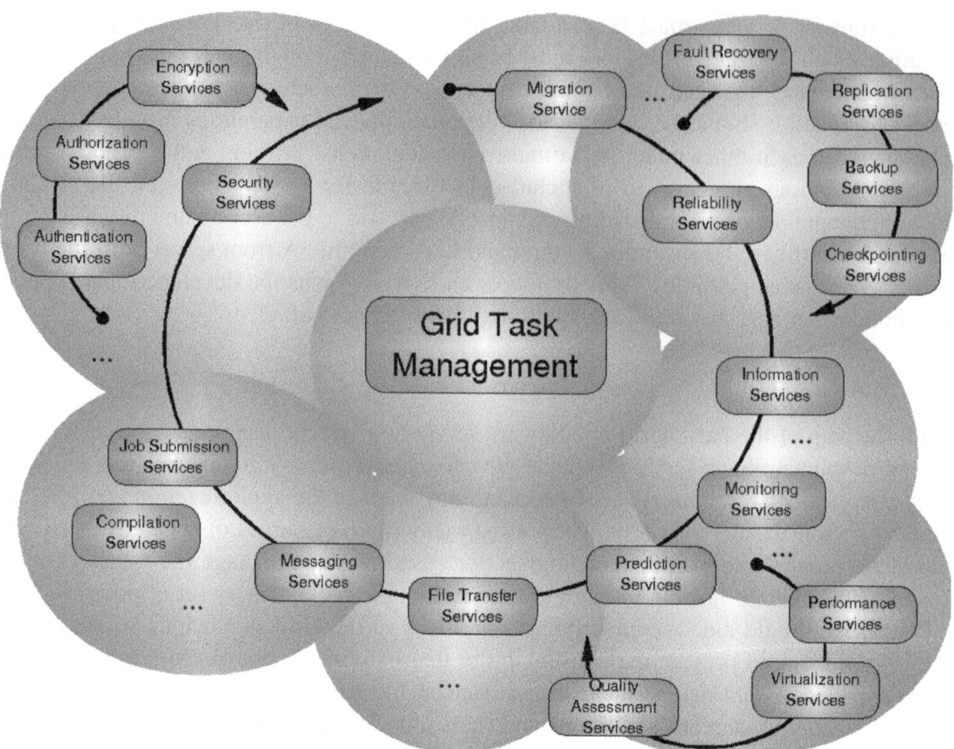

Figure 5.5 GridTask management is a complex issue. It involves the coordinated effort of integrating a number of concepts reaching from Migration, to security. Such concepts can be provided as part of task management services.

development, facilitating the maintenance of accounts and the monitoring of the services running at remote locations.

5.7 Grid Middleware Toolkits

Currently, the community is building Grid middleware toolkits that address some of the complex issues related to the Grid approach. These efforts are still in their infancies. However, they have made great progress in the past years. In this section, we briefly outline a subset of these developments in order to introduce several different toolkits.

5.7.1 Globus Toolkit

The Globus Project has contributed significantly to the Grid effort through (a) conducting research on Grid-related issues such as resource management, security, information services, data management, and application development environments; (b) developing the Globus Toolkit as part of an open-source development project; (c) assisting in the planning and building of large-scale testbeds, both for research and for production use by scientists and engineers; (d) contributing as part of collaborations in a large number of application-oriented efforts that develop large-scale Grid-enabled applications in collaboration with scientists and engineers; and (e) participating in community activities that include educational outreach and Grid standards as part of the Global Grid Forum.

The Globus Toolkit has evolved from a project developing APIs (Nexus and DUROC) to a project that also develops protocols and services (MDS, GRAM, and their uses through the Java CoG Kit). During the past few years, the Globus Toolkit has evolved from an API-oriented solution to a more protocol and service-oriented architecture.

The Globus Toolkit includes the following features:

Security is an important aspect of the Globus Toolkit [2]. The Grid Security Infrastructure (GSI) uses public key cryptography as the basis for its functionality. It enables key security services such as mutual authentication, confidential communication, delegation, and single sign-on. GSI builds the core for implementing other Globus Toolkit services.

Communication (in the C-based Globus Toolkit) is handled through the GlobusIO API that provides TCP, UDP, IP multicast, and file I/O services with support for security, asynchronous communication, and Quality-of-Service. An important tool provided by the Globus Project is MPICH-G2, which supports Message Passing Interface (MPI) across several distributed computers.

Information about a Grid is handled in the Globus Toolkit version 2 through the Metacomputing Directory Service (MDS). The concept of a distributed directory service for the Grid was first defined in [22] and later refined in [23]. The Metacomputing Directory Service manages information about entities in a Grid in a distributed fashion. The implementation of MDS in GT2 is based on the Lightweight Directory Access Protocol (LDAP). In [17], we have shown that the Globus architecture can be significantly simplified while combining the information services with a job submission service. We expect this concept to be introduced into future versions of the Globus Toolkit. In version 3 of the toolkit, this protocol is being replaced by XML while providing similar functionalities as the original toolkit.

Resource management within the Globus Toolkit is handled through a layered system in which high-level global resource management services are built on top of local resource allocation services [2]. The current Globus Toolkit resource management system comprises three components: (1) an extensible resource specification language for exchanging information about resource requirements among the components in the Globus Toolkit resource management architecture; (2) a standardized interface to local resource management tools including, for example, PBS, LSF, and Condor; and (3) a resource coallocation API that may allow the construction of sophisticated coallocation strategies that enable use of multiple resources concurrently (DUROC) [15, 24, 25].

Data management is supported by integration of the GSI protocol to access remote files through, for example, the HTTP and the FTP protocols.

Data Grids are supported through replica catalog services in the newest release of the Globus Toolkit. These services allow copying of the most relevant portions of a data set to local storage for faster access. Installation of the extensive toolkit is enabled through a packaging software that can generate custom-designed installation distributions.

5.7.2 Commodity Grid Kits

The Globus Project provides an elementary set of Grid middleware. Unfortunately, these services may not be compatible with the commodity technologies used for application development by other Grid middleware developers.

To overcome this difficulty, the Commodity Grid project is creating *Co*mmodity *G*rid Toolkits (CoG Kits) that define mappings and interfaces between Grid services and particular commodity frameworks [26]. Technologies and frameworks of interest include Java, Python, CORBA, Perl, Web services, .NET, and JXTA.

Existing Python and Java CoG Kits provide the best support for a subset of the services within the Globus Toolkit. The Python CoG Kit uses the Simplified Wrapper and Interface Generator (SWIG) in order to wrap the Globus C-API, while the Java CoG Kit is a complete reimplementation of the Globus protocols in Java. Although the Java CoG Kit can be classified as middleware for integrating advanced Grid services, it can also be viewed both as a system providing advanced services currently not available in the Globus Toolkit and as a framework for designing computing portals [27]. Both the Java and Python CoG Kits are popular with Grid programmers and have been used successfully in many community projects. The usefulness of the Java CoG Kit has been proven during the design and implementation of the Globus Toolkit version 3 (GT3) and version 4 (GT4). Today the Java CoG Kit has been made an integral part of the GT3 release. Some features that are not included in the GT3 release but can be downloaded from the Java CoG Kit Web pages [28] are the availability of prototype user interfaces that are designed as JavaBeans and can be easily integrated in interface definition environments for rapid prototyping. As these components work on GT2, GT3 and GT4, the CoG Kit provides a new level of abstraction for Grid middleware developers (see Figures 5.6 and 5.7). In contrast to the Globus Toolkit, the Java CoG Kit has extended the notion of middleware while not only providing middleware for elementary Grid services but also middleware for Grid portal development. A variety of other tools are built on top of the Java CoG Kit.

5.7.3 Open Grid Services Architecture

The Open Grid Services Architecture (OGSA) [29] initiative of the GGF defines the artifacts for a standard service-oriented Grid framework on the basis of the emerging W3C-defined Web services technology [30]. A service-oriented Grid framework provides a loosely coupled technology- and platform-independent integration environment that allows different vendors to provide Grid-enabled services in a variety of technologies, yet conforming to the GGF-defined OGSA standards.

Web services, as defined by W3C, comprise "a software system identified by a URI, whose public interfaces and bindings are defined and described using XML. Its definition can be discovered by other software systems. These systems may then interact with the Web service in a manner prescribed by its definition, using XML based messages conveyed by Internet protocols." In other words, a Web service is a network-enabled software component that can be described, published, discovered, and invoked.

Web services are described through the Web services description language (WSDL), providing an XML format for defining Web services as set of endpoints operating on messages. Web service descriptions can be published and discovered on centrally available

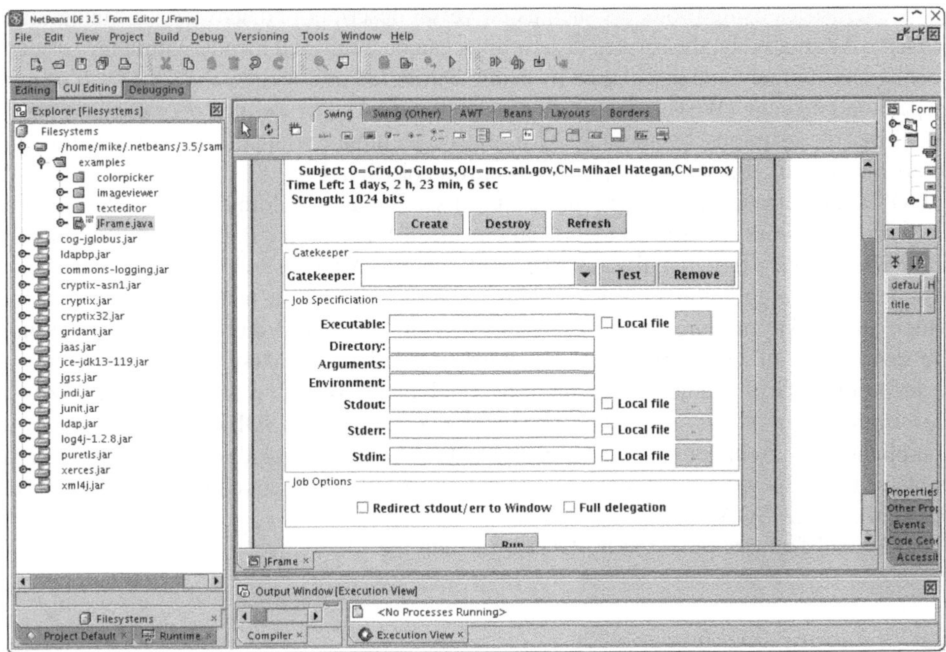

Figure 5.6 The Java CoG Kit contains a number of JavaBeans that provides the next level of Grid middleware to develop simple, yet powerful components for Grid users. These beans can be reused in commodity interface Development environments (IDE), providing a Grid IDE

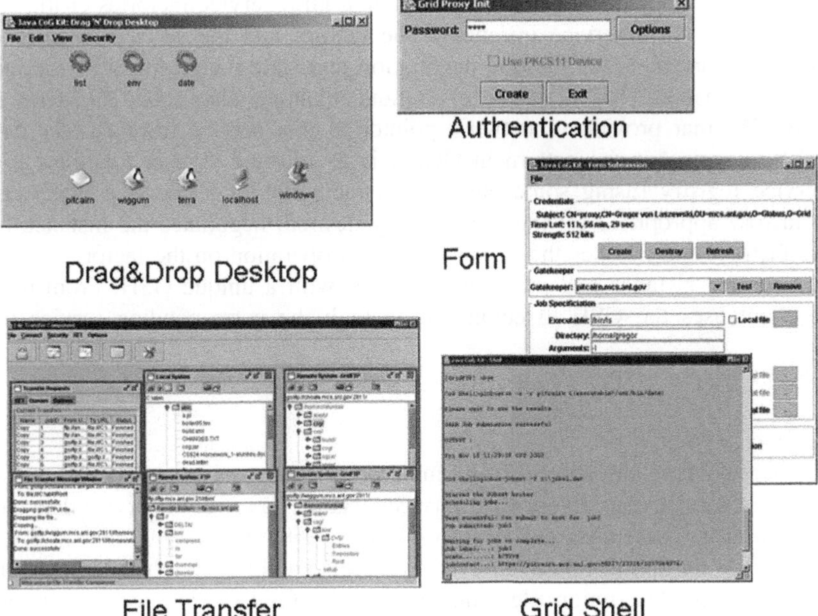

Figure 5.7 Selected Java CoG Kit GUI components include convenient prototype interfaces to perform file transfer, job submissions, and service access while interfacing with Grid middleware

service registries using protocols such as universal description, discovery, and integration (UDDI). Alternatively, the service descriptions can be published locally within the service-hosting environment and discovered by using distributed protocols such as the Web service inspection language (WSIL). Web services can be invoked by defined operations as a part of service descriptions by passing XML message inputs wrapped in standard Internet protocols such as the simple object access protocol (SOAP).

OGSA has introduced the concept of a Grid service as a building block of the service-oriented framework. A Grid service is an enhanced Web service that extends the conventional Web service functionality into the Grid domain. A Grid service handles issues such as state management, global service naming, reference resolution, and Grid-aware security, which are the key requirements for a seamless, universal Grid architecture. To facilitate the description of Grid services, GGF has extended the conventional WSDL schema to incorporate Grid-enabled attributes. The Grid-aware WSDL is called as *Grid Service Description LanguageGrid Service Description Language* (GSDL). Hence, an OGSA-compliant Grid architecture will be composed of Grid services described using GSDL, published and discovered on specialized Grid registries, and invoked by using an Internet protocol such as SOAP.

Additional features provided by OGSA include the following:

Service Factories and Instances.

The Web services paradigm addresses the discovery and invocation of persistent services. The Grid paradigm supports the creation, management, and termination of transient service instances handled at runtime. The OGSA framework defines entities and patterns between these entities to support such dynamic management of transient instances. OGSA implements the factory pattern, whereby persistent Grid service factories enable the creation of transient Grid service instances. The factory and the service instance collectively provide constructs for lifetime negotiation and state management of the instance. Every service instance has a universally unique identity, also called the *Grid service handle* (GSH), that provides a network pointer to that service instance. As shown in Figure 5.8, a typical usage pattern in OGSA is as follows. A user identifies a central Grid service registry (using some out-of-band methodology), selects a persistent Grid service factory appropriate for the functionality desired, negotiates the instance lifetime with the factory, and invokes the instance-creation operation on the factory. The factory creates a service instance and provides the client with a unique GSH. From that point on, the client uses the GSH to communicate with the corresponding service instance directly.

Service Data.

To support discovery, introspection, and monitoring of Grid services, OGSA introduced the concept of service data. Every Grid service instance has a set of service data elements associated with it. These elements refer to metadata used internally by the OGSA framework, as well as the runtime data retrieved by clients and other external services. In other words, the Service Description Elements (SDEs) of a Grid service instance represent its internal and external state. In an object-oriented terminology, the SDEs in a Grid service are equivalent to instance variables in an object.

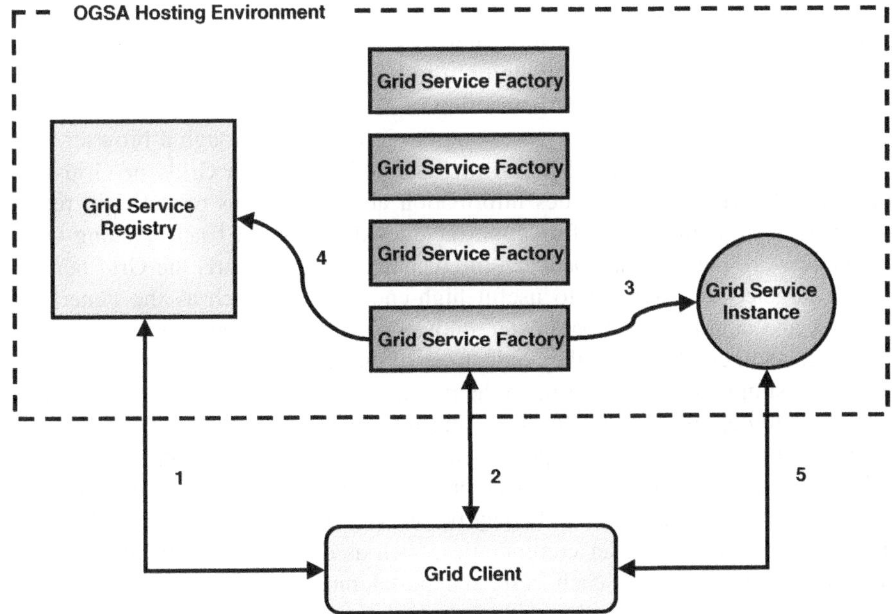

Figure 5.8 Typical usage pattern for OGSA services. (1) The client accesses the central Grid registry and discovers the handle for a Grid service factory capable of providing the required functionality. (2) It then invokes the instance-creation operation of the Grid service factory. (3) The factory creates a Grid service instance and provides the client a unique network handle (GSH) for this instance. (4) The factory then registers the instance handle with the Grid registry so that other clients can discover it. This step is optional. (5) The client uses the GSH to communicate with the Grid service instance on the basis of the operations promised in its description (GSDL)

Notification.

OGSA also provides the classic *publish/subscribe* style communication mechanism with Grid services. Hence, clients can subscribe to receive notification messages based for any activity on a predefined notification topic or change in the data value of SDEs. OGSA provides mechanisms for a *notification source* and *notification sink* to allow asynchronous, one-way delivery of messages from the source to the sink. Hence, OGSA shifts from a client-server Web service model to a peer-to-peer Grid service model.

WS-RF.

Most recently the standards proposed to the GGF have been revised and broken up into a smaller set of standards that are closer in nature to the commodity Web service model. A number of standards that address these new issues have been proposed and are now also under review by OASIS.

5.8 Portal Middleware for Grids

The term "portal" is not uniformly defined within the computer science community. Sometimes it is used interchangeably for integrated desktops, electronic market places, or information hubs [31, 32]. We use the term portal in the more general sense of a community access point to information and services. A portal typically is most useful when designed for a particular community in mind.

Web portals build on the current generation of Web-based commodity technologies on the basis of the HTTP protocol for accessing the information through a browser.

A *Grid portal* is a specialized portal for users of production Grids or Grid-enabled applications. A Grid portal provides information about the status of the Grid resources and services. Commonly, this information includes the status of batch queuing systems, load, and network performance between the resources. Furthermore, the Grid portal may provide a targeted access point to useful high-end services, such as the generation of a compute- and data-intensive parameter study for climate change. Grid portals provide communities another advantage: they hide much of the complex logic to drive Grid-related services with simple interaction through the portal interface. Furthermore, they reduce the effort needed to deploy software for accessing resources on production Grids.

In contrast to Web portals, Grid portals may not be restricted to simple browser technologies but may use specialized plug-ins or executables to handle the data visualization requirements. These advanced portals may be the Web portals of tomorrow. A Grid portal may deal with different user communities, such as developers, application scientists, administrators, and users. In each case, the portal must support a personal view that remembers the preferred interaction with the portal at time of entry. To meet the needs of this diverse community, sophisticated Grid portal middleware must provide commodity collaborative tools such as news-readers, e-mail, chat, and video conferencing, and event scheduling.

It is important to recognize the fact that Grid middleware must also include software that allows the creation of portals for Grids. Such middleware has in the past been relying on the Java CoG Kit and can, through abstractions, continue to build on the convenient functions provided. Several efforts are using Jetspeed as part of this middleware to enable Grid portals [33–35].

Grid Middleware to Deploy Virtual Organization Portals.
Compute center portals provide a collective view of and access to a distributed set of high-performance computing resources as part of a high-performance computing center. Typical candidates are HotPage [36], and UNICORE [11].

HotPage enables researchers to easily find information about each of the resources in the computational Grid. This information (which is stored in HTML) includes technical documentation, operational status, load and current usage, and queued jobs. Hence, it combines information provided by Grid middleware services and other informational databases maintained as part of the virtual organization. Additionally, HotPage enables users to access and manipulate files and data and to submit, monitor, and delete jobs. Grid access is through the Globus Toolkit [37] or the Network Weather Service [38]. The HotPage backend is accessed through Perl CGI scripts that create the pages requested. HotPage has been installed on a variety of production Grids, such as NPACI and NASA IPG [2]. Future versions of HotPage are expected to be implemented in Java by using the Java CoG Kit, OGSA, and Jetspeed.

UNICORE (UNiform Interface to COmputing REsources) provides a vertical integration environment for Grids, including access to resources through a Java Swing framework. It is designed to assist in the workflow management of tasks to be scheduled on resources part of supercomputing centers similar to [10]. A UNICORE workflow comprises hierarchical assemblies of interdependent tasks, with dependencies that are mapped to actions such as execution, compilation, linking, and scripting according to resource requirements on target machines on the Grid. Besides strong authentication, UNICORE assists in compiling and running applications and in transferring input and output data. One of the main components of UNICORE is the preparation and modification of structured jobs through a graphical user interface that supports workflows. It allows the submission, monitoring, and control of the execution as part of a client that gets installed on the user's machine. New functionality is developed to handle system administration and management, modeling of dynamic and extensible resources, creation of application-specific client and server extensions, improved data and file management functions, and runtime control of complex job chains. The ability to utilize Globus Toolkit–enabled resources within UNICORE [34] while reusing the Java CoG Kit is under development. Future developments will also include GT3.

5.9 Applications Using and Enhancing Grid Middleware

In this section, we focus on three applications representative of current Grid activities reusing and enhancing Grid middleware.

5.9.1 Astrophysics

The Astrophysics Simulation Collaboratory is an example of a Grid application requiring large numbers of tightly coupled resources that are used as part of large parameter studies. The Astrophysics Simulation Collaboratory (ASC) was originally developed in support of numerical simulations in astrophysics. It has led to the development of a general-purpose code for partial differential equations in three dimensions [39].

The astrophysics simulation collaboratory (ASC) pursues the following objectives: (a) Promote the creation of a community for sharing and developing simulation codes and scientific results; (b) enable transparent access to remote resources, including computers, data storage archives, information servers, and shared code repositories; (c) enhance domain-specific component and service development supporting problem-solving capabilities, such as the development of simulation codes for the astrophysical community or the development of advanced Grid services reusable by the community; (d) distribute and install programs onto remote resources while accessing code repositories, compilation, and deployment services; (e) enable collaboration during program execution to foster interaction during the development of parameters and the verification of the simulations; (f) enable shared control and steering of the simulations to support asynchronous collaborative techniques among collaboratory members; and (g) provide access to domain-specific clients that, for example, enable access to multimedia streams and other data generated during the execution of the simulation.

To communicate these objectives as part of a collaboratory, ASC uses a Grid portal on the basis of Java Server Pages (JSP) for thin-client access to Grid services. Specialized services support community code development through online code repositories [2].

5.9.2 Earthquake Engineering

The intention of NEESgrid is to build a national-scale distributed virtual laboratory for earthquake engineering. The initial goals of the project are to (1) extend the Globus Information Service to meet the specialized needs of the community and (2) develop a set of application-specific services, reusing existing Grid services. Ultimately, the system will include a collaboration and visualization environment, specialized servers to handle and manage the environment, and access to external system and storage provided by NCSA [40].

One of the objectives of NEESgrid is to enable observation and data access to experiments in real time. Both centralized and distributed data repositories will be created to share data between different locations on the Grid. These repositories will have data management software to assist in rapid and controlled publication of results. A software library will be created to distribute simulation software to users. This will allow users with NEESgrid-enabled desktops to run remote simulations on the Grid.

NEESgrid comprises a layered architecture, with each component being built on core Grid services that handle authentication, information, and resource management but are customized to fit the needs of earthquake engineering community.

5.9.3 High-energy Physics Grids

A number of large projects related to high-energy research have recognized the potential and necessity of Grids is part of their sophisticated problem-solving infrastructure. Such projects include Particle Physics Data Grid (PPDG) [41], international Virtual Data Grid Laboratory (iVDGL) [42], and the European DataGrid [43].

PPDG is a collaboratory project concerned with providing next-generation infrastructure for current and future high-energy and nuclear physics experiments. One of the important requirements of PPDG is to deal with the enormous amount of data that is created during high-energy physics experiment and must be analyzed by large groups of specialists. Data storage, replication, job scheduling, resource management, and security components supplied by the Globus Toolkit, Condor, STACS, SRB, and EU DataGrid projects all will be integrated for easy use by the physics collaborators. Development of PPDG is supported under the DOE SciDAC initiative (Particle Physics Data Grid Collaboratory Pilot) [20, 41].

The goal of iVDGL is to establish and utilize a virtual laboratory comprising heterogeneous computing and storage resources in the United States, Europe, and other regions linked by high-speed networks and operated as a single system for the purposes of interdisciplinary experimentation in Grid-enabled, data-intensive scientific computing. iVDGL is aiming at the physics community while providing future production services, which may also be designed in the Grid Physics Network (GriPhyN) [44]. Integration of the two projects will provide a community production Grid.

In general, the communities addressed by such Grids require sharing and modifying large amounts of data.

Many of these projects reuse the concept of virtualization, which is well understood by the processor design community in addressing the design of memory hierarchies. In Grid projects, however, the virtualization goes beyond the concept of tertiary storage space and includes not only file systems but also the annotation of data with metadata. This metadata allows the scientists to formulate complex context-sensitive queries returning the appropriate information. As a result, some of these queries have been generated by state-of-the-art algorithms that are too resource-intensive to be run redundantly by many users. Thus, it is beneficial to store the results in a cache-like system that can be queried rather than generated. On the other hand, some of the calculations performed on the data may actually be cheaper and less resource-intense than storing the data and their intermediate results. In both cases, it is important to annotate the data and the way it has or can be created with metadata. Efforts such as [45]–[47] are reporting on progress in this area. In many such efforts, lessons learned from parallel computing while mapping an abstract direct acyclic graph specification onto a physical environment are used. Through late mapping that postpones the mapping in successive steps, the dynamical of the actual Grid resources are optimized. At present, however, these efforts do not address the more formal aspects such as deadlocks or resource over- and underprovision.

5.10 Concluding Remarks

In this chapter, we have concentrated on several aspects of middleware for Grids. We have seen that the Grid approach and the creation of Grid middleware are a natural continuation of established parallel and distributed computing research. We highlighted a number of projects that address some – but not all – of the issues that must be resolved before the Grid is truly universal. In addition to the development of middleware, interfaces are needed that can be used by application scientists to access Grids. These interfaces are provided by portal middleware toolkits that allow the easy integration of Grid middleware in ongoing application development efforts. On the basis of the efforts in the community, we observe the trend that many advanced tools will be part of the middleware of tomorrow. Hence, the level of abstraction included in middleware to enable collaborative research is expected to rise significantly over the next years. Also critical are commodity Grid toolkits, enabling access to Grid functionality in the area of Web services, programming paradigms and frameworks, as well as the integration with other programming languages. The tools and technologies discussed in this chapter are the first step in the creation of a global computing Grid.

Acknowledgments

This work was funded in part by the US Department of Energy, Office of Advanced Scientific Computing, "CoG Kits: Enabling Middleware for Designing Science Applications, Web Portals, and Problem-Solving Environments," under Contract W-31-109-ENG-38. Globus Project is a trademark held by the University of Chicago. Globus Toolkit is a registered trademark held by the University of Chicago. We thank Gail Pieper and Jens Vöckler for their comments on this chapter.

Bibliography

[1] Whitehead, A. N. *Science and the Modern World*, Free Press, New York, 1967, p. 58.

[2] von Laszewski, G. and Wagstrom, P. (2004) Gestalt of the Grid, *Performance Evaluation and Characterization of Parallel and Distributed Computing Tools, Series on Parallel and Distributed Computing*, pp. 149–187; Wiley, (to be published); http://www.mcs.anl.gov/~gregor/papers/vonLaszewski–gestalt.pdf

[3] Smarr, L. and Catlett, C. (1992) Metacomputing. *Communications of the ACM*, **35**(6), 44–52.

[4] Foster, I. (2002) The Grid: A New Infrastructure for 21st Century Science. *Physics Today*, **55**(22), 42; http://www.aip.org/pt/vol-55/iss-2/p42.html

[5] von Laszewski, G., Su, M.-H., Insley, J. A., Foster, I., Bresnahan, J., Kesselman, C., Thiebaux, M., Rivers, M. L., Wang, S., Tieman, B., and McNulty, I. (1999) Real-Time Analysis, Visualization, and Steering of Microtomography Experiments at Photon Sources. *Ninth SIAM Conference on Parallel Processing for Scientific Computing*, San Antonio, TX, March 22-24; http://www.mcs.anl.gov/~gregor/papers/vonLaszewski–siamCmt99.pdf

[6] Oram, A., ed. (2001) *Peer-To-Peer*, O'Reiley.

[7] Foster, I., Kesselman, C., and Tuecke, S. (2002) The Anatomy of the Grid: Enabling Scalable Virtual Organizations. *International Journal of High Performance Computing Applications*, **15**(3), 200–222, Fall 2001; http://www.globus.org/research/papers/anatomy.pdf

[8] von Laszewski, G., Foster, I., Gawor, J., Smith, W., and Tuecke, S. (2000) CoG Kits: A Bridge Between Commodity Distributed Computing and High-Performance Grids. *ACM Java Grande 2000 Conference*, San Francisco, CA, June 3-5 , pp. 97–106; http://www.mcs.anl.gov/~gregor/papers/vonLaszewski–cog-final.pdf

[9] von Laszewski, G., Amin, K., Hategan, M., and Zaluzec, N. J. (2004) GridAnt: A Client-Controllable Grid Workflow System. *37th Hawai'i International Conference on System Science*, Island of Hawaii, Big Island, January 5-8; http://www.mcs.anl.gov/~gregor/papers/vonLaszewski–gridant-hics.pdf

[10] von Laszewski, G. (1999) A Loosely Coupled Metacomputer: Cooperating Job Submissions Across Multiple Supercomputing Sites. *Concurrency, Experience, and Practice*, **11**(5), 933–948; http://www.mcs.anl.gov/~gregor/papers/vonLaszewski–CooperatingJobs.ps

[11] Unicore, Web Page, http://www.unicore.de/. Accessed 4/22/2004.

[12] Grimshaw, A. S. and Wulf, W. A. (1997) The Legion Vision of a Worldwide Virtual Computer. *Communications of the ACM*, **40**(1), 39–45; http://legion.virginia.edu/copy-cacm.html

[13] Condor: High Throughput Computing, Web Page, http://www.cs.wisc.edu/condor/. Accessed 4/22/2004.

[14] The Global Grid Forum Web Page, Web Page, http://www.gridforum.org. Accessed 4/22/2004.

[15] Portable Batch System, Web Page, Veridian Systems, http://www.openpbs.org/. Accessed 4/22/2004.

[16] Butler, R., Engert, D., Foster, I., Kesselman, C., Tuecke, S., Volmer, J., and Welch, V. (2000) A National-Scale Authentication Infrastructure. *IEEE Computer*, **33**(12), 60–66.

[17] von Laszewski, G., Gawor, J., Peña, C. J., and Foster, I. InfoGram: A Peer-to-Peer Information and Job Submission Service. *Proceedings of the 11th Symposium on High Performance Distributed Computing*, Edinburgh, UK, July 24-26, pp. 333–342; http://www.mcs.anl.gov/~gregor/papers/vonLaszewski–infogram.ps

[18] Allcock, B. and Madduri, R. Reliable File Transfer Service, Web Page, http://www-unix.globus.org/ogsa/docs/alpha3/services/reliable_transfer.html

[19] NSF Middleware Initiative, Web Page, http://www.nsf-middleware.org/Middleware/

[20] Scientific Discovery through Advanced Computing (SciDAC) (2001) Web Page, http://scidac.org/

[21] Information Power Grid Engineering and Research Site (2001) Web Page, http://www.ipg.nasa.gov/

[22] von Laszewski, G., Fitzgerald, S., Foster, I., Kesselman, C., Smith, W., and Tuecke, S. (1997) A Directory Service for Configuring High-Performance Distributed Computations. *Proceedings of the 6th IEEE Symposium on High-Performance Distributed Computing*, August 5-8, Portland, OR, pp. 365–375; http://www.mcs.anl.gov/~gregor/papers/fitzgerald–hpdc97.pdf

[23] Czajkowski, K., Fitzgerald, S., Foster, I., and Kesselman, C. (2001) Grid Information Services for Distributed Resource Sharing. *Proceedings of the Tenth IEEE International Symposium on High-Performance Distributed Computing*, San Francisco, CA, August 7-9; IEEE Press, pp. 181–184, http://www.globus.org/research/papers/MDS-HPDC.pdf.

[24] Load Sharing Facility, Web Page, Platform Computing, Inc, http://www.platform.com/

[25] Czajkowski, K., Foster, I., and Kesselman, C. (1999) Co-Allocation Services for Computational Grids. *Proceedings of the 8th IEEE Symposium on High Performance Distributed Computing*.

[26] von Laszewski, G., Foster, I., Gawor, J., and Lane, P. (2001) A Java Commodity Grid Kit. *Concurrency and Computation: Practice and Experience*, **13**(8-9), 643–662. http://www.mcs.anl.gov/~gregor/papers/vonLaszewski–cog-cpe-final.pdf

[27] von Laszewski, G., Foster, I., Gawor, J., Lane, P., Rehn, N., and Russell, M. (2001) Designing Grid-based Problem Solving Environments and Portals. *34th Hawaiian International Conference on System Science*, Maui, Hawaii, January 3-6; http://www.mcs.anl.gov/~gregor/papers/vonLaszewski–cog-pse-final.pdf

[28] The Commodity Grid Project, Web Page, http://www.globus.org/cog

[29] Foster, I., Kesselman, C., Nick, J., and Tuecke, S. (2002) *The Physiology of the Grid: An Open Grid Services Architecture for Distributed Systems Integration*, Web Page, http://www.globus.org/research/papers/ogsa.pdf

[30] World Wide Web Consortium, Web Page, http://www.w3.org/. Accessed 4/22/2004.

[31] Smarr, L. (2001) *Infrastructures for Science Portals*, http://www.computer.org/internet/v4n1/smarr.htm

[32] Fox, G. C. and Furmanski, W. (1999) High Performance Commodity Computing, in *The Grid: Blueprint for a New Computing Infrastructure* (eds I. Foster and C. Kesselman), Morgan Kaufman.

[33] The Jetspeed Webpage, Web Page, http://jakarta.apache.org/jetspeed/. Accessed 4/22/2004.

[34] Grid Interoperability Project, Web Page, http://www.grid-interoperability.org/. Accessed 4/22/2004.

[35] Snelling, D., van den Berghe, S., von Laszewski, G., Wieder, P., MacLaren, J., Brooke, J., Nicole, D., and Hoppe, H.-C. (2001) *A Unicore Globus Interoperability Layer*, Web Page, http://www.grid-interoperability.org/D4.1b_draft.pdf

[36] NPACI HotPage (2001) Web Page, https://hotpage.npaci.edu/

[37] The Globus Project, Web Page, http://www.globus.org

[38] Gaidioz, B., Wolski, R., and Tourancheau, B. (2000) Synchronizing Network Probes to Avoid Measurement Intrusiveness with the Network Weather Service. *Proceedings of 9th IEEE High-Performance Distributed Computing Conference*, August, pp. 147-154; http://www.cs.ucsb.edu/ rich/publications/

[39] von Laszewski, G., Russell, M., Foster, I., Shalf, J., Allen, G., Daues, G., Novotny, J., and Seidel, E. (2002) Community Software Development with the Astrophysics Simulation Collaboratory. *Concurrency and Computation: Practice and Experience*, **14**, 1289-1301; http://www.mcs.anl.gov/~gregor/papers/vonLaszewski–cactus5.pdf

[40] NEESgrid Homepage, Web Page, http://www.neesgrid.org/. Accessed 4/22/2004.

[41] Particle Physics Data Grid (2001) Web Page, http://www.ppdg.net/. Accessed 4/22/2004.

[42] The International Virtual Data Grid Laboratory, Web Page, http://www.ivdgl.org/. Accessed 4/22/2004.

[43] The DataGrid Project (2000) http://www.eu-datagrid.org/. Accessed 4/22/2004.

[44] GriPhyN - Grid Physics Network, Web Page, http://www.griphyn.org/index.php

[45] Foster, I., Vöckler, J.-S., Wilde, M., and Zhao, Y. (2002) Chimera: A Virtual Data System for Representing, Querying and Automating Data Derivation. *14th International Conference on Scientific Database Management*, Edinburgh, Scotland.

[46] von Laszewski, G., Ruscic, B., Amin, K., Wagstrom, P., Krishnan, S., and Nijsure, S. (2003) A Framework for Building Scientific Knowledge Grids Applied to Thermochemical Tables. *The International Journal of High Performance Computing Application*, **17**, http://www.mcs.anl.gov/~gregor/papers/vonLaszewski–knowledge-grid.pdf

[47] Pancerella, C., Myers, J. D., Allison, T. C., Amin, K., Bittner, S., Didier, B., Frenklach, M., William, J., Green, H., Ho, Y.-L., Hewson, J., Koegler, W., Lansing, C., Leahy, D., Lee, M., McCoy, R., Minkoff, M., Nijsure, S., von Laszewski, G., Montoya, D., Pinzon, R., Pitz, W., Rahn, L., Ruscic, B., Schuchardt, K., Stephan, E., Wagner, A., Wang, B., Windus, T., Xu, L., and Yang, C. (2003) Metadata in the Collaboratory for Multi-Scale Chemical Science. *2003 Dublin Core Conference: Supporting Communities of Discourse and Practice-Metadata Research and Applications*, Seattle, WA, September 28-October 2; http://www.mcs.anl.gov/~gregor/papers/vonLaszewski–dublin-core.pdf

6

QoS-enabled Middleware

Nanbor Wang, Christopher D. Gill[1], Douglas C. Schmidt, Aniruddha Gokhale, Balachandran Natarajan[2], Joseph P. Loyall, Richard E. Schantz, and Craig Rodrigues[3]

[1] *Washington University*
[2] *Vanderbilt University*
[3] *BBN Technologies*

6.1 Introduction

6.1.1 Emerging Trends

Commercial off-the-shelf (COTS) middleware technologies, such as The Object Management Group (OMG)'s Common Object Request Broker Architecture (CORBA) [28], Sun's Java RMI [56], and Microsoft's COM+ [25], have matured considerably in recent years. They are being used to reduce the time and effort required to develop applications in a broad range of information technology (IT) domains. While these middleware technologies have historically been applied to *enterprise applications* [15], such as online catalog and reservation systems, bank asset management systems, and management planning systems, over 99% of all microprocessors are now used for distributed real-time and embedded (DRE) systems [2] that control processes and devices in physical, chemical, biological, or defense industries.

These types of DRE applications have distinctly different characteristics than conventional desktop or back-office applications in that *the right answer delivered too late can become the wrong answer*, that is, failure to meet key QoS requirements can lead to catastrophic consequences. Middleware that can satisfy stringent Quality-of-Service (QoS) requirements, such as predictability, latency, efficiency, scalability, dependability, and security, is therefore increasingly being applied to DRE application development. Regardless of the domain in which middleware is applied, however, its goal is to help

Middleware for Communications. Edited by Qusay H. Mahmoud
 ISBN 0-470-86206-8

expedite the software process by (1) making it easier to integrate parts together and (2) shielding developers from many inherent and accidental complexities, such as platform and language heterogeneity, resource management, and fault tolerance.

Component middleware [50] is a class of middleware that enables reusable services to be composed, configured, and installed to create applications rapidly and robustly. In particular, component middleware offers application developers the following reusable capabilities:

- *Connector mechanisms between components*, such as remote method invocations and message passing
- *Horizontal infrastructure services*, such as request brokers
- *Vertical models of domain concepts*, such as common semantics for higher-level reusable component services ranging from transaction support to multilevel security.

Examples of COTS component middleware include the CORBA Component Model (CCM) [26], Java 2 Enterprise Edition (J2EE) [49], and the Component Object Model (COM) [4], each of which uses different APIs, protocols, and component models.

As the use of component middleware becomes more pervasive, DRE applications are increasingly combined to form distributed systems that are joined together by the Internet and intranets. Examples of these types of DRE applications include *industrial process control systems*, such as hot rolling mill control systems that process molten steel in real time, and *avionics systems*, such as mission management computers [46]; [47] that help aircraft pilots plan and navigate through their routes. Often, these applications are combined further with other DRE applications to create "systems of systems," which we henceforth refer to as "large-scale DRE systems."

The *military command and control systems* shown in Figure 6.1 exemplifies such a large-scale DRE system, where information is gathered and assimilated from a diverse collection of devices (such as unmanned aerial vehicles and wearable computers), before being presented to higher-level command and control functions that analyze the information and coordinate the deployment of available forces and weaponry. In the example, both the reconnaissance information such as types and locations of threats, and the tactical decisions such as retasking and redeployment, must be communicated within specified timing constraints across system boundaries to all involved entities.

6.1.2 Key Technical Challenges and Solution Approaches

The following key technical challenges arise when developing and deploying applications in the kinds of large-scale DRE systems environments outlined in Section 6.1.1:

1. **Satisfying multiple QoS requirements in real time.** Most DRE applications have stringent QoS requirements that must be satisfied simultaneously in real time. Example QoS requirements include processor allocation and network latency, jitter, and bandwidth. To ensure that DRE applications can achieve their QoS requirements, various types of *QoS provisioning* must be performed to allocate and manage system computing and communication resources end-to-end. QoS provisioning can be performed in the following ways:

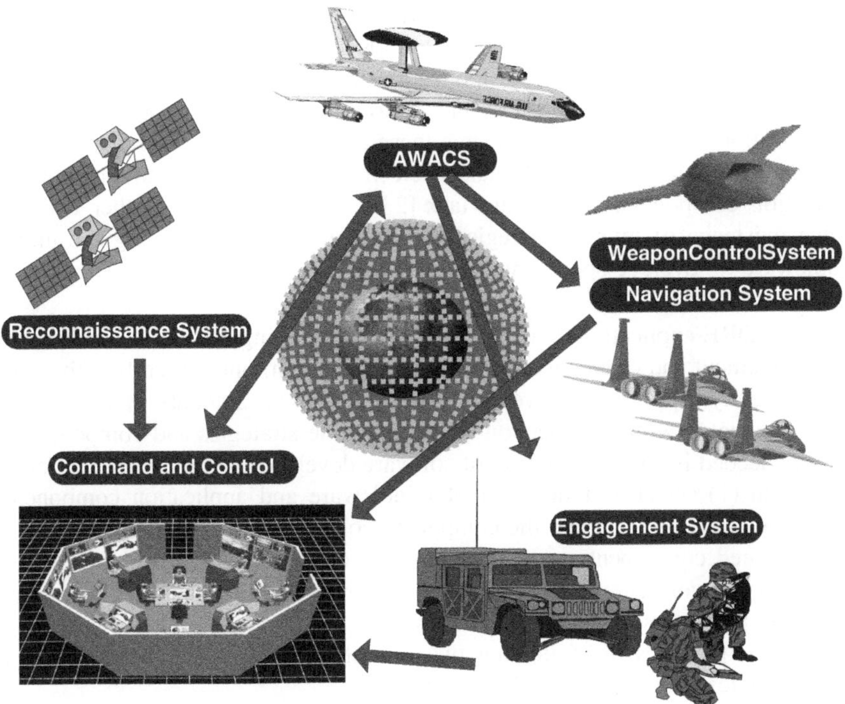

Figure 6.1 An example large-scale DRE system

- *Statically*, where ensuring adequate resources required to support a particular degree of QoS is preconfigured into an application. Examples of static QoS provisioning include task prioritization and communication bandwidth reservation. Section 6.4.1.1 describes a range of QoS resources that can be provisioned statically.
- *Dynamically*, where the resources required are determined and adjusted on the basis of the runtime system status. Examples of dynamic QoS provisioning include runtime reallocation reprioritization to handle bursty CPU load, primary, secondary, and remote storage access, and competing network traffic demands. Section 6.4.2.1 describes a range of QoS resources that can be provisioned dynamically.

QoS provisioning in large-scale DRE systems crosscuts multiple layers and requires end-to-end enforcement. Conventional component middleware technologies, such as CCM, J2EE, and COM+, were designed largely for applications with business-oriented QoS requirements such as data persistence, confidentiality, and transactional support. Thus, they do not effectively address enforcing the stringent, multiple, and simultaneous QoS requirements of DRE applications end-to-end. What is therefore needed is *QoS-enabled component middleware* that preserves existing support for heterogeneity in standard component middleware, yet also provides multiple dimensions of QoS provisioning and enforcement [35] to meet those end-to-end real-time QoS requirements of DRE applications.

2. **Alleviating the complexity of composing and managing large-scale DRE software.** To reduce lifecycle costs and time-to-market, application developers today largely assemble and deploy DRE applications by manually selecting a set of compatible common-off-the-shelf (COTS) and custom-developed components. To compose an application successfully requires that these components have compatible interfaces, semantics, and protocols, which makes selecting and developing a compatible set of application components a daunting task [23]. This problem is further compounded by the existence of myriad strategies for configuring and deploying the underlying middleware to leverage special hardware and software features.

Moreover, DRE application demands for QoS provisioning (which in turn require end-to-end enforcement and often pervade an entire application) only exacerbate the complexity. Consequently, application developers spend nontrivial amounts of time debugging problems associated with the selection of incompatible strategies and components. What is therefore needed is an integrated set of software development processes, platforms, and tools that can (1) select a suitable set of middleware and application components, (2) analyze, synthesize, and validate the component configurations, (3) assemble and deploy groups of related components to their appropriate execution platforms, and (4) dynamically adjust and reconfigure the system as operational conditions change, to maintain the required QoS properties of DRE applications.

This chapter provides the following two contributions to research and development efforts that address the challenges described above:

- We illustrate how enhancements to standard component middleware can simplify the development of DRE applications by composing QoS provisioning policies statically with applications. Our discussion focuses on a QoS-enabled enhancement of standard CCM [26] called the *Component-Integrated ACE ORB* (CIAO), which is being developed at Washington University in St Louis and the Institute for Software Integrated Systems (ISIS) at Vanderbilt University.
- We describe how dynamic QoS provisioning and adaptation can be addressed using middleware capabilities called *Qoskets*, which are enhancements to the Quality Objects (QuO) [57] middleware developed by BBN Technologies. Our discussion focuses on how Qoskets can be combined with CIAO to compose adaptive QoS assurance into DRE applications dynamically. In particular, Qoskets manage modular QoS *aspects*, which can be combined with CIAO and woven to create an integrated QoS-enabled component model.

A companion chapter in this book [16] on Model Driven Middleware discusses how QoS-enabled component middleware can be combined with Model Driven Architecture (MDA) tools to rapidly develop, generate, assemble, and deploy flexible DRE applications that can be tailored readily to meet the needs of multiple simultaneous QoS requirements.

All the material presented in this chapter is based on the CIAO and QuO middleware, which can be downloaded in open-source format from deuce.doc.wustl.edu/Download.html and quo.bbn.com/quorelease.html, respectively. This middleware is being applied to many projects worldwide in a wide range of domains, including telecommunications, aerospace, financial services, process control, scientific computing, and distributed interactive simulations.

6.1.3 Chapter Organization

The remainder of this chapter is organized as follows: Section 6.2 gives a brief overview of middleware technologies and describes limitations of conventional object-based middleware technologies; Section 6.3 describes how component middleware addresses key limitations of object-oriented middleware and then explains how current component middleware does not yet support DRE application development effectively; Section 6.4 illustrates how the CIAO and QuO middleware expands the capabilities of conventional component middleware to facilitate static and dynamic QoS provisioning and enforcement for DRE applications; Section 6.5 compares our work on CIAO and Qoskets with related research; and Section 6.6 presents concluding remarks.

6.2 The Evolution of Middleware

In the early days of computing, software was developed from scratch to achieve a particular goal on a specific hardware platform. Since computers were themselves much more expensive than the cost to program them, scant attention was paid to systematic software reuse and composition of applications from existing software artifacts. Over the past four decades, the following two trends have spurred the transition from hardware-centric to software-centric development paradigms:

- **Economic factors** – Because of advances in VLSI and the commoditization of hardware, most computers are now *much* less expensive than the cost to program them.
- **Technological advances** – With the advent of software development technologies, such as object-oriented programming languages and distributed object computing technologies, it has become easier to develop software with more capabilities and features.

A common theme underlying the evolution of modern software development paradigms is the desire for reuse, that is, to compose and customize applications from preexisting software building blocks [11]. Major modern software development paradigms all aim to achieve this common goal, but differ in the type(s) and granularity of building blocks that form the core of each paradigm. The development and the evolution of middleware technologies also follow the similar goal to capture and reuse design information learned in the past, within various layers of software.

This section provides an overview of middleware and describes the limitations of conventional distributed object computing (DOC) middleware, which has been the dominant form of middleware during the 1990s. Section 6.3 then presents an overview of how component middleware overcomes the limitations of conventional DOC middleware.

6.2.1 Overview of Middleware

Middleware is reusable software that resides between applications and the underlying operating systems, network protocol stacks, and hardware [38]. Middleware's primary role is to bridge the gap between application programs and the lower-level hardware and software infrastructure to coordinate how parts of applications are connected and how they interoperate. Middleware focuses especially on issues that emerge when such programs are used across physically separated platforms. When developed and deployed properly, middleware can reduce the cost and risk of developing distributed applications and systems by helping

- simplify the development of distributed applications by providing a consistent set of capabilities that is closer to the set of application design-level abstractions than to the underlying computing and communication mechanisms;
- provide higher-level abstraction interfaces for managing system resources, such as instantiation and management of interface implementations and provisioning of QoS resources;
- shield application developers from low-level, tedious, and error-prone platform details, such as socket-level network programming idioms;
- amortize software lifecycle costs by leveraging previous development expertise and capturing implementations of key patterns in reusable frameworks, rather than rebuilding them manually for each use;
- provide a wide array of off-the-shelf developer-oriented services, such as transactional logging and security, which have proven necessary to operate effectively in a distributed environment;
- ease the integration and interoperability of software artifacts developed by multiple technology suppliers, over increasingly diverse, heterogeneous, and geographically separated environments [6];
- extend the scope of portable software to higher levels of abstraction through common industry-wide standards.

The emergence and rapid growth of the Internet, beginning in the 1970s, brought forth the need for distributed applications. For years, however, these applications were hard to develop because of a paucity of methods, tools, and platforms. Various technologies have emerged over the past 20+ years to alleviate complexities associated with developing software for distributed applications and to provide an advanced software infrastructure to support it. Early milestones included the advent of Internet protocols [33]; [34], interprocess communication and message passing architectures [9], micro-kernel architectures [1], and Sun's Remote Procedure Call (RPC) model [48]. The next generation of advances included OSF's Distributed Computing Environment (DCE) [37], CORBA [28], and DCOM [4]. More recently, middleware technologies have evolved to support DRE applications (e.g., Real-time CORBA [27]), as well as to provide higher-level abstractions, such as component models (*e.g.*, CCM [26], J2EE [49]), and model-driven middleware [16].

The success of middleware technologies has added the middleware paradigm to the familiar operating system, programming language, networking, and database offerings used by previous generations of software developers. By decoupling application-specific functionality and logic from the accidental complexities inherent in a distributed infrastructure, middleware enables application developers to concentrate on programming application-specific functionality, rather than wrestling repeatedly with lower-level infrastructure challenges.

6.2.2 Limitations of Conventional Middleware

Section 6.2.1 briefly described the evolution of middleware over the past 20+ years. One of the watershed events during this period was the emergence of distributed object computing (DOC) middleware in the late 1980s/early 1990s [41]. DOC middleware represented the confluence of two major areas of software technology: *distributed computing systems* and *object-oriented design and programming*. Techniques for developing distributed systems focus on integrating multiple computers to act as a unified scalable computational resource. Likewise, techniques for developing object-oriented systems focus on reducing complexity by creating reusable frameworks and components that reify successful patterns and software architectures [13]; [5]; [45]. DOC middleware therefore uses object-oriented techniques to distribute reusable services and applications efficiently, flexibly, and robustly over multiple, often heterogeneous, computing and networking elements.

The Object Management Architecture (OMA) in the CORBA 2.x specification [29] defines an advanced DOC middleware standard for building portable distributed applications. The CORBA 2.x specification focuses on *interfaces*, which are essentially contracts between clients and servers that define how clients *view* and *access* object services provided by a server. Despite its advanced capabilities, however, the CORBA 2.x standard has the following limitations [53]:

1. Lack of functional boundaries. The CORBA 2.x object model treats all interfaces as client/server contracts. This object model does not, however, provide standard *assembly* mechanisms to decouple dependencies among collaborating object implementations. For example, objects whose implementations depend on other objects need to discover and connect to those objects explicitly. To build complex distributed applications, therefore, application developers must explicitly program the connections among interdependent services and object interfaces, which is extra work that can yield brittle and nonreusable implementations.

2. Lack of generic server standards. CORBA 2.x does not specify a generic server framework to perform common server configuration work, including initializing a server and its QoS policies, providing common services (such as notification or naming services), and managing the runtime environment of each component. Although CORBA 2.x standardized the interactions between object implementations and object request brokers (ORBs), server developers must still determine how (1) object implementations are installed in an ORB and (2) the ORB and object implementations interact. The lack of a generic component server standard yields tightly coupled, *ad hoc* server implementations, which increase the complexity of software upgrades and reduce the reusability and flexibility of CORBA-based applications.

3. Lack of software configuration and deployment standards. There is no standard way to distribute and start up object implementations remotely in CORBA 2.x specifications. Application administrators must therefore resort to in-house scripts and procedures to deliver software implementations to target machines, configure the target machine and software implementations for execution, and then instantiate software implementations to make them ready for clients. Moreover, software implementations are often modified to accommodate such *ad hoc* deployment mechanisms. The need of most reusable software implementations to interact with other software implementations and services further aggravates the problem. The lack of higher-level software management standards results in systems that are harder to maintain and software component implementations that are much harder to reuse.

6.3 Component Middleware: A Powerful Approach to Building DRE Applications

This section presents an overview of component middleware and the CORBA Component Model. It then discusses how conventional component middleware lacks support for the key QoS provisioning needs of DRE applications.

6.3.1 Overview of Component Middleware and the CORBA Component Model

Component middleware [50] is a class of middleware that enables reusable services to be composed, configured, and installed to create applications rapidly and robustly. Recently, component middleware has evolved to address the limitations of DOC middleware described in Section 6.2.2 by

- creating a virtual boundary around larger application component implementations that interact with each other only through well-defined interfaces,
- defining standard container mechanisms needed to execute components in generic component servers, and
- specifying the infrastructure to assemble, package, and deploy components throughout a distributed environment.

The CORBA Component Model (CCM) [26] is a current example of component middleware that addresses limitations with earlier generations of DOC middleware. The CCM specification extends the CORBA object model to support the concept of components and establishes standards for implementing, packaging, assembling, and deploying component implementations. From a client perspective, a CCM component is an extended CORBA object that encapsulates various interaction models via different interfaces and connection operations. From a server perspective, components are units of implementation that can be installed and instantiated independently in standard application server runtime environments stipulated by the CCM specification. Components are larger building blocks than objects, with more of their interactions managed to simplify and automate key aspects of construction, composition, and configuration into applications.

A *component* is an implementation entity that exposes a set of *ports*, which are named *interfaces* and connection points that components use to collaborate with each other. Ports include the following interfaces and connection points shown in Figure 6.2:

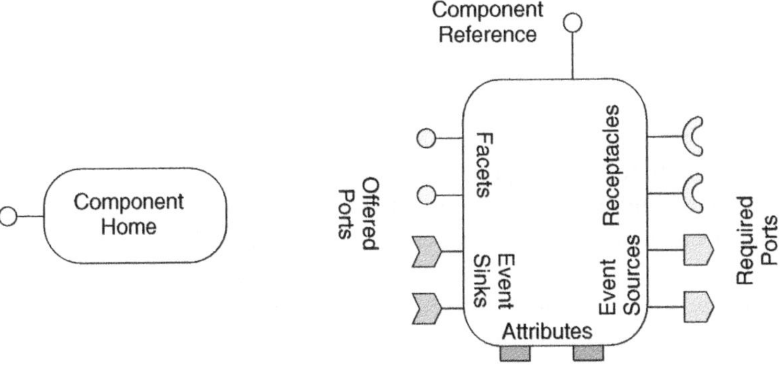

Figure 6.2 Client view of CCM components

- **Facets**, which define a named interface that services method invocations from other components synchronously.
- **Receptacles**, which provide named connection points to synchronous facets provided by other components.
- **Event sources/sinks**, which indicate a willingness to exchange event messages with other components asynchronously.

Components can also have *attributes* that specify named parameters that can be configured later via metadata specified in component property files.

Figure 6.3 shows the server-side view of the runtime architecture of the CCM model. A *container* provides the server runtime environment for component implementations called *executors*. It contains various predefined hooks and operations that give components access

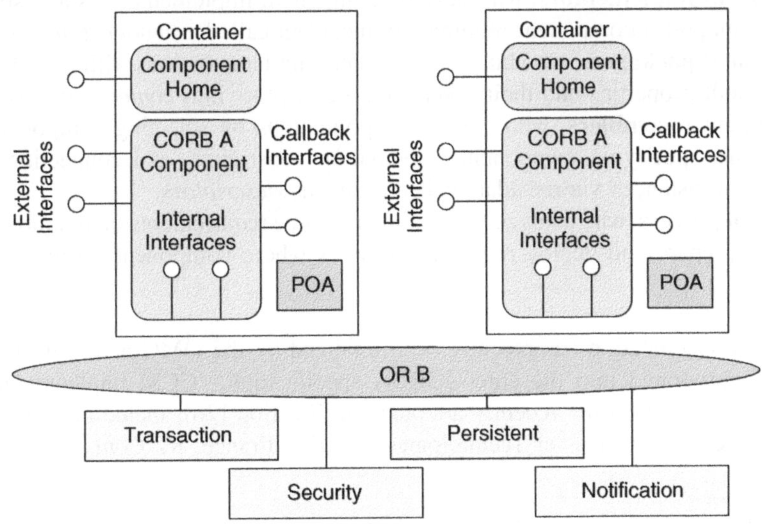

Figure 6.3 Overview of the CCM runtime architecture

to strategies and services, such as persistence, event notification, transaction, replication, load balancing, and security. Each container defines a collection of runtime strategies and policies, such as an event delivery strategy and component usage categories, and is responsible for initializing and providing runtime contexts for the managed components. Component implementations have associated metadata, written in XML, that specify the required container strategies and policies.

In addition to the building blocks outlined above, the CCM specification also standardizes various aspects of stages in the application development lifecycle, notably component implementation, packaging, assembly, and deployment as shown in Figure 6.4, where each stage of the lifecycle adds information pertaining to these aspects. The numbers in the discussion below correspond to the labels in Figure 6.4. The CCM Component Implementation Framework (CIF) (1) automatically generates component implementation skeletons and persistent state management mechanisms using the Component Implementation Definition Language (CIDL). CCM packaging tools (2) bundle implementations of a component with related XML-based component metadata. CCM assembly tools (3) use XML-based metadata to describe component compositions, including component locations and interconnections among components, needed to form an assembled application. Finally, CCM deployment tools (4) use the component assemblies and composition metadata to deploy and initialize applications.

The tools and mechanisms defined by CCM collaborate to address the limitations described in Section 6.2.2. The CCM programming paradigm separates the concerns of composing and provisioning reusable software components into the following development roles within the application lifecycle:

- **Component designers**, who define the component features by specifying what each component does and how components collaborate with each other and with their clients. Component designers determine the various types of ports that components offer and/or require.
- **Component implementors,** who develop component implementations and specify the runtime support a component requires via metadata called *component descriptors*.
- **Component packagers**, who bundle component implementations with metadata giving their default properties and their component descriptors into *component packages*.
- **Component assemblers**, who configure applications by selecting component implementations, specifying component instantiation constraints, and connecting ports of component instances via metadata called *assembly descriptors*.
- **System deployers**, who analyze the runtime resource requirements of assembly descriptors and prepare and deploy required resources where component assemblies can be realized.

The CCM specification has recently been finalized by the OMG and is in the process of being incorporated into the core CORBA specification.[1] CCM implementations are now available based on the recently adopted specification [26], including *OpenCCM* by the Universite des Sciences et Technologies de Lille, France, *K2 Containers* by iCMG,

[1] The CORBA 3.0 specification [28] released by the OMG only includes changes in IDL definition and Interface Repository changes from the Component specification.

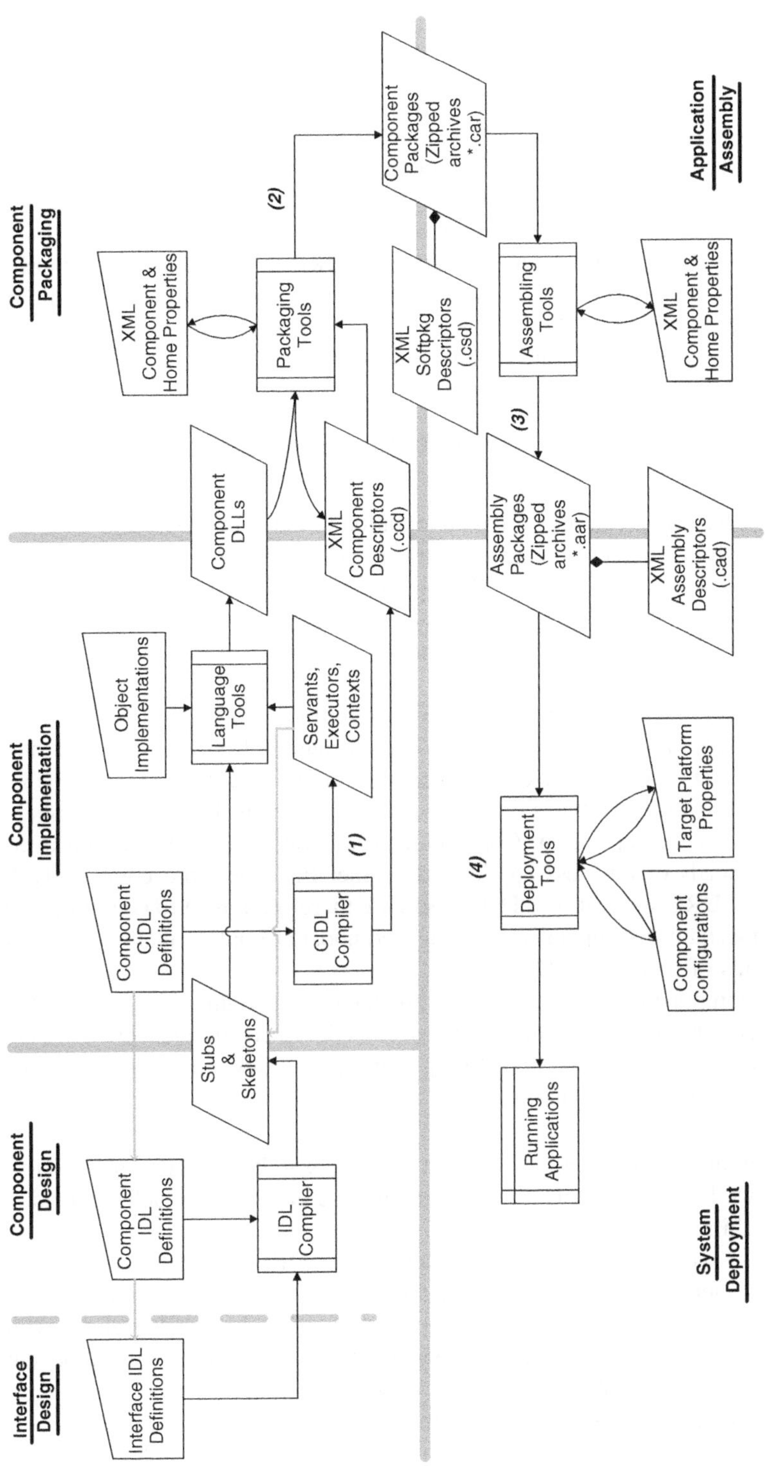

Figure 6.4 Overview of the CCM development lifecycle

MicoCCM by FPX, *Qedo* by Fokus, and *CIAO* by the DOC groups at Washington University in St Louis and the Institute for Software Integrated Systems (ISIS) at Vanderbilt University. The architectural patterns used in CCM [52] are also used in other popular component middleware technologies, such as J2EE [24]; [3] and .NET.

Among the existing component middleware technologies, CCM is the most suitable for DRE applications since the current base-level CORBA specification is the only standard COTS middleware that has made substantial progress in satisfying the QoS requirements of DRE systems. For example, the OMG has adopted several DRE-related specifications, including **Minimum CORBA**, **Real-time CORBA**, **CORBA Messaging**, and **Fault-tolerant CORBA**. These QoS specification and enforcement capabilities are essential for supporting DRE systems. Our work therefore focuses on CCM as the basis for developing QoS-enabled component models that are able to support DRE systems.

6.3.2 Limitations with Conventional Component Middleware for Large-scale DRE Systems

Large-scale DRE applications require seamless integration of many hardware and software systems. Figure 6.5 shows a representative air traffic control system that collects and processes real-time flight status from multiple regional radars. On the basis of the real-time flight data, the system then reschedules flights, issues air traffic control commands to airplanes in flight, notifies airports, and updates the displays in an airport's flight bulletin boards.

The types of systems shown in Figure 6.5 require complicated provisioning, where developers must connect numerous distributed or collocated subsystems together and define the functionality of each subsystem. Component middleware can reduce the software development effort for these types of large-scale DRE systems by enabling application development through composition. Conventional component middleware, however, is designed for the needs of business applications, rather than the more complex QoS provisioning needs of DRE applications. Developers are therefore forced to configure and control these QoS mechanisms imperatively in their component implementations to meet the real-time demands of their applications.

It is possible for component developers to take advantage of certain middleware or OS features to implement QoS-enabled components by embedding certain QoS provisioning code within a component implementation. Many QoS capabilities, however, cannot be implemented *solely* within a component because

- QoS provisioning must be done end-to-end, that is, it needs to be applied to many interacting components. Implementing QoS provisioning logic internally in each component hampers its reusability.
- Certain resources, such as thread pools in Real-time CORBA, can be provisioned only within a broader execution unit, that is, a component server rather than a component. Since component developers often have no *a priori* knowledge about with which other components a component implementation will collaborate, the component implementation is not the right level at which to provision QoS.
- Certain QoS assurance mechanisms, such as configuration of nonmultiplexed connections between components, affect component interconnections. Since a reusable component

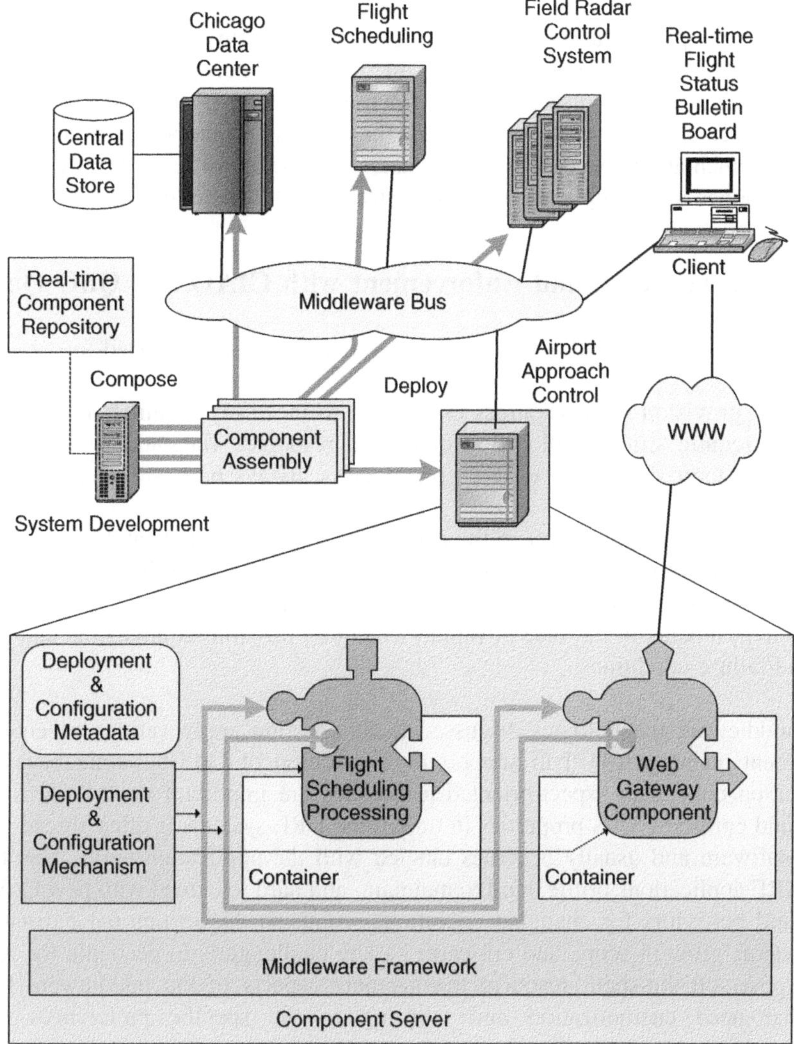

Figure 6.5 Integrating DRE applications with component middleware

implementation may not know how it will be composed with other components, it is not generally possible for component implementations to perform QoS provisioning in isolation.

- Many QoS provisioning policies and mechanisms require the installation of customized infrastructure modules to work correctly in meeting their requirements. However, some of the policies and mechanisms in support of controlled behaviors such as high throughput and low latency, may be inherently incompatible. It is hard for QoS provisioning mechanisms implemented within components to manage these incompatibilities without knowing the end-to-end QoS context *a priori*.

In general, isolating QoS provisioning functionality into each component prematurely commits every implementation to a specific QoS provisioning scenario in a DRE application's lifecycle. This tight coupling defeats one of the key benefits of component middleware: *separating component functionality from system management.* By creating dependencies between application components and the underlying component framework, component implementations become hard to reuse, particularly for large-scale DRE systems whose components and applications possess stringent QoS requirements.

6.4 QoS Provisioning and Enforcement with CIAO and QuO Qoskets

This section describes middleware technologies that we have developed to

1. statically provision QoS resources end-to-end to meet key requirements. Some DRE systems require strict preallocation of critical resources via static QoS provisioning for closed loop systems to ensure critical tasks always have sufficient resources to complete;
2. monitor and manage the QoS of the end-to-end functional application interactions;
3. enable the adaptive and reflective decision-making needed to dynamically provision QoS resources robustly and enforce the QoS requirements of applications for open loop environments in the face of rapidly changing mission requirements and environmental/failure conditions.

The middleware technologies discussed in this section apply various aspect-oriented development techniques [19] to support the separation of QoS systemic behavior and configuration concerns. Aspect-oriented techniques are important since code for provisioning and enforcing QoS properties in traditional DRE systems is often spread throughout the software and usually becomes tangled with the application logic. This tangling makes DRE applications brittle, hard to maintain, and hard to extend with new QoS mechanisms and behaviors for changing operational contexts. As Section 6.3.2 discusses, as DRE systems grow in scope and criticality, a key challenge is to decouple the reusable, multipurpose, off-the-shelf, resource management aspects of the middleware from aspects that need customization and tailoring to the specific preferences of each application.

On the basis of our experience in developing scores of large-scale research and production DRE systems over the past two decades, we have found that it is most effective to separate the programming and provisioning of QoS concerns along the two dimensions shown in Figure 6.6 and discussed below:

Functional paths, which are flows of information between client and remote server applications. The middleware is responsible for ensuring that this information is exchanged efficiently, predictably, scalably, dependably, and securely between remote nodes. The information itself is largely application-specific and determined by the functionality being provided (hence the term "functional path").

QoS systemic paths, which are responsible for determining how well the functional interactions behave end-to-end with respect to key DRE QoS properties, such as (1) when,

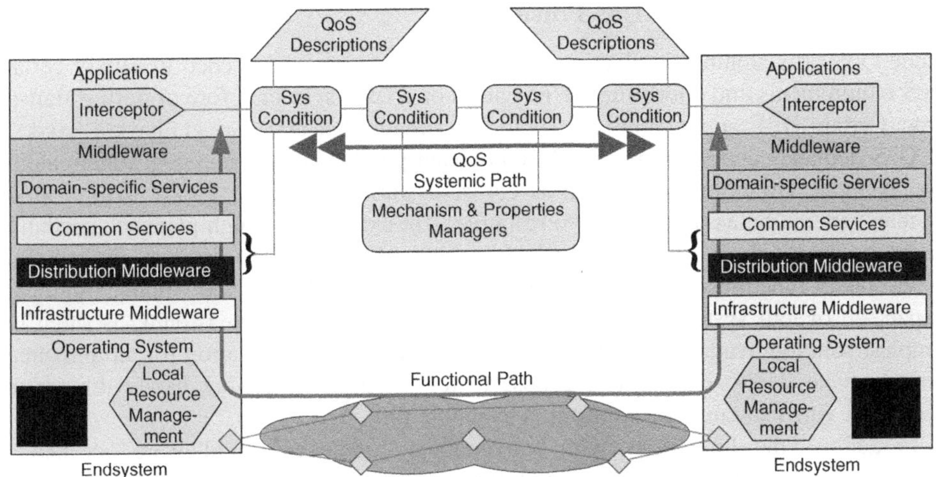

Figure 6.6 Decoupling the functional path from the systemic QoS path

how, and what resources are committed to client/server interactions at multiple levels of distributed systems, (2) the proper application and system behavior if available resources are less than expected, and (3) the failure detection and recovery strategies necessary to meet end-to-end dependability requirements.

In next generation of large-scale DRE systems, the middleware – rather than operating systems or networks alone – will be responsible for separating QoS systemic properties from functional application properties and coordinating the QoS of various DRE system and application resources end-to-end. The architecture shown in Figure 6.6 enables these properties and resources to change independently, for example, over different distributed system configurations for the same application.

The architecture in Figure 6.6 assumes that QoS systemic paths will be provisioned by a different set of *specialists* – such as systems engineers, administrators, operators, and possibly automated computing agents – and *tools* – such as MDA tools [16] – than those customarily responsible for programming functional paths in DRE systems. In conventional component middleware, such as CCM described in Section 6.3.1, there are multiple software development roles, such as component designers, assemblers, and packagers. QoS-enabled component middleware identifies yet another development role: the *Qosketeer* [57]. The Qosketeer is responsible for performing QoS provisioning, such as preallocating CPU resources, reserving network bandwidth/connections, and monitoring/enforcing the proper use of system resources at runtime to meet or exceed application and system QoS requirements.

6.4.1 Static Qos Provisioning via QoS-enabled Component Middleware and CIAO

Below, we present an overview of static QoS provisioning and illustrate how QoS-enabled component middleware, such as CIAO, facilitates the composition of static QoS provisioning into DRE applications.

6.4.1.1 Overview of Static QoS Provisioning

Static QoS provisioning involves predetermining the resources needed to satisfy certain QoS requirements and allocating the resources of a DRE system before or during start-up time. Certain DRE applications use static provisioning because they (1) have a fixed set of QoS demands and (2) require tightly bounded predictability for system functionality. For example, key commands (such as collision warning alerts to a pilot avionic flight systems) should be assured access to resources, for example, through planned scheduling of those operations or assigning them the highest priority [14]. In contrast, the handling of secondary functions, such as flight path calculation, can be delayed without significant impact on overall system functioning. In addition, static QoS provisioning is often the simplest solution available, for example, a video streaming application for an unmanned aerial vehicle (UAV) may simply choose to reserve a fixed network bandwidth for the audio and video streams, assuming it is available [40].

To address the limitations with conventional component middleware outlined in Section 6.3.2, it is necessary to make QoS provisioning specifications an integral part of component middleware and apply the aspect-oriented techniques to decouple QoS provisioning specifications from component functionality. This separation of concerns relieves application component developers from tangling the code to manage QoS resources within the component implementation. It also simplifies QoS provisioning that crosscut multiple interacting components to ensure proper end-to-end QoS behavior.

To perform QoS provisioning end-to-end throughout a component middleware system robustly, static QoS provisioning specifications should be decoupled from component implementations. Static QoS requirements should be specified instead using metadata associated with various application development lifecycles supported by the component middleware. This separation of concerns helps improve component reusability by preventing premature commitment to specific QoS provisioning parameters. QoS provisioning specifications can also affect different scopes of components and their behaviors, for example, thread-pools that are shared among multiple components, versus a priority level assigned to a single-component instance.

Different stages of the component software development lifecycle involve commitments to certain design decisions that affect certain scopes of a DRE application. To avoid constraining the development stage unnecessarily, it is therefore important to ensure that we do not commit to specific QoS provisioning prematurely. We review these stages in the component software development lifecycle below and identify the QoS specifications appropriate for each of these stages:

- QoS provisioning specifications that are part of a component implementation need to be specified in the metadata associated with the component implementation, that is, component descriptors. Examples of QoS-supporting features include (1) a real-time ORB on which a component implementation depends, or (2) QoS provisioning parameters, such as the required priority model that a component implementation elects to make explicitly configurable.
- Configurable specifications of a component, such as its priority model and priority level, can be assigned default values in the component package by associating a component implementation with component property files as described in Section 6.3.1.

- Resources, such as thread-pools and prioritized communication channels, that need to be shared among multiple components in a component server should be allocated separately as logical resource policies. Component instances within the same component server can then share these logical resources by associating them with common logical resource policies in the application assembly metadata described in Section 6.3.1.
- Component assembly metadata must also be extended to provision QoS resources for component interconnections, such as defining private connections, performing network QoS reservations, or establishing preconnections to minimize runtime latencies.
- Generic component servers provide the execution environment for components, as described in Section 6.3.1. To ensure a component server is configured with the mechanisms and resources needed to support the specified QoS requirements, deployment tools should be extended to include middleware modules that can configure the component servers. Examples include customized communication mechanisms and custom priority mappings.

Figure 6.7 illustrates several types of static QoS provisioning that are common in DRE applications, including

1. **CPU resources**, which need to be allocated to various competing tasks in a DRE system to make sure these tasks finish on time
2. **Communication resources**, which the middleware uses to pass messages around to "connect" various components in a distributed system together
3. **Distributed middleware configurations**, which are plug-ins that a middleware framework uses to realize QoS assurance.

6.4.1.2 Static QoS Provisioning with CIAO

Figure 6.8 shows the key elements of the Component-Integrated ACE ORB (CIAO), which is a QoS-enabled implementation of CCM being developed at Washington University, St Louis, and the Institute for Software Integrated Systems (ISIS) at Vanderbilt University. CIAO extends the The ACE ORB (TAO) [21] to support components and

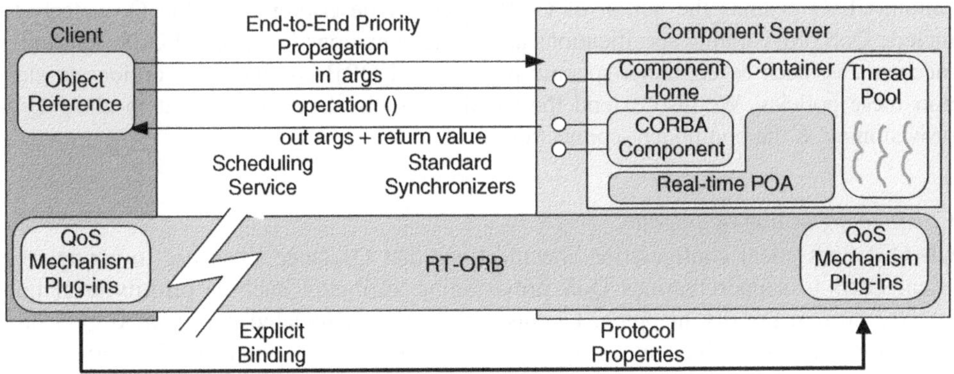

Figure 6.7 Examples of static QoS provisioning

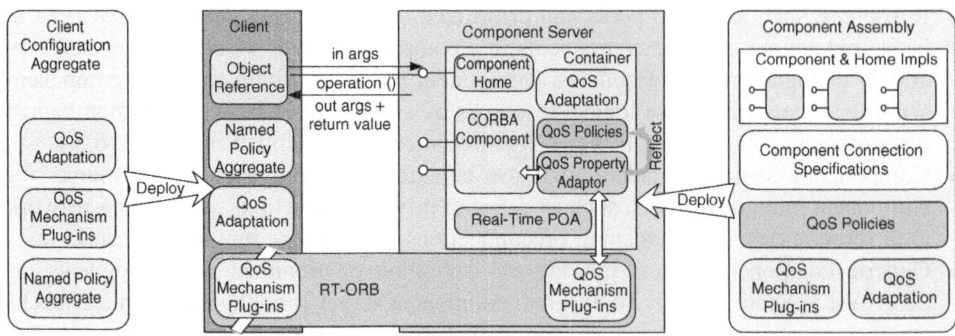

Figure 6.8 Key elements in CIAO

simplify the development of DRE applications by enabling developers to declaratively provision QoS policies end-to-end when assembling a system. TAO is an open-source, high-performance, highly configurable Real-time CORBA ORB that implements key patterns [45] to meet the demanding QoS requirements of DRE systems. TAO supports the standard OMG CORBA reference model [30] and Real-time CORBA specification [27], with enhancements designed to ensure efficient, predictable, and scalable QoS behavior for high-performance and real-time applications.

TAO is developed atop lower-level middleware called ACE [43, 44], which implements core concurrency and distribution patterns [45] for communication software. ACE provides reusable C++ wrapper facades and framework components that support the QoS requirements of high-performance, real-time applications. ACE and TAO run on a wide range of OS platforms, including Windows, most versions of UNIX, and real-time operating systems such as Sun/Chorus ClassiX, LynxOS, and VxWorks.

To support the role of the Qosketeer, CIAO extends CCM to support static QoS provisioning as follows:

Application assembly descriptor.
A CCM *assembly descriptor* specifies how components are composed to form an application. CIAO extends the notion of CCM assembly descriptor to enable Qosketeers to include QoS provisioning specifications and implementations for required QoS supporting mechanisms, such as container-specific policies and ORB configuration options to support these policies. We also extend the CCM assembly descriptor format to allow QoS provisioning at the component-connection level.

Client configuration aggregates.
CIAO defines client configuration specifications that Qosketeers can use to configure a client ORB to support various QoS provisioning attributes, such as priority-level policy and custom priority mapping. Clients can then be associated with named QoS provisioning policies (defined in an aggregate) that are used to interact with servers and

provide end-to-end QoS assurance. CIAO enables the transparent installation of client configuration aggregates into a client ORB.

QoS-enabled containers.
CIAO enhances CCM containers to support QoS capabilities, such as various server-specified Real-time CORBA policies. These QoS-enabled containers provide the central ORB interface that Qosketeers can use to provision component QoS policies and interact with ORB endsystem QoS assurance mechanisms, such as Real-time POA and ORB proprietary internal interfaces, required by the QoS policies.

QoS adaptation.
CIAO also supports installation of meta-programming hooks, such as portable interceptors, servant managers, smart proxies, and skeletons, that can be used to inject dynamic QoS provisioning transparently. Section 6.4.2 describes these QoS adaptation mechanisms in detail.

To support the capabilities described above, CIAO extends the CCM metadata framework so that developers can bind the QoS requirements and the supporting mechanisms during various stages of the development lifecycle, including component implementation, component packaging, application assembly, and application deployment. These capabilities enable CIAO to statically provision the types of QoS resources outlined in Section 6.4.1.1 as follows:

CPU resources.
These policies specify how to allocate CPU resources when running certain tasks, for example, by configuring the priority models and priority levels of component instances.

Communication resources.
These policies specify strategies for reserving and allocating communication resources for component connections, for example, an assembly can request a private connection between two critical components in the system and reserve bandwidth for the connection using the RSVP protocol [40].

Distributed middleware configuration.
These policies specify the required software modules that control the QoS mechanisms for:

- **ORB configurations.** To enable the configuration of higher-level policies, CIAO needs to know how to support the required functionality, such as installing and configuring customized communication protocols.
- **Meta-programming mechanisms.** Software modules, such as those developed with the QuO Qosket middleware described in Section 6.4.2 that implement dynamic QoS provisioning and adaptation, can be installed statically at system composition time via meta-programming mechanisms, such as smart proxies and interceptors [55].

In summary, DRE application developers can use CIAO at various points of the development cycle to (1) decouple QoS provisioning functionality from component implementations and (2) compose static QoS provisioning capabilities into the application via the component assembly and deployment phases. This approach enables maximum reusability of components and robust composition of DRE applications with diverse QoS requirements.

6.4.2 Dynamic QoS Provisioning via QuO Adaptive Middleware and Qoskets

Section 6.4.1 described the static QoS provisioning capabilities provided by CIAO. We now present an overview of dynamic QoS provisioning and describe how the QuO Qosket middleware framework [39] can be used to manage dynamic QoS provisioning for DRE applications.

6.4.2.1 Overview of Dynamic QoS Provisioning

Dynamic QoS provisioning involves the allocation and management of resources at runtime to satisfy application QoS requirements. Certain events, such as fluctuations in resource availability due to temporary overload, changes in QoS requirements, or reduced capacity due to failures or external attacks, can trigger reevaluation and reallocation of resources. Dynamic QoS provisioning requires the following middleware capabilities:

- To detect changes in available resources, middleware must *monitor* DRE system status to determine if reallocations are required. For example, network bandwidth is often shared by multiple applications on computers. Middleware for bandwidth-sensitive applications, such as video conferencing or image processing, is therefore responsible for determining changes in total available bandwidth and bandwidth usable by a specific application.
- To *adapt* to the change by adjusting the use of resources required by an application, when available resources change. For instance, middleware that supports a video conferencing application [40] may choose to (1) lower the resolution temporarily when there is less available network bandwidth to support the original resolution and (2) switch back to the higher resolution when sufficient bandwidth becomes available. Other examples (such as changing the degree of fault tolerance, substituting a simple text interface in place of imagery intensive applications with short real-time constraints, or temporarily ceasing operation in deference to higher priority activities) illustrate the breadth of DRE application adaptive behavior.

Although applications can often implement dynamic QoS provisioning functionality themselves using conventional middleware, they must work around the current structures intended to simplify the development of DRE systems, which often becomes counter productive. Such *ad hoc* approaches lead to nonportable code that depends on specific OS features, tangled implementations that are tightly coupled with the application software, and other problems that make it hard to adapt the application to changing requirements, and even harder to reuse an implementation. It is therefore essential to separate the functionality of dynamic QoS provisioning from both the lower-level distribution middleware *and* the application functionality.

Figure 6.9 illustrates the types of dynamic QoS provisioning abstractions and mechanisms that are necessary in large-scale DRE systems:

1. **Design formalisms** to specify the level of service desired by a client, the level of service a component expects to provide, operating regions interpreting the ranges of possible measured QoS, and actions to take when the level of QoS changes.
2. **Runtime capabilities** to adapt application behavior on the basis of the current state of QoS in the system.
3. **Management mechanisms** that keep track of the resources in the protocol and middleware infrastructure that need to be measured and controlled dynamically to meet application requirements and mediate total demand on system resources across and between the platforms.

6.4.2.2 Overview of QuO

The Quality Objects (QuO) [57] framework is adaptive middleware developed by BBN Technologies that allows DRE system developers to use aspect-oriented software

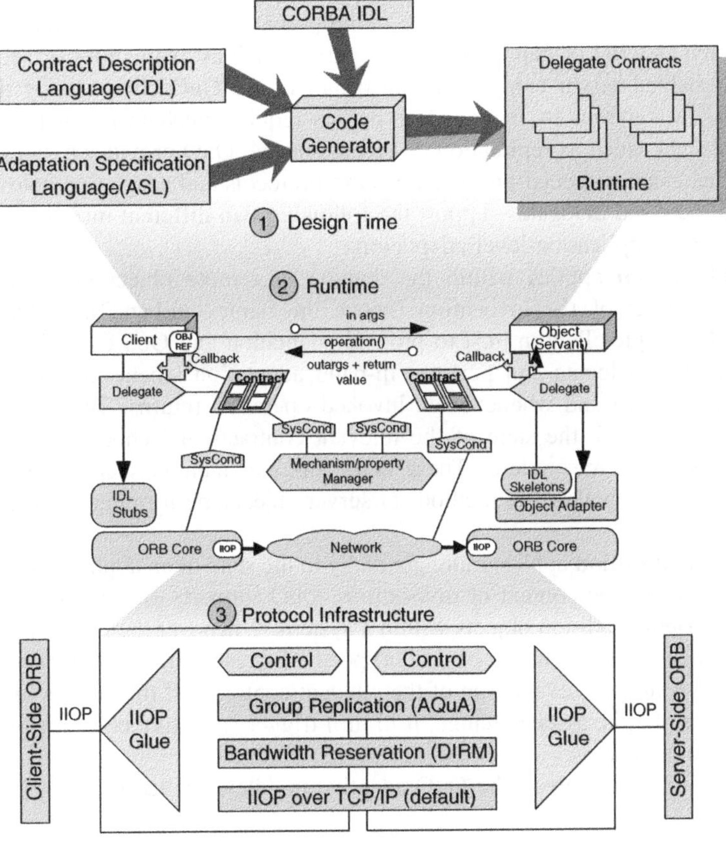

Figure 6.9 Examples of dynamic QoS provisioning

development [19] techniques to separate the concerns of QoS programming from "business" logic in DRE applications. The QuO framework allows DRE developers to specify (1) their QoS requirements, (2) the system elements that must be monitored and controlled to measure and provide QoS, and (3) the behavior for adapting to QoS variations that occur at runtime.

The runtime architecture shown in Figure 6.9 illustrates how the elements in QuO support the following dynamic QoS provisioning needs:

- **Contracts** specify the level of service desired by a client, the level of service a component expects to provide, operating regions indicating possible measured QoS, and actions to take when the level of QoS changes.
- **Delegates** act as local proxies for remote components. Each delegate provides an interface similar to that of the remote component stub, but adds locally adaptive behavior on the basis of the current state of QoS in the system, as measured by the contract.
- **System Condition objects** provide interfaces to resources, mechanisms, and ORBs in the DRE system that need to be measured and controlled by QuO contracts.

QuO applications can also use resource or property managers that manage given QoS resources such as CPU or bandwidth, or properties such as availability or security, for a set of QoS-enabled server components on behalf of the QuO clients using those server components. In some cases, managed properties require mechanisms at lower levels in the protocol stack, such as replication or access control. QuO provides a gateway mechanism [42] that enables special-purpose transport protocols and adaptation below the ORB.

QuO contracts and delegates support the following two different models for triggering middleware- and application-level adaptation:

In-band adaptation applies within the context of remote object invocations, and is a direct side effect of the invocation (hence the name "in-band"). A QuO delegate applies the Interceptor Pattern [45] to provide adaptation and QoS control for the object access. The QuO delegate can perform in-band adaptation whenever a client makes a remote operation call and whenever an invoked operation returns. The delegates on the client and server check the state of the relevant contracts and choose behaviors on the basis of the state of the system. These behaviors can include shaping or filtering the method data, choosing alternate methods or server objects, performing local functionality, and so on.

Out-of-band adaptation applies more generally to the state of an application, a system, or a subsystem outside the context of invocations. QuO supports out-of-band adaptation by monitoring system condition objects within a system. Whenever the monitored conditions change (or whenever they change beyond a specified threshold), a system condition object triggers an asynchronous evaluation of the relevant contracts. If this results in a change in contract region, that is, a state change, it in turn triggers adaptive behavior asynchronous to any object interactions.

For more information about the QuO adaptive middleware, see [57]; [42]; [22]; [51].

6.4.2.3 Qoskets: QuO Support for Reusing Systemic Behavior

One goal of QuO is to separate the role of a systemic QoS programmer from that of an application programmer. A complementary goal of this separation of programming roles is that systemic QoS behaviors can be encapsulated into reusable units that are not only developed separately from the applications that use them but that can be reused by selecting, customizing, and binding them to new application programs. To support this goal, we have defined a *Qosket* [39] as a unit of encapsulation and reuse of systemic behavior in QuO applications. A Qosket is each of the following, simultaneously:

- **A collection of crosscutting implementations**, that is, a Qosket is a set of QoS specifications and implementations that are woven throughout a DRE application and its constituent components to monitor and control QoS and systemic adaptation.
- **A packaging of behavior and policy**, that is, a Qosket generally encapsulates elements of an adaptive QoS behavior and a policy for using that behavior, in the form of contracts, measurements, and code to provide adaptive behavior.
- **A unit of behavior reuse**, largely focused on a single property, that is, a Qosket can be used in multiple DRE applications, or in multiple ways within a single application, but typically deals with a single property (*for example*, performance, dependability, or security).

Qoskets are an initial step towards individual behavior packaging and reuse, as well as a significant step toward the more desirable (and much more complex) ability to compose behaviors within an application context. They are also a means toward the larger goal of flexible design trade-offs at runtime among properties (such as real-time performance, dependability, and security) that vary with the current operating conditions. Qoskets are used to bundle in one place all of the specifications for controlling systemic behavior, independent of the application in which the behavior might end up being used.

In practice, a Qosket is a collection of the interfaces, contracts, system condition objects, callback components, unspecialized adaptive behavior, and implementation code associated with a reusable piece of systemic behavior. A Qosket is fully specified by defining the following:

1. The contracts, system condition objects, and callback components it encapsulates
2. The Adaptation Specification Language (ASL) template code, defining partial specifications of adaptive behavior,
3. Implementation code for instantiating the Qosket's encapsulated components, for initializing the Qosket, and for implementing the Qosket's defined systemic measurement, control, and adaptation, and
4. The interfaces that the Qosket exposes.

The general structure of Qoskets, including modules they encapsulate and interfaces they expose, is illustrated in Figure 6.10. The two interfaces that Qoskets expose correspond to these two use cases:

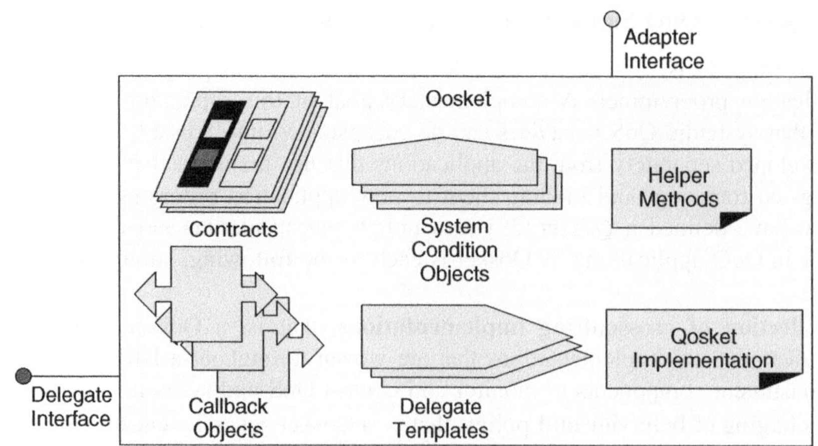

Figure 6.10 Qoskets encapsulate QuO components into reusable behaviors

- **The adapter interface**, which is an application programming interface. This interface provides access to QoS measurement, control, and adaptation features in the Qosket (such as the system condition objects, contracts, and so forth) so that they can be used anywhere in an application.
- **The delegate interface**, which is an interface to the in-band method adaptation code. In-band adaptive behaviors of delegates are specified in the QuO ASL language. The adaptation strategies of the delegate are encapsulated and woven into the application using code generation techniques.

6.4.3 Integrated QoS provisioning via CIAO and Qoskets

As discussed in Section 6.4.2.3, Qoskets provide abstractions for dynamic QoS provisioning and adaptive behaviors. The current implementation of Qoskets in QuO, however, requires application developers to modify their application code manually to "plug in" adaptive behaviors into existing applications. Rather than retrofitting DRE applications to use Qosket-specific interfaces, it would be more desirable to use existing and emerging COTS component technologies and standards to encapsulate QoS management, both static and dynamic.

Conversely, CIAO allows system developers to compose static QoS provisioning, adaptation behaviors, and middleware support for QoS resources allocating and managing mechanisms into DRE applications transparently, as depicted in Section 6.4.1.2. CIAO did not, however, initially provide an abstraction to model, define, and specify dynamic QoS provisioning. We are therefore leveraging CIAO's capability to configure Qoskets transparently into component servers to provide an integrated QoS provisioning solution, which enables the composition of both static and dynamic QoS provisioning into DRE applications.

The static QoS provisioning mechanisms in CIAO enable the composition of Qoskets into applications as part of component assemblies. As in the case of provisioning static QoS policies, developers can compose and weave together Qosket modules of different

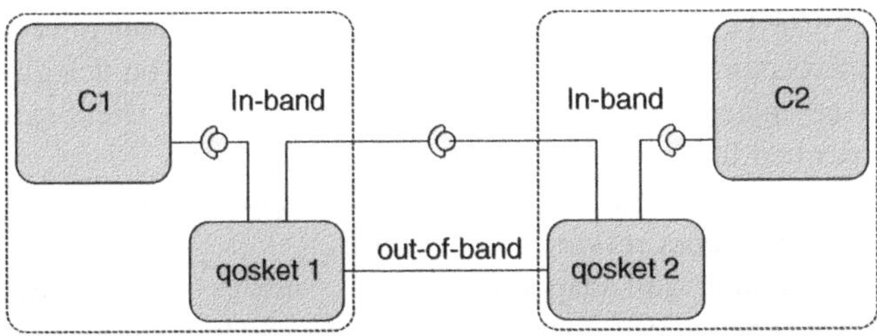

Figure 6.11 Composing Qosket components in an application

granularities into component-based applications at various stages of the development lifecycle. The most straightforward approach is to implement Qosket behaviors as CCM components, as shown in Figure 6.11. Each Qosket component offers interception facets and receptacles that can be used to provide in-band adaptation along the functional path. Qosket components can also provide other ports to support out-of-band adaptation and interact with related QoS management mechanisms.

Implementing Qoskets as CCM components takes advantage of the CCM assembly framework and is thus suitable for prototyping a dynamic QoS provisioning capability. It requires modifying the application assembly directly, however, and must therefore be woven into an application at application composition time. Inserting Qoskets as conventional CCM components requires the use of regular CCM interfaces to define interception facets and receptacles. This exposure could in theory impose significant overhead in network communication because it permits Qosket delegate components to be deployed remotely from the targeting components, which defeats the purpose of using Qoskets to closely manage local behavior to achieve end-to-end QoS management.

To resolve these concerns, CIAO provides meta-mechanisms to weave in Qosket delegates and integrate Qosket modules into a CCM application. As shown in Figure 6.12, CIAO can install a Qosket using the following mechanisms:

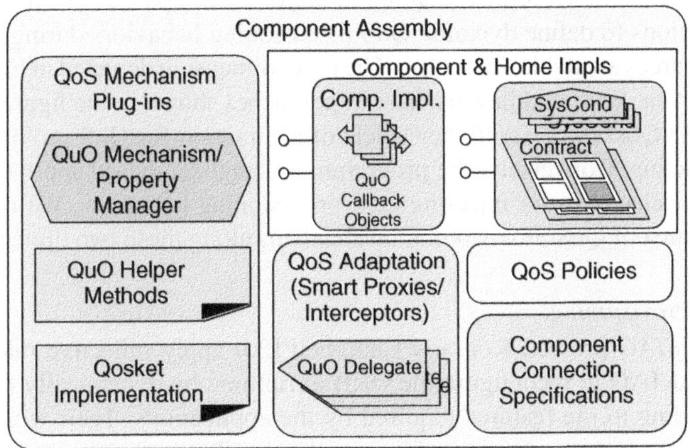

Figure 6.12 Composing a Qosket using CIAO

- QuO delegates can be implemented as smart proxies [55] and injected into components using interceptors by providing hooks in containers. The delegate configuration metadata can be injected into assembly descriptors or the client-side configuration aggregates described in Section 6.4.1.2.
- Developers can specify a Qosket-specific ORB configuration and assemble QoS mechanisms into the component server or client ORB.
- Out-of-band provisioning and adaptation modules, such as contracts, system conditions, and callback components, can continue to be implemented and assembled as separate CCM components into servers.
- Interactions between delegates and system condition objects can then be specified by additional assembly descriptors that, for example, connect the delegate metadata with the packaging-time metadata for the system condition components.

The approach described above requires support from the component middleware, but provides better control over the integration of Qosket modules. Since this approach does not require the modification of application assembly at a *functional* level, it more thoroughly decouples QoS management from the application logic written by component developers.

Although using CIAO to compose Qoskets into component assemblies simplifies retrofitting, a significant problem remains: *component crosscutting*. Qoskets are useful for separating concerns between systemic QoS properties and application logic, as well as implementing limited crosscutting between a single client/component pair. Neither Qoskets nor CIAO yet provide the ability to crosscut application components, however. Many QoS-related adaptations will need to modify the behavior of several components at once, likely in a distributed way. Some form of dynamic aspect-oriented programming might be used in this context, which is an area of ongoing research [31].

6.5 Related Work

This section reviews work on QoS provisioning mechanisms using the taxonomy shown in Figure 6.13. One dimension depicted in Figure 6.13 is *when QoS provisioning is performed*, that is, static versus dynamic QoS provisioning, as described in Section 6.1. Some enabling mechanisms allow static QoS provisioning before the start-up of a system, whereas others provide abstractions to define dynamic QoS provisioning behaviors during runtime on the basis of the resources available at the time. The other dimension depicted in Figure 6.13 is the *level of abstraction*. Both middleware-based approaches shown in the figure, that is, CIAO and BBN's QuO Qoskets, offer higher levels of abstraction for QoS provisioning specification and modeling. Conversely, the programming language-based approach offers metaprogramming mechanisms for injecting QoS provisioning behaviors. We review previous research in the area of QoS provisioning mechanisms along these two dimensions.

Dynamic QoS Provisioning.

In their *dynamicTAO* project, Kon and Campbell [20] apply reflective middleware techniques to extend TAO to reconfigure the ORB at runtime by dynamically linking selected modules, according to the features required by the applications. Their work falls into the same category as *Qoskets* (shown in Figure 6.13), in that both provide the mechanisms for realizing *dynamic* QoS provisioning at the middleware level. Qoskets offer a more

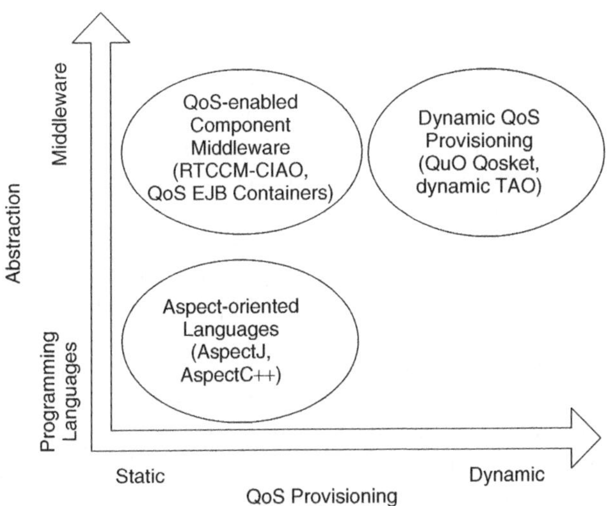

Figure 6.13 Taxonomy of QoS provisioning enabling mechanisms

comprehensive QoS provisioning abstraction, however, whereas Kon and Campbell's work concentrates on configuring middleware capabilities.

Moreover, although Kon and Campbell's work can also provide QoS adaptation behavior by dynamically (re)configuring the middleware framework, their research may not be as suitable for DRE applications, since dynamic loading and unloading of ORB components can incur significant and unpredictable overheads and thus prevent the ORB from meeting application deadlines. Our work on CIAO relies upon Model Driven Architecture (MDA) tools [16] to analyze the required ORB components and their configurations. This approach ensures that the ORB in a component server contains only the required components, without compromising end-to-end predictability.

QoS-enabled Component Middleware.
Middleware can apply the Quality Connector pattern [8] to meta-programming techniques for specifying the QoS behaviors and configuring the supporting mechanisms for these QoS behaviors. The container architecture in component-based middleware frameworks provides the vehicle for applying meta-programming techniques for QoS assurance control in component middleware, as previously identified in [54]. Containers can also help apply aspect-oriented software development [19] techniques to plug in different systemic behaviors [7]. These projects are similar to CIAO in that they provide a mechanism to inject "aspects" into applications statically at the middleware level.

Miguel de Miguel further develops the work on QoS-enabled containers by extending a QoS EJB container interface to support a `QoSContext` interface that allows the exchange of QoS-related information among component instances [10]. To take advantage of the QoS container, a component must implement `QoSBean` and `QoSNegotiation` interfaces. This requirement, however, adds an unnecessary dependency to component implementations. Section 6.3.2 examines the limitations of implementing QoS behavior logic in component implementations.

QoS-Enabled Distributed Objects (Qedo).
The Qedo project [12] is another ongoing effort to make QoS support an integral part of CCM. Qedo targets applications in the telecommunication domain and supports information streaming. It defines a metamodel that defines multiple categories of QoS requirements for applications. To support the modeled QoS requirements, Qedo defines extensions to CCM's container interface and the Component Implementation Framework (CIF) to realize the QoS models [36].

Similar to QuO's Contract Definition Language (CDL), Qedo's contract metamodel provides mechanisms to formalize and abstract QoS requirements. QuO (Qosket) contracts are more versatile, however, because they not only provide high levels of abstraction for QoS status but also define actions that should take place when state transitions occur in contracts. Qedo's extensions to container interfaces and the CIF also require component implementations to interact with the container QoS interface and negotiate the level of QoS contract directly. While this approach is suitable for certain applications where QoS is part of the functional requirements, it inevitably tightly couples the QoS provisioning and adaptation behaviors into the component implementation, and thus hampers the reusability of component. In comparison, our approach explicitly avoids this coupling and tries to *compose* the QoS provision behaviors into the component systems.

Aspect-Oriented Programming Languages.
Aspect-oriented programming (AOP) [19] languages provide language-level abstractions for weaving different aspects that crosscut multiple layers of a system. Examples of AOP languages include AspectJ [18] and Aspect $C++$ [32]. Similar to AOP, CIAO supports injection of aspects into systems at the middleware level using meta-programming techniques. Both CIAO and AOP weave aspects statically, that is, before program execution, and neither defines an abstraction for dynamic QoS provisioning behaviors. In contrast, Qoskets use AOP techniques dynamically to organize and connect the various dimensions of the QoS middleware abstractions with each other and with the program to which the behavior is being attached.

6.6 Concluding Remarks

DRE applications constitute an increasingly important domain that requires stringent support for multiple simultaneous QoS properties, such as predictability, latency, and dependability. To meet these requirements, DRE applications have historically been custom-programmed to implement their QoS provisioning needs, making them so expensive to build and maintain that they cannot adapt readily to meet new functional needs, different QoS provisioning strategies, hardware/software technology innovations, or market opportunities. This approach is increasingly infeasible, since the tight coupling between custom DRE software modules increases the time and effort required to develop and evolve DRE software. Moreover, QoS provisioning crosscuts multiple layers in applications and requires end-to-end enforcement that makes DRE applications even harder to develop, maintain, and adapt.

One way to address these coupling issues is by refactoring common application logic into *object-oriented application frameworks* [17]. This solution has limitations, however, since application objects can still interact directly with each other, which encourages tight coupling. Moreover, framework-specific bookkeeping code is also required within the applications to manage the framework, which can also tightly couple applications to the framework they are developed upon. It becomes nontrivial to reuse application objects and port them to different frameworks.

Component middleware [50] has emerged as a promising solution to many of the known limitations with object-oriented middleware and application frameworks. Component middleware supports a higher-level packaging of reusable software artifacts that can be distributed or collocated throughout a network. Existing component middleware, however, does not yet address the end-to-end QoS provisioning needs of DRE applications, which transcend component boundaries. QoS-enabled middleware is therefore necessary to separate end-to-end QoS provisioning concerns from application functional concerns.

This chapter describes how the Component Integrated ACE ORB (CIAO) middleware developed by Washington University in St Louis and the Institute for Software Integrated Systems at Vanderbilt University is enhancing the standard CCM specification to support static QoS provisioning by preallocating resources for DRE applications. We also describe how BBN's QuO Qosket middleware framework provides powerful abstractions that help define and implement reusable dynamic QoS provisioning behaviors. By combining QuO Qoskets with CIAO, we are developing an integrated end-to-end QoS provisioning solution for DRE applications.

When augmented with Model Driven Middleware tools, such as CoSMIC [16], QoS-enabled component middleware and applications can be provisioned more effectively and efficiently at even higher levels of abstraction. CIAO, QuO Qoskets, and CoSMIC are open-source software that is currently available and can be obtained from www.dre.vanderbilt.edu/CIAO/, quo.bbn.com/, and www.dre.vanderbilt.edu/cosmic, respectively.

Bibliography

[1] Accetta, M., Baron, R., Golub, D., Rashid, R., Tevanian, A., and Young, M. (1986) Mach: A New Kernel Foundation for UNIX Development. *Proceedings of the Summer 1986 USENIX Technical Conference and Exhibition*, Atlanta, GA, pp. 93–112.

[2] Alan, B. and Andy, W. (2001) *Real-Time Systems and Programming Languages, 3rd ed., Addison Wesley Longmain*, Reading, MA.

[3] Alur, D., Crupi, J., and Malks, D. (2001) *Core J2EE Patterns: Best Practices and Design Strategies*, Prentice Hall.

[4] Box, D. (1998) *Essential COM*, Addison-Wesley, Reading, MA.

[5] Buschmann, F., Meunier, R., Rohnert, H., Sommerlad, P., and Stal, M. (1996) *Pattern-Oriented Software Architecture—A System of Patterns*, Wiley & Sons, New York.

[6] Cemal, Y., Adam, P., and Schmidt, D. C. (2003) Distributed Continuous Quality Assurance: The Skoll Project *Workshop on Remote Analysis and Measurement of Software Systems (RAMSS)* IEEE/ACM, Portland, OR.

[7] Conan, D., Putrycz, E., Farcet, N., and DeMiguel, M. (2001) Integration of Non-Functional Properties in Containers. *Proceedings of the Sixth International Workshop on Component-Oriented Programming (WCOP).*

[8] Cross, J. K. and Schmidt, D. C. (2002) Applying the Quality Connector Pattern to Optimize Distributed Real-time and Embedded Middleware, in *Patterns and*

Skeletons for Distributed and Parallel Computing (eds F. Rabhi and S. Gorlatch), Springer-Verlag.

[9] Davies, D., Holler, E., Jensen, E., Kimbleton, S., Lampson, B., LeLann, G., Thurber, K., and Watson, R. (1981) *Distributed Systems- Architecture and Implementation – An Advanced Course*, Springer-Verlag.

[10] de Miguel, M. A. (2002) QoS-Aware Component Frameworks. *The* 10th *International Workshop on Quality of Service (IWQoS 2002)*, Miami Beach, FL.

[11] Schmidt, D. C. and Frank B. (2003) Patterns, Frameworks, and Middleware: Their Synergistic Relationships, *International Conference on Software Engineering (ICSE)*, IEEE/ACM, Portland, OR, pp. 694–704.

[12] FOKUS n.d, (2003) Qedo Project Homepage http://qedo.berlios.de/.

[13] Gamma, E., Helm, R., Johnson, R., and Vlissides, J. (1995) *Design Patterns: Elements of Reusable Object-Oriented Software*, Addison-Wesley, Reading, MA.

[14] Gill, C., Schmidt, D. C., and Cytron, R. (2003) Multi-Paradigm Scheduling for Distributed Real-Time Embedded Computing. *IEEE Proceedings, Special Issue on Modeling and Design of Embedded Software.*

[15] Gokhale, A., Schmidt, D. C., Natarajan, B., and Wang, N. (2002) Applying Model-Integrated Computing to Component Middleware and Enterprise Applications. *The Communications of the ACM Special Issue on Enterprise Components, Service and Business Rules.*

[16] Gokhale, A., Schmidt, D. C., Natarajan, B., Gray, J., and Wang, N. (2003) Model Driven Middleware, in *Middleware for Communications* (ed. Q. Mahmoud), Wiley & Sons, New York.

[17] Johnson, R. (1997) Frameworks = Patterns + Components. *Communications of the ACM*, **40**(10), 39–42.

[18] Kiczales, G., Hilsdale, E., Hugunin, J., Kersten, M., Palm, J., and Griswold, W. G. (2001) An Overview of AspectJ. *Lecture Notes in Computer Science*, **2072**, 327–355.

[19] Kiczales, G., Lamping, J., Mendhekar, A., Maeda, C., Lopes, C. V., Loingtier, J. M., and Irwin, J. (1997) Aspect-Oriented Programming. *Proceedings of the 11th European Conference on Object-Oriented Programming.*

[20] Kon, F., Costa, F., Blair, G., and Campbell, R. H. (2002) The Case for Reflective Middleware. *Communications of the ACM*, **45**(6), 33–38.

[21] Krishna, A. S., Schmidt, D. C., Klefstad, R., and Corsaro, A. (2003) Real-time Middleware, in *Middleware for Communications* (ed. Q. Mahmoud), Wiley & Sons, New York.

[22] Loyall, J. P., Bakken, D. E., Schantz, R. E., Zinky, J. A., Karr, D., Vanegas, R., and Anderson, K. R. (1998) QoS Aspect Languages and Their Runtime Integration. *Proceedings of the Fourth Workshop on Languages, Compilers and Runtime Systems for Scalable Components (LCR98)*, pp. 28–30.

[23] Luis, I., José, M. T., and Antonio, V. (2002) Selecting Software Components with Multiple Interfaces, *Proceedings of the* 28th *Euromicro Conference (EUROMICRO'02)*, IEEE, Dortmund, Germany, pp. 26-32.

[24] Marinescu, F. and Roman, E. (2002) *EJB Design Patterns: Advanced Patterns, Processes, and Idioms*, John Wiley & Sons, New York.

[25] Morgenthal, J. P. (1999) Microsoft COM+ Will Challenge Application Server Market, www.microsoft.com/com/wpaper/complus-appserv.asp.

[26] Obj (2002a) *CORBA Components* OMG Document formal/2002-06-65 edn.

[27] Obj (2002b) *Real-time CORBA Specification* OMG Document formal/02-08-02 edn.

[28] Obj (2002c) *The Common Object Request Broker: Architecture and Specification* 3.0.2 edn.

[29] Obj 2002d *The Common Object Request Broker: Architecture and Specification* 2.6.1 edn.

[30] Obj (2002e) *The Common Object Request Broker: Architecture and Specification* 3.0.2 edn.

[31] Office DIE n.d, (2003) Program Composition for Embedded Systems (PCES) www.darpa.mil/ixo/.

[32] Olaf, S., Andreas, G., and Schröder-Preikschat, W. (2002) Aspect C++: An Aspect-Oriented Extension to C++. *Proceedings of the 40th International Conference on Technology of Object-Oriented Languages and Systems (TOOLS Pacific 2002).*

[33] Postel, J. (1980) *User Datagram Protocol. Network Information Center RFC 768*, pp. 1–3.

[34] Postel, J. (1981) *Transmission Control Protocol. Network Information Center RFC 793*, pp. 1–85.

[35] Rajkumar, R., Lee, C., Lehoczky, J. P., and Siewiorek, D. P. (1998) Practical Solutions for QoS-based Resource Allocation Problems, *IEEE Real-Time Systems Symposium*, IEEE, Madrid, Spain, pp. 296–306.

[36] Ritter, T., Born, M., Unterschütz, T., and Weis, T. (2003) A QoS Metamodel and its Realization in a CORBA Component Infrastructure, *Proceedings of the* 36th *Hawaii International Conference on System Sciences, Software Technology Track, Distributed Object and Component-based Software Systems Minitrack, HICSS 2003* HICSS, Honolulu, HW, on CD-ROM.

[37] Rosenberry, W., Kenney, D., and Fischer, G. (1992) *Understanding DCE*, O'Reilly and Associates, Inc.

[38] Schantz, R. E. and Schmidt, D. C. (2002) Middleware for Distributed Systems: Evolving the Common Structure for Network-centric Applications, in *Encyclopedia of Software Engineering* (ed. J. Marciniak and G. Telecki), Wiley & Sons, New York.

[39] Schantz, R., Loyall, J., Atighetchi, M., and Pal, P. (2002) Packaging Quality of Service Control Behaviors for Reuse, *Proceedings of the* 5th *IEEE International Symposium on Object-Oriented Real-time Distributed Computing (ISORC)*, IEEE/IFIP, Crystal City, VA, pp. 375–385.

[40] Schantz, R., Loyall, J., Schmidt, D., Rodrigues, C., Krishnamurthy Y., and Pyarali, I. (2003) Flexible and Adaptive QoS Control for Distributed Real-time and Embedded Middleware. *Proceedings of Middleware 2003, 4th International Conference on Distributed Systems Platforms*, IFIP/ACM/USENIX, Rio de Janeiro, Brazil.

[41] Schantz, R. E., Thomas, R. H., and Bono, G. (1986) The Architecture of the Cronus Distributed Operating System, *Proceedings of the 6th International Conference on Distributed Computing Systems*, IEEE, Cambridge, MA, pp. 250–259.

[42] Schantz, R. E., Zinky, J. A., Karr, D. A., Bakken, D. E., Megquier, J., and Loyall, J. P. (1999) An Object-level Gateway Supporting Integrated-Property Quality of Service. *Proceedings of The 2nd IEEE International Symposium on Object-oriented Real-time Distributed Computing (ISORC 99).*

[43] Schmidt, D. C. and Huston, S. D. (2002a) *C++ Network Programming, Volume 1: Mastering Complexity with ACE and Patterns*, Addison-Wesley, Boston, MA.

[44] Schmidt, D. C. and Huston, S. D. (2002b) *C++ Network Programming*, Volume 2: Systematic Reuse with ACE and Frameworks, Addison-Wesley, Reading, MA.

[45] Schmidt, D. C., Stal, M., Rohnert, H., and Buschmann, F. (2000) *Pattern-Oriented Software Architecture: Patterns for Concurrent and Networked Objects*, Volume 2, Wiley & Sons, New York.

[46] Sharp, D. C. (1998) Reducing Avionics Software Cost Through Component Based Product Line Development. *Proceedings of the 10th Annual Software Technology Conference.*

[47] Sharp, D. C. (1999) Avionics Product Line Software Architecture Flow Policies. *Proceedings of the 18th IEEE/AIAA Digital Avionics Systems Conference (DASC).*

[48] Sun Microsystems (1988) RPC: Remote procedure call protocol specification, Technical Report RFC-1057, Sun Microsystems, Inc, Mountain View, CA.

[49] Sun Microsystems (2001) Java^{TM} 2 Platform Enterprise Edition, http://java.sun.com/j2ee/index.html.

[50] Szyperski, C. (1998) *Component Software—Beyond Object-Oriented Programming*, Addison-Wesley, Santa Fe, NM.

[51] Vanegas, R., Zinky, J. A., Loyall, J. P., Karr, D., Schantz, R. E., and Bakken, D. (1998) QuO's Runtime Support for Quality of Service in Distributed Objects. *Proceedings of Middleware 98, the IFIP International Conference on Distributed Systems Platform and Open Distributed Processing (Middleware 98).*

[52] Volter, M., Schmid, A., and Wolff, E. (2002) *Server Component Patterns: Component Infrastructures Illustrated with EJB, Wiley Series in Software Design Patterns*, Wiley & Sons, West Sussex, England.

[53] Wang, N., Schmidt, D. C., and O'Ryan, C. (2000) An Overview of the CORBA Component Model, in *Component-Based Software Engineering* (ed. G. Heineman and B. Councill), Addison-Wesley, Reading, MA, pp. 557–572.

[54] Wang, N., Schmidt, D. C., Kircher, M., and Parameswaran, K. (2001a) Towards a Reflective Middleware Framework for QoS-enabled CORBA Component Model Applications. *IEEE Distributed Systems Online*, **2**(5), http://dsonline.computer.org/0105/features/wan0105_print.htm.

[55] Wang, N., Schmidt, D. C., Othman, O., and Parameswaran, K. (2001b) Evaluating Meta-Programming Mechanisms for ORB Middleware. *IEEE Communication Magazine, special issue on Evolving Communications Software: Techniques and Technologies*, **39**(10), 102–113.

[56] Wollrath, A., Riggs, R., and Waldo, J. (1996) A Distributed Object Model for the Java System. *Proceedings of 2nd Conference on Object-oriented Technologies and Systems (COOTS)*, pp. 219–232; USENIX Association, Toronto, Ontario, Canada.

[57] Zinky, J. A., Bakken, D. E., and Schantz, R. (1997) Architectural Support for Quality of Service for CORBA Objects. *Theory and Practice of Object Systems*, **3**(1), 1–20.

7

Model Driven Middleware

Aniruddha Gokhale, Douglas C. Schmidt, Balachandran Natarajan[1], Jeff Gray[2], Nanbor Wang[3]

[1] *Vanderbilt University*
[2] *University of Alabama*
[3] *Washington University*

7.1 Introduction

Emerging trends and technology challenges.
A growing number of computing resources are being expended to control distributed real-time and embedded (DRE) systems, including medical imaging, patient monitoring equipment, commercial and military aircraft and satellites, automotive braking systems, and manufacturing plants. Mechanical and human control of these systems are increasingly being replaced by DRE software controllers [44]. The real-world processes controlled by these DRE applications introduce many challenging Quality-of-Service (QoS) constraints, including

- **Real-time requirements**, such as low latency and bounded jitter
- **High-availability requirements**, such as fault propagation/recovery across distribution boundaries and
- **Physical requirements**, such as limited weight, power consumption, and memory footprint.

DRE software is generally harder to develop, maintain, and evolve [52, 20] than mainstream desktop and enterprise software because of conflicting QoS constraints, for example, bounded jitter versus fault tolerance and high-throughput versus minimal power consumption.

The tools and techniques used to develop DRE applications have historically been highly specialized. For example, DRE applications have traditionally been scheduled using

Middleware for Communications. Edited by Qusay H. Mahmoud
 ISBN 0-470-86206-8

fixed-priority periodic algorithms [29], where time is divided into a sequence of identical frames at each processor, and the processor executes each task for a uniform interval within each frame. DRE applications also often use frame-based interconnects, such as 1553, VME, or TTCAN buses where the traffic on an interconnect is scheduled at system design time to link the processors. Moreover, highly specialized platforms and protocols, such as cyclic executives [30] and time-triggered protocols [26], have been devised to support the development of DRE applications.

Highly specialized technologies have been important for developing traditional real-time and embedded systems, such as statically scheduled single-processor avionics mission computing systems [18]. These special-purpose technologies often do not scale up effectively, however, to address the needs of the new generation of large-scale DRE systems, such as air traffic and power grid management, which are inherently network-centric and dynamic. Moreover, as DRE applications grow in size and complexity, the use of highly specialized technologies can make it hard to adapt DRE software to meet new functional or QoS requirements, hardware/software technology innovations, or emerging market opportunities.

A candidate solution: QoS-enabled component middleware.

During the past decade, a substantial amount of R&D effort has focused on developing *QoS-enabled component middleware* as a means to simplify the development and reuse of DRE applications. As shown in Figure 7.1, QoS-enabled component middleware is systems software that resides between the applications and the underlying operating systems, network protocol stacks, and hardware and is responsible for providing the following capabilities:

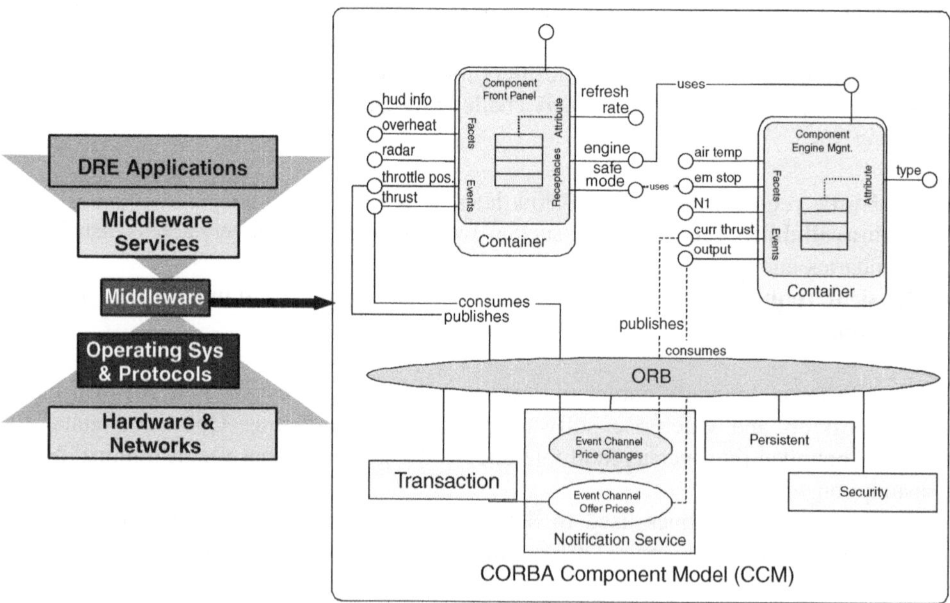

Figure 7.1 Component middleware layers and architecture

1. **Control over key end-to-end QoS properties.** One of the hallmarks of DRE applications is their need for strict control over the end-to-end scheduling and execution of CPU, network, and memory resources. QoS-enabled component middleware is based on the expectation that QoS properties will be developed, configured, monitored, managed, and controlled by a different set of specialists (such as middleware developers, systems engineers, and administrators) and tools than those responsible for programming the application functionality in traditional DRE systems.
2. **Isolation of DRE applications from the details of multiple platforms.** Standards-based QoS-enabled component middleware defines a communication model that can be implemented over many networks, transport protocols, and OS platforms. Developers of DRE applications can therefore concentrate on the application-specific aspects of their systems and leave the communication and QoS-related details to developers of the middleware.
3. **Reduction of total ownership costs.** QoS-enabled component middleware defines crisp boundaries between the components in the application, which reduces dependencies and maintenance costs associated with replacement, integration, and revalidation of components. Likewise, core components of component architectures can be reused, thereby helping to further reduce development, maintenance, and testing costs.

A companion chapter in this book [61] examines how recent enhancements to standard component middleware–particularly Real-time CORBA [38] and the CORBA Component Model [37]–can simplify the development of DRE applications by composing static QoS provisioning policies and dynamic QoS provisioning behaviors and adaptation mechanisms into applications.

Unresolved challenges.

Despite the significant advances in QoS-enabled component middleware, however, applications in important domains (such as large-scale DRE systems) that require simultaneous support for multiple QoS properties are still not well supported. Examples include shipboard combat control systems [51] and supervisory control and data acquisition (SCADA) systems that manage regional power grids. These types of large-scale DRE applications are typified by the following characteristics:

- **Stable applications and labile infrastructures** – Most DRE systems have a longer life than commercial systems [7]. In the commercial domain, for instance, it is common to find applications that are revised much more frequently than their infrastructure. The opposite is true in many large-scale DRE systems, where the application software must continue to function properly across decades of technology upgrades. As a consequence, it is important that the DRE applications interact with the changing infrastructure through well-managed interfaces that are *semantically stable*. In particular, if an application runs successfully on one implementation of an interface, it should behave equivalently on another version or implementation of the same interface.
- **End-to-end timeliness and dependability requirements** – DRE applications have stringent timeliness (i.e., end-to-end predictable time guarantees) and dependability requirements [49]. For example, the timeliness in DRE systems is often expressed as an upper bound in response to external events, as opposed to enterprise systems where

it is expressed as events-per-unit time. DRE applications generally express the dependability requirements as a probabilistic assurance that the requirements will be met, as opposed to enterprise systems, which express it as availability of a service.

- **Heterogeneity** – Large-scale DRE applications often run on a wide variety of computing platforms that are interconnected by different types of networking technologies with varying performance properties. The efficiency and predictability of execution of the different infrastructure components on which DRE applications operate varies on the basis of the type of computing platform and interconnection technology.

Despite the advantages of QoS-enabled component middleware, the unique requirements of large-scale DRE applications described earlier require a new generation of sophisticated tools and techniques for their development and deployment. Recent advances in QoS-enabled component middleware technology address many requirements of DRE applications, such as heterogeneity and timeliness. However, the remaining challenges discussed below impede the rapid development, integration, and deployment of DRE applications using COTS middleware:

i. Accidental complexities in identifying the right middleware technology. Recent improvements in middleware technology and various standardization efforts, as well as market and economical forces, have resulted in a multitude of middleware stacks, such as those shown in Figure 7.2. This heterogeneity often makes it hard, however, to identify the right middleware for a given application domain. Moreover, there exist limitations on how much application code can be factored out as reusable patterns and components in various layers for each middleware stack. This limit on refactoring in turn affects the optimization possibilities that can be implemented in different layers of the middleware. The challenge for DRE application developers is thus to choose the right middleware technology that can provide the desired levels of end-to-end QoS.

ii. Accidental complexities in configuring middleware. In QoS-enabled component middleware, both the components and the underlying component middleware framework may have a large number of configurable attributes and parameters that can be set at various stages of development lifecycle, such as composing an application or deploying

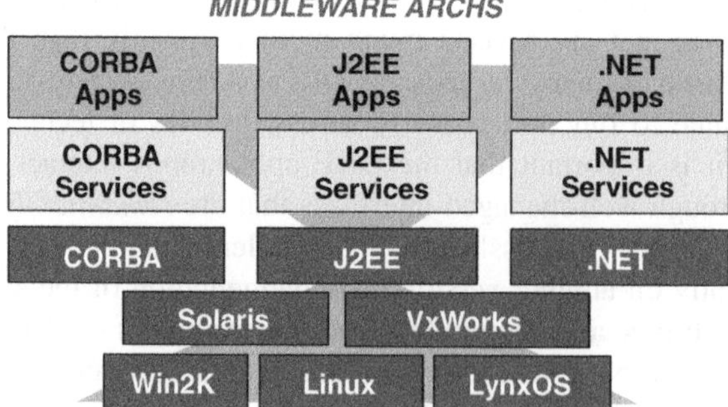

Figure 7.2 Multiple middleware stacks

an application in a specific environment. It is tedious and error-prone, however, to manually ensure that all these parameters are semantically consistent throughout an application. Moreover, such *ad hoc* approaches have no formal basis for validating and verifying that the configured middleware will indeed deliver the end-to-end QoS requirements of the application. An automated and rigorous tool-based approach is therefore needed that allows developers to formally analyze application QoS requirements and then synthesize the appropriate set of configuration parameters for the application components and middleware.

iii. Accidental complexities in composing and integrating software systems. Composing an application from a set of components with syntactically consistent interface signatures simply ensures they can be connected together. To function correctly, however, collaborating components must also have compatible semantics and invocation protocols, which are hard to express via interface signatures alone. *Ad hoc* techniques for determining, composing, assembling, and deploying the right mix of semantically compatible, QoS-enabled COTS middleware components do not scale well as the DRE application size and requirements increase. Moreover, *ad hoc* techniques, such as manually selecting the components, are often tedious, error-prone, and lack a solid analytical foundation to support verification and validation.

iv. Satisfying multiple QoS requirements simultaneously. DRE applications often possess multiple QoS requirements that the middleware must help to enforce simultaneously. Owing to the uniqueness and complexity of these QoS requirements, the heterogeneity of the environments in which they are deployed, and the need to interface with legacy systems and data, it is infeasible to develop a one-size-fits-all middleware solution that can address these requirements. Moreover, it is also hard to integrate highly configurable, flexible, and optimized components from different providers while still ensuring that application QoS requirements are delivered end-to-end.

v. Lack of principled methodologies to support dynamic adaptation capabilities. To maintain end-to-end QoS in dynamically changing environments, DRE middleware needs to be adaptive. Adaptation requires instrumenting the middleware to reflect upon the runtime middleware, operating systems, and network resource usage data and adapting the behavior on the basis of the collected data. DRE application developers have historically defined middleware instrumentation and program adaptation mechanisms in an *ad hoc* way and used the collected data to maintain the desired QoS properties. This approach creates a tight coupling between the application and the underlying middleware, while also scattering the code that is responsible for reflection and adaptation throughout many parts of DRE middleware and applications, which makes it hard to configure, validate, modify, and evolve complex DRE applications consistently.

To address the challenges described above, we need principled methods for specifying, programming, composing, integrating, and validating software throughout these DRE applications. These methods must enforce the physical constraints of the system. Moreover, they must satisfy stringent functional and systemic QoS requirements within an entire system. What is required is a set of standard integrated tools that allow developers to specify application requirements at higher levels of abstraction than that provided by low-level mechanisms, such as conventional general-purpose programming languages, operating systems, and middleware platforms. These tools must be able to analyze the requirements and synthesize the required metadata that will compose applications from the right set of middleware components.

A promising solution: Model Driven Middleware.
A promising way to address the DRE software development and integration challenges described above is to develop *Model Driven Middleware* by combining the Object Management Group (OMG)'s *Model Driven Architecture* (MDA) technologies [35, 2] with *QoS-enabled component middleware* [9, 50, 60, 61]. The OMG MDA is an emerging paradigm for expressing application functionality and QoS requirements at higher levels of abstraction than is possible using conventional third-generation programming languages, such as Visual Basic, Java, C++, or C#. In the context of DRE middleware and applications, MDA tools can be applied to

1. **Model** different functional and systemic properties of DRE applications in separate platform-independent models. Domain-specific aspect model weavers [16] can integrate these different models into composite models that can be further refined by incorporating platform-specific aspects.
2. **Analyze** different – but interdependent – characteristics and requirement of application behavior specified in the models, such as scalability, predictability, safety, schedulability, and security. Tool-specific model interpreters [27] translate the information specified by models into the input format expected by model checking and analysis tools [19, 55]. These tools can check whether the requested behavior and properties are feasible given the specified application and resource constraints.
3. **Synthesize** platform-specific code and metadata that is customized for particular component middleware and DRE application properties, such as end-to-end timing deadlines, recovery strategies to handle various runtime failures in real time, and authentication and authorization strategies modeled at a higher level of abstraction than that provided by programming languages (such as C, C++, and Java) or scripting languages (such as Perl and Python).
4. **Provision** the application by assembling and deploying the selected application and middleware components end-to-end using the configuration metadata synthesized by the MDA tools.

The initial focus of MDA technologies were largely on enterprise applications. More recently, MDA technologies have emerged to customize QoS-enabled component middleware for DRE applications, including aerospace [1], telecommunications [34], and industrial process control [48]. This chapter describes how MDA technologies are being applied to QoS-enabled CORBA component middleware to create Model Driven Middleware frameworks.

Chapter organization.
The remainder of this chapter is organized as follows: Section 7.2 presents an overview of the OMG Model Driven Architecture (MDA) effort; Section 7.3 describes how the Model

Driven Middleware paradigm, which is an integration of MDA and QoS-enabled component middleware, resolves key challenges associated with DRE application integration; Section 7.4 provides a case study of applying Model Driven Middleware in the context of real-time avionics mission computing; Section 7.5 compares our work on Model Driven Middleware with related efforts; and Section 7.6 presents concluding remarks.

7.2 Overview of the OMG Model Driven Architecture (MDA)

The OMG has adopted the Model Driven Architecture (MDA) shown in Figure 7.3 to standardize the integration of the modeling, analysis, simulation and synthesis paradigm with different middleware technology platforms. MDA is a development paradigm that applies domain-specific modeling languages systematically to engineer computing systems, ranging from small-scale real-time and embedded systems to large-scale distributed enterprise applications. It is *model-driven* because it uses models to direct the course of understanding, design, construction, deployment, operation, maintenance, and modification. MDA is a key step forward in the long road of converting the art of developing software into an engineering process. This section outlines the capabilities and benefits of OMG's MDA.

7.2.1 Capabilities of the MDA

The OMG MDA approach is facilitated by domain-specific modeling environments [59], including model analysis and model-based program synthesis tools [27]. In the MDA

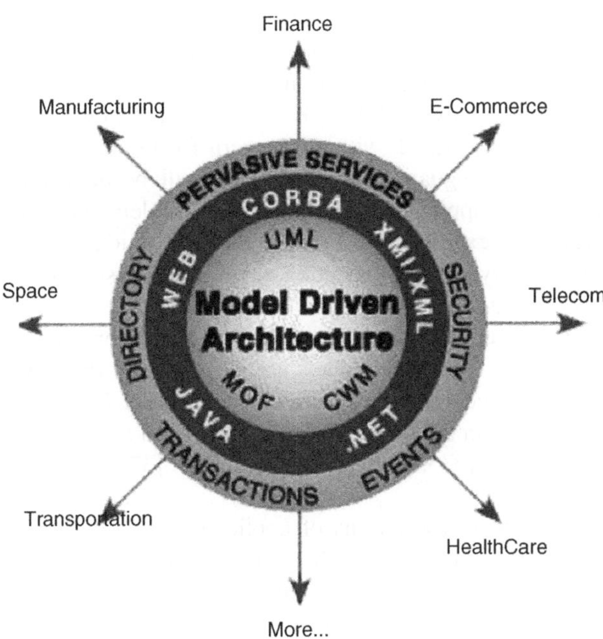

Figure 7.3 Roles and relationships in OMG model-driven architecture

paradigm, application developers capture integrated, end-to-end views of entire applications in the form of models, including the interdependencies of components. Rather than focusing on a single custom application, the models may capture the essence of a *class* of applications in a particular domain. MDA also allows domain-specific modeling languages to be formally specified by *metamodels* [54, 22].

A metamodel defines the abstract and concrete syntax, the static semantics (i.e., well-formedness rules), and semantic mapping of the abstract syntax into a semantic domain for a domain-specific modeling language (DSML), which can be used to capture the essential properties of applications. The (abstract) metamodel of a DSML is translated into a (concrete) domain-specific modeling paradigm, which is a particular approach to creating domain models supported by a custom modeling tool. This paradigm can then be used by domain experts to create the models and thus the applications.

The MDA specification defines the following types of models that streamline platform integration issues and protect investments against the uncertainty of changing platform technology:

- **Computation-independent models (CIMs)** that describe the computation-independent viewpoint of an application. CIMs provide a domain-specific model that uses vocabulary understandable to practitioners of a domain and hides nonessential implementation details of the application from the domain experts and systems engineers. The goal of a CIM is to bridge the gap between (1) domain experts who have in-depth knowledge of the subject matter, but who are not generally experts in software technologies, and (2) developers who understand software design and implementation techniques, but who are often not domain experts. For example, the Unified Modeling Language (UML) can be used to model a GPS guidance system used in avionics mission computing [53] without exposing low-level implementation artifacts.
- **Platform-independent models (PIMs)** that describe at a high level how applications will be structured and integrated, without concern for the target middleware/OS platforms or programming languages on which they will be deployed. PIMs provide a formal definition of an application's functionality implemented on some form of a virtual architecture. For example, the PIM for the GPS guidance system could include artifacts such as priority of the executing thread or worst-case execution time for computing the coordinates.
- **Platform-specific models (PSMs)** that are *constrained* formal models that express platform-specific details. The PIM models are mapped into PSMs via *translators*. For example, the GPS guidance system that is specified in the PIM could be mapped and refined to a specific type in the underlying platform, such as a QoS-enabled implementation [61] of the CORBA Component Model (CCM) [37].

The CIM, PIM, and PSM descriptions of applications are formal specifications built using modeling standards, such as UML, that can be used to model application functionality and system interactions. The MDA enables the application requirements captured in the CIMs to be traced to the PIMs/PSMs and vice versa. The MDA also defines a platform-independent meta-modeling language, which is also expressed using UML, that allows platform-specific models to be modeled at an even higher level of abstraction.

Figure 7.3 also references the Meta-Object Facility (MOF), which provides a framework for managing any type of metadata. The MOF has a layered metadata architecture with a meta-metamodeling layer and an object modeling language–closely related to UML–that ties together the metamodels and models. The MOF also provides a repository to store metamodels.

The Common Warehouse Model (CWM) shown in Figure 7.3 provides standard interfaces that can manage many different databases and schemas throughout an enterprise. The CWM interfaces are designed to support management decision making and exchange of business metadata between diverse warehouse tools to help present a coherent picture of business conditions at a single point in time. The OMG has defined the XML Metadata Interchange (XMI) for representing and exchanging CWM metamodels using the Extended Markup Language (XML).

The OMG partitions the architecture of a computing system into the following three levels where MDA-based specifications are applicable:

1. The **Pervasive services** level constitutes a suite of PIM specifications of essential services, such as events, transactions, directory and security, useful for large-scale application development.
2. The **Domain facilities** level constitutes a suite of PIM specifications from different domains such as manufacturing, healthcare and life science research within the OMG.
3. The **Applications** level constitutes a suite of PIM specifications created by software vendors for their applications.

The three levels outlined above allow a broad range of services and application designs to be reused across multiple platforms. For instance, some of the domain-specific services from the OMG could be reused for other technology platforms, rather than designing them from scratch.

As shown in Figure 7.4, MDA uses a set of tools to

- model the application in a computation-independent and platform-independent style;
- analyze the interdependent features of the system captured in a model; and
- determine the feasibility of supporting different system quality aspects, such as QoS requirements, in the context of the specified constraints.

Another set of tools translate the PIMs into PSMs. As explained in Section 7.2.1, PSMs are executable specifications that capture the platform behavior, constraints, and interactions

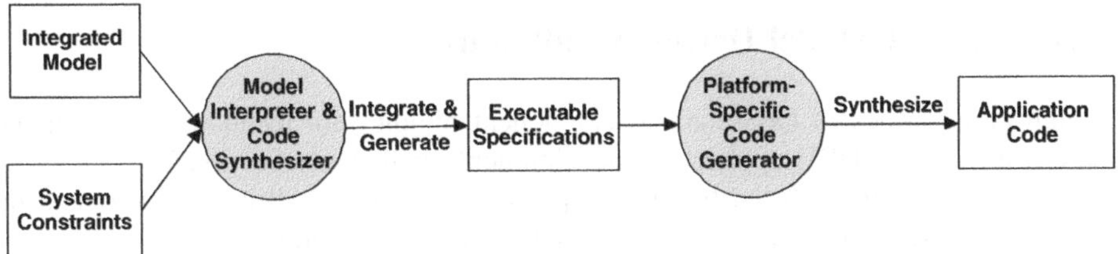

Figure 7.4 The model driven architecture computing process

with the environment. These executable specifications can in turn be used to synthesize various portions of application software.

7.2.2 Benefits of the MDA

When implemented properly, MDA technologies help to:

- Free application developers from dependencies on particular software APIs, which ensures that the models can be used for a long time, even as existing software APIs become obsolete and replaced by newer ones.
- Provide correctness proofs for various algorithms by analyzing the models automatically and offering refinements to satisfy various constraints.
- Synthesize code that is highly dependable and robust since the tools can be built using provably correct technologies.
- Rapidly prototype new concepts and applications that can be modeled quickly using this paradigm, compared to the effort required to prototype them manually.
- Save companies and projects significant amounts of time and effort in design and maintenance, thereby also reducing application time-to-market.

Earlier generations of computer-aided software engineering (CASE) technologies have evolved into sophisticated tools, such as *objectiF* and *in-Step* from MicroTool and *Paradigm Plus*, *VISION*, and *COOL* from Computer Associates. This class of products has evolved over the past two decades to alleviate various complexities associated with developing software for enterprise applications. Their successes have added the MDA paradigm to the familiar programming languages and language processing tool offerings used by previous generations of software developers. Popular examples of MDA or MDA-related tools being used today include the Generic Modeling Environment (GME) [27], Ptolemy [5], and MDA-based UML/XML tools, such as *Codagen Architect* or *Metanology*.

As described in Section 7.2.1, MDA is a platform-independent technology that aims to resolve the complexities involved with COTS obsolescence and the resulting application transition to newer technologies. The design and implementation of MDA tools for a given middleware technology, however, is itself a challenging problem that is not completely addressed by the standards specifications. The next section addresses these challenges by describing how MDA technology can be effectively combined with QoS-enabled middleware to create *Model Driven Middleware*.

7.3 Overview of Model Driven Middleware

Although rapid strides in QoS-enabled component middleware technology have helped resolve a number of DRE application development challenges, Section 7.1 highlighted the unresolved challenges faced by DRE application developers. It is in this context that the OMG's Model Driven Architecture can be effectively combined with these QoS-enabled component middleware technologies to resolve these challenges. We coined the term *Model Driven Middleware* to describe integrated suites of MDA tools that can be applied to the design and runtime aspects of QoS-enabled component middleware.

This section first outlines the limitations of prior efforts to use modeling and synthesis techniques for lifecycle management of large-scale applications, including DRE applications. These limitations resulted from the lack of integration between modeling techniques and QoS-enabled middleware technologies. We then describe how to effectively integrate MDA with QoS-enabled middleware.

7.3.1 Limitations of Using Modeling and Middleware in Isolation

Earlier efforts in model-driven synthesis of large-scale applications and component middleware technologies have evolved from different perspectives, that is, modeling tools have largely focused on design issues (such as structural and behavioral relationships and associations), whereas component middleware has largely focused on runtime issues (such as (re)configuration, deployment, and QoS enforcement). Although each of these two paradigms have been successful independently, each also has its limitations, as discussed below:

Complexity due to heterogeneity.
Conventional component middleware is developed using separate tools and interfaces written and optimized manually for each middleware technology (such as CORBA, J2EE, and .NET) and for each target deployment (such as various OS, network, and hardware configurations). Developing, assembling, validating, and evolving *all* this middleware manually is costly, time-consuming, tedious, and error-prone, particularly for runtime platform variations and complex application use cases. This problem is exacerbated as more middleware, target platforms, and complex applications continue to emerge.

Lack of sophisticated modeling tools.
Previous efforts at model-based development and code synthesis attempted by CASE tools generally failed to deliver on their potential for the following reasons [2]:

- They attempted to generate entire applications, including the middleware infrastructure and the application logic, which often led to inefficient, bloated code that was hard to optimize, validate, evolve, or integrate with existing code.
- Because of the lack of sophisticated domain-specific languages and associated metamodeling tools, it was hard to achieve *round-trip engineering*, that is, moving back and forth seamlessly between model representations and the synthesized code.
- Since CASE tools and modeling languages dealt primarily with a restricted set of platforms (such as mainframes) and legacy programming languages (such as COBOL), they did not adapt well to the distributed computing paradigm that arose from advances in PC and Internet technology and newer object-oriented programming languages, such as Java, C++, and C#.

7.3.2 Combining Model Driven Architecture and QoS-enabled Component Middleware

The limitations with modeling techniques and component middleware outlined above can largely be overcome by integrating OMG MDA and component middleware as follows:

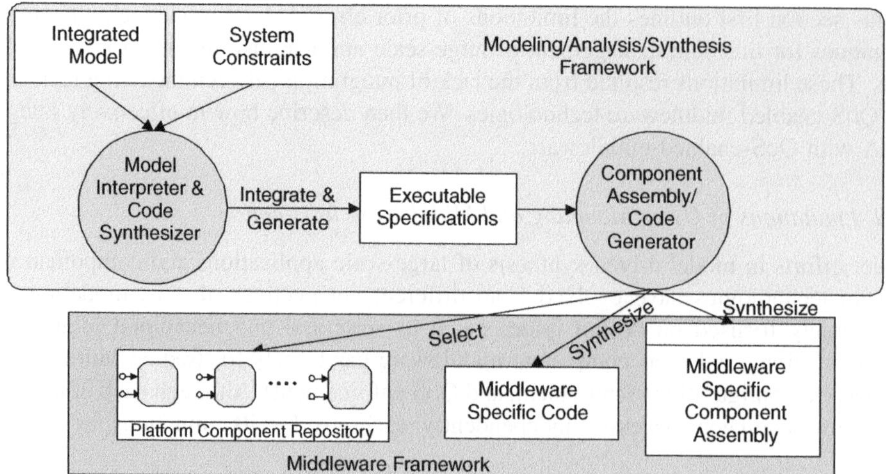

Figure 7.5 Integrating MDA and component middleware to create model driven middleware

- Combining MDA with component middleware helps overcome problems with earlier-generation CASE tools since it does not require the modeling tools to generate all the code. Instead, large portions of applications can be *composed* from reusable, prevalidated middleware components, as shown in Figure 7.5.
- Combining MDA and component middleware helps address environments where control logic and procedures change at rapid pace, by synthesizing and assembling newer extended components that implement the new procedures and processes.
- Combining component middleware with MDA helps make middleware more flexible and robust by automating the configuration of many QoS-critical aspects, such as concurrency, distribution, resource reservation, security, and dependability. Moreover, MDA-synthesized code can help bridge the interoperability and portability problems between different middleware for which standard solutions do not yet exist.
- Combining component middleware with MDA helps to model the interfaces among various components in terms of standard middleware interfaces, rather than language-specific features or proprietary APIs.
- Changes to the underlying middleware or language mapping for one or many of the components modeled can be handled easily as long as they interoperate with other components. Interfacing with other components can be modeled as constraints that can be validated by model checkers, such as Cadena [19].

Figure 7.6 illustrates seven points at which MDA can be integrated with component middleware architectures and applied to DRE applications. Below, we present examples of each of these integration points in the Model Driven Middleware paradigm:

1. Configuring and deploying application services end-to-end. Developing complex DRE applications requires application developers to handle a variety of configuration, packaging and deployment challenges, such as

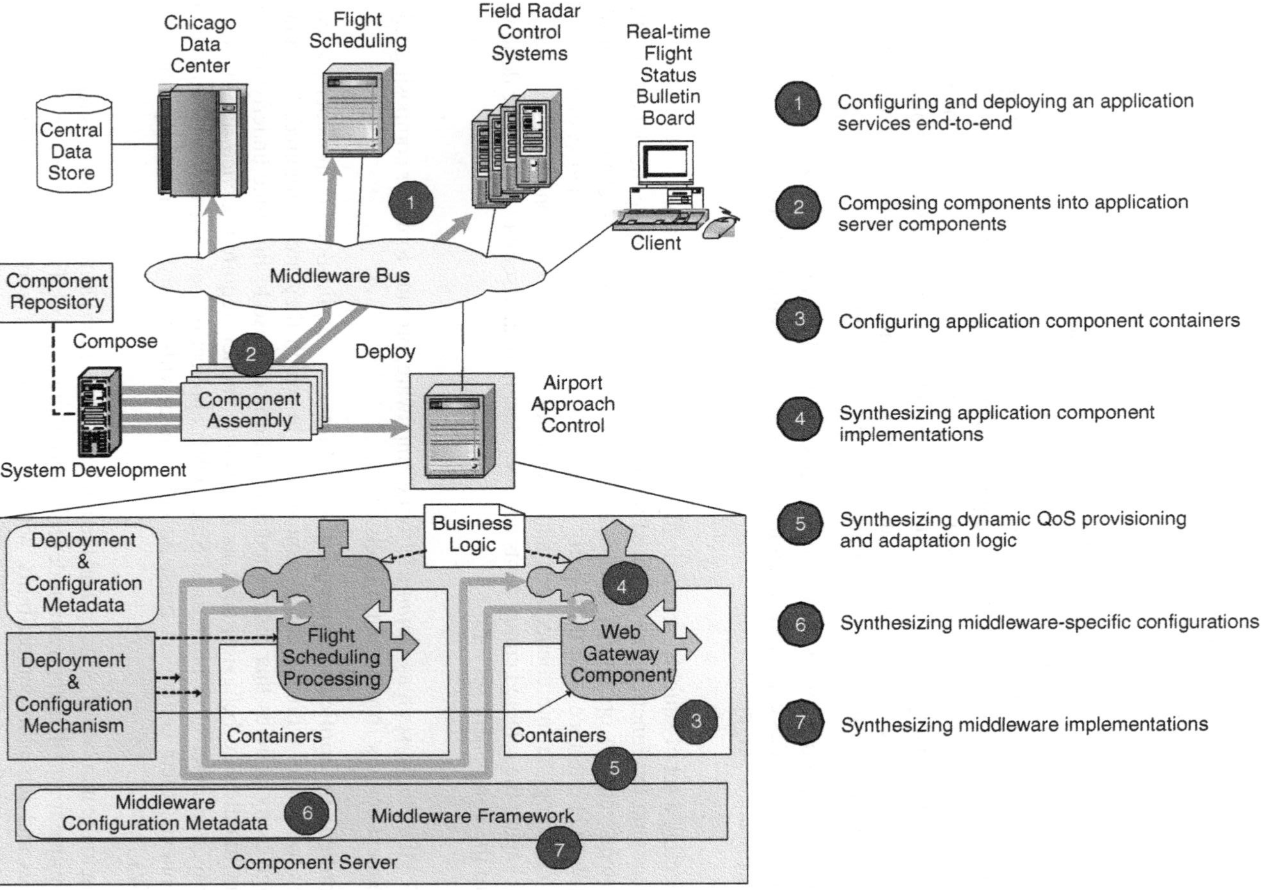

Figure 7.6 Integrating model driven architecture with component middleware

- configuring appropriate libraries of middleware suites tailored to the QoS and footprint requirements of the DRE application;
- locating the appropriate existing services;
- partitioning and distributing application processes among component servers using the same middleware technologies; and
- provisioning the QoS required for each service that comprises an application end-to-end.

It is a daunting task to identify and deploy all these capabilities into an efficient, correct, and scalable end-to-end application configuration. For example, to maintain correctness and efficiency, services may change or migrate when the DRE application requirements change. Careful analysis is therefore required for large-scale DRE systems to effectively partition collaborating services on distributed nodes so the information can be processed efficiently, dependably, and securely. The OMG's Deployment and Configuration specification [40] addresses these concerns and describes the mechanisms by which distributed component-based applications are configured and deployed.

2. Composing components into component servers. Integrating MDA with component middleware provides capabilities that help application developers to compose components into application servers by

- selecting a set of suitable, semantically compatible components from reuse repositories;
- specifying the functionality required by new components to isolate the details of DRE systems that (1) operate in environments where DRE processes change periodically and/or (2) interface with third-party software associated with external systems;
- determining the interconnections and interactions between components in metadata;
- packaging the selected components and metadata into an assembly that can be deployed into the component server.

OMG MDA tools, such as *OptimalJ* from Compuware, provide tools for composing J2EE component servers from visual models.

3. Configuring application component containers. Application components use containers to interact with the component servers in which they are configured. Containers manage many policies that distributed applications can use to fine-tune underlying component middleware behavior, such as its priority model, required service priority level, security, and other QoS properties. Since DRE applications consist of many interacting components, their containers must be configured with consistent and compatible QoS policies.

Owing to the number of policies and the intricate interactions among them, it is tedious and error-prone for a DRE application developer to *manually* specify and maintain component policies and semantic compatibility with policies of other components. MDA tools can help automate the validation and configuration of these container policies by allowing system designers to specify the required system properties as a set of models. Other MDA tools can then analyze the models and generate the necessary policies and ensure their consistency.

The Embedded Systems Modeling Language (ESML) [21] developed as part of the DARPA MoBIES program uses MDA technology to model the behavior of, and interactions between, avionics components. Moreover, the ESML model generators synthesize fault management and thread policies in component containers.

4. Synthesizing application component implementations. Developing complex DRE applications involves programming new components that add application-specific functionality. Likewise, new components must be programmed to interact with external systems and sensors (such as a machine vision module controller) that are not internal to the application. Since these components involve substantial knowledge of application domain concepts (such as mechanical designs, manufacturing process, workflow planning, and hardware characteristics) it would be ideal if they could be developed in conjunction with systems engineers and/or domain experts, rather than programmed manually by software developers.

The shift toward high-level design languages and modeling tools is creating an opportunity for increased automation in generating and integrating application components. The goal is to bridge the gap between specification and implementation via sophisticated aspect weavers [24] and generator tools [27] that can synthesize platform-specific code customized for specific application properties, such as resilience to equipment failure, prioritized scheduling, and bounded worst-case execution under overload conditions.

The Constraint Specification Aspect Weaver (C-SAW) [16] and Adaptive Quality Modeling Environment (AQME) [33] tools developed as part of the DARPA PCES program use MDA technology to provide a model-driven approach for weaving in and synthesizing QoS adaptation logic DRE application components. In particular, AQME is used in conjunction with QuO/Qoskets [57] to provide adaptive QoS policies for an unmanned aerial vehicle (UAV) real-time video distribution application [47].

5. Synthesizing dynamic QoS provisioning and adaptation logic. On the basis of the overall system model and constraints, MDA tools may decide to plug in existing dynamic QoS provisioning and adaptation modules, using appropriate parameters. When none is readily available, the MDA tools can assist in creating new behaviors by synthesizing appropriate logic, for example, using QoS-enabled aspect languages [46]. The generated dynamic QoS behavior can then be used in system simulation dynamically to verify its validity. It can then be composed into the system as described above.

The AQME [33] modeling language mentioned at the end of integration point 4 above models the QuO/Qosket middleware by modeling system conditions and service objects. For example, AQME enables modeling the interactions between the sender and receiver of the UAV video streaming applications, as well as parameters that instrument the middleware and application components.

6. Synthesizing middleware-specific configurations. The infrastructure middleware technologies used by component middleware provide a wide range of policies and options to configure and tune their behavior. For example, CORBA Object Request Brokers (ORBs) often provide many options and tuning parameters, such as various types of transports and protocols, various levels of fault tolerance, middleware initialization options, efficiency of (de)marshaling event parameters, efficiency of demultiplexing incoming method calls, threading models, and thread priority settings, and buffer sizes, flow control, and buffer overflow handling. Certain combinations of the options provided by the middleware may be semantically incompatible when used to achieve multiple QoS properties.

For example, a component middleware implementation could offer a range of security levels to the application. In the lowest security level, the middleware exchanges all the messages over an unsecure channel. The highest security level, in contrast, encrypts and decrypts messages exchanged through the channel using a set of dynamic keys. The same middleware could also provide an option to use zero-copy optimizations to minimize latency. A modeling tool could automatically detect the incompatibility of trying to compose the zero-copy optimization with the highest security level (which makes another copy of the data during encryption and decryption).

Advanced meta-programming techniques, such as adaptive and reflective middleware [25, 14, 11, 7] and aspect-oriented programming [24], are being developed to configure middleware options so they can be tailored for particular DRE application use cases.

7. Synthesizing middleware implementations. MDA tools can also be integrated with component middleware by using generators to synthesize custom middleware implementations. This integration is a more aggressive use of modeling and synthesis than integration point 6 described above since it affects middleware *implementations*, rather than just their configurations. For example, application integrators could use these capabilities to generate highly customized implementations of component middleware so that it (1) only includes the features actually needed for a particular application and (2) is carefully fine-tuned to the characteristics of particular programming languages, operating systems, and networks.

The customizable middleware architectural framework *Quarterware* [6] is an example of this type of integration. Quarterware abstracts basic middleware functionality and allows application-specific specializations and extensions. The framework can generate core facilities of CORBA, Remote Method Invocation (RMI), and Message Passing Interface (MPI). The framework-generated code is optimized for performance, which the authors demonstrate is comparable–and often better–than many commercially available middleware implementations.

7.4 Model Driven Middleware Case Study: Integrating MDA with QoS-enabled Middleware for Avionics Mission Computing

The Model Driven Middleware tool suite we are developing is called CoSMIC (Component Synthesis with Model Integrated Computing) [13, 31, 12]. Our research on CoSMIC is manifested in the integration of OMG MDA with QoS-enabled component middleware, such as CIAO [61], along the seven points illustrated in Figure 7.6. This section illustrates how the Model Driven Middleware concept manifested in CoSMIC is being operationalized in practice for the avionics mission computing domain.

Within the CoSMIC framework, we are integrating the OMG MDA tools and processes with the CIAO QoS-enabled component middleware platform [61] and applying them in the context of Boeing's Bold Stroke [52, 53] avionics mission computing platform. Bold

Stroke is a large-scale DRE platform that is based heavily on QoS-enabled component middleware. The CoSMIC tools we are developing for avionics mission computing are designed to model and analyze both application functionality and end-to-end application QoS requirements. With CIAO's support for QoS-enabled, reusable CCM components, it is then possible for CoSMIC to

- model the QoS requirements of avionics applications using domain-specific modeling languages that we have developed
- associate the models with different static and dynamic QoS profiles
- simulate and analyze dynamic behaviors
- synthesize appropriate middleware configuration parameters including different policies of the CIAO containers
- synthesize the QoS-enabled application functionality in component assemblies and
- weave in crosscutting runtime adaptation logic metadata.

Figure 7.7 illustrates the interaction between CoSMIC and CIAO that enables the resolution of the challenges outlined in Section 7.1. Below, we describe how we are using Model Driven Middleware in the context of avionics mission computing.

Handling avionics mission computing middleware configuration.
The CIAO CCM implementation provides a large number of configuration parameters to fine-tune its performance to meet the QoS needs of Bold Stroke avionics applications. CIAO imposes several constraints on what combinations of these options are valid for a given specification of application QoS requirements. To handle these complexities, CoSMIC provides a modeling paradigm [22] called the *Options Configuration Modeling Language* (OCML) [12]. CIAO middleware developers can use OCML to model the available configuration parameters and the different constraints involving these options. Likewise, the Bold Stroke avionics mission computing application developers can use the OCML modeling, analysis, and synthesis framework to (1) model the desired QoS

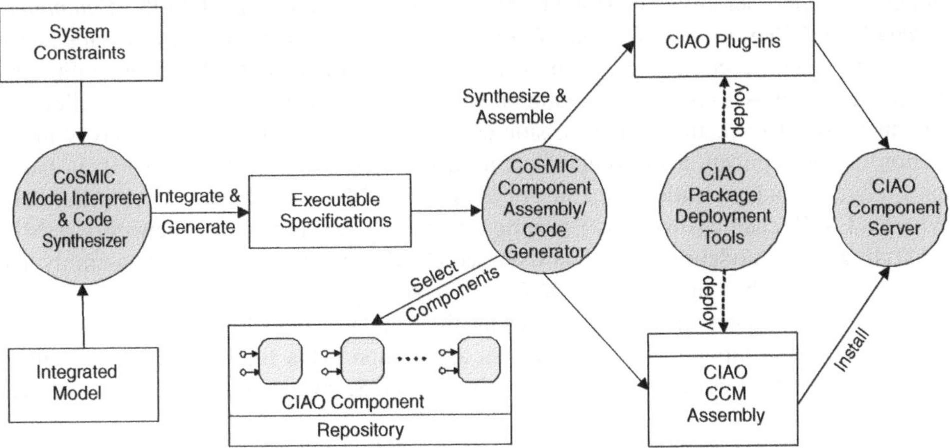

Figure 7.7 Interactions between CoSMIC and CIAO

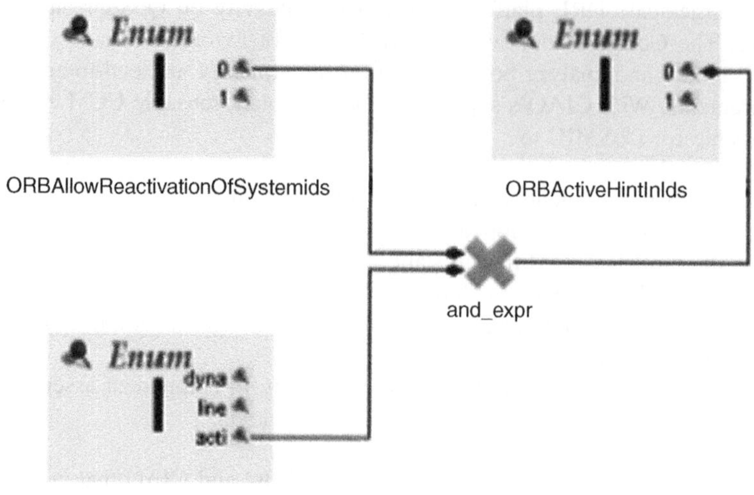

Figure 7.8 Example using OCML modeling paradigm

requirements and (2) synthesize the appropriate middleware configuration metadata that is then used by CIAO to fine-tune its performance.

Figure 7.8 illustrates an example of using the OCML modeling paradigm to model a rule for combining configuration options in CIAO. DRE application developers using the OCML paradigm to choose the CIAO middleware configuration parameters will be constrained by the rules imposed by the suite of OCML models, such as those illustrated in Figure 7.8.

Handling avionics middleware component assembly and deployment.
The CoSMIC tool suite provides modeling of DRE systems, their QoS requirements, and component assembly and deployment. A modeling paradigm called the *Component Assembly and Deployment Modeling Language* (CADML) [12] has been developed for this purpose. In the context of avionics mission computing, the CADML assembly and deployment models convey information on which Bold Stroke components are collocated on which processor boards of the mission computer. CADML models also convey information on the replication of components used to enhance reliability. CADML is based on a related paradigm called the Embedded Systems Modeling Language (ESML) [21]. Whereas ESML enables modeling a proprietary avionics component middleware, CoSMIC CADML enables modeling the standards-based CCM components, and their assembly and deployment described in the OMG D&C specification [40].

Figure 7.9 illustrates an example of using the CADML modeling paradigm to model an assembly of components of an avionics application. This figure illustrates a simple avionics application comprising an assembly of a GPS, an airframe, and a navigational display component.

The CoSMIC tool also provides synthesis tools targeted at the CIAO QoS-enabled component assembly and deployment. As described in our companion chapter in this

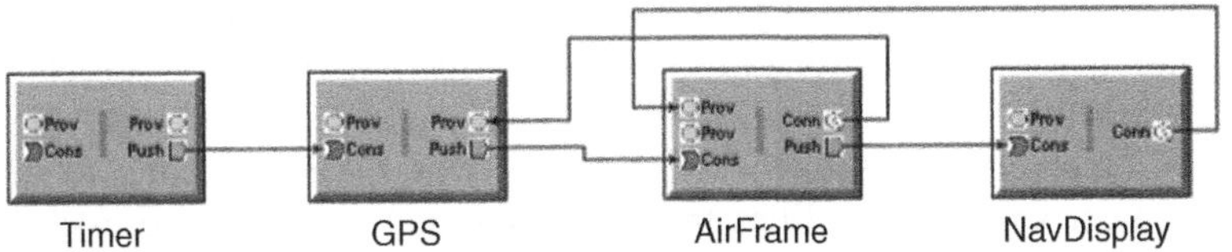

Figure 7.9 Using CADML to model avionics component assembly

book [61], QoS-enabled component middleware, such as CIAO, abstracts component QoS requirements into metadata that can be specified in a component assembly after a component has been implemented. Decoupling QoS requirements from component implementations greatly simplifies the conversion and validation of an application model with multiple QoS requirements into CCM deployment of DRE applications. The synthesis tools use the CADML models to generate the assembly and deployment data.

The CoSMIC component assembly and deployment tools described in this section have been applied successfully for modeling and synthesis of a number of Bold Stroke product scenarios. To address scalability issues for Bold Stroke component assemblies comprising large number of components, for example, 50 or more, CADML allows partitioning the assemblies into manageable logical units that are then themselves assembled. Moreover, using CADML to synthesize the assembly metadata for such large assemblies is a big win over manually handcrafting the same both in terms of the scalability as well as correctness of the synthesized assembly metadata.

7.5 Related Work

This section reviews related work on model driven architectures and describes how modeling, analysis, and generative programming techniques are being used to model and provision QoS capabilities for DRE component middleware and applications.

Model-based software development.

Research on Model Driven Middleware extends earlier work on Model-Integrated Computing (MIC) [56, 17, 28, 15] that focused on modeling and synthesizing embedded software. MIC provides a unified software architecture and framework for creating Model-Integrated Program Synthesis (MIPS) environments [27]. Examples of MIC technology used today include GME [27] and Ptolemy [5] (used primarily in the real-time and embedded domain) and MDA [35] based on UML [36] and XML [10] (which have been used primarily in the business domain). Our work on CoSMIC combines the GME tool and UML modeling language to model and synthesize QoS-enabled component middleware for use in provisioning DRE applications. In particular, CoSMIC is enhancing GME to produce domain-specific modeling languages and generative tools for DRE applications, as well as developing and validating new UML profiles (such as the UML profile for CORBA [39], the UML profile for quality of service [41], and UML profile for schedulability, performance and time [42]) to support DRE applications.

As part of an ongoing collaboration [47] between ISIS, University of Utah, and BBN Technologies, work is being done to apply GME techniques to model an effective resource management strategy for CPU resources on the TimeSys Linux real-time OS [58]. TimeSys Linux allows applications to specify CPU reservations for an executing thread, which guarantee that the thread will have a certain amount of CPU time, regardless of the priorities of other threads in the system. Applying GME modeling to develop the QoS management strategy simplifies the simulation and validation necessary to assure end-to-end QoS requirements for CPU processing.

The Virginia Embedded System Toolkit (VEST) [55] is an embedded system composition tool that enables the composition of reliable and configurable systems from COTS component libraries. VEST compositions are driven by a modeling environment that uses the GME tool [27]. VEST also checks whether certain real-time, memory, power, and cost constraints of DRE applications are satisfied.

The Cadena [19] project provides an MDA tool suite with the goal of assessing the effectiveness of applying static analysis, model-checking, and other light-weight formal methods to CCM-based DRE applications. The Cadena tools are implemented as plug-ins to IBM's Eclipse integrated development environment (IDE) [43]. This architecture provides an IDE for CCM-based DRE systems that ranges from editing of component definitions and connections information to editing and debugging of auto-generated code templates.

Commercial successes in model-based software development include the Rational Rose [32] suite of tools used primarily in enterprise applications. Rose is a model-driven development tool suite that is designed to increase the productivity and quality of software developers. Its modeling paradigm is based on the Unified Modeling Language (UML). Rose tools can be used in different application domains including business and enterprise/IT applications, software products and systems, and embedded systems and devices. In the context of DRE applications, Rose has been applied successfully in the avionics mission computing domain [52].

Program transformation technologies.

Program transformation is used in many areas of software engineering, including compiler construction, software visualization, documentation generation, and automatic software renovation. The approach basically involves changing one program to another. Program transformation environments provide an integrated set of tools for specifying and performing semantic-preserving mappings from a source program to a new target program.

Program transformations are typically specified as rules that involve pattern matching on an abstract syntax tree (AST). The application of numerous transformation rules evolves an AST to the target representation. A transformation system is much broader in scope than a traditional generator for a domain-specific language. In fact, a generator can be thought of as an instance of a program transformation system with specific hard-coded transformations. There are advantages and disadvantages to implementing a generator from within a program transformation system. A major advantage is evident in the preexistence of parsers for numerous languages [4]. The internal machinery of the transformation system may also provide better optimizations on the target code than could be done with a stand-alone generator.

Generative Programming (GP) [8] is a type of program transformation concerned with designing and implementing software modules that can be combined to generate specialized and highly optimized systems fulfilling specific application requirements. The goals are to (1) decrease the conceptual gap between program code and domain concepts (known as achieving high intentionality), (2) achieve high reusability and adaptability, (3) simplify managing many variants of a component, and (4) increase efficiency (both in space and execution time).

GenVoca [3] is a generative programming tool that permits hierarchical construction of software through the assembly of interchangeable/reusable components. The GenVoca model is based upon stacked layers of abstraction that can be composed. The components can be viewed as a catalog of problem solutions that are represented as pluggable components, which then can be used to build applications in the catalog domain.

Yet another type of program transformation is aspect-oriented software development (AOSD). AOSD is a new technology designed to more explicitly separate concerns in software development. The AOSD techniques make it possible to modularize crosscutting aspects of complex DRE systems. An aspect is a piece of code or any higher-level construct, such as implementation artifacts captured in a MDA PSM, that describes a recurring property of a program that crosscuts the software application, that is, aspects capture crosscutting concerns). Examples of programming language support for AOSD constructs include AspectJ [23] and AspectC++ [45].

7.6 Concluding Remarks

Large-scale distributed real-time and embedded (DRE) applications are increasingly being developed using QoS-enabled component middleware [61]. QoS-enabled component middleware provides policies and mechanisms for provisioning and enforcing large-scale DRE application QoS requirements. The middleware itself, however, does not resolve the challenges of choosing, configuring, and assembling the appropriate set of syntactically and semantically compatible QoS-enabled DRE middleware components tailored to the application's QoS requirements. Moreover, a particular middleware API does not resolve all the challenges posed by obsolescence of infrastructure technologies and its impact on long-term DRE system lifecycle costs.

The OMG's Model Driven Architecture (MDA) is emerging as an effective paradigm for addressing the challenges described above. The MDA is a software development paradigm that applies domain-specific modeling languages systematically to engineer computing systems. This chapter provides an overview of the emerging paradigm of *Model Driven Middleware*, which applies MDA techniques and tools to help configure and deploy QoS-enabled component middleware and DRE applications and large-scale systems of systems.

To illustrate recent progress on Model Driven Middleware, we describe a case study of applying the CoSMIC tool suite in the domain of avionics mission computing. CoSMIC is designed to simplify the integration of DRE applications that consist of QoS-enabled component middleware, such as the CIAO QoS enhancements to the CORBA Component Model (CCM) [61]. CoSMIC provides platform-dependent metamodels that describe middleware and container configurations, as well as platform-independent metamodels to describe DRE application QoS requirements. These metamodels can be used to provision static and dynamic resources in CIAO. By extending CIAO to support component deployment metadata for QoS policies, such as real-time priorities, DRE applications can be composed from existing components while applying various QoS policies. This capability not only reduces the cost of developing DRE applications but also makes it easier to analyze the consistency of QoS policies applied throughout a system using MDA tools.

All the material presented in this book chapter is based on the CoSMIC Model Driven Middleware tools available for download at www.dre.vanderbilt.edu/cosmic. The associated component middleware CIAO can be downloaded in open-source format from www.dre.vanderbilt.edu/Download.html.

Bibliography

[1] Aeronautics, L. M. (2003) Lockheed Martin (MDA Success Story) http://www.omg.org/mda/mda_files/LockheedMartin.pdf.

[2] Allen, P. (2002) Model Driven Architecture. *Component Development Strategies*, **12**(1).

[3] Batory, D., Singhal, V., Thomas, J., Dasari, S., Geraci, B., and Sirkin, M. (1994) The GenVoca Model of Software-System Generators. *IEEE Software*, **11**(5), 89–94.

[4] Baxter, I., Pidgeon, C., and Mehlichm, M. (2004) DMS: Program Transformation for Practical Scalable Software Evolution. *International Conference on Software Engineering (ICSE)*, Edinburgh, Scotland, May 2004.

[5] Buck, J. T., Ha, J. T., S., Lee, E. A., and Messerschmitt, D. G. (1994) Ptolemy: A Framework for Simulating and Prototyping Heterogeneous Systems. *International Journal of Computer Simulation*, special issue on Simulation Software Development, **4**(2), 155–182.

[6] Campbell, R., Singhai, A., and Sane, A. (1998) Quarterware for Middleware. *Proceedings of the 18th International Conference on Distributed Computing Systems (ICDCS)*. 136–150.

[7] Cross, J. K. and Schmidt, D. C. (2002) Applying the Quality Connector Pattern to Optimize Distributed Real-time and Embedded Middleware, in *Patterns and Skeletons for Distributed and Parallel Computing* (eds F. Rabhi and S. Gorlatch), Springer-Verlag, London.

[8] Czarnecki, K. and Eisenecker, U. (2000) *Generative Programming: Methods, Tools, and Applications*, Addison-Wesley, Boston, MA.

[9] de Miguel, M. A. (2002) QoS-Aware Component Frameworks. *The* 10th *International Workshop on Quality of Service (IWQoS 2002)*, Miami Beach, FL.

[10] Domain, W. A. (2004) *Extensible Markup Language (XML)*, http://www.w3c.org/XML.

[11] Fábio, M. C. and Gordon, S. B. (1999) A Reflective Architecture for Middleware: Design and Implementation. *ECOOP'99, Workshop for PhD Students in Object Oriented Systems*.

[12] Gokhale, A. (2003) *Component Synthesis using Model Integrated Computing*, www.dre.vanderbilt.edu/cosmic.

[13] Gokhale, A., Natarajan, B., Schmidt, D. C., Nechypurenko, A., Gray, J., Wang, N., Neema, S., Bapty, T., and Parsons, J. (2002) CoSMIC: An MDA Generative Tool for Distributed Real-time and Embedded Component Middleware and Applications.

Proceedings of the OOPSLA 2002 Workshop on Generative Techniques in the Context of Model Driven Architecture, ACM, Seattle, WA.

[14] Gordon, S. B., Coulson, G., Robin, P., and Papathomas, M. (1998) An Architecture for Next Generation Middleware, *Proceedings of the IFIP International Conference on Distributed Systems Platforms and Open Distributed Processing*, Springer-Verlag, London, pp. 191–206.

[15] Gray, J., Bapty, T., and Neema, S. (2001) Handling Crosscutting Constraints in Domain-Specific Modeling. *Communications of the ACM*, **44**(10), 87–93.

[16] Gray, J., Sztipanovits, J., Bapty, T., Neema, S., Gokhale, A., and Schmidt, D. C. (2003) Two-level Aspect Weaving to Support Evolution of Model-Based Software, in *Aspect-Oriented Software Development* (eds R. Filman, T. Elrad, M. Aksit, and S. Clarke), Addison-Wesley, Reading, MA.

[17] Harel, D. and Gery, E. (1996) Executable Object Modeling with Statecharts, *Proceedings of the 18th International Conference on Software Engineering*, IEEE Computer Society Press, Berlin, pp. 246–257.

[18] Harrison, T. H., Levine, D. L., and Schmidt, D. C. (1997) The Design and Performance of a Real-time CORBA Event Service. *Proceedings of OOPSLA '97*, ACM, Atlanta, GA, pp. 184–199.

[19] Hatcliff, J., Deng, W., Dwyer, M., Jung, G., and Prasad, V. (2003) Cadena: An Integrated Development, Analysis, and Verification Environment for Component-Based Systems. *Proceedings of the International Conference on Software Engineering*, Portland, OR.

[20] Joseph, K. C. and Patrick, L. (2001) Proactive and Reactive Resource Reallocation in DoD DRE Systems. *Proceedings of the OOPSLA 2001 Workshop "Towards Patterns and Pattern Languages for OO Distributed Real-time and Embedded Systems"*.

[21] Karsai, G., Neema, S., Bakay, A., Ledeczi, A., Shi, F., and Gokhale, A. (2002) A Model-Based Front-end to ACE/TAO: The Embedded System Modeling Language. *Proceedings of the Second Annual TAO Workshop*, Arlington, VA.

[22] Karsai, G., Sztipanovits, J., Ledeczi, A., and Bapty, T. (2003) Model-Integrated Development of Embedded Software. *Proceedings of the IEEE*, **91**(1), 145–164.

[23] Kiczales, G., Hilsdale, E., Hugunin, J., Kersten, M., Palm, J., and Griswold, W. G. (2001) An overview of AspectJ. *Lecture Notes in Computer Science*, **2072**, 327–355.

[24] Kiczales, G., Lamping, J., Mendhekar, A., Maeda, C., Lopes, C. V., Loingtier, J. M., and Irwin, J. (1997) Aspect-Oriented Programming. *Proceedings of the 11th European Conference on Object-Oriented Programming*.

[25] Kon, F., Costa, F., Blair, G., and Campbell, R. H. (2002) The Case for Reflective Middleware. *Communications of the ACM*, **45**(6), 33–38.

[26] Kopetz, H. (1997) *Real-Time Systems: Design Principles for Distributed Embedded Applications*, Kluwer Academic Publishers, Norwell, MA.

[27] Ledeczi, A., Bakay, A., Maroti, M., Volgysei, P., Nordstrom, G., Sprinkle, J., and Karsai, G. (2001) Composing Domain-Specific Design Environments. *IEEE Computer*, **34**(11), 44–51.

[28] Lin, M. (1999) Synthesis of Control Software in a Layered Architecture from Hybrid Automata, *HSCC*, pp. 152–164.

[29] Liu, C. and Layland, J. (1973) Scheduling Algorithms for Multiprogramming in a Hard-Real-Time Environment. *Journal of the ACM*, **20**(1), 46–61.

[30] Locke, C. D. (1992) Software Architecture for Hard Real-Time Applications: Cyclic Executives vs. Fixed Priority Executives. *The Journal of Real-Time Systems*, **4**, 37–53.

[31] Lu, T., Turkaye, E., Gokhale, A., and Schmidt, D. C. (2003) CoSMIC: An MDA Tool Suite for Application Deployment and Configuration. *Proceedings of the OOPSLA 2003 Workshop on Generative Techniques in the Context of Model Driven Architecture*, ACM, Anaheim, CA.

[32] Matthew, D. (1999) *Rose RealTime–A New Standard for RealTime Modeling: White Paper*, Rose Architect Summer Issue, 1999 edn, Rational (IBM).

[33] Neema, S., Bapty, T., Gray, J., and Gokhale, A. (2002) Generators for Synthesis of QoS Adaptation in Distributed Real-Time Embedded Systems. *Proceedings of the ACM SIGPLAN/SIGSOFT Conference on Generative Programming and Component Engineering (GPCE'02)*, Pittsburgh, PA.

[34] Networks, L. G. (2003) Optical Fiber Metropolitan Network, http://www.omg.org/mda/mda_files/LookingGlassN.pdf.

[35] Obj (2001a) *Model Driven Architecture (MDA)*, OMG Document ormsc/2001-07-01 edn.

[36] Obj (2001b) *Unified Modeling Language (UML) v1.4*, OMG Document formal/2001-09-67 edn.

[37] Obj (2002a) *CORBA Components*, OMG Document formal/2002-06-65 edn.

[38] Obj (2002b) *Real-time CORBA Specification*, OMG Document formal/02-08-02 edn.

[39] Obj (2002c) *UML Profile for CORBA*, OMG Document formal/02-04-01 edn.

[40] Obj (2003a) *Deployment and Configuration Adopted Submission*, OMG Document ptc/03-07-08 edn.

[41] Obj (2003b) *UML Profile for Modeling Quality of Service and Fault Tolerance Characteristics and Mechanisms Joint Revised Submission*, OMG Document realtime/03-05-02 edn.

[42] Obj (2003c) *UML Profile for Schedulability*, Final Draft OMG Document ptc/03-03-02 edn.

[43] Object Technology International, Inc. (2003) *Eclipse Platform Technical Overview: White Paper*, Updated for 2.1, Original publication July 2001 edn, Object Technology International, Inc.

[44] Ogata, K. (1997) *Modern Control Engineering*, Prentice Hall, Englewood Cliffs, NJ.

[45] Olaf, S., Gal, A., and Schröder-Preikschat, W. (2002) Aspect C++: An Aspect-Oriented Extension to C++. *Proceedings of the 40th International Conference on Technology of Object-Oriented Languages and Systems (TOOLS Pacific 2002).*

[46] Pal, P., Loyall, J., Schantz, R., Zinky, J., Shapiro, R., and Megquier, J. (2000) Using QDL to Specify QoS Aware Distributed (QuO) Application Configuration. *Proceedings of the International Symposium on Object-Oriented Real-time Distributed Computing (ISORC)*, IEEE/IFIP, Newport Beach, CA.

[47] Schantz, R., Loyall, J., Schmidt, D., Rodrigues, C., Krishnamurthy, Y., Pyarali, I. (2003) Flexible and Adaptive QoS Control for Distributed Real-time and Embedded Middleware. *Proceedings of Middleware 2003, 4th International Conference on Distributed Systems Platforms*, IFIP/ACM/USENIX, Rio de Janeiro, Brazil.

[48] Railways, A. (2003) Success Story OBB, http://www.omg.org/mda/mda_files/SuccessStory_OeBB.pdf/.

[49] Rajkumar, R., Juvva, K., Molano, A., and Oikawa, S. (1998) Resource Kernel: A Resource-Centric Approach to Real-Time Systems. *Proceedings of the SPIE/ACM Conference on Multimedia Computing and Networking*, ACM.

[50] Ritter, T., Born, M., Unterschütz T., and Weis, T. (2003) A QoS Metamodel and its Realization in a CORBA Component Infrastructure. *Proceedings of the* 36th *Hawaii International Conference on System Sciences, Software Technology Track, Distributed Object and Component-based Software Systems Minitrack, HICSS 2003*, HICSS, Honolulu, HI.

[51] Schmidt, D. C., Schantz, R., Masters, M., Cross, J., Sharp, D., and DiPalma, L. (2001) Towards Adaptive and Reflective Middleware for Network-Centric Combat Systems. *CrossTalk*.

[52] Sharp, D. C. (1998) Reducing Avionics Software Cost Through Component Based Product Line Development. *Proceedings of the 10th Annual Software Technology Conference*.

[53] Sharp, D. C. and Roll, W. C. (2003) Model-Based Integration of Reusable Component-Based Avionics System. *Proceedings of the Workshop on Model-Driven Embedded Systems in RTAS 2003*.

[54] Sprinkle, J. M., Karsai, G., Ledeczi, A., and Nordstrom, G. G. (2001) The New Metamodeling Generation. *IEEE Engineering of Computer Based Systems*, IEEE, Washington, DC.

[55] Stankovic, J. A., Zhu, R., Poornalingam, R., Lu, C., Yu, Z., Humphrey, M., and Ellis, B. (2003) VEST: An Aspect-based Composition Tool for Real-time Systems. *Proceedings of the IEEE Real-time Applications Symposium*, IEEE, Washington, DC.

[56] Sztipanovits, J. and Karsai, G. (1997) Model-Integrated Computing. *IEEE Computer*, **30**(4), 110–112.

[57] Technologies, B n.d. Quality Objects (QuO), www.dist-systems.bbn.com/papers.

[58] TimeSys (2001) TimeSys Linux/RT 3.0, www.timesys.com.

[59] van Deursen, A., Klint, P., and Visser, J. (2002) Domain-Specific Languages, http://homepages.cwi.nl/ jvisser/papers/dslbib/index.html.

[60] Wang, N., Schmidt, D. C., Gokhale, A., Gill, C. D., Natarajan, B., Rodrigues, C., Loyall, J. P., and Schantz, R. E. (2003a) Total Quality of Service Provisioning in Middleware and Applications. *The Journal of Microprocessors and Microsystems*, **27**(2), 45–54.

[61] Wang, N., Schmidt, D. C., Gokhale, A., Rodrigues, C., Natarajan, B., Loyall, J. P., Schantz, R. E., and Gill, C. D. (2003b) QoS-enabled Middleware, in *Middleware for Communications* (ed. Q. Mahmoud), Wiley & Sons, New York.

8

High-Performance Middleware-Based Systems

Shikharesh Majumdar
Carleton University

8.1 Introduction

The ability to run distributed applications over a set of diverse platforms is crucial for achieving scalability as well as gracefully handling the evolution in hardware and platform design. Alterations to an existing system are made at the application level as well. Additional components are added to an existing system for handling an increase in workload or the incorporation of new features in a telecommunication or an embedded application for example. Because of the continuous improvement in computing technology, the newly added components are often built using a technology that is different from that used for implementing the legacy components. Moreover, the new feature in the embedded system may require a special platform for execution. An effective middleware system is required to provide the glue that holds such a heterogeneous distributed system together, and achieve high system performance. Using such middleware software, it is possible for two application components written in different languages and implemented on top of different operating systems to communicate with one another. Early work on middleware for process-oriented systems are discussed in [34]. This chapter, however, focuses on Distributed Object Computing (DOC), which is currently one of the most desirable paradigms for application implementation [30]. It is concerned with the performance of middleware systems that provide this communication infrastructure and interoperability in a heterogeneous distributed environment.

A number of middleware standards such as CORBA, DCOM, and DCE are discussed in the literature. We focus on CORBA that is widely used by distributed system builders.

Middleware for Communications. Edited by Qusay H. Mahmoud
 ISBN 0-470-86206-8

There is a great deal of interest in client-server systems that employ DOC [3]. Heterogeneity often appears in client-server systems that span multiple different platforms. DOC combines the advantages of distributed computing such as concurrency and reliability with the well-known attributes of Object Oriented (OO) technology: encapsulation, polymorphism, and inheritance. It promotes reusability of software and enhances the use of Commercial Off-The-Shelf software (COTS) components for the construction of large applications as well as provides high performance by distributing application functionality over multiple computing nodes. The Common Object Request Broker Architecture (CORBA) [20] is a standard proposed by the Object Management Group for the construction of object- oriented client-server systems in which clients can receive service from servers that may run on diverse platforms. Both the clients and servers use a common standard Interface Definition Language (IDL) for interfacing with the Object Request Broker (ORB) that provides client-server intercommunication as well as a number of other facilities such as location and trading services through the ORB agent [21]. A high-level description of the CORBA architecture and its use in heterogeneous OO environments are discussed in [36].

A large body of research exists in the field of client-server systems. A representative set of previous work that has focused on Distributed Object Computing is discussed. Different coordination models are possible for effectively handling the diverse design issues of distributed systems. A number of models for coordination of the interactions between client, server, and the middleware agent is presented in [3]. The agent provides a dynamic directory service, which keeps track of all object implementations and locates the appropriate server. In many COTS ORB products such as Visibroker [6] and Orbix [12], the agent also performs load balancing. In certain middleware products, this entity is called the *locator* or *name server* but we will refer to the entity as an *agent* in this paper. Various techniques that can be employed in developing the client-side in a distributed application are described in [11].

There is a rapid growth of interest in CORBA middleware. This is reflected in several conferences on the topic and a special issue of the Communications of the ACM on CORBA [28]. A description of the CORBA broker architectural framework in terms of its concepts, roles, and behaviors of its components are discussed in [32]. A number of different architectures for multithreaded systems and their impact on performance of CORBA-based systems are described in [27]. The use of CORBA in enterprise computing and network management is discussed in [31] and [10] respectively. Extension of CORBA to ubiquitous computing is presented in [15], whereas reflective middleware architectures using component-based models are described in [24]. Comparatively less work seems to have been done on the performance impact of middleware systems that this chapter focuses on.

Three approaches to enhancing the performance of CORBA-based systems are presented. These approaches are a result of our research on CORBA performance and are summarized in this chapter. Detailed discussions are provided in the relevant papers included in the bibliography. In the first approach, performance improvement is achieved through the deployment of an effective client-agent-server interaction architecture. The second approach advocates the design of performance-enhanced middleware that exploits limited heterogeneity in systems for performance improvement. The third approach is aimed at the application level and concerns the use of effective techniques for design and implementation of the application. The applicability of an approach depends on the user of the techniques. The first two are useful for middleware system builders, whereas the last is important to application designers and programmers. A brief outline of the rest of the chapter is presented. Section 8.2 discusses the performance issues related to the middleware-based systems. The three approaches to engineering performance into middleware-based systems are presented next. Section 8.3 presents a discussion of the use of an appropriate interaction architecture for improving performance. Section 8.4 outlines the performance-optimization techniques that effectively utilize limited heterogeneity in systems. Performance-improvement techniques that can be deployed at the application level are discussed in Section 8.5, whereas Section 8.6 presents the summary and conclusions.

The chapter is useful for builders and users of high-performance interoperable distributed systems that are built using the general CORBA specifications. A special CORBA specification for real-time system called RT-CORBA is available. These systems need to be predictable, often requiring that hard deadlines be met for transactions. Such systems are considered by other researchers (see [15] for example) but are beyond the scope of the discussion presented in this chapter.

8.2 Performance of CORBA Middleware

A variety of different classes of systems demand high performance. These include real-time avionics applications, process control systems, and image processing applications as well as different types of telecommunication products such as voice and data switches. Two performance metrics are of special interest in such applications and are defined next.

Response Time is a measure of latency. It is the difference in time between the arrival of a request or an event and the time at which response to the request is completed.

Throughput is a measure of system capacity. It is typically measured in number of requests completed per unit time.

Performance-sensitive applications demand low latency and high throughput. A desirable attribute of most systems is *scalability*. A system is called scalable if its throughput grows in proportion with an increase in workload.

Although middleware provides interoperability, if not designed carefully, middleware-based systems can incur a severe performance penalty. A number of papers have captured the performance limitations of several CORBA-based middleware products. The limitations of the original CORBA standard in providing real-time QoS, and in handling partial failures are described in [16]. The paper also presents a CORBA-based framework that uses the Virtual Synchrony model to build reliable distributed systems through synchronous method invocations and asynchronous message passing. Using load-balancing techniques on systems that consist of replicated objects can lead to an improvement in system performance. The integrating of load balancing with name service is proposed in [5]. CORBA has been used in the context of various applications. The performance of a group membership protocol based on CORBA is discussed in [19]. This paper provides a performance comparison of three implementations of a group membership protocol: one implementation uses UDP sockets, while the other two use CORBA for communication. The paper demonstrates that CORBA can be used to implement high-performance group membership protocols for heterogeneous, distributed computing environments. Although there is some performance degradation due to CORBA, the performance degradation can be reduced by carefully choosing an appropriate design.

The latency and scalability problems in conventional middleware products are discussed in [9]. The key sources of overhead and techniques for reducing these are discussed. An investigation of the latency and scalability of two commercial products Orbix-2.1 and Visibroker 2.0 on systems employing high-speed ATM networks is presented in [8]. The authors observe high latency for these CORBA implementations that also give rise to poor system scalability. The nonoptimized internal buffering in the middleware products as well as a large presentation layer overhead leads to a high variability in latency that is unacceptable for many real-time applications. Scalability is a desirable attribute of most systems. Measurements on the ATM-based system also revealed that none of these products could handle a large number of objects per server. New versions of these products with better performance have been introduced subsequently. An important characteristic of some of these products is the use of multithreading in their implementations.

A multithreaded middleware allows the system to effectively handle multiple client requests that occur simultaneously. Since a ready thread can use the CPU when the previously executing thread is blocked for I/O, average waiting times for requests can be minimized. A performance comparison of a number of products with a multithreaded architecture, CORBAplus, HP Orb Plus, miniCOOL, MT-Orbix, TAO, and Visibroker is available in [27]. Applications such as Intelligent Networks, WWW applications, and Collaborative Virtual Reality employ millions of objects globally distributed across a large number of hosts. These systems demand a very high throughput and low-latency method invocations. Scalable partial replication–based middleware and OO approaches for constructing such middleware are discussed in [18]. Performance issues in the context of fault tolerant CORBA-based systems are discussed in [7].

8.3 Impact of Client-Server Interaction Architectures

The client-agent-server interaction architecture is observed to have a strong impact on system performance [1]. A detailed investigation of a number of such architectures that are inspired by [3] is presented in [2]. These architectures refer to the way a request from the client is routed to the server and a reply from the server is sent back to the client and are briefly described.

8.3.1 Three Interaction Architectures

Handle-driven ORB (HORB). When a client wants to request a service, it sends the server name to the agent in the middleware system (see Step 1 in Figure 8.1). The agent performs a name to object reference (IOR) mapping and sends an IOR or handle back to the client (Step 2). The client uses the handle to contact the server (Step 3) and receive the result of the desired service (Step 4). A number of COTS ORB that include Visibroker [6] (previously known as ORBeline) and Orbix-MT [12] and E2ASP [13] use such a handle-driven architecture.

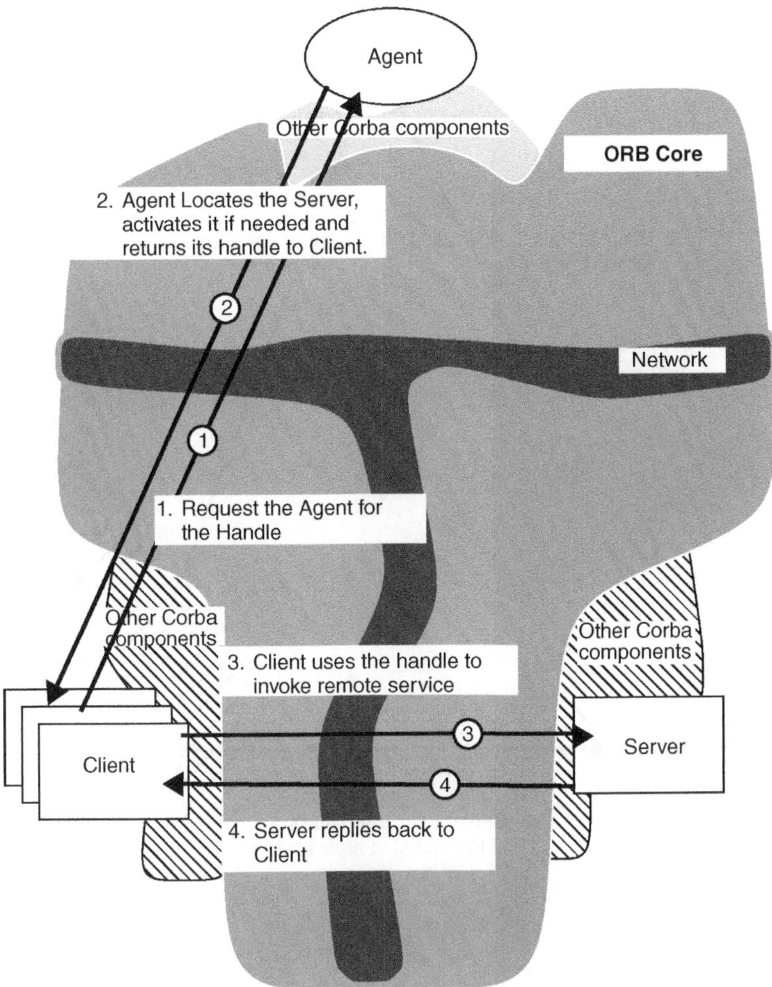

Figure 8.1 The handle-driven architecture (from Abdul-Fatah, I. and Majumdar, S. (2002) Performance of CORBA-Based Client-Server Architectures. *IEEE Transactions on Parallel & Distributed Systems*, **13**(2), 111–127.)

Forwarding ORB (F-ORB). The client sends the entire request for service to the agent (see Step 1 in Figure 8.2). The agent locates the appropriate server and forwards the request as well as the stringified client handle to it (Step 2). The server performs the desired service and sends a response back to the client (Step 3).

Process planner ORB (PORB). In a number of situations, the client needs to invoke methods in multiple servers to complete a given job. A PORB architecture that uses concurrent server invocations is useful in such a situation. An example system in which a job consists of concurrent requests to N servers is presented in Figure 8.3. The client sends a composite request in which all the services (require to complete the entire job) with their respective parameters to the agent (see Step 1 in Figure 8.3). The P-ORB

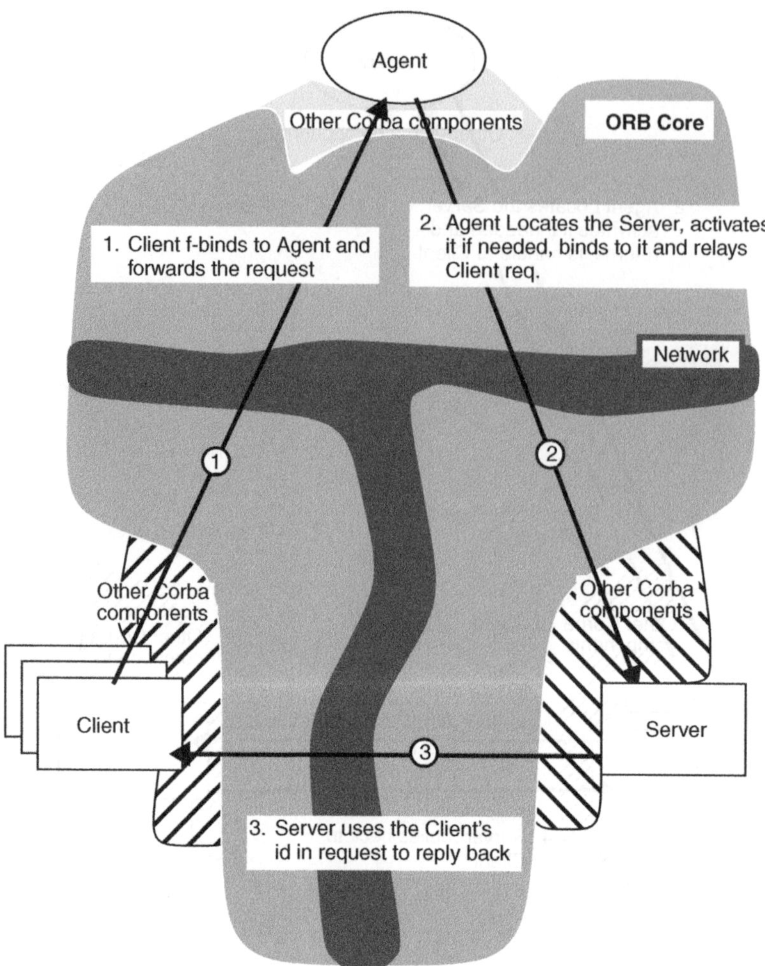

Figure 8.2 The forwarding architecture (from Abdul-Fatah, I. and Majumdar, S. (2002) Performance of CORBA-Based Client-Server Architectures. *IEEE Transactions on Parallel & Distributed Systems*, **13**(2), 111–127.)

agent extracts each individual request from the composite request and determines their precedence relationships. If multiple servers can be activated concurrently, the agent sends a request to each server (Step 2). Each server performs the desired service and sends the result of the operation back to the agent (Step 3). When responses to all the component requests are available, the agent sends a consolidated reply to the client (Step 4).

8.3.2 Performance Comparison

A comparison of the performances of these architectures is presented in Figure 8.4. The system throughput is measured on a network of SPARC workstations running Solaris 2.5. The prototypes for the three architectures are based on ORBeline, a CORBA-compliant

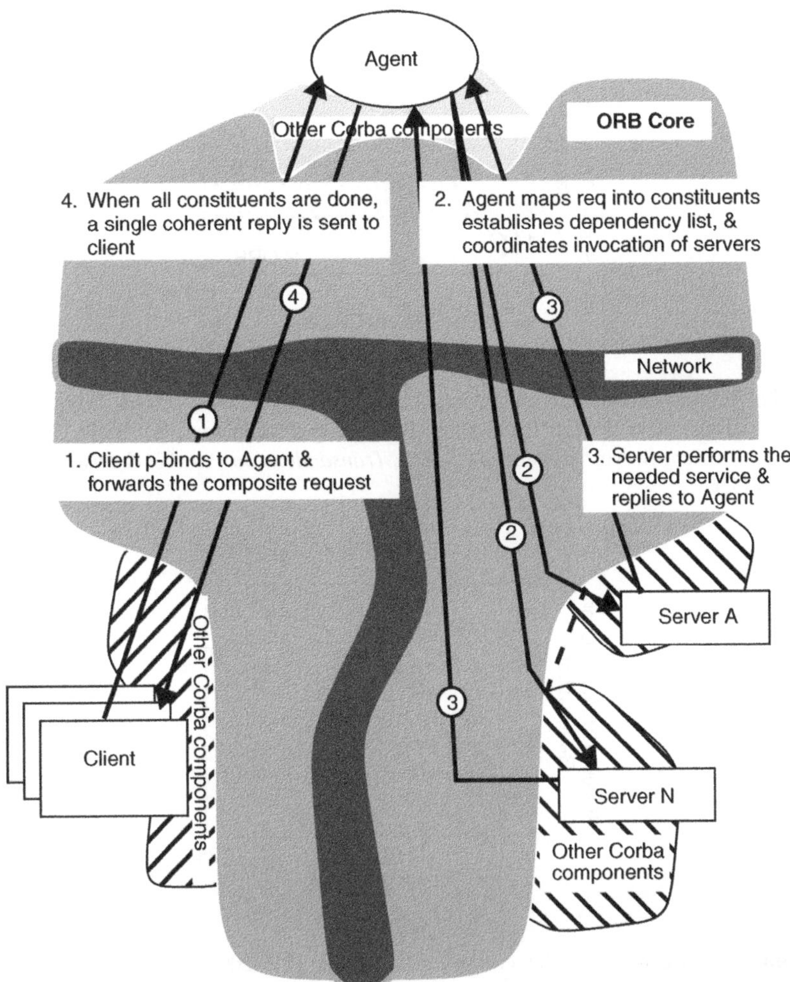

Figure 8.3 The process planner architecture (from Abdul-Fatah, I. and Majumdar, S. (2002) Performance of CORBA-Based Client-Server Architectures. *IEEE Transactions on Parallel & Distributed Systems*, **13**(2), 111–127.)

middleware [23] that is a predecessor of Visibroker [6]. A synthetic workload consisting of N clients, two replicated servers with fixed CPU demands of S1 and S2, message length L, and a simulated internetwork delay of D was used. Clients operate cyclically with two server invocations in each cycle. Two servers are invoked sequentially in case of the H-ORB and the F-ORB, whereas the two servers are concurrently invoked in case of the P-ORB. A closed system is used: upon completion of a cycle, the client repeats the cycle. Figure 8.4 is representative of the results observed. More data is available in [2]. As shown in Figure 8.4b(a), at low system load (smaller N), the F-ORB and the P-ORB perform better than the H-ORB. At higher loads (high N), the F-ORB and the P-ORB saturate, and the H-ORB demonstrates a superior performance. The synchronous

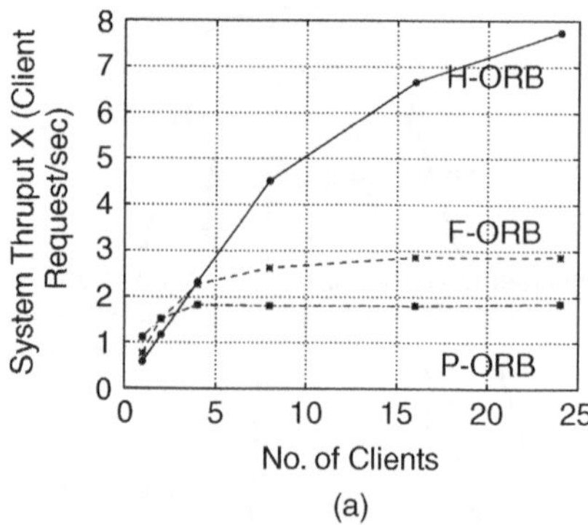

Figure 8.4a Performance of the architectures with no cloning. D = 200 ms, MsgSz(L) = 4800 Bytes, S1 = 10 ms, S2 = 15 ms (from Abdul-Fatah, I. and Majumdar, S. (2002) Performance of CORBA-Based Client-Server Architectures. *IEEE Transactions on Parallel & Distributed Systems*, **13**(2), 111–127.)

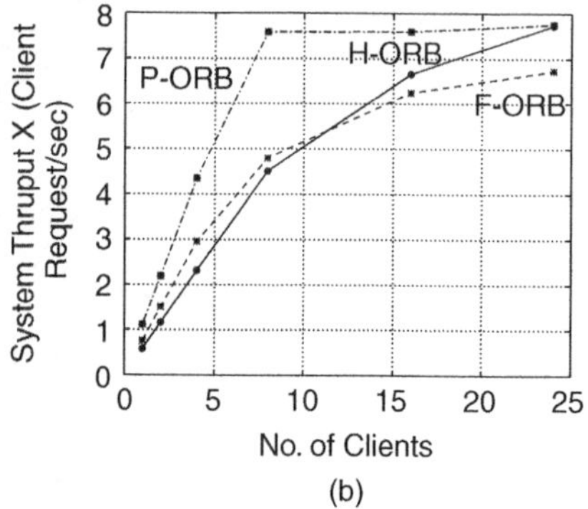

Figure 8.4b Performance of architectures with 8 clones. D = 200 ms, MsgSz(L) = 4800 Bytes, S1 = 10 ms, S2 = 15 ms (from Abdul-Fatah, I. and Majumdar, S. (2002) Performance of CORBA-Based Client-Server Architectures. *IEEE Transactions on Parallel & Distributed Systems*, **13**(2), 111–127.)

communication used leads to a software bottleneck and the throughputs of the F-ORB and the P-ORB flatten off at higher values of N. The software bottleneck can be alleviated by using agent cloning that deploys multiple processes at the agent (see Figure 8.4b(b)).

Each of the three architectures H-ORB, F-ORB, and P-ORB has a special feature that makes it more attractive than the others in a specific situation. The marshaling/unmarshaling cost is small in the H-ORB for messages between the agent and the clients. Four messages are required, however, to complete a method invocation by the client. Both the P-ORB and F-ORB result in a smaller number of message transfers per method invocation in comparison to the H-ORB. The P-ORB also uses concurrency in server execution to improve performance. Note that the client in an F-ORB is more complex than a client in the H-ORB. This is because the request is sent by the client to the agent, but the reply is received from the server. Research is required for hiding this complexity from the application programmer. Analytic models for evaluating the performance of these middleware architectures is presented in [22]. A detailed discussion of the performance of these architectures is presented in [2] and some of the major observations are summarized.

- Latency-Scalability Trade-off: There seems to be a trade-off between latency obtained at low load and throughput obtained at higher loads when a single agent process is used. The architecture that exhibits best latency characteristics at low load may rank lower when scalability at high load is considered. The P-ORB that uses concurrency in server execution and the F-ORB that reduces the number of messages perform much better than the H-ORB at lower system loads. The H-ORB, however, demonstrates a much better scalability in comparison with the two other architectures in the absence of agent cloning. It achieves a higher throughput at higher system loads when the two other architectures are saturated (see Figure 8.4b(a)).
- Software Bottlenecks and Agent Cloning: The agent in the F-ORB and the P-ORB are more complex than an HORB agent. Moreover, they are observed to spend a greater amount of time in a "blocked waiting for a message to be received" state in comparison to an HORB agent [2]. This can lead to a software bottleneck at the agent. The fact that the performance bottleneck is a software process and not a hardware device is an interesting feature of these client-server systems that use a synchronous mechanism for message transfer. The detrimental effects of software bottlenecks are effectively reduced by process cloning, which uses multiple copies of the agent process.
- Concurrency versus Additional Waiting: The P-ORB uses concurrency in server execution for improving system performance. However, this architecture results in a larger number of interactions between servers and the agent in comparison to the two other architectures. Since the agent communicates with a large number of entities (all clients and all servers), the probability of a server waiting when returning the computation results to the agent increases with the system load. The client in an F-ORB, on the other hand, waits for the server and releases the server as soon as the reply is received. This extra waiting experienced by the server counteracts the performance improvement that accrues from parallelism in server execution. As a result, in some situations the overall performance of the P-ORB can be worse than that of the F-ORB (see [2] for details).

The combination of the good attributes of the FORB and the HORB into an *Adaptive ORB (A-ORB)* is described in [29]. Depending on the system load, the interaction architecture switches between a handle-driven and a forwarding mode. The adaptive architecture is observed to perform significantly better than a pure H-ORB or a pure F-ORB for a broad range of system load. A detailed report on the investigation of adaptive middleware is presented in [17].

8.4 Middleware Performance Optimization

A number of issues that concern the performance of CORBA-based middleware are described in the literature. Several sources of overhead in IIOP implementations such as excessive marshaling/unmarshaling operations, data copying, and high-level function calling are identified in [27]. The performance-optimization techniques suggested by the author include optimizing for the common case, eliminating waste of CPU cycles that occurs during data copying, for example, replacing general purpose methods with specialized efficient methods, as well as precomputation and caching of values. Applying these techniques to SunSoft IIOP produced a significant performance improvement for various types of messages transmitted over an ATM network. Optimization techniques for the collocation of CORBA objects are described in [37].

Techniques that exploit the increasing availability of parallel platforms are discussed in [14]. Using moderately priced Symmetric Multiprocessors (SMP's) for hosting CORBA servers can lead to high-performance systems. The paper demonstrates the viability of using multithreaded systems using a thread per request scheme and discusses how locality-based techniques can improve server performance on such multithreaded systems.

8.4.1 Systems with Limited Heterogeneity

Many distributed systems that include a variety of different embedded systems are characterized by limited heterogeneity. That is, only a few components are built using different technologies (programming language/operating system), whereas the majority of the components are implemented in the same programming language and run on top of the same operating system. A mobile wireless access server researched by Nortel Networks to support mobile users in accessing enterprise applications when they are traveling is an example of such a system and is described by Parsons in a personal communication [25]. The system provides a number of functionalities that include controlling of incoming calls, personalized interactive voice response for callers, and web-based access for wireless devices. Most core functionalities are provided by Java components, whereas C++ components are used for communicating with backend systems using third party software that provide runtime libraries written in C++. Most of the intercommunications occur between Java components with occasional interactions between a Java and a C++ component.

It is possible to exploit such limited heterogeneity in systems for achieving high system performance [4]. The basic idea is to use a CORBA-compliant ORB when two dissimilar components (Java and C++ component in the previous example) communicate and bypass the CORBA-compliant ORB and use a faster private path when two similar components (two Java components in the previous example) interact with each other. Three performance-enhancement techniques for CORBA-based middleware that exploit limited heterogeneity in systems are described in [4] and are exploited in building a system called *flyover* that is described next.

8.4.2 Flyover

A flyover is a fast path for communication among similar components in a system that is characterized by limited heterogeneity. With this architecture, the client/servers that are developed using different technologies communicate through the CORBA-compliant

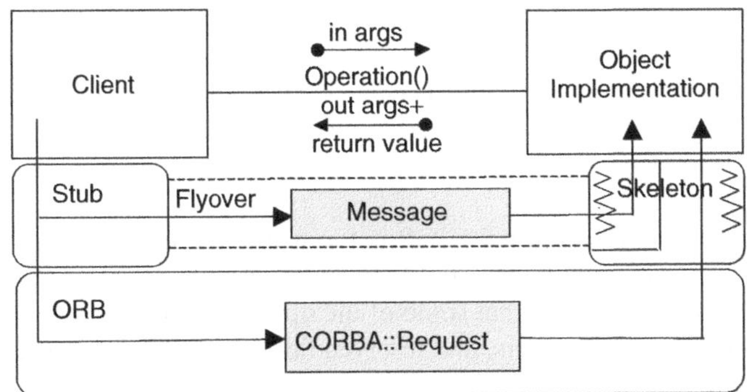

Figure 8.5 General architecture of the flyover (from Wu, W.-K. and Majumdar, S. (2002) Engineering CORBA-Based Systems for High Performance. *Proceedings of the International Conference on Parallel Processing (ICPP'02)*, Vancouver, Canada, August 2002.)

ORB, whereas components built using the same programming language and operating system use a *flyover*, which incorporates private protocols and bypasses a number of standard CORBA operations. The flyover not only achieves a higher performance by exploiting by saving in terms of CORBA overheads but also preserves both the client/server application transparency and location transparency, without ORB source code modification. Using a flyover, the application designers can focus on their core applications rather than developing their own proprietary method for performance optimization.

Figure 8.5 shows the general architecture of the flyover. There are two paths through which the client/server pair can communicate. For clients and servers built using different programming languages or different operating systems, the interprocess communication is mediated through the ORB. However, if the client and server use the same programming language, operating system, and hardware platforms, they can communicate directly by a low-level communication interface and proprietary protocols to bypass the ORB (see Figure 8.5). This avoids the additional overhead introduced by the CORBA-compliant middleware and achieves a performance gain.

The main goal of the flyover is to achieve performance enhancement on systems with limited heterogeneity. The following techniques are used to achieve such a performance gain.

- Preventing CDR Conversion: If the nodes on which the client and server are located use the same kind of hardware architecture, the overhead of data conversion can be avoided.
- Removing Padding Bytes: Padding bytes are used in CDR to fill up the gap between primitive data types that can start at specific byte boundaries. Padding bytes do not have any semantic value, but they consume communication bandwidth. The primitive data types are not aligned with specific byte boundaries within the "optimized GIOP" message used in the flyover. As a result, this strategy can minimize the size of GIOP messages and save communication time. A more detailed discussion is provided in [38]. Removal of padding bytes are explained further in Section 8.6.2.

- Direct Socket Communication: The intercommunication between the client and server is handled by a private BSD socket [33].
- Efficient Request Demultiplexing: An *Object ID* replaces the object key to uniquely identify the target object in the server during request demultiplexing. Moreover, an *Operation ID* replaces the operation name to uniquely identify the static function of the target object.
- GIOP Optimization: The original GIOP is too heavyweight for high-speed communication. To provide a smaller message footprint, the protocol used in the flyover is a highly optimized GIOP. Note that some of the optimization techniques such as removing magic field, version numbers, and reserved fields in the optimized GIOP are similar to those described in [26].

8.4.2.1 The Interface Definition Language Processor

To incorporate the flyover into CORBA-based applications, we have developed a tool prototype called the *Interface Definition Language Processor (IDLP)*[38]. The application developer uses this tool instead of the original IDL compiler for compiling the IDL interface specification files. Figure 8.6 describes the steps involved in generating client and server executables using the IDLP (the current version of which supports only C++). IDLP consists of two main components: the IDL parser and the C++ parser. The IDL parser analyzes the IDL interfaces defined in the IDL interface specification file and modifies this specification file. It also writes relevant information (e.g., interface name and interface function name) into data structures that are shared by the IDL parser and the C++ parser. This information is useful in analyzing the structures of the client stub, server skeleton, and the header file. In the next step, IDLP calls the CORBA IDL compiler (supplied by the ORB vendor) to generate the header file, client stub, and the server skeleton.

The C++ parser then analyzes the structure of the header file, client stub, and server skeleton and performs the necessary modifications required for the incorporation of the

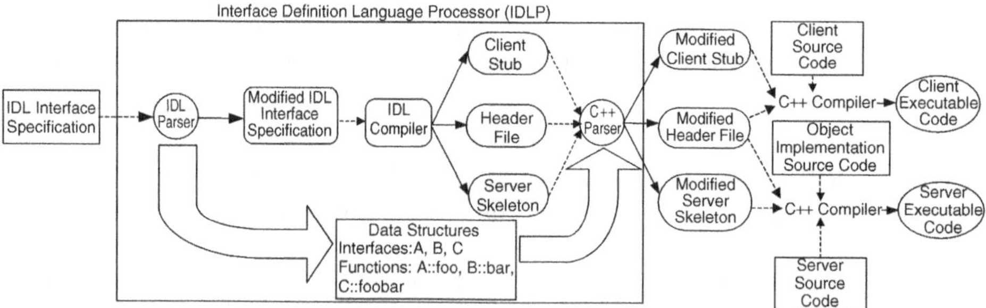

Figure 8.6 Generation of client and server executable codes using the IDLP (from Wu, W.-K. and Majumdar, S. (2002) Engineering CORBA-Based Systems for High Performance. *Proceedings of the International Conference on Parallel Processing (ICPP'02)*, Vancouver, Canada, August 2002.)

flyover into the CORBA-based applications. From the application programmer's point of view, there is no change in the CORBA application development process except that the IDLP is used instead of the vendor-supplied IDL compiler in compiling IDL specification files. The IDLP tool is implemented by using the scripting language Perl.

8.4.3 Performance of Flyover

The flyover prototype described in [38] is based on Iona's Orbix-MT2.3. Using a synthetic workload, we ran a number of experiments to compare the performance of the flyover, Orbix-MT 2.3 and Orbix 2000 with different system and workload parameters. The experiments were run on a network of SPARC workstations running under Solaris 2.6. A representative set of results is presented in this chapter. A complete set of experimental results can be found in [38]. Figure 8.7 and Table 8.1 show the performance of the flyover for CORBA arrays of long. These results correspond to experiments in which a client invokes a method in a server with an array of specific size as argument. The server has a zero service time and the called method returns immediately upon invocation. The response time is the difference between the time at which the response from the server is received by the client and the time at which a method was invoked by the

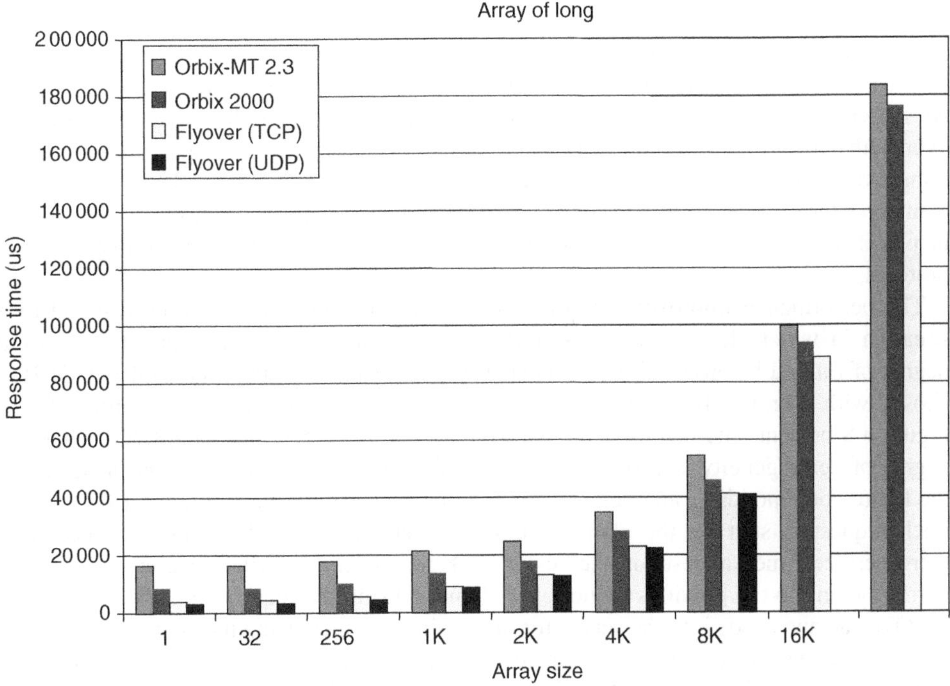

Figure 8.7 Performance of flyover (from Wu, W.-K. and Majumdar, S. (2002) Engineering CORBA-Based Systems for High Performance. *Proceedings of the International Conference on Parallel Processing (ICPP'02)*, Vancouver, Canada, August 2002.)

Table 8.1 Performance improvement for CORBA array of long (from Wu, W.-K. and Majumdar, S. (2002) Engineering CORBA-Based Systems for High Performance. *Proceedings of the International Conference on Parallel Processing (ICPP'02)*, Vancouver, Canada, August 2002.)

Array size	Improvement of flyover (TCP) over Orbix-MT 2.3	Improvement of flyover (TCP) over Orbix 2000	Improvement of flyover (UDP) Flyover (TCP)
1	77.06%	55.37%	19.99%
32	73.88%	49.30%	24.53%
256	69.77%	45.91%	17.49%
1K	58.51%	34.37%	3.16%
2K	47.94%	26.58%	3.72%
4K	34.44%	18.81%	2.39%
8K	24.05%	9.84%	1.08%
16K	10.94%	5.32%	–
32K	5.94%	2.11%	–

client. The array sizes are chosen in such a way that the performance comparisons can be done for a range of response times. The array sizes used in the experiment vary from 1 to 128 K. Note that because the upper bound on the size of UDP data packet is 65535 bytes, results for Flyover (UDP) are not available for array sizes of 16 K and 32 K. As shown in the figure and table, the flyover exhibits a substantially higher performance over both Orbix-MT 2.3 and Orbix 2000 in sending an array of long. For example, in transferring an array long [33], the flyover achieves a 73.88% and a 49.30% improvement over Orbix-MT 2.3 and Orbix 2000 respectively. A significant further improvement in performance can be achieved by using UDP instead of TCP as the transport layer protocol.

The performance improvements discussed in the previous paragraph correspond to a situation in which the flyover is used 100% of the time. A system characterized by a degree of limited heterogeneity of p is analyzed. On such a system, the CORBA ORB is chosen with a probability of p, whereas the flyover is chosen with a probability of 1-p. Figure 8.8 presents the results of transferring arrays long[256] and long[8192] when the degree of heterogeneity is varied from 0 to 1. The results of the experiment presented in the figure corresponds to an open system in which requests arrive and join a FIFO queue. Each request is sent by the client to a server that processes the request immediately (zero service time) and sends the result back to the client. After receiving the result corresponding to the previous request, the client picks up the request from the head of the FIFO queue and sends it to the server for processing. By increasing the arrival rate, we can increase the system load and investigate the impact of contention for system resources. When the degree of heterogeneity p is 1, the system is totally heterogeneous. The response time curve shown in ($p = 1$) corresponds to the performance achieved with Orbix-MT 2.3. However, p equal to 0 indicates the system is totally homogeneous and corresponds to a system in which the flyover is taken 100% of the time. When p is between 0 and 1, the response time curve is between the lower bound (the flyover prototype) and the

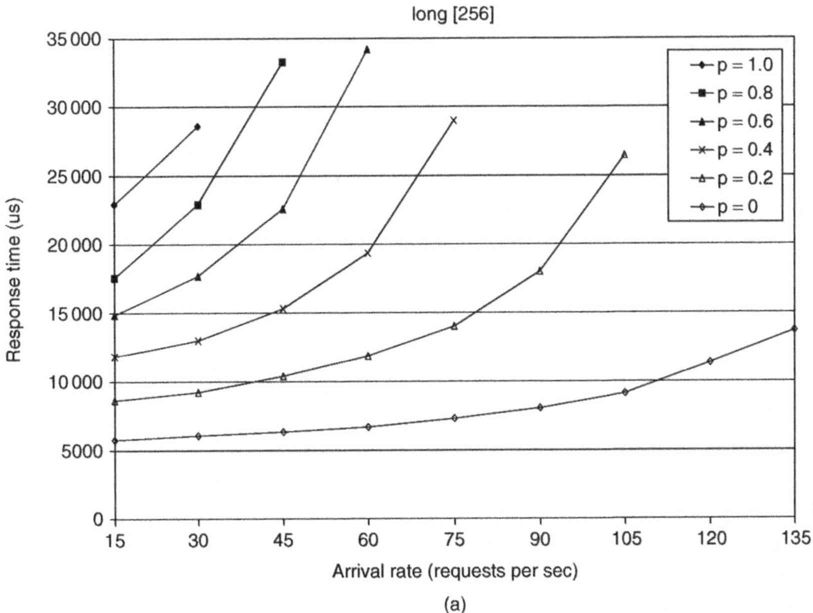

Figure 8.8a Performance of system with limited heterogeneity for an array of long[256] (from Tao, W. and Majumdar, S. (2002) Application Level Performance Optimizations for CORBA-Based Systems. *Proceedings of the International Workshop on Software and Performance (WOSP'02)*, Rome, Italy, July 2002.)

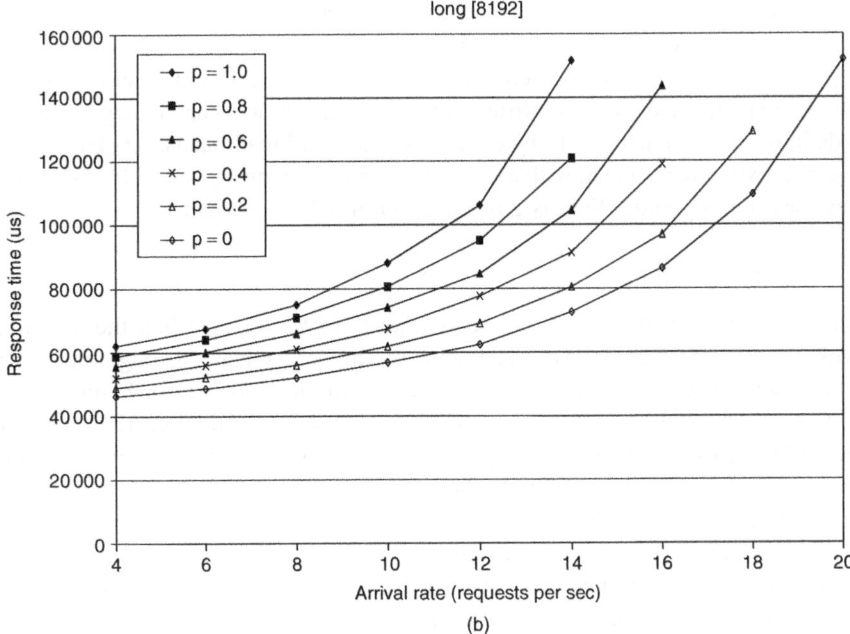

Figure 8.8b Performance of system with limited heterogeneity for an array of long[8192] (from Tao, W. and Majumdar, S. (2002) Application Level Performance Optimizations for CORBA-Based Systems. *Proceedings of the International Workshop on Software and Performance (WOSP'02)*, Rome, Italy, July 2002.)

upper bound (Orbix-MT 2.3). As the degree of heterogeneity decreases, a higher number of CORBA invocations are passing through the flyover. Hence, the response time curve shifts down and moves close to the curve corresponding to a totally homogeneous system.

This experiment shows that even when a small subset of similar communicating components exists within the system (e.g., with a degree of heterogeneity $p = 0.8$), the overall system performance can still be improved substantially. For example, in sending an array long[256] at a rate of 30 requests per second, the flyover shows a 21% improvement in response time approximately when p is equal to 0.8 (see Figure 8.8a). The performance improvement introduced by the flyover is more pronounced at higher system loads.

Overheads

The performance optimization techniques improve system performance in different ways, but their incorporation into the system through the flyover introduces an additional overhead. The primary impact of this overhead is the increase in binding time in comparison to a regular CORBA ORB. In many systems, the location of servers is static. This increase in binding time can be amortized over multiple invocations on the same object reference by the client. Thus, the flyover can give rise to a substantial performance improvement in systems in which clients make multiple calls to the same server. A detailed discussion of the overhead is presented in [38].

8.5 Application Level Performance Optimizations

The performance optimization techniques discussed in the previous sections require engineering of the middleware product. The techniques presented in this section are at the application level. A set of guidelines for performance enhancement that can be used during application design and implementation is described in [35] and is summarized. The guidelines are concerned with four different issues: connection setup latency, placement of methods into servers, packing of objects into servers, and load balancing. The guidelines are based on experiments made with a synthetic workload running on a network of SPARC workstations running under Solaris 2.6. Iona's Orbix- MT2.3 was used.

8.5.1 Connection Setup Latency

A connection between the client and server needs to be set up when the client wants to invoke a method in the server. The state of the server has a strong impact on the connection setup latency: setting up a connection with an active server is much faster in comparison to setting up a connection with a server that is currently inactive. In resource-constrained systems in which it is not possible to keep all the servers active, maintaining an active state for at least the popular servers that are used often can lead to a large reduction in the connection setup delay.

Clients and servers are typically compiled into separate processes. When the client and server are allocated on the same node, collocating the client-server pair in the same address space is recommended: an order of magnitude improvement in performance is achieved when the client-server collocation is used [35]. Note that some middleware vendors provide special functions for achieving such a collocation.

8.5.2 Parameter Passing

Complex data structures are often passed between a client and its server. Significant performance differences are observed among different ways of parameter passing. The performance differences are observed to increase with the size of the parameter list. Unraveling complex structures into the component primitive types is observed to give rise to the best performance. The performance benefits arise primarily from the reduction of the number of padding bytes used in a CDR representation of the parameters transmitted on the wire. An example that captures the performance benefit is presented next.

Figure 8.9 displays three different ways of parameter passing involving an array of structures. Option I (see Figure 8.9a) is the most natural in which the array of structures is directly passed. In Option II (see Figure 8.9b), the array of structures is unraveled into two component arrays of primitive data types, whereas in Option III (se Figure 8.9c), two methods each with a separate parameter are invoked. We assume that the methods in three different cases are implemented in such a way that the same functional result is achieved with each option. The relative performances of the three options are presented in Figure 8.10 that displays the response time for a zero-server execution time observed with each option normalized with respect to that achieved with Option I.

Option II demonstrates the best performance. Its relative performance with respect to the most natural Option I increases with the array size. A performance improvement of over 100% is observed, for example, for arrays with 1000 or more elements. It is

Option I : Method 1 with one parameter
struct CD [size] {
char c; double d}
(a)

Figure 8.9a Different ways of parameter passing–option I (from Tao, W. and Majumdar, S. (2002) Application Level Performance Optimizations for CORBA-Based Systems. *Proceedings of the International Workshop on Software and Performance (WOSP'02)*, Rome, Italy, July 2002.)

Option II : Method 1 with two parameters
char [size] c; double [size] d;
(b)

Figure 8.9b Different ways of parameter passing–option II (from Tao, W. and Majumdar, S. (2002) Application Level Performance Optimizations for CORBA-Based Systems. *Proceedings of the International Workshop on Software and Performance (WOSP'02)*, Rome, Italy, July 2002.)

Option III : Method 1 with parameter char [size] c;
Method 2 with parameter double [size] d;
(c)

Figure 8.9c Different ways of parameter passing–option III (from Tao, W. and Majumdar, S. (2002) Application Level Performance Optimizations for CORBA-Based Systems. *Proceedings of the International Workshop on Software and Performance (WOSP'02)*, Rome, Italy, July 2002.)

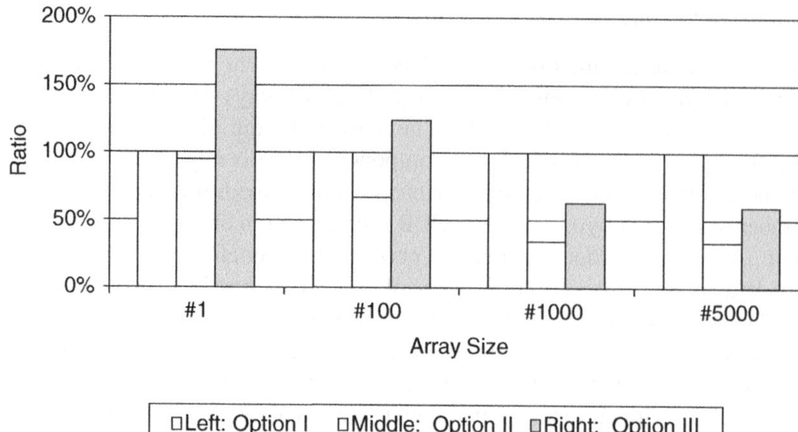

Figure 8.10 The effect of parameter passing on response time with zero server execution time (from Tao, W. and Majumdar, S. (2002) Application Level Performance Optimizations for CORBA-Based Systems. *Proceedings of the International Workshop on Software and Performance (WOSP'02)*, Rome, Italy, July 2002.)

interesting to note that although at lower array sizes Option III is inferior in performance to Option I, Option III assumes the position of the second best performer at higher array sizes.

The rationale behind system behavior lies in the data alignment specified in CORBA's CDR format that is used by the ORB for transferring data over the wire. Since data types such as long and double have to start at specific byte boundaries, padding bytes may be used between two different types of data elements. For example, if a double is to follow a char and the char starts at position 0, the double will start at position 8, resulting in padding bytes between the two elements. Padding bytes do not have any semantic value, but consume bandwidth. By unraveling the structure, Option II leads to an efficient packing of the elements of the same type together and reduces the number of padding bytes. The concomitant savings are two fold: in terms of communication time as well as in terms of CPU time consumed in marshaling and unmarshaling. For higher array sizes, Option III performs better than Option I. Although Option III incurs an additional method invocation overhead, this additional overhead is offset at higher array sizes by the reduction in the number of padding bytes.

Although the unraveling of complex data structures (Option II) is observed to produce a performance benefit for all parameter sizes, a larger performance benefit is expected to accrue for communication-bound systems characterized by larger messages exchanged between the client and the server.

8.5.3 Combination of Methods and Performance Recovery

A secondary, nevertheless important observation is made in the context of parameter passing and discussed in [35]. Additional passes through existing code are sometimes made for reducing system overheads for improving performance. Combining multiple methods into one is often performed during such performance recovery operations for reducing the method invocation overheads. Combining two methods into one makes intuitive sense during system redesign and modification aimed at performance improvement. System modification for performance enhancement is common in the software industry and was observed by the author in the context of telephone switches built by Nortel. The relative performances of Option I and Option III demonstrate that an unexpected result may accrue if this redesigning operation is not performed carefully.

The empirical data in Figure 8.10 demonstrates a nonintuitive result: the system with two method invocations performs better than the system with a single method invocation passing an array of structures when the size of the array is large. The occurrence of such a nonintuitive result indicates that a careful consideration of the application is required before performing such operations.

8.5.4 Method Placement and Object Packing

Compared to the other factors discussed in this section, method placement is observed to have the least impact on performance. The method dispatching latency is reduced by packing a larger number of methods into an object [35]. The position of the method in the interface has a small impact on performance: methods called more frequently should thus be defined as close to the beginning of the interface declaration as possible. Note that this observation is dependent on the search method used by a particular ORB product and may/may not be important in the context of different products.

When all the objects are allocated on the same CPU, the designer should avoid placing the objects on separate servers. Collocating the objects into the same server process or packing all the methods into one object demonstrated a higher performance in comparison to the system in which a separate process is created for each object [35].

8.5.5 Load Balancing

Application-level load balancing on systems with replicated servers can improve system performance significantly. Static load balancing based on *a priori* knowledge of client-server workload characteristics that classified clients into a "busy" and a "nonbusy" class on the basis of their frequency of server invocations is described in [35]. Such a static load-balancing approach can lead to a significant performance improvement in comparison to an "unbalanced" system. Although the availability of such *a priori* knowledge is possible for dedicated applications, it is unlikely to be available during earlier stages of system design. Moreover, in many applications, a client may not remain in one "busy" or "nonbusy" class and may change its behavior, depending on system state. A number of dynamic strategies are useful in such a situation. These strategies are based on the well-known round-robin strategy used in load balancing and scheduling. Both local and global strategies are investigated [35]. The local strategies in which each client selects a server instance independently using a round-robin strategy is preferable for smaller systems. For more complex systems, using a centralized dispatcher-based global round-robin strategy can significantly improve system throughput.

8.6 Summary and Conclusions

Three different approaches to engineering performance into middleware-based systems are presented in this chapter. The first two concern the engineering of the middleware product itself, whereas performance-optimization techniques at the application level are advocated by the third. Thus, the first two techniques are important for middleware system builders, whereas the third is important in application design and implementation.

The first technique concerns the client-agent-server interaction architecture. Three pure and one hybrid architecture are discussed. The handle-driven ORB is observed to be lightweight in construction. The forwarding ORB leads to a smaller number of messages per method invocation, whereas the process planner ORB uses parallel server invocations for improving performance. Hybrid architectures demonstrated superior performance. An adaptive ORB that switches between a handle-driven and a forwarding mode is observed to outperform a pure handle-driven or a pure forwarding architecture.

The second approach exploits limited heterogeneity in systems. The regular CORBA-compliant path is used for client-server interaction when two dissimilar system components interact with one another. A flyover that uses a private socket and an efficient communication protocol is used when two similar components interact with one another. By saving the overheads associated with the CORBA ORB, a flyover-based system can lead to a large improvement in performance. The performance benefit clearly depends on the degree of heterogeneity that determines the relative frequency with which the flyover is chosen. Experimental results demonstrate that a significant performance improvement can accrue even when a large degree of heterogeneity exists on the system.

The third approach concerns techniques to be used by the application system developer. Methods to reducing connection setup latency and for effective packing of objects into servers are described. The way parameters are passed during method invocation is observed to have a strong impact on performance. Unraveling of complex data structures is observed to give rise to substantial performance savings. Performance enhancement through application-level load balancing is also discussed. Both static and dynamic techniques, each of which is useful in a different situation, are discussed.

8.7 Acknowledgments

The material presented in this chapter is based on the research performed by graduate students, Istabrak Abdul-Fatah, E-Kai Shen, Imran Ahmad, Wai-Keung Wu, and Weili Tao and the resulting material described in various papers referenced in the text. Their efforts, without which this chapter will not be possible, are gratefully acknowledged. Financial support for this research came from various sources including Natural Sciences and Engineering Research Council of Canada, the Department of National Defense, Communications and Information Technology, Ontario, and Nortel Networks.

Bibliography

[1] Abdul-Fatah, I. and Majumdar, S. (1998) Performance Comparison of Architectures for Client-Server Interactions in CORBA. *Proceedings of the International Conference on Distributed Computing Systems (ICDCS '98)*, Amsterdam, Netherlands, May 1998.

[2] Abdul-Fatah, I. and Majumdar, S. (2002) Performance of CORBA-Based Client-Server Architectures. *IEEE Transactions on Parallel & Distributed Systems*, **13**(2), 111–127.

[3] Adler, R. (1995) Distributed Coordination Models for Client/Server Computing. *IEEE Computer*, **28**(4), 14–22.

[4] Ahmad, I. and Majumdar, S. (2001) Achieving High Performance in CORBA-Based Systems with Limited Heterogeneity. *Proceedings of the IEEE International Symposium on Object Oriented Real Time Computing (ISORC 2001)*, Magdeburg Germany, May 2001.

[5] Barth, T., Flender, B., Freisleben, B., and Thilo, F. (1999) Load Distribution in a CORBA Environment. *Proceedings of the International Symposium on Distributed Objects and Applications (DOA'99)*, Edinburgh, UK, September 1999.

[6] Borland Inprise (1999) *Visibroker: CORBA Technology from Inprise*, Borlan Inprise.

[7] Friedman, R. and Hadad, E. (2002) FTS: a High-Performance CORBA Fault-Tolerance Service. *Proceedings of the Seventh IEEE International Workshop on Object-Oriented Real-Time Dependable Systems (WORDS 2002)*, San Diego, CL, January 2002.

[8] Gokhale, A. S. and Schmidt, D. C. (1997) Evaluating CORBA Latency and Scalability Over High-Speed ATM Networks. *Proceedings of the IEEE 17th International Conference on Distributed Systems (ICDCS '97)*, Baltimore, MD, May 1997.

[9] Gokhale, A. S. and Schmidt, D. C. (1998) Measuring and Optimizing CORBA Latency and Scalability Over High-speed Networks. *IEEE Transaction on Computers*, **47**(4), 391–413.

[10] Haggerty, P. and Seetharaman, K. (1998) The Benefits of CORBA-Based Network Management. *Communications of the ACM*, **41**(10), 73–80.

[11] Henning, M. and Vinoski, S. (1999) *Advanced CORBA Programming with C++*, Addison-Wesley, Boston, MA.

[12] Iona Technologies (1997) *Orbix Programmers' Guide*, Iona, Dublin, Ireland.

[13] Iona Technologies (2002) *E2 ASP Users Guide*, Iona, Dublin, Ireland.

[14] Jacobsen, H.-A. and Weissman, B. (1998) Towards High-Performance Multithreaded CORBA Servers. *Proceedings of the International Conference on Parallel and Distributed Processing Techniques and Applications (PDPTA'98)*, Las Vegas, NV, July 1998.

[15] Kon, F., Roman, M., Liu, P., Mao, J., Yamane, T., Magalha, C., and Campbell, R. H. (2000) Monitoring, Security, and Dynamic Configuration with the Dynamic TAO Reflective ORB. *Proceedings of the IFIP International Conference on Distributed Systems Platforms and Open Distributed Processing (Middleware 2000)*, Pallisades, New York, April 2000.

[16] Maffeis, S. and Schmidt, D. C. (1997) Constructing Reliable Distributed Communication Systems with CORBA. *IEEE Communications Magazine*, **35**(2), 72–78.

[17] Majumdar, S., Shen, E.-K., and Abdul-Fatah, I. (2004) Performance of Adaptive CORBA Middleware. *Journal of Parallel and Distributed Systems*, **64**(2), 201–218.

[18] Martin, P., Callaghan, V., and Clark, A. (1999) High Performance Distributed Objects Using Caching Proxies for Large Scale Applications. *Proceedings of the International Symposium on Distributed Objects and Applications (DOA'99)*, Edinburgh, UK, September.

[19] Mishra, S. and Liu, X. (2000) Design, Implementation and Performance Evaluation of a High Performance CORBA Group Membership Protocol. *Proceedings of the 7th International Conference on High Performance Computing*, Bangalore, India, December.

[20] Object Management Group (2002) *The Common Object Request Broker: Architecture and Specification*, ver 3.0.22 http://www.omg.org.

[21] Otte, R., Patrick, P., and Roy, M. (1996) *Understanding CORBA The Common Object Request Broker Architecture*, Prentice Hall, Upper Saddle River, NJ.

[22] Petriu, D. C., Amer, H., Majumdar, S., and Abdul-Fatah, I. (2000) Analytic Modelling of Middleware. *Proceedings* of the Workshop on Software and Performance (WOSP 2000), Ottawa, Canada, September 2000.

[23] PostModern Computing Technologies Inc. (1994) *ORBeline Reference Guide*, Mountain View, CA.

[24] Parlavantzas, N., Coulson, G., and Blair, G. S. (2000) On the Design of Reflective Middleware Platforms. *Proceedings of the RM-2000 Workshop on Reflective Middleware*, New York, USA, April 2000.

[25] Parsons, E.W., *Personal Communication*, Nortel Networks, Ottawa, ON, September 2000.

[26] Schmidt D. C. and Cleeland, C. (1999) Applying Patterns to Develop Extensible ORB Middleware. *IEEE Communications Magazine*, **37**(4), 54–63.

[27] Schmidt, D. C. (1998) Evaluating Architectures for Multi-threaded CORBA Object Request Brokers. *Communications of the ACM*, **41**(10), 54–61.

[28] Seetharaman, K. (1998) Introduction. *Communications of the ACM*, **41**(10), 34–36.

[29] Shen, E.-K., Majumdar, S., and Abdul-Fatah, I. (2000) The Performance of Adaptive Middleware Systems. *Proceedings of the ACM Principles of Distributed Computing Conference (PODC)*, Portland, ME, July 2000.

[30] Shokri, E. and Sheu, P. (2000) Real-Time Distributed Computing: An Emerging Field. *IEEE Computer*, **33**(6), 45–46.

[31] Siegel, J. (1998) OMG Overview: CORBA and OMA in Enterprise Computing. *Communications of the ACM*, **41**(10), 37–43.

[32] Stal, M. (1995) The Broker Architectural Framework. *Proceedings of the Object-Oriented Programming Systems, Languages and Applications Conference (OOPSLA'95)*, Austin, MN, October 1995.

[33] Stevens, W. R. (1990) *UNIX Network Programming*, Prentice Hall, Upper Saddle River, NJ.

[34] Tanenbaum, A. (1995) *Distributed Operating Systems*, Prentice Hall, Englewood Cliffs, NJ.

[35] Tao, W. and Majumdar, S. (2002) Application Level Performance Optimizations for CORBA-Based Systems. *Proceedings of the International Workshop on Software and Performance (WOSP'02)*, Rome, Italy, July 2002.

[36] Vinoski, S. (1997) CORBA: Integrating Diverse Applications Within Distributed Heterogeneous Environments. *IEEE Communications Magazine*, **35**(2), 46–55.

[37] Wang, N., Vinoski, S., and Schmidt, D.C. (1999) *Collocation Optimizations for CORBA*, C++ Report, September 1999.

[38] Wu, W.-K. and Majumdar, S. (2002) Engineering CORBA-Based Systems for High Performance. *Proceedings of the International Conference on Parallel Processing (ICPP'02)*, Vancouver, Canada, August 2002.

9

Concepts and Capabilities of Middleware Security

Steven Demurjian, Keith Bessette, Thuong Doan[1], Charles Phillips[2]

[1]*The University of Connecticut*
[2]*United States Military Academy*

9.1 Introduction

Distributed computing applications for the twenty-first century, constructed from legacy, commercial-off-the-shelf (COTS), database, and new client/server applications, require stakeholders (i.e., software architects, system designers, security officers, etc.) to architect and prototype solutions that facilitate the interoperation of new and existing applications in a network-centric environment. In these solutions, security must play a fundamental part, considered at early and all stages of the design and development lifecycle. The emergence of distributed computing technology such as DCE [6, 13], CORBA [4, 16, 18], DCOM/OLE [3], J2EE/EJB [12, 15], JINI [1, 17], and .NET [11, 14] has enabled the parallel and distributed processing of large, computation-intensive applications. Historically, the incorporation of security has often been an afterthought, dependent on programmatic effort rather than a cohesive mechanism seamlessly incorporated into the underlying technology. However, there has been a dramatic turnaround in the support of security, particularly in modern middleware platforms such as Common Object Request Broker Architecture (CORBA), .NET, and Java 2 Platform, Enterprise Edition (J2EE). The objective of this chapter is a two-fold examination of the concepts and capabilities of middleware security: exploring the security capabilities in three popular middleware platforms, namely, CORBA, .NET, and J2EE; and utilizing middleware concepts to realize complex and critical security approaches, namely, role-based and mandatory access control.

Toward this objective, in Section 9.2 of this chapter, we discuss the state of the art in support of middleware security, focusing on the security abilities of CORBA, .NET,

Middleware for Communications. Edited by Qusay H. Mahmoud
 ISBN 0-470-86206-8

and J2EE. For CORBA, we explore its security features for confidentiality, integrity, accountability, and availability [5], keeping in mind that CORBA security is a *meta-model* that characterizes a breadth of security capabilities representing different security models, paradigms, and techniques. On the other hand, .NET and J2EE provide actual security capabilities via their respective runtime environments, and the *application programmer interfaces* (APIs), which provide security functionality. As such, .NET and J2EE can be considered, at some level, a realization of the CORBA security meta-model, implementing a subset of its features and capabilities. Thus, for both .Net and J2EE, we present and discuss their security potential and features with a five-prong approach: *code-based access control* for security embedded in actual program code, *role-based access control* for user-oriented security, *secure code verification and execution* that focuses on runtime security support, *secure communication* of the exchange of messages and information, and *secure code and data protection* to detail cryptographic capabilities. When relevant, we compare and contrast the features of .Net and J2EE, and we also note their similarities to different portions of the CORBA meta-model. Overall, in Section 9.2, we cover a range of security capabilities from conceptual (meta-model of CORBA) to practical (.NET and J2EE).

In Section 9.3 of this chapter, we demonstrate the utilization of middleware to realize complex security capabilities for distributed applications, namely, role-based access control (RBAC) and mandatory access control (MAC). We report on an RBAC/MAC security model and enforcement framework for a distributed environment comprised of software resources interacting via middleware [7, 8, 9, 10]. Our approach concentrates on the services (APIs) of software resources, providing the means to customize and restrict by time intervals, data values, and clearance levels, allowing different users to have limited access to resource APIs based on role and security level. For enforcement of privileges, the Unified Security Resource (USR) provides the security infrastructure via middleware services: *Security Policy Services* to manage roles and their privileges; *Security Authorization Services* to authorize roles to users; and, *Security Registration Services* to identify clients and track security behavior. These services are used by administrative tools to manage roles by granting/revoking privileges and setting classification levels, to assign clearances and authorize roles to end users, and to handle the delegation of responsibilities from user to user. We report on the current prototype of the USR enforcement framework, which utilizes both CORBA and JINI as the underlying middleware [25], to provide insight on utilizing middleware to realize sophisticated and complex security capabilities. Finally, Section 9.4 offers concluding and summary remarks.

9.2 Security in CORBA, .NET, and J2EE

In this section, we explore the security capabilities of CORBA [5], .NET [21], and J2EE [24], highlighting their features and providing a context that explains their approaches. The major difference between the support for security in CORBA (as opposed to its realization in an actual CORBA product, e.g., Visibroker) and security in .NET/J2EE is that the CORBA security specification is a *meta-model*, akin to the fact that UML is a meta-model with various implementations (e.g., Together, Rational, etc.). As a meta-model, the CORBA security specification is attempting to generalize many security models and associated security principles, to arrive at a specification that is comprehensive (wide variety of security capabilities at the model level – RBAC, MAC, encryption, etc.) while simultaneously being language independent (not tied to Java, C++, .NET, etc.). As such, the CORBA security specification is a step above .NET/J2EE, each of which, in fact, can be characterized as an implementation or instance of the CORBA security meta-model. The reality is that the CORBA security specification itself [5] emphasizes that there is no single security model and associated implementation that satisfies all of the capabilities and features of the CORBA security meta-model. Our objective in this section is to introduce and highlight the security capabilities of CORBA (Section 9.2.1), .NET (Section 9.2.2), and J2EE (Section 9.2.3). When relevant, we accentuate the similarities and differences of the three approaches.

9.2.1 CORBA Security Capabilities

In this section, we examine the CORBA Security Service Specification [5] for the CORBA middleware framework [4]. Unlike its .NET and J2EE counterparts, the CORBA Security Service Specification is a meta-model, containing capabilities that span a broad spectrum of security features and techniques. The CORBA Security Service Specification focuses on four key aspects of security: confidentiality, integrity, accountability, and availability. *Confidentiality* is concerned with access to information, to limit access to those individuals (programs) that have been given explicit permission. *Integrity* has a similar meaning to the database concept of integrity (correct and consistent data values), but in the context of security, is focused on the idea that only authorized users are allowed to modify information in specific ways (e.g., read, update, delete, etc.), and that the delegation of this authorization between users is tightly controlled. *Accountability* involves the fact that users must be responsible for all of their actions, and that security mechanisms must be able to monitor and track the accountability. Lastly, if users have been appropriately authorized, then their authorizations require the system's *availability*. These four aspects in turn lead to six key features of the CORBA security model: *identification* and *authentication* of users (are they who they say they are), where users can be clients, software artifacts, and so on, who must provide proof to verify themselves prior to access; *authorization* and *access control* of users to objects (does a user have permission to do what s/he wants?), with authorization involving the granting, denying, and revocation of access to objects, and access control the means to realize the privilege definition process and to support enforcement; *security auditing* for both the accountability of a user's actions and to track those actions in an audit (security log); *security of communication* between the users and the objects that they interact with, to ensure that not just the permissions (access control) are satisfied, but that the exchange of information and messages is secure; *nonrepudiation* to provide proof of access, both on the transmission from the client, and on the receipt by the target object; and, *administration* of all of the security requirements for an application, which includes the security policy, users, roles, permissions, and so on.

The remainder of this section reviews the CORBA security model, details the access control process in CORBA, explores the different security views that are supported, and concludes by briefly examining the execution process of a secure access request. Note that the material in this section is from [5], adapted and rewritten to be consistent with the discussion on security middleware. Since the CORBA security specification [5] is extensive (400+ pages), it is impossible in this context to present its breadth and depth. Thus, the objective is to introduce the reader to its core underlying concepts.

9.2.1.1 The CORBA Security Model

In Figure 9.1, we present the interactions of a client and target object using CORBA (and Object Request Broker – ORB) in the presence of active security services and capabilities. The structural model of CORBA is comprised of different levels used to facilitate secure object invocation by clients. First, there are application-level components, such as a client request services and a target object providing services (see Figure 9.1). The degree that application-level components are aware of security can vary; the visibility of security can be dependent on programmatic actions whether the client (or user) and/or the target object (application) is cognizant of security. Second, as shown in Figure 9.1, there are many different components that realize security, including the *ORB core* and its services, the *ORB security services*, and the *policy objects* that contain the actual security requirements for an application (not shown). In Figure 9.1, there are two ORB security services: the *Access Control Service*, which verifies from the client perspective if the operation being requested is permitted, and if so, enforces the audit actions that track the actual access; and, the *Secure Invocation Service* that on the client side is used to connect the client to the target object and on the target side is used to protect the target object in its interactions with the client. Third, their actual security services must be implemented at some point by the middleware; in Figure 9.1, this is referring to the fact that both the Access Control and Secure Invocation Service specifications must be available in actual CORBA platforms (implementations). These implementations, which are platform specific, must interact with the security and other features that are supported by the OS and hardware.

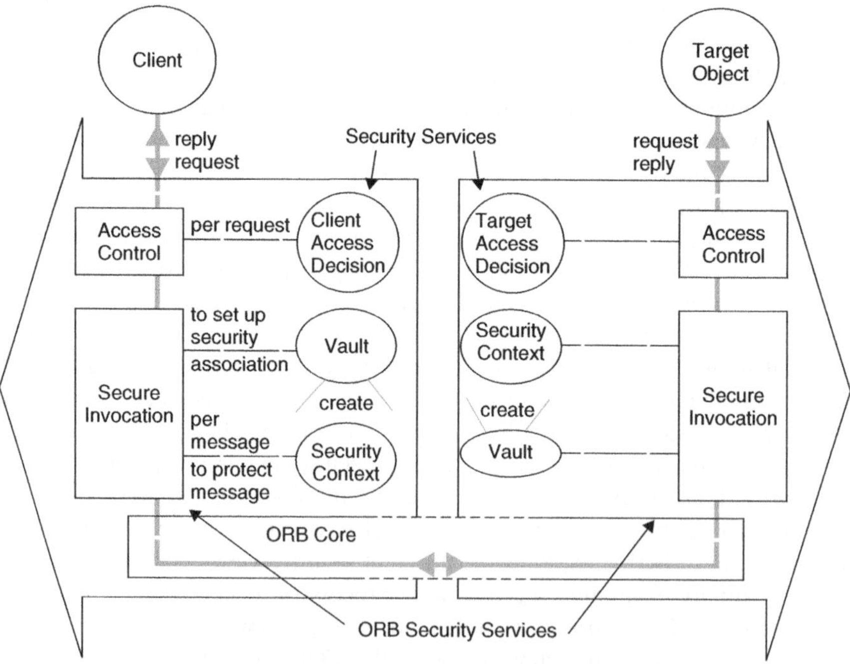

Figure 9.1 CORBA security approach (Figure 2–47 in Object Management Group (2002) *Security Service Specification – Version 1.8*, http://www.omg.org.)

9.2.1.2 Access Control in CORBA

In this section, we begin to focus on the details of CORBA security (see Figure 9.1 again) by investigating the privilege definition process as it relates to the client and the target object, and as it involves the interactions of the client and the target object. In the CORBA security reference model, the access control process must verify the characteristics of a subject's permissions (via *privilege attributes*) against the target objects that are themselves managed via *control attributes* (grouped as *domains*) and operations (grouped as *rights*). The combination of privilege attributes, control attributes, and domains provides the means to define security requirements, and provides the basis for enforcing those requirements by actual clients against target objects, as shown in Figure 9.1.

Privilege attributes are associated with the user (client or other software artifact), referred to as the *principal*, and are used to capture all of the various security permissions for access control, including the identity (e.g., user id) of the principal, the role(s) of the principal, the group(s) that the principal belongs to in the organization, the security clearance (e.g., secret, classified, etc.) of the principal, the target objects and operations to which the principal has been granted access, and any other enterprise-wide privileges. Clearly, privilege attributes contain the security requirements for each user (principal). Control attributes are associated with each target object, to track the security privileges for an object's perspective. While privilege attributes focused on capabilities of individual principals, control attributes track all of the principals (and their access) on a target-object-by-target-object basis. For example, an access control list entry for each target object would track the list of principals who have been authorized. In addition, control attributes can also track security information on the target object itself. For example, a target object might have a security classification (e.g., secret, classified, etc.), or may be limited in access to certain time periods. Finally, note that the *rights* of a target object are the set of operations that are available for assignment to each principal. For example, this may be the right to invoke a method, modify a portion of the target object, and so on.

Defining privileges for principals and target objects, while not a difficult process, can be time consuming and complex, when one considers, for even a moderate sized organization, the number of principals, and more importantly, the tens or hundreds of thousands (or more) of target objects for which security must be defined. To assist in this process, the idea of a *domain* can be leveraged, which provides a context to define common characteristics and capabilities related to security. We focus on the *security policy domain* that represents the scope over which each security policy is enforced, assuming that an organization may have multiple policies. A security policy domain permits the definition of security requirements for a group of target objects, allowing this group to be managed as a whole, thereby reducing the needed administrative effort. To address the structural aspects of an organization in the security definition process, policy domain hierarchies can be employed. A *policy domain hierarchy* allows a security administrator to design a hierarchy of policy domains, and then delegate subsets of the hierarchy (subdomain) to different individuals. For example, in a health care organization, they could be subdomains for patient objects, test-result objects, employee objects, and so on, and the security for these subdomains could be delegated to different people in hospital administration/IT. This separation into subdomains allows both access control and auditing to also be administered at these levels, doling out the responsibility for these tasks. There are also other abstraction techniques to organize the security definition process, including federated policy domains, system- and application-enforced policies, and overlapping policy domains [5].

9.2.1.3 Security User Views

The CORBA-compliant security model can be viewed differently by five types of users: the enterprise manager, the end user, the application developer, the operation system administrator, and the object system implementer. We briefly review each of these views:

- The *Enterprise Management View* encompasses an organization-wide perspective on the treatment of security requirements. The objective of this view is to protect all of the information assets of the organization (enterprise), and to accomplish this at a reasonable cost. Risk assessment, protection against malicious and inadvertent security breaches, countermeasures, and so on, are all part of this view. The end result will be the definition of security policies, as appropriate, for different portions of an organization.
- The *End User View* focuses on the actual, individual principal (see Section 9.2.1.2), the privileges that are authorized, and the authentication that must take place to permit the user to perform his/her tasks against specific target objects according to a particular job's responsibilities. As such, the end user view involves the privilege attributes (e.g., user id, role, clearance, etc.).
- The *Application Developer View* focuses on the degree that stakeholders (e.g., software engineers, programmers, developers, etc.) must be aware of the security capabilities of the enterprise. In some situations, the enterprise may wish to define security, but make those definitions transparent to the majority of stakeholders. In this case, the ORB security services could be automatically called to enforce security of the principals against target objects. In other situations, the security may be the strict responsibility of all stakeholders, who would interact explicitly and programmatically with the security services.
- The *Administrator's View* is the security management perspective, providing all of the capabilities to administer and control security policy definition and enforcement, including creating and maintaining the domains, assigning the privilege attributes to end users, administrating the security policies, monitoring the control attributes of target objects, and so on.
- The *Object System Implementer's View* differs from the application developers' view, since these stakeholders are actually responsible for prototyping and implementing the ORB, and for our purposes, the ORB security services. Remember, it may be the case that a given enterprise decides to implement its own unique security services; this view is intended for this situation.

Overall, these various views are all critical, since they collectively provide the means and context to detail and realize all of the security requirements for an organization.

9.2.1.4 CORBA Security Execution Model

To complete our discussion, we review the execution process related to security for the situation where a client attempts to access a target object, which involves the concept

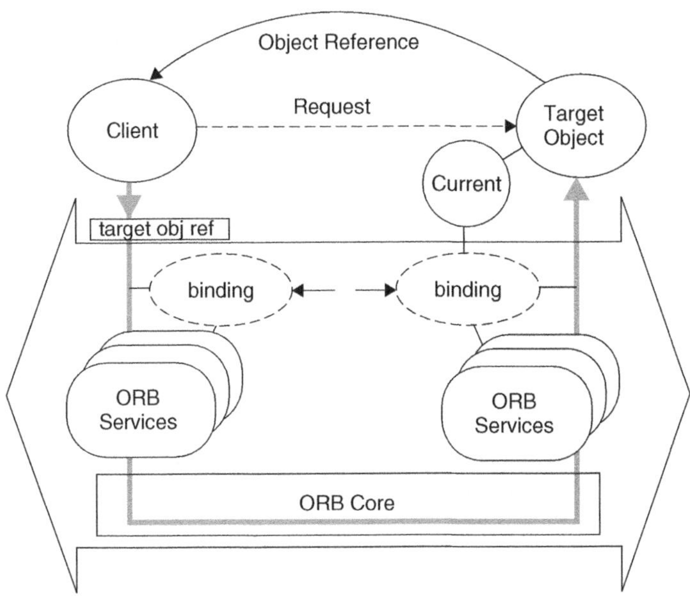

Figure 9.2 Security binding (Figure 2–24 in Object Management Group (2002) *Security Service Specification – Version 1.8*, http://www.omg.org.)

of binding and object references. To accomplish this in a secure manner, there are many different steps that must occur. First, as shown in Figure 9.2, there must be interactions from the client to the target object, with the client making a request to access the target object. For this request to be successful, the client must obtain a *binding* to the target object, which requires a check to see if the client has the permissions (via the privilege attributes) to invoke an operation on the target object (via the control attributes). If so, the binding is established, and a reference to the target object is returned, thereby allowing the invocation.

In some situations, the ability to obtain an object reference may involve interaction with policy object and the domain manager, as shown in Figure 9.3. Recall that as discussed in Section 9.2.1.2, a policy domain can be defined for groups of target objects to capture common security features in a single policy. If this is true, then each security policy domain has its own policy domain manager (see Figure 9.3) that controls and manages the policy for all involved target objects. Thus, the binding as discussed in Figure 9.2 may need to be more complex, specifically, it may require the ORB to create the reference and associate the reference with multiple security policy domains.

9.2.2 .NET Security Capabilities

This section discusses the security capabilities of .NET, focusing on the breadth and depth of security features that the platform supports in order to maintain secure development and execution environments. The following key definitions are briefly reviewed to understand this section: *assembly*, which refers to compiler-generated code, specifically, Microsoft intermediate language (MSIL); *evidence*, which refers to the "proof" that is

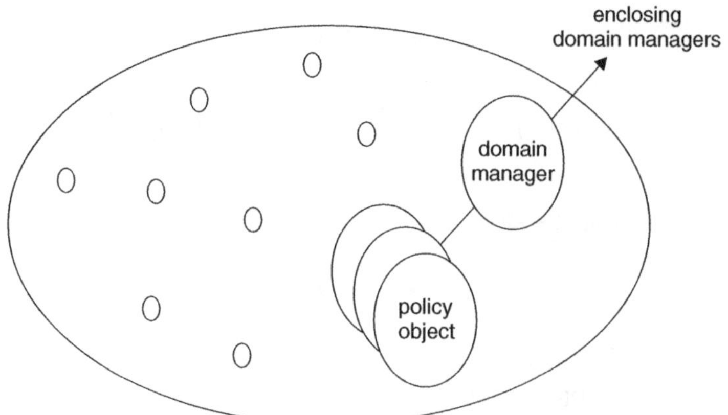

Figure 9.3 Domain objects (Figure 2–25 in Object Management Group (2002) *Security Service Specification – Version 1.8*, http://www.omg.org.)

supplied regarding identity; and *permissions*, which refers to a privilege that is given to perform an operation on a resource, and in the case of an assembly, the permission set for all allowable privileges on all required resources.

To understand the security capabilities of .NET, we begin by reviewing the structural model for security, which consists of three components: the Common Language Runtime (CLR), the Hosting Environment, and the Security Settings, all of which are shown, along with their interactions, in Figure 9.4. To start, the Hosting Environment is attempting to execute an application, and to do so, it must provide both the code (via assembly) and its identity (via evidence) in its interactions with CLR. CLR is the core component of

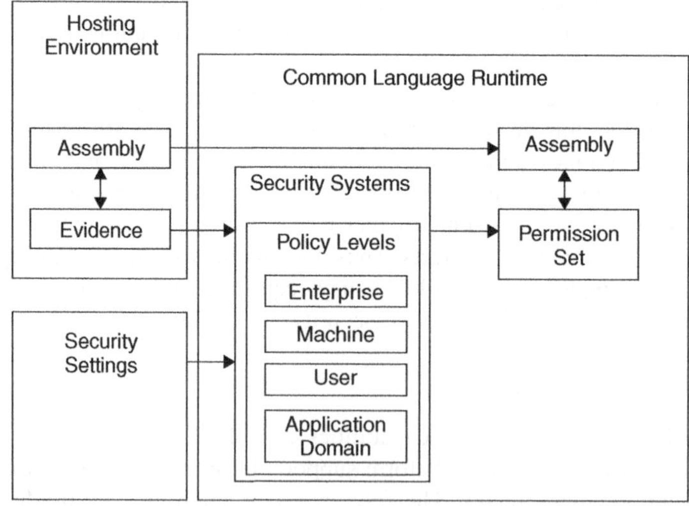

Figure 9.4 .NET security structural model Open Web Application Security Project (2002) *J2EE and .NET Security*, http://www.owasp.org/downloads/J2EEandDotNetsecurityByGerMulcahy.pdf.

the .Net Framework, providing a secure execution environment through the principles of what Microsoft describes as managed code and code access security [23]. CLR contains the Security System, which realizes the security policy at different levels, namely, the enterprise, machine, user, and application domain. To do so, policy files comprised of rules are defined by a security administrator to allow permission sets to different types of code or users on the basis of information provided by the evidence. For an actual application, the different parameters related to security must be set within the Security System, as shown by the input from the Security Settings box in Figure 9.4, thereby establishing the security requirements at one or more policy levels. In order for execution to occur within CLR, the assembly is utilized to identify the required permission set (i.e., allowances given to a piece of code to execute a certain method or access a certain resource) and be provided with evidence from the Host to the Security System. These interactions in .NET, while different from the ones proposed for CORBA in Figure 9.1 and Figure 9.2, do share a similar philosophical approach in requiring specific actions to be taken before actual access can occur. One focus of this section is to explore the interactions among these components.

Given this background, there are five main security capabilities of .NET that are intended to keep the code, data, and systems safe from inadvertent or malicious errors [21]:

1. *Code-based access control* gives permissions at the code level to access resources on the basis of the application or protection domain that the code is assigned to, which must also include the evidences of the code to verify identity.
2. *Role-based access control*, a popular approach based on a user's responsibilities and objectives for an application, gives permissions to a user to access resources on the basis of the user's role in the system.
3. *Secure code verification and execution*, similar in concept to bytecode verification in Java, analyzes bytecode (MSIL) and ensures that the executing code is staying within its allowed domain.
4. *Secure communication* provides the ability to pass data and messages locally or remotely in a secure manner, to avoid both intentional and inadvertent data/message modification.
5. *Secure code and data protection* ensures that code has not been modified without authorization by utilizing cryptographic solutions and signed distribution files.

Clearly, the first three in this list involve security requirements, as realized within the Security System, Security Settings, Evidence, and Permission Set from Figure 9.4; the last two focus on programmatic techniques to ensure the secure execution and exchange of information, code, and messages. The remainder of this section reviews .NET's security features using these five different capabilities.

9.2.2.1 Code-Based Access Control

Code-based access control, a main component of CLR's security, focuses on the ability of code to access resources, and in .NET, this corresponds to under what situations access by a code segment is permitted (prevented). The determination of what a piece of code is allowed to do is decided by the origins and intentions of the code itself, which can be decomposed into *evidence-based security*, *permissions*, and a *security policy*. During execution, the CLR reviews evidence of an assembly, determines an identity for the assembly, and looks up and grants permissions on the basis of the security policy for that assembly identity [23]. To understand code-based access control, we start with *evidence-based security*, which is used by the CLR to determine the origin(s) of an assembly. At runtime, the CLR examines the meta-data of an assembly to determine the origin of the code, the creator of the assembly, and the URL and zone that the assembly came from. In .NET, a *zone* represents the domain that the assembly is from, for example, Internet, LAN, local machine, and so on. The association of the meta-data and its corresponding assembly is verified by the CLR [23].

The successful verification of evidence leads to the identification of *permissions*, which represent the privileges that are given to code and code segments. A permission assigned to a code segment represents the ability to execute a certain method or access a certain resource. An assembly will request permissions to execute, and these requests are answered at runtime by the CLR, assuming that the assembly has provided apropos evidence. If not, CLR throws a security exception and an assembly's request is denied. It is also possible that a request may be only partially denied, with the CLR dynamically assigning the assembly a lower level of permission than requested. Since there are numerous different permissions that can be requested, permissions are grouped into sets where each set has the same level of security and trust. For example, an assembly that has originated from the Internet zone may be granted an Internet permission set that pertains to the execution of untrusted code, allowing the behavior of nonlocal code to be tightly controlled. Thus, assemblies, like permissions, can be grouped on the basis of similar requests and zones. Permissions and permission sets in .NET are similar to, respectively, privilege/control attributes and domains in CORBA (see Section 9.2.1.2).

The grouping of assemblies on the basis of different criteria is essentially establishing different *security policies* for different code groupings [21, 23]. In .NET, here are the three different security policies that are supported, realized in *security policy files*, and administered by *security personnel*: enterprise level for a cohesive and comprehensive policy for the entire enterprise (i.e., all assemblies from all machines for all users); machine level for individual hardware platforms to permit different policies for different machines; and user level to allow permissions that capture individual responsibilities. Clearly, these different security policy levels are a subset of the CORBA views (see Section 9.2.1.3). It is important that the different policies (enterprise, machine, and user) are consistent with one another, since they must all effectively interact at runtime to grant and verify permissions of users executing applications on machines across the enterprise. The .NET framework provides the means to organize security policy groups of assemblies (code) into hierarchical categories based on the identity that the CLR determines from the evidence in the meta-data. Once related assemblies have been grouped and categorized, the actual security policy can be specified as permissions for all assemblies in a group. This is accomplished using *security policy files* to capture security requirements, with the proviso that a policy file may limit the permissions of another policy file, but cannot entirely restrict all permissions. The actual decisions and definition of permission for a given enterprise are typically made by the security administrator for the assemblies and domains. To accomplish this, the .NET configuration tool or the Code Access Security Tool (Caspol.exe) [23] can be utilized.

9.2.2.2 Role-Based Access Control

Role-based access control extends the policies and permissions concepts of code-based access control to apply to a user or role. A *role* represents a logical grouping of users that all have similar capabilities, which we see at the OS level (administrator and guest roles) but can also be domain-specific (e.g., in a health care application, there could be physician, nurse, and technician roles). Roles allow the assignment of specific sets of privileges to perform certain operations. .NET uses role-based security to *authenticate* an identity and to pass on that identity to resources, thereby *authorizing* the users playing roles access to resources according to policies and permissions. .NET applies the term 'principal' to role membership, and the permissions of role-based security are managed by the PrincipalPermission object. In addition, there are numerous authentication mechanisms [22] such as: *windows authentication* used by applications through the *Security Support Provider Interface (SSPI)*; *form-based authentication* is from HTTP, with the form requesting user log-on credentials to be passed on to the application server submitted in the form of a cookie; *passport authentication* is a centralized service that requires only a single log-on for members of sites that subscribe to it; *Internet Information Services (IIS)server authentication*, which includes Basic and Digest authentication and X.509 Certificates with SSL, for authentication of applications hosted on the IIS server; and *impersonation authentication*, which allows an application to access another application using a different identity while maintaining responsibility to the original user. .NET offers specific control of role-based security by supporting role permission checking declaratively and imperatively. .NET authentication mechanisms are expected to run in a Windows environment, and as a result, IIS is the only supported server of the .NET framework. This expectancy limits the flexibility of .NET to maintain the fine granularity of role-based access control on other platforms [19]. For instance, IIS does not support application-level authentication since authentication is done at the OS level, thereby limiting flexibility.

9.2.2.3 Secure Code Verification and Execution

The most significant security concern in using middleware-based approaches is that remote code has the potential to execute in a local environment, which is a worrisome concept for system and network administrators. Thus, starting with Java, and continuing with .NET, verifying code and executing it within a safe environment must be supported to prevent system weaknesses from being exposed by an application error, malicious or not. .NET performs security checks during the code execution process; stack integrity, bytecode structure, buffer overflows, method parameter types, semantics, and accessibility policies are all verified. These code checks are inserted into certain locations as .NET compiles programs into MSIL [22]. Like Java, .NET uses the concept of a 'sandbox' for code execution. The sandbox can be described by the analogy of a child who can only play with the objects in the sandbox unless given permission to use objects outside of the sandbox. The sandbox of .NET is called the *application domain*, which applies static boundaries to its execution environment and contains all of the loaded classes of an assembly, where multiple assemblies may be loaded into the same application domain. There are limitations that are imposed. For example, there is no way for an object to directly reference another object in a different application domain; any such access must be performed remotely programmatically. Overall, the application domain supplies a fixed solution to the executing environment, taking this privilege and burden away from the developer, including the need to deal with unmanaged code [20]. To accomplish this, all of the security checks to verify code are done on managed code in a managed environment (the CLR). In addition, .NET allows for unmanaged code to bypass the CLR and provides the means to access legacy applications and code outside the CLR by supplying managed wrappers for unmanaged code [21, 23]. From a security perspective, this bypass is problematic, since it provides the means for unauthorized and unauthenticated access (malicious or unintentional) to occur, thereby compromising any of the defined security policies (enterprise, machine, and user).

9.2.2.4 Secure Communication

The transmission and communication of sensitive data across both local and remote systems must be securely accomplished. In .NET, secure communications occurs at the application level via the Secure Sockets Layer (SSL) and Transport Layer Security (TLS). These two protocols determine what cryptographic algorithm and session keys are to be used. .NET applications can use the Windows SSPI, but only as unmanaged code, which as discussed in Section 9.2.2.3, has the potential to open a system to unauthorized access. In using .NET, Microsoft is promoting the exclusive use of IIS, and while IIS does support SSL and TLS, it usage of files to transfer messages is inefficient by today's standards, and another potential security hole. Thus, .NET developers have a choice of either running unmanaged code via Windows SSPI, which has its own security problems, or utilizing IIS for creating secure communications, which itself is insecure, based on the history of attacks on IIS servers [19].

9.2.2.5 Secure Code and Data Protection

The last aspect of .NET's security involves assurances regarding secure code and data exchange. To provide a confidence level to developers and users, .NET (and other such systems) must uphold code and data reliability and validity, thereby insuring secure execution of code and access/exchange of data. This means that any code loaded by the system must supply evidence of its source, version signature, and proof that there has not been any unauthorized modification; the same is applicable to data. .NET provides a number of different techniques for maintaining code and data integrity [21]. Certificate management for the signing of code and data sources is provided by .NET, which uses strong-named assemblies that include the assembly name and version information. The assembly is signed with an RSA keypair nullifying the chance of unauthorized modification, and the version information is included in order to avoid DLL conflicts during execution [21]. .NET utilizes and extends the Windows Cryptographic API to make cryptographic solutions very configurable. However, since .NET's encryption functionality is tied to the Windows API, this could pose a problem if the developer is using another OS, which is likely in enterprise applications that involve heterogeneous hardware and software. The major criticism of code and data protection in .NET is that it is not difficult to reverse engineer the cryptographic algorithms and security protocols since both are based on published portable executable standards (bytecode).

9.2.3 J2EE Security Capabilities

This section mirrors Section 9.2.2 on .NET, to discuss the security capabilities of J2EE, focusing on its ability to keep code, data, and systems safe from inadvertent or malicious errors. As shown in Figure 9.5, the compilation of Java source code creates platform-independent Java bytecode that is able to execute either locally or remotely. The execution process of Java bytecode involves a number of different components, including the class loader (with bytecode verifier), the Java class libraries (APIs), and the Java Virtual Machine (JVM), which interacts with the operating system. The root of the hierarchy is the primordial class loader that loads the base classes. The JVM provides a secure runtime environment in combination with the class loader and Java class libraries. JVM manages memory by dynamically allocating different areas for use by different programs, isolating executing code, and performing runtime checks (e.g., array bounds), which each has a role in ensuring security. The block labeled Runtime System as shown in Figure 9.5, contains the *Security Manager*, *Access Controller*, and other features that all interact to maintain security of executing code. The Security Manager and Access Controller examine and implement the security policy [23]. Java programs can be either standalone *applications* or *applets* that can be downloaded from the Internet and displayed/manipulated safely within a Web browser. Security considerations in J2EE are important for both applications and applets, but applets are of particular concern for security, since they represent remote code that is brought in and executed on a local machine. To control applet behavior, Java uses a sandbox, which forces downloaded applets to run in a confined portion of the system, and allows the software engineer to customize a security policy. As presented, Figure 9.5 is at a much lower level than the meta-model of CORBA (see Section 9.2.1 again), and as such, this clearly illustrates the assertion that security in Java and J2EE is a partial realization of CORBA capabilities. In the remainder of this

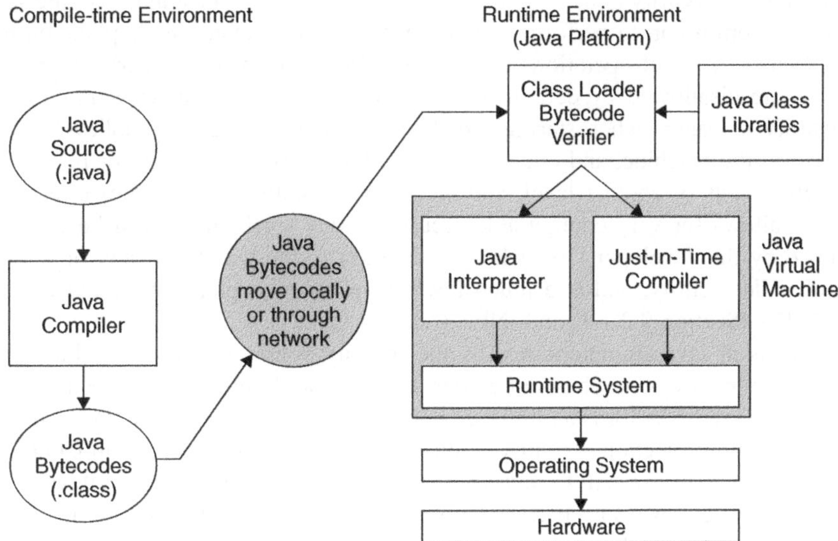

Figure 9.5 The Java 2 platform – compile and execute

section, we explore J2EE's support for code-based access control, role-based access control, secure code verification and execution, secure communication, and secure code and data protection.

9.2.3.1 Code-based Access Control

Code-based access control in J2EE is implemented through the JVM, the class loader, and the Security Manager and Access Controller [23]. The hierarchy of the class loader is supposed to prevent unauthorized and untrusted code from replacing any code in the base classes [23]. In the case of applets, the class loader has additional capabilities: it must determine how and when applets can load classes; and it is responsible for fetching the applet's code from the remote machine, creating and enforcing a namespace hierarchy, and preventing applets from invoking methods that are part of the system's class loader. An executing Java environment permits multiple class loaders, each with its own namespace, to be simultaneously active. Namespaces allow the JVM to group classes on the basis of where they originated (e.g., local or remote). Java applications are free to create their own class loaders. In fact, the Java Development Kit (JDK) provides a template for a class loader to facilitate customization. Before a class loader may permit execution, its code must be checked by the *bytecode verifier*. The verifier ensures that the code (application or applet), which may not have been generated by a Java compiler, adheres to all of the rules of the language. In fact, in order to do its job, the verifier assumes that all code is meant to crash or penetrate the system's security measures. Using a bytecode verifier means that Java validates all untrusted code before permitting execution within a namespace. Thus, namespaces ensure that one applet/application cannot affect the rest of the runtime environment, and code verification ensures that an applet/application cannot violate its own namespace.

The *Security Manager* enforces the boundaries around the sandbox by implementing and imposing the security policy for applications. All classes in Java must ask the security manager for permission to perform certain operations. By default, an application has no security manager, so all operations are allowed. But, if there is a security manager, all operations are disallowed by default. Existing browsers and applet viewers create their own security managers when starting up. Java has only two security policy levels, one for the executing machine, and one for the user. Each level can expand or restrict on all of the permissions of another level, and there can be multiple policy files at each level. While this allows for very configurable security policy levels, they must be created and maintained by the developer [19], which is contrary to the way that many organizations wish to handle security, and at a much lower level of conceptualization and realization than CORBA (Section 9.2.1.3) and .NET (Section 9.2.2.1).

Permissions in Java are determined by the security policy at runtime, and are granted by the security policy on the basis of evidence, in a more limited manner than .NET. The evidence that Java looks for is a publisher signature and a location origin. Permissions are also grouped into protection domains (similar to security policy domains in CORBA and to security policy files in .NET) and associated with groups of classes in Java in much the way they are grouped into permission sets and associated with code groups in .NET, and classes are grouped by their origins. However, in Java, code-based access control is not automatic, but requires programmatic effort by the software engineer. Thus, the permissions capabilities in .NET are more robust as compared to J2EE, allowing a more fine-grained level of permissions to be established for code.

9.2.3.2 Role-Based Access Control

In support of role-based access control, J2EE uses the Java Authentication and Authorization Service (JAAS). JAAS is an integrated package that implements a Java version of the Pluggable Authentication Module (PAM) framework. Using JAAS, software engineers are allowed to modify and then plug-in domain/application-specific authentication modules [19]. JAAS currently supports authentication methods including Unix, JNDI, and Kerberos, which is at a coarser level of granularity that roles in .NET (see Section 9.2.2.2), more akin to OS level security. JAAS can only provide limited impersonation authentication because the user identity is different for the application and OS levels. Java servlets also support authentication through all of the HTTP methods (Basic, Digest, and form) [24], which is similar to .NET. User access checking can be done both declaratively and imperatively within different components of J2EE: Java servlets provide user access checking declaratively at the servlet level, Enterprise JavaBeans (EJB) provide user access checking declaratively down to the method level, and JAAS provides user access checking imperatively within methods [23]. Essentially, J2EE and .NET offer very similar capabilities, with their realization slightly differing in approach, which is most likely tracked to the differences in the underlying paradigm philosophies of Java and Microsoft.

9.2.3.3 Secure Code Verification and Execution

J2EE and .NET share a similar approach to secure code verification and execution, and given the timeline of their development (Java predates .NET), it would not be surprising if this approach was strongly influenced by Java. J2EE security checks are performed during the code execution process, and as discussed in Section 9.2.3.1, have their roots in the JVM and runtime environment. Java interprets bytecodes and has a bytecode verifier traverse the bytecodes before it goes to the Just-in-Time (JIT) compiler or JVM [24]. Stack integrity, bytecode structure, buffer overflows, method parameter types, semantics, and accessibility policies are all checked and verified against the executing code. Java also uses the concept of a 'sandbox' for code execution, to restrict the actions of an applet. The applet may do anything it wants within its sandbox, but cannot read or alter any data outside it. The sandbox model supports the running of untrusted code in a trusted environment so that if a user accidentally imports a hostile applet, that applet cannot damage the local machine. One result of this approach is that the security policy is hard coded as part of the application, providing little or no flexibility either to modify the policy or to have discretionary access control. The sandbox of Java is called the *protection domain*, which applies dynamic boundaries to its sandbox. Java's *protected domains*, which is similar to application domains in .NET (Section 9.2.2.3), constitute an extension of the sandbox, and determine the domain and scope in which an applet can execute. Two different protected domains can interact only through trusted code, or by explicit consent of both parties.

Because of the hierarchical class loader structure, an object can access another object in another protection domain as long as they were both loaded from the same class loader. The protection domain is flexible as far as what executing code can do, but this flexibility depends on the sophistication of the developer in using techniques that do not result in a security breach [23]. The JVM is a managed environment, constantly running security checks to verify executing code. To implement sandboxes, the Java platform relies on three major components: the class loader, the bytecode verifier, and the Security Manager. Each component plays a key role in maintaining the integrity of the system, assuring that only the correct classes are loaded (class loader), that the classes are in the correct format (bytecode verifier), and that untrusted classes will neither execute dangerous instructions nor access protected system resources (Security Manager). The security provided by the sandbox has been present in the earliest versions of Java, and as such, the level of assurance of behavior is likely superior in J2EE as compared to .NET.

9.2.3.4 Secure Communication

Secure communications in J2EE are done at the application level. Like .NET, J2EE supports Secure Sockets Layer (SSL) and Transport Layer Security (TLS). In addition, Java provides Java Secure Sockets Extensions (JSSE) for implementing secure communications. JSSE is a configurable and flexible solution that uses SSL and TLS to create a secure connection using sockets (SSLSocketFactory), and can use this connection for remote method invocations (RMI). J2EE transcends .NET in that it is not limited in its usage with an application server, as in .NET's IIS. As a rule, the stability and insulation from external attacks has been much higher in J2EE; this is likely traced to the fact that .NET continues to build upon the Microsoft platform that has numerous security holes.

9.2.3.5 Secure Code and Data Protection

Java both provides ways of maintaining code and data integrity. J2EE provides Java Cryptography Extensions (JCE) and the Java Cryptography Architecture (JCA) for cryptographic functionality. JCE supports key exchange, uses message authentication algorithms, and makes it easier for developers to implement different encryption algorithms [23]. In addition, cryptographers have developed a way to generate a short, unique representation of a message, a *message digest* that can be encrypted and then used as a digital signature. Message digests in Java provides the functionality of a message digest algorithm. Message digests are secure one-way hash functions that take arbitrary-sized data and output a fixed-length hash value. In addition to message digests, Java provides the ability for signed distribution files via certificate management that can be utilized to verify code and data sources. A supplier bundles Java code (and any related files) into a JAR (a Java Archive), and then signs the file with a digital signature. The JAR is released as a version, and the client can verify the authenticity of the supplier by verifying the signature. An unsigned class may be added to a JAR file, but not to a package within a JAR file, in order to maintain the security of private data and methods of a package [24]. However, since JAR manifest files do not require version information, there is the potential for DLL-like conflicts; recall that .NET (Section 9.2.2.5) permitted version information.

9.3 RBAC and MAC using CORBA and JINI

Our research over the past years, has focused on an approach that seeks to integrate role-based access control (RBAC) and mandatory access control (MAC) into a security model and enforcement framework for a distributed environment comprised of software artifacts/resources interacting via middleware [2, 7, 8, 9, 10]. In support of our effort, in the top half of Figure 9.6, the typical scenario facing government and corporations today is presented, where various software artifacts (i.e., legacy applications, COTS, databases, clients, servers, etc.) must all interoperate with one another via middleware (i.e., Lookup Services) to provide the computing infrastructure to support day-to-day operations. In such a scenario, one can conceptualize each of the software artifacts in terms of *resources* that provide *services* (methods) for use within the environment, and as such, each artifact publishes an *application programmer interface* (API). The problem with these APIs is that they contain all of the public methods needed by all users without regard to security. If one user (e.g., a physician) needs access to a method (e.g., prescribe_medicine) via a patient tool, then that method must be part of the API, and as such, the responsibility would be on the software engineer to ensure that the method is only accessible via the patient tool to users who are physicians, and not all users of the patient tool (which may include nurses, administrators, billing, etc.). Thus, in many applications, the ability to control the visibility of APIs on the basis of user role would be critical to ensure security.

Our approach concentrates on the APIs of software resources, the services, providing the means for them to be customizable and restricted by time intervals, data values, and clearance levels to define the portions of APIs that can be invoked on the basis of the responsibilities of a user role and the security level of user. For enforcement, as shown in Figure 9.6, there is the Unified Security Resource (USR), which provides the security infrastructure via middleware services: *Security Policy Services* to manage roles; *Security Authorization Services* to authorize roles to users; and, *Security Registration Services* to identify clients and track security behavior. These services are utilized by a set of administrative and management tools, namely, the *Security Policy Client* (SPC) to manage user roles, the *Security Authorization Client* (SAC) to authorize roles to end users, and the *Security Delegation Client* (SDC) to handle the delegation of responsibilities from user to user. We note that when the work on this effort began in the 1999 timeframe, explicit middleware support for security, as discussed in Sections 9.2.1, 9.2.2, and 9.2.3, were in their infancy or not even released. Hence, we developed our own middleware security services in support of our approach, which is the focus of this section.

In the remainder of this section, we begin in Section 9.3.1 by providing an overview of the security modeling capabilities and features that are supported in our approach. Then, in Section 9.3.2, we explain the processing from the perspective of the middleware services that are provided. Finally, in Section 9.3.3, we report on the current prototype of the USR security middleware, which utilizes both CORBA and JINI as the underlying middleware [25]. Lastly, please note that USR works with both JINI and Visibroker (CORBA), but it is hardwired; it is not set up to work with any middleware platform as of this time. USR is a research software that illustrates the way that middleware can be leveraged to design custom security; it is not a commercially viable platform. The USR middleware is not pluggable; however, the source code is freely available for

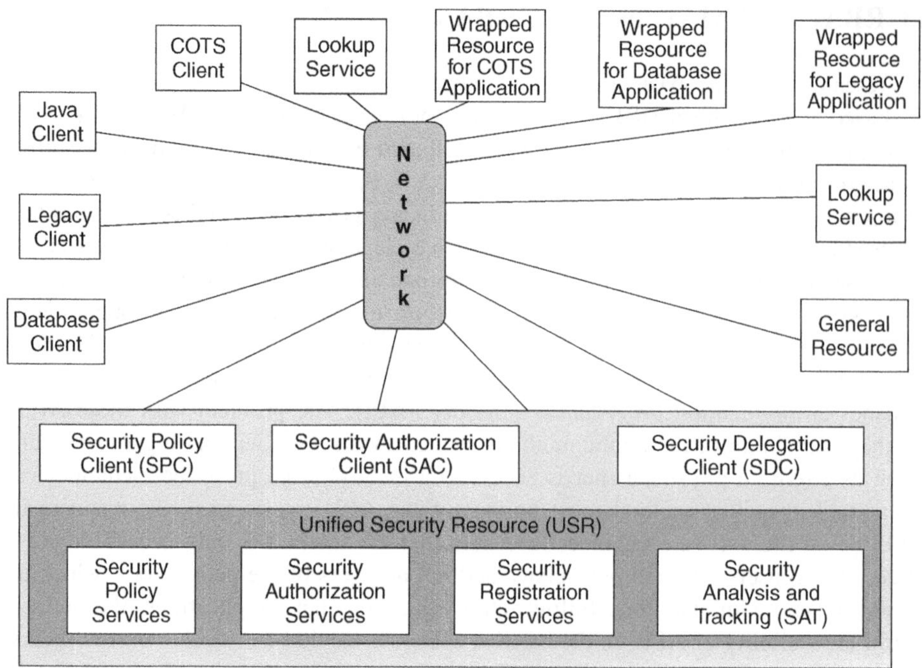

Figure 9.6 The security framework

download. The intent is to demonstrate the feasibility and potential of advanced security with middleware.

9.3.1 Overview of the RBAC/MAC Security Model

This section reviews the definitions for our RBAC/MAC security model [7, 8] with role delegation [2] for customized access to artifact API methods. We focus on the relevant concepts rather than formal definitions, which can be found elsewhere [9, 10]. To begin, to constrain access on the basis of time, we define a *lifetime*, LT, as a time interval with start time (st) and end time (et) of the form (mo., day, yr., hr., min., sec.). In support of MAC, we define *sensitivity levels* unclassified (U), confidential (C), secret (S), and top secret (T) forming a hierarchy: U < C < S < T, with c*learance (CLR)* given to users and *classification (CLS)* given to entities (roles, methods, etc.). To characterize the distributed environment (top half of Figure 9.6), we define a distributed application as a set of unique *software resources* (e.g., legacy, COTS, DB, etc.), each composed of unique *service set*, each composed of a unique *method set*. Methods are defined by their name, LT of access, CLS level, and parameters; Services and resources by their name, LT of access, and CLS level, where CLS of service is minimum (least secure) of its methods and the CLS of a resource is the minimum of its services. Note that tracking methods in this fashion is similar to control attributes for target objects in CORBA (Section 9.2.1.2).

To begin to represent privileges, we define a user role, UR, to be a triple of name, LT of access, and CLS level, and maintain a list of all URs for an application. Next, a user is uniquely identified by UserId (typically name), LT of access, and CLR level, with an associated user list for the application. The objective is to assign URs various methods to represent the privileges that a role is allowed. To facilitate this, we allow this assignment to be constrained on the basis of a method's actual parameter values (called a *signature constraint*, SC) and limit when a method can be active for a UR in time (called a *time constraint*, TC). With these definitions in hand, a *user-role authorization*, URA, associates a UR with a method M constrained by SC (what values can a method be invoked) and a TC (when can a method be invoked). Then, a *user authorization*, UA, associates a user with a UR and a TC (when can a user be authorized to a role). Finally, at runtime, a *client, C,* is authorized to a user, identified by a unique *client token* comprised of user, UR, IP address, and client creation time. All of the capabilities and concepts of roles and users are the privilege attributes for principals in CORBA (Section 9.2.1.2).

To illustrate the concepts, consider a health care application could have URs for Nurse, Physician, Biller, and so on, and these roles could be assigned methods for manipulating a resource Patient Record that has methods Record_Patient_History assigned to Nurse, Set_Vital_Signs assigned to Nurse and Physician, Prescribe_Medicine assigned to Physician, and Send_Bill assigned to Biller. Actual users would be assigned URs (e.g., Steve to Physician, Lois to Nurse, Charles to Biller, etc.), thereby acquiring the ability to invoke methods. There could also be limits placed on the methods assignment to a role to control values (e.g., Nurse Lois can only access patients Smith and Jones for the Record_Patient_History and Set_Vital_Signs methods) or to control time (e.g., Biller Charles can only access Send_Bill after patient has been discharged).

9.3.2 The Security Services of USR

The Unified Security Resource (USR) consists of three sets of services: *Security Policy Services* managing roles and their privileges; *Security Authorization Services* to authorize roles to users; and *Security Registration Services* to identify clients and track security behavior. The USR is a repository for all static and dynamic security information on roles, clients, resources, authorizations, etc., and is organized into a set of services, as given in Figure 9.7. Our security services in USR are counterparts of the ORB security services (Access Control and Secure Invocation in Figure 9.1) for CORBA.

Security Policy Services (see Figure 9.7) are utilized to define, track, and modify user roles, to allow resources to register their services and methods (and signatures), and to grant/revoke access by user roles to resources, services, and/or methods with optional time and signature constraints. These services are used by a security officer to define a policy, and by the resources (e.g., database, Java server, etc.) to dynamically determine if a client has permission to execute a particular [resource, service, method] under a time and/or signature constraint. There are five different Security Policy Services: Register for allowing a resource to (un)register itself, its services, and their methods (and signatures), which is used by a resource for secure access to its services; Query Privilege for verification of privileges; User Role to allow the security officer to define and delete user roles; Constraint to allow time and signature constraints to be defined by the security officer, and for these constraints to be dynamically verified at runtime; and Grant-Revoke for establishing privileges.

Security Authorization Services (Figure 9.7) are utilized to maintain profiles on the clients (e.g., users, tools, software agents, etc.) that are authorized and actively utilizing

```
SECURITY POLICY SERVICES

Register Service
Register_Resource(R_Id);
Register_Service(R_Id, S_Id);
Register_Method(R_Id, S_Id, M_Id);
Register_Signature(R_Id, S_Id, M_Id, Signat);
UnRegister_Resource(R_Id);
UnRegister_Service(R_Id, S_Id);
UnRegister_Method(R_Id, S_Id, M_Id);
UnRegister_Token(Token)

Query Privileges Service
Query_AvailResource();
Query_AvailMethod(R_Id);
Query_Method(Token, R_Id, S_Id, M_Id);
Check_Privileges(Token, R_Id,
                 S_Id, M_Id, ParamValueList);
User Role Service
Create_New_Role(UR_Name, UR_Disc, UR_Id);
Delete_Role(UR_Id);

Constraint Service
DefineTC(R_Id, S_Id, M_Id, SC);
DefineSC(R_Id, S_Id, M_Id, SC);
CheckTC(Token, R_Id, S_Id, M_ID);
CheckSC(Token, R_Id, S_Id, M_ID, ParamValueList);

Grant_Revoke Service
Grant_Resource(UR_Id, R_Id);
Grant_Service(UR_Id, R_Id, S_Id);
Grant_Method(UR_Id, R_Id,S_Id, M_Id);
Grant_SC(UR_Id, R_Id, S_Id, M_Id, SC);
Grant_TC(UR_Id, R_Id, S_Id, M_Id, TC);
Revoke_Resource(UR_Id, R_Id);
Revoke_Service(UR_Id, R_Id, S_Id);
Revoke_Method(UR_Id, R_Id,S_Id, M_Id);
Revoke_SC(UR_Id, R_Id, S_Id, M_Id, SC);
Revoke_TC(UR_Id, R_Id, S_Id, M_Id, TC);

SECURITY AUTHORIZATION SERVICES

Authorize Role Service
Grant_Role(UR_Id, User_Id);
Revoke_Role(UR_Id, User_Id);

Client Profile Service
Verify_UR(User_Id, UR_Id);
Erase_Client(User_Id);
Find_Client(User_Id);
Find_All_Client();

SECURITY REGISTRATION SERVICES

Register Client Service
Create_Token(User_Id, UR_Id, Token);
Register_Client(User_Id, IP_Addr, UR_Id);
UnRegister_Client(User_Id, IP_Addr, UR_Id);
IsClient_Registered(Token);
Find_Client(User_Id, IP_Addr);

Security Tracking and Analysis Services
Tracking Service: Logfile (Log String)
Analysis Service: Analyze   (Java Class File)
```

Figure 9.7 The services of USR

nonsecurity services, allowing a security officer to authorize users to roles. There are two services: Authorize Role for the security officer to grant and revoke a role to a user with the provision that a user may be granted multiple roles, but must play only a single role when utilizing a client application; and Client Profile for the security officer to monitor and manage the clients that have active sessions. *Security Registration Services* are utilized by clients at start-up for identity registration (client id, IP address, and user role), which allows a unique Token to be generated for each session of a client. Finally, the *Global Clock Resource* (GCR) (Figure 9.8) is used by Security Policy Services to verify a TC when a client (via a UR) is attempting to invoke a method, and Security Registration Services to obtain the time, which is then used in the generation of a unique Token.

To illustrate the utilization of the services as given in Figure 9.7, consider the processing as shown in Figure 9.8, which illustrates two different paths: the middleware interactions required by a client joining the distributed environment – Steps 1 to 8; and various steps that are then taken in order to attempt to access a resource (invoke a method) – Steps 9 to 12. In Steps 1 to 4, the client is authenticated. In Step 5, the client selects a role to play for the session. In Steps 6 to 8, a token is generated and assigned to the user for the session via a name and password verification of the USR. The user chooses a role and registers via the RegisterClient method, which requests a global time from the global clock resource and returns a token via CreateToken. In Step 9, the client discovers the desired method from the lookup service (JINI or CORBA), and attempts to invoke the method with its parameters and the Token. In Step 10, the resource uses the

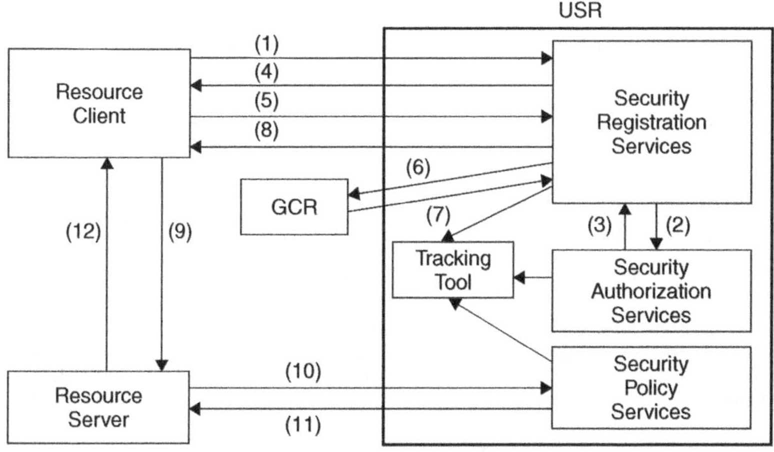

1. Login (user_id, pass_wd)
2. isClientAuthorized()
3. getRoleList(user_id)
4. Return Role List
5. Register_Client (user_id, pass_wd, role)
6. Request Time from GRC
7. Return Current Global Time
8. If Client time not expired, return Token
9. Make Resource Request
10. Check Authentication and Authorization Constraints via hasClientRight(role_id, R_id, S_id, m_id, Token, paramlist)
11. Return Results of constraint Check
12. If Validated Request, Return Results to Client

Figure 9.8 Client interactions and service invocations

hasClientRight method (Step 11) to check whether the user/client meets all of the MACC, time, and signature constraints required to invoke the method (Step 12). If the handle for the method invocation is returned, the client can then invoke the method. The flow as given in Figure 9.8 and described in this paragraph is an example of the attainment of CORBA's identity and authentication, and authorization and access control, as given in the paragraph in Section 9.2.1.

9.3.3 Prototyping/Administrative and Management Tools

This section reviews the prototyping efforts for our constraint-based model and enforcement framework. As shown in Figure 9.9, we have designed and implemented the entire security framework as given in Figure 9.6, which includes the USR and administrative/management tools, capable of interacting with either CORBA or JINI as the middleware. We have also prototyped a University DB resource and client and a hospital application where a client can access information in a Patient DB (Figure 9.9). The hospital portion of Figure 9.9 (left side) interacts with CORBA as the middleware; the university portion (right side) uses JINI. Note that since the middleware lookup service is transparent to the user, a client could be constructed to use CORBA and/or JINI depending on which one is active in the network. From a technology perspective, the university application in Figure 9.9 (right side) is realized using Java 1.3, JINI 1.1, Windows NT 4.0 and Linux, and Oracle 8.1.7. The hospital application (left side) uses the same technologies except for Visibroker 4.5 for Java as middleware. Both of the resources (Patient DB and University DB) operate in a single environment using shared USR transparently, and are designed

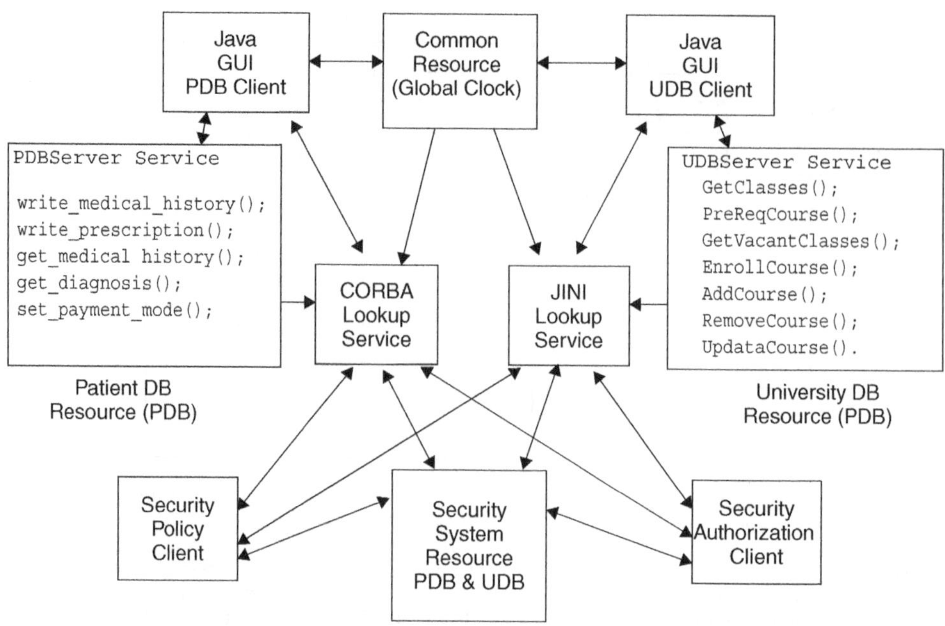

Figure 9.9 Prototype security and resource architecture

to allow them to register their services with CORBA, JINI, or both. The University DB Java Client allows students to query course information and enroll in classes, and faculty to query and modify the class schedule. The Patient DB Java Client supports similar capabilities in a medical domain.

In addition, the prototyping environment supports administrative and management tools, as shown in the bottom portion of Figure 9.6. The *Security Policy Client (SPC)*, shown in Figure 9.10, can be used to define and remove roles, and to grant and revoke privileges (i.e., CLR/CLS, time constraints, resources, services, methods, and/or signature constraints). The security officer can also inspect and monitor security via the tracking capabilities of SPC. Also, SPC is used to establish whether a role is delegatable. In Figure 9.10, the tab for defining a user role, assigning a classification, and determining the delegation status, is shown. The *Security Authorization Client (SAC)*, shown in Figure 9.11, supports authorization of role(s) to users. A user may hold more than one role, but can act in only one role at a time. SAC provides the security officer with the ability to create a new User as discussed in Section 9.3.1. SAC is also used to authorize role delegation. In Figure 9.11, a user is being granted access to role for a specific time period. Not shown is the *Security Delegation Client (SDC)* (see [2, 20]), which is intended to handle the delegation of responsibilities from user to user.

9.4 Conclusion

In this chapter, we have explored the ability of middleware to support security from a number of different perspectives. First, in Section 9.2.1, we examined the CORBA

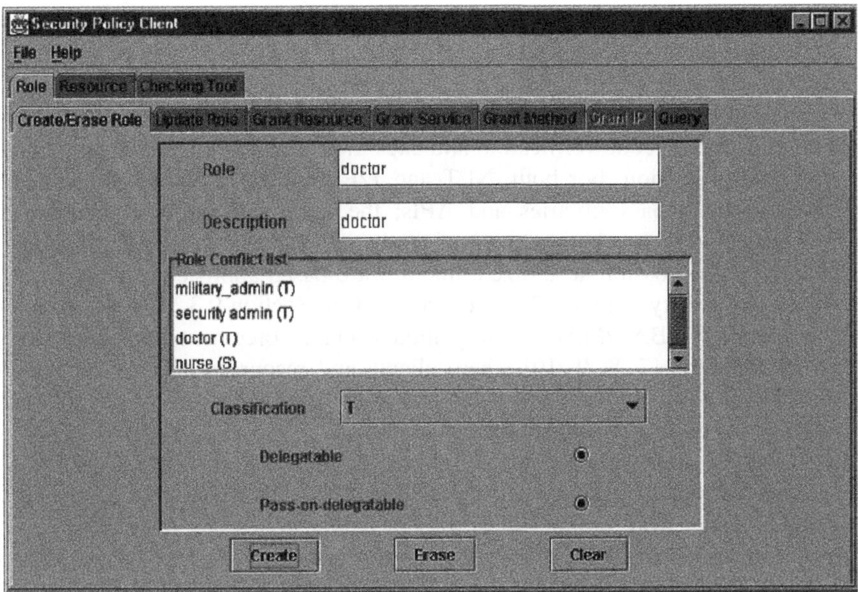

Figure 9.10 Security policy client – defining a role

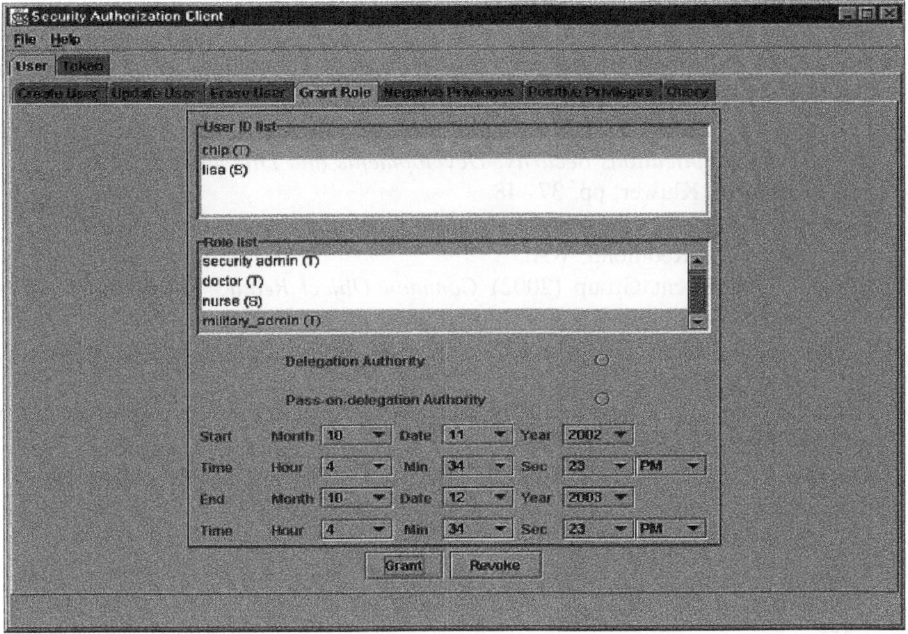

Figure 9.11 Security authorization client – granting a role

security specification [4] meta-model, which is extremely comprehensive, and attempts to capture a wide range of security models, features, techniques, and capabilities. Then, in Sections 9.2.2 and 9.2.3, we reviewed the security capabilities of .NET and J2EE respectively, with a consistent presentation based on code-based access control, role-based access control, secure code verification and execution, secure communication, and secure code and data protection. For both .NET and J2EE, we have avoided implementation-level discussion of class libraries and APIs; the interested reader is referred to [21] and [24], respectively. In addition to out-of-the-box security as presented in Sections 9.2.2 and 9.2.3, it is also possible to utilize a middleware-based approach to design and prototype advanced security features. Towards that end, in Section 9.3, we presented ongoing work on a unified RBAC/MAC security model and enforcement framework for a distributed environment [7, 8, 9, 10] where clients and resources interact via middleware (JINI and Visibroker). In the approach, client interactions to resource APIs are controlled on a role-by-role basis, constrained by security level, time, and data values. As presented, the RBAC/MAC security model realizes many of the features of the CORBA security specification, including privilege attributes (for users and roles) and control attributes (for methods). This work is interesting since it demonstrates that middleware can be an extremely effective tool to easily and seamlessly integrate complex security capabilities so that they can be applied to legacy and modern resources. Overall, we believe that the reader can use the material presented in this chapter as a basis to design and integrate security features and capabilities into distributed applications.

Bibliography

[1] Arnold, K., Osullivan, B., Scheifler, R. W., Waldo, J., Wollrath, A., O'Sullivan, B., and Scheifler, R. (1999) *The JINI Specification*, Addison-Wesley.

[2] Liebrand, M. *et al.* (2003) Role Delegation for a Resource-Based Security Model, in *Data and Applications Security: Developments and Directions II* (eds E. Gudes and S. Shenoi), Kluwer, pp. 37–48.

[3] Microsoft Corporation (1995) *The Component Object Model (Technical Overview)*, Microsoft Press, Redmond, WA.

[4] Object Management Group (2002) *Common Object Request Broker Architecture: Core Specification – Version 3.0.2*, http://www.omg.org.

[5] Object Management Group (2002) *Security Service Specification – Version 1.8*, http://www.omg.org.

[6] Open Software Foundation (1994) OSF DCE Application Development Guide - revision 1.0, OSF, Cambridge, MA.

[7] Phillips, C. *et al.* (2002) Security Engineering for Roles and Resources in a Distributed Environment. *Proceedings of the 3rd Annual ISSEA Conference*, Orlando, FL, March 2002.

[8] Phillips, C. *et al.* (2002) Towards Information Assurance in Dynamic Coalitions. *Proceedings of 2002 IEEE Information Assurance Workshop*, West Point, NY, June 2002.

[9] Phillips, C. *et al.* (2003) Security Assurance for an RBAC/MAC Security Model. *Proceedings of 2003 IEEE Information Assurance Workshop*, West Point, NY, June 2003.

[10] Phillips, C. *et al.* (2003) Assurance Guarantees for an RBAC/MAC Security Model. *Proceedings of the 17th IFIP 2002 11.3 WG Conference*, Estes Park, CO, August 2003.

[11] Riordan, R. (2002) *Microsoft ADO.NET Step by Step*, Microsoft Press.

[12] Roman, E. (1999) *Mastering Enterprise JavaBeans and the Java 2 Platform, Enterprise Edition*, John Wiley & Sons.

[13] Rosenberry, W., Kenney, D., and Fischer, G. (1991) *Understanding DCE*, O'Reilly & Associates.

[14] Sceppa, D. (2002) *Microsoft ADO.NET (Core Reference)*, Microsoft Press.

[15] Valesky, T. (1999) *Enterprise JavaBeans: Developing Component-Based Distributed Applications*, Addison-Wesley.

[16] Vinoski, S. (1997) CORBA: Integrating Diverse Applications Within Distributed Heterogeneous Environments. *IEEE Communications Magazine*, **14**(2), 46–55.

[17] Waldo, J. (1999) The JINI Architecture for Network-Centric Computing. *Communications of the ACM*, **42**(7), 76–82.

[18] Yang, Z. and Duddy, K. (1996) CORBA: A Platform for Distributed Object Computing. *ACM Operating Systems Review*, **30**(2), 4–31.

[19] DevX Enterprise Zone (2002) *Software Engineers Put .NET and Enterprise Java Security to the Test*, http://www.devx.com/enterprise/articles/dotnetvsjava/GK0202-1.asp.

[20] Microsoft Corporation (2003) Microsoft Developer's Network. *Microsoft .NET Security*, http://msdn.microsoft.com/library/default.asp?url=/nhp/Default.asp?contentid=28001369.

[21] Open Web Application Security Project (2002) *J2EE and .NET Security*, http://www.owasp.org/downloads/J2EEandDotNetsecurityByGerMulcahy.pdf.

[22] Sun Microsystems (2003) J2EE Security Model. *Java 2 Platform Security*, http://java.sun.com/j2se/1.4.1/docs/guide/security/spec/security-spec.doc.html.

[23] University of Connecticut's Distributed Security (2003) http://www.engr.uconn.edu/~steve/DSEC/dsec.html.

[24] Microsoft Corporation (2003) Microsoft TechNet. *Security in the Microsoft .NET Framework*, http://www.microsoft.com/technet/treeview/default.asp?url=/technet/itsolutions/net/evaluate/fsnetsec.asp.

[25] Got Dot Net (2002) *Application Domain FAQ*, http://www.gotdotnet.com/team/clr/AppdomainFAQ.aspx.

10

Middleware for Scalable Data Dissemination

Panos K. Chrysanthis[1], Vincenzo Liberatore[2], Kirk Pruhs[1]

[1] *University of Pittsburgh*
[2] *Case Western Reserve University*

10.1 Introduction

A major problem on the Internet is the scalable dissemination of information. This problem is particularly acute exactly at the time when the scalability of data delivery is most important, for example, election results on the night of the 2000 US presidential election, and news during 9/11/2001. The current unicast pull framework simply does not scale up to these types of workloads. One proposed solution to this scalability problem is to use multicast communication. Multicast introduces many nontrivial data management problems, which are common to various data dissemination applications. Multicast data management issues can be addressed by a middleware such that

- the middleware would free application developers from the need of implementing in each application a common set of data;
- the middleware would free application developers from the details of the underlying multicast transport, management algorithms;
- the middleware would implement and unify high-performance state-of-the-art data management methods and algorithms while hiding their complexity from the application developer.

This chapter describes the design, architecture, implementation, and research issues of such a middleware. Particular emphasis will be placed on previous and current research to improve the performance of the middleware implementation. Current applications of

Middleware for Communications. Edited by Qusay H. Mahmoud
 ISBN 0-470-86206-8

the middleware include the scalable dissemination of Web contents and a healthcare alert system, and are described in this chapter.

The middleware addresses the following cored data management issues in multicast-enabled data dissemination applications:

Document selection. Select documents for which broadcast is appropriate from those documents for which different dissemination methods are preferable.
Scheduling. Decide how frequently and in which order documents are multicast.
Consistency. Support currency and transactional semantics for updated contents.
Cache Replacement. Broadcast data can be cached at client site, and local caches are managed by replacement policies tailored to the broadcast environment.
Hybrid schemes. Support client-initiated requests to complement the multicast approach.
Indexing. Reduce client waiting and tuning time by advertising transmission times.

The middleware architecture is built from individual components that can be selected or replaced, depending on the underlying multicast transport mechanism or on the application needs. A prototype will shortly be made available at `http://dora.cwru.edu/mware/` and `http://db.cs.pitt.edu/mware/`.

This chapter is organized as follows. Section 10.2 overviews the middleware components, their relationship, and gives an example of the middleware operations. Section 10.3 reviews previous work on data dissemination systems. Section 10.4 discusses each building block within the middleware in terms of implementation, state-of-the-art research, and open research problems. Section 10.5 discusses the integration of the various data management operations. Section 10.6 describes an application of middleware for data dissemination. Section 10.7 concludes the chapter.

10.2 Architecture Overview

The outline of a possible configuration of the middleware and its relationship with the application and transport layers is shown in Figure 10.1. The configuration demonstrates the main building blocks, each tied to a particular data management function in multicast environments (document selection, scheduling, consistency, caching, and indexing). The middleware has server-side and client-side components; most of the server-side building blocks have a corresponding module on the clients and vice versa. Sections 10.4 and 10.5 will describe in detail each one of these components and their integration. The rest of this section gives a motivating example and demonstrates the purpose of each one of the middleware building blocks.

The application is a highly scalable Web server and, in the example, we assume that the underlying transport is IP multicast. The following discussion closely follows the original work of Almeroth *et al.* [8]. The objective of the Web server application is to scale to a large client population, and scalability is accomplished by using the data dissemination middleware. The configuration in Figure 10.1 enables hybrid dissemination schemes, whereby the server can disseminate data by choosing any combination of the following three schemes: *multicast push*, *multicast pull*, and *unicast push*. The end user should not perceive that Web resources are downloaded with a variety of methods, as the

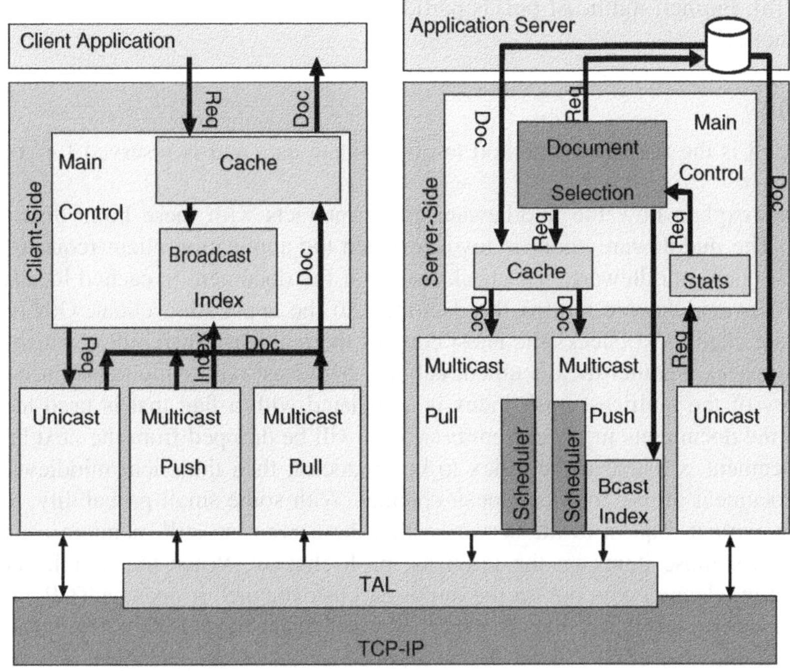

Figure 10.1 Middleware data dissemination architecture

browser and the middleware shield the user from the details of the hybrid dissemination protocol.

Multicast Push.
In multicast push, the server repeatedly sends information to the clients without explicit client requests. (For example, television is a classic multicast push system.) Multicast push is an ideal fit for asymmetric communication links, such as satellites and base station methods, where there is little or no bandwidth from the client to the server. For the same reason, multicast push is also ideal to achieve maximal scalability of Internet hot spots. Hence, generally multicast push should be restricted to hot resources.

Multicast Pull.
In multicast pull, the clients make explicit requests for resources, and the server broadcasts the responses to all members of the multicast group. If the number of pending requests for a document is over some threshold, then the server transmits the document on the multicast pull channel, instead of individually transmitting the document to each of the requesting clients. One would expect that this possibility of aggregation would improve user-perceived performance for the same reason that proxy caches improve user-perceived performance, that is, it is common for different users to make requests to the same resource. Multicast pull is a good fit for "warm resources" for which repetitive multicast push cannot be justified, while there is an advantage in aggregating concurrent client

requests [8]. Further, multicast pull is particularly helpful when some documents quickly become hot [17].

Unicast Pull.
Unicast pull is the traditional method to disseminate data and is reserved for cold documents.

We now explain how the middleware client interacts with these three dissemination methods. The middleware client is invoked when the application client requests a document. The client middleware first checks to see if the document is cached locally, and if so, the client middleware returns the document to the application client. Otherwise, the middleware client next checks the latest copy of the multicast push index, which it stores locally, to check whether the document is being broadcast on the multicast push channel. Each entry of the multicast push index is associated with a flag that is used to indicate which of the documents in the current broadcast will be dropped from the next broadcast. If the document is listed in the index to be broadcast, then the client middleware waits for the document on the multicast push channel. With some small probability, the client makes an explicit request to the server so that the server can still maintain estimates of popularity for those items on the multicast push channel. When the middleware client learns that the document is not on the multicast push channel, it opens a TCP connection to the server and explicitly requests the document using the HTTP. After making this request, the client needs to listen for a response on all of the channels. The document may be returned on the multicast push channel if the document had recently become hot. The document may be returned on the multicast pull channel if several requests for this document arrived nearly simultaneously at the server. And the document may also be returned over the TCP connection that the client has with the server. The client might also observe the TCP connection being closed without the document being returned in the case that the server has determined that the document will not be served to this client via unicast.

The hybrid data dissemination scheme is supported by additional components.

Document Selection.
In the Web server application, the *document selection* unit periodically gathers statistics on document popularity. Once statistics have been collected, the server partitions the resources into *hot*, *warm*, and *cold* documents. When a client wishes to request a Web document, it either downloads it from a multicast group or it requests the document explicitly. In the former case, the client needs to find the multicast address on which the server transmits hot resources. Multicast address determination can be accomplished with a variety of schemes including an explicit http request followed by redirection to the appropriate multicast address, hashing the URI to a multicast group [8], using a well-known multicast address paired to the IP address of the origin server in single-source multicast [39], or application-level discovery in the case of end-to-end multicast.

Bandwidth Allocation.
This component determines what fraction of the server bandwidth is devoted to servicing documents via the multicast push channel, and what fraction of the server bandwidth

is devoted to servicing documents through the pull channels. This component is closely related to the document selection component. We use an adaptation of an algorithm proposed by Azar *et al.* [11] for the case that the server only has a multicast push channel and unicast. In [11], it is shown that if one wants to minimize the server bandwidth required to obtain a particular delay L for all the documents, then the multicast push channel should consist of those documents with probability of access greater than $1/(\lambda L)$, where λ is the arrival rate of document requests. By binary search, we can then find the minimum delay L obtainable by our server's fixed known bandwidth.

Indexing.
The server also broadcasts an *index* of sorted URIs or URI digests, which quickly allows the client to determine whether the requested resource is in the hot broadcast set [43, 52, 81, 6]. On the whole, the client determines the multicast group, downloads the appropriate portions of the index, and determines whether the resource is upcoming along the cyclic broadcast. If the request is not in the hot broadcast set, the client makes an explicit request to the server, and simultaneously starts listening to the warm multicast group if one is available. If the page is cold, the requested resource is returned on the same connection.

Multicast Pull Scheduling.
The *multicast pull scheduling* component resolves contention among client request for the use of the warm multicast channel and establishes the order in which pages are sent over that channel.

Multicast Push Scheduling.
In multicast push, the server periodically broadcasts hot resources to the clients. The server chunks hot resources into nearly equal-sized pages that fit into one datagram and then cyclically sends them on the specified multicast group along with index pages. The frequency and order in which pages are broadcast is determined by the *multicast push scheduling* component. Pages are then injected at a specified rate that is statically determined from measurements of network characteristics [8]. Alternatively, different connectivity can be accommodated with a variety of methods: the multicast can be replicated across multiple layers [16, 21, 80], it can be supported by router-assisted congestion control [62], or it can use application-level schemes in end-to-end multicast [23]. Client applications can recover from packet loss by listening to consecutive repetitions of the broadcast [8], or pages can be encoded redundantly with a variety of schemes that allow the message to be reconstructed [21].

Caching and consistency.
Upon receipt of the desired pages, the client buffers them to reconstruct the original resource and can *cache* resources to satisfy future request [51, 58]. The set of hot pages is cyclically multicast, and so received pages are current in that they cannot be more than one cycle out-of-date. Furthermore, certain types of *consistency* semantics can be guaranteed by transmitting additional information along with the control pages [74, 68].

Transport Adaptation Layer.
The TAL (Transport Adaptation Layer) renders middleware and applications independent of the particular choice of a multicast transport protocol.

10.3 Background and Historical Notes

10.3.1 Multicast Data Dissemination

An early example of scalable data dissemination is the teletext system, which uses the spare bandwidth allocated in the European television channels to implement multicast push data dissemination. Television-based teletext systems have been deployed since the early 80s by most national television companies in Western Europe. Teletext has since attained nationwide diffusion and reaches most households. It provides a continuous information source and has deeply influenced the lifestyle of the countries where it is operational [36, 75].

In the context of data networks, early examples of multicast push include high-throughput multiprocessors database systems over high-bandwidth networks [20] and a community information system over wireless [36]. Multicast push is a good fit for *infostations*, a critical component in a wireless information system where connectivity is spotty [42]. Early results and directions in the field are discussed in [33]. The DBIS-Toolkit [9] introduces a gateway for data delivery. The toolkit provides a set of components to adapt various data dissemination methods with each other. It builds an overlay network where different logical links use either multicast or unicast, push or pull, and synchronous or asynchronous communication.

10.3.2 Multicast

In *broadcast* communication, packets are delivered to all the hosts in a network. In *multicast* communication, packets are delivered to a specified set of hosts. In many types of networks, broadcasting is available natively at the physical or at the data link layer. For example, broadcasting is the basic communication mode in shared media, such as wireless and optical. In these networks, broadcast is the foundation for other communication methods (multicast and unicast), which are implemented by selectively discarding frames at the receiver site if that frame was not directed to that host. Since broadcasting is available natively, it is the natural method to disseminate data on such media. Broadcast data dissemination can be used over wireless links, including satellites [4] and base station schemes [42], and optical networks.

Broadcast does not scale to multihop networks and, consequently, is only used for special applications (e.g., DHCP) in these networks. The lack of a broadcast primitive complicates the implementation of multicast as well. In the context of the Internet, multicast is supported by the network layer [27, 39] but is seldom enabled for reasons including security, performance, scalability, difficulty of management, and accounting [23]. As a result, multicast is more commonly supported at the application layer (*overlay multicast*) through protocols such as HyperCast [61], OverCast [45], Narada [23], or Yoid. Multicast protocols, whether based on IP multicast or on overlays, differ in the type of functionality that they offer. For example, reliable multicast [38] ensures error recovery and single-source multicast [39] allows for the added scalability of IP multicast. Additionally, a peer-to-peer system such as Chord [24] can support some multicast-like functionality.

In unicast communication, the sender typically regulates the rate at which it injects packets into the network so as to avoid overwhelming the receiver (*flow control*) or intermediate network nodes (*congestion control*) while simultaneously achieving maximum throughput [65]. In a multicast environment, flow and congestion control are significantly complicated by the heterogeneity of the receiver population. In particular, each receiver can have a receive buffer of different size and different available end-to-end bandwidth. A highly scalable solution to flow and congestion control is *layered multicast* [63]. In a layered multicast scheme, data is multicast simultaneously on L different channels (*layers*). In the most common scheme, the transmission rate r_l at layer l is $r_0 = r_1 = 1$ and $r_l = 2r_{l-1}$ for $1 < l < L$ (basically, the rates increase by a factor of 2 from one layer to the next). A receiver can subscribe to a prefix of layers $0, 1, \ldots, l$, thereby selecting to receive data at a rate, which is at least 1/2 of the bandwidth available from the source to the receiver.

10.4 Middleware Components

In this section, we discuss all the components of the middleware. We will describe known algorithms for the implementation of the components and their analytical and experimental evaluation. We will also discuss open research issues pertaining to each one of the components. Figure 10.1 gives a broad architectural overview of how the following components can fit together in a typical configuration of the middleware.

10.4.1 Transport Adaptation Layer

A first middleware component is the Transport Adaptation Layer (TAL). Its objective is to enable the middleware to interact with different types of multicast transport within a uniform interface. Different multicast protocols often present different APIs and different capabilities. It is unlikely that a single multicast mechanism would be able to satisfy the requirements of all applications [38], and so the middleware must be able to interact with various underlying multicast transport protocols. As a result, the TAL allows us to write the middleware with a unique multicast API while retaining the flexibility as to the exact multicast transport. The TAL is the lowest common layer within the middleware (Figure 10.1) and provides transport-like functionality to all other communicating middleware modules. It possesses a server component (mostly used to stream pages out on a multicast channel) and a client component (mostly used to receive pages from such stream).

The Java Reliable Multicast (JRM) library [72] is an existing implementation that contains a TAL-like interface, the *Multicast Transport API (MTAPI)*. The MTAPI supports multiple underlying multicast protocols and allows for new protocols to be seamlessly added. It should be noted that the TAL's objective is not to restrict a given application to a set of specific multicast protocols. An additional functionality of TAL is to integrate various multicast packages so that clients supporting different multicast protocols can communicate with each other through the middleware.

An application follows the following steps to employ the TAL. Before activating the middleware, the application will call the channel management API (e.g., in JRM) to:

- Choose the most appropriate multicast transport among those that are available.
- Interface with transport layer functions that resolve channel management issues, including the creation, configuration, destruction, and, possibly, address allocation of multicast channels upon requests.
- Interface with transport layer functions that support session management, including session advertising, discovery, join, and event notification.

The application can then invoke a security API (as in JRM) to establish data confidentiality and integrity as well as message, sender, and receiver authentication. At this stage, the application has a properly configured multicast channel, which will be passed to the data management middleware along with a description of the target data set. Additionally, if the application also desires a multicast pull channel for the warm pages, it creates and configures a second multicast channel through analogous invocations of the channel management and security API's and passes it to the middleware. Other middleware modules use the TAL (e.g., MTAPI) to:

- Send and receive data from multicast groups.
- Obtain aggregate information from clients if the multicast transport layer supports cumulative feedback from the clients.

The TAL is a thin layer that does *not* implement features that are missing or are inappropriate for the underlying transport. The purpose of the TAL is to provide a common interface to existing protocols and not to replace features that are not implemented in the given protocols. For example, the TAL does not provide any additional security, but it simply interfaces with existing security modules in the underlying multicast layer.

While the transport adaptation layer aims at supporting diverse transport modules, the nature of the middleware and of the target applications that will run on it impose certain constraints on the types of transport layers that can be supported. The most important one is that data dissemination needs to scale to a very large receiver group. Consequently, the transport must avoid the problem of ACK/NACK implosion, through, for example, ACK aggregation [38] or NACK suppression [32, 37]. Alternatively, a reliable multicast transport can adopt open-loop reliability solutions such as cyclical broadcast [8] or error-correcting codes [21]; open-loop solutions are an especially good fit for asymmetric communication environments such as satellites or heavily loaded servers. On the other hand, the middleware does not necessarily need a multicast transport that provides Quality-of-Service guarantees, that accommodates multiple interacting senders, that supports intermittent data flows, or that provides secure delivery. Receivers can join the multicast at different start times, but will not receive copies of data sent before their start time.

10.4.2 Document Selection

As discussed in Section 10.2, the middleware operates according to a hybrid scheme whereby multicast push, multicast pull, and unicast are used depending on whether individual pages are hot, warm, or cold. The server must partition the set of pages into these three groups, and such operation is accomplished by the document selection module. The document selection is one of the primary entry points to the middleware at the server side: the application dynamically specifies the target document set to the document selection unit through an appropriate API and the unit will autonomously partition such data set for use of downstream middleware components (Figure 10.1). The document selection unit uses statistics of document popularity collected within the main control module, caches documents in main memory if possible, and requests to the application documents that are not cached. Document selection is a server-only module: the client must conform to the selection choices made by the server.

The implementation of the document selection unit ultimately depends on the following consideration: pages should be classified as hot, warm, or cold so that overall middleware performance is improved [76]. Save this general consideration, however, the document selection unit can be implemented in various ways. Document selection can be accomplished by having client request all documents, including those in the multicast push broadcast [8]. The advantage of this solution is that the server obtains precise estimates of document popularity, at the price of forcing client requests even when such requests are not needed. As a result, client operations could be delayed and dissemination does not achieve maximal scalability. Another solution is to temporarily remove documents from the broadcast cycle and count the number of ensuing requests. Then, if the server receives a large number of requests, it assumes that the document is still hot and it is reinserted in the hot resource set [76]. In addition to these established methods, alternatives include collecting aggregate statistics from multicast feedback, as in single-source multicast, and randomly sampling the client population [17].

10.4.3 Multicast Push Scheduling

In multicast push, a server cyclically multicasts a set of hot pages. The objective of the multicast push scheduling component is to decide the frequency and order in which those pages are disseminated. The component obtains the multicast page set from the main control unit (so that no further decision as to the membership in that set is made downstream of the document selection component, Figure 10.1). Multicast push scheduling also interacts with the broadcast index component (Section 10.4.5). The multicast push scheduling components are present only at the server side.

The implementation of multicast push scheduling can proceed according to different algorithms. The simplest broadcast strategy is to adopt a *flat broadcast schedule*, whereby each page is transmitted once every n ticks. A sample flat schedule is depicted in Figure 10.2(a). Nonflat schemes are desirable when some pages are more popular than others, in which case the most popular pages should be devoted a larger fraction of the available bandwidth. A simple way to differentiate pages is through *frequency multiplexing*, where the data source partitions the data set across several physical channels according to their popularity [55, 70]. Differentiated treatment arises from aggregating a smaller amount of hotter data on one channel and a larger amount of cooler data on another channel. Since channel bandwidth is the same, the fewer hotter pages receive a proportionally larger amount of bandwidth than cooler pages. An alternative is *time-division multiplexing*, whereby pages are partitioned into a set of *logical channels*, and the logical channels alternate over the same physical channel [4]. The broadcast schedule is flat within a single logical channel, but hotter channels contain less pages or are

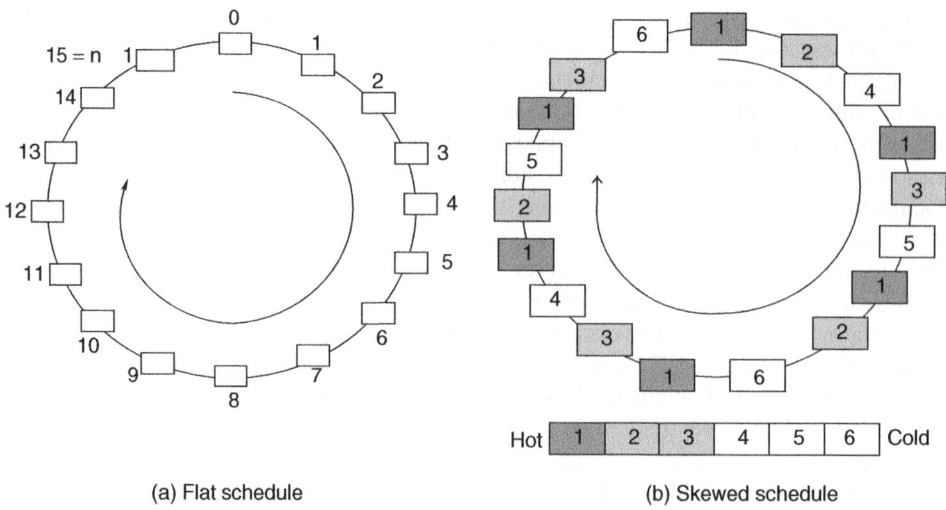

Figure 10.2 Figure (a) is an example of a flat broadcast program. Pages are numbered from 0 to $n-1$, and are cyclically transmitted by the server in that order. Figure (b) is a skewed broadcast schedule that is obtained by multiplexing on the same physical channel the logical channels {1}, {2, 3}, {4, 5, 6}

scheduled for broadcast more frequently than cooler channels. Thus, hotter pages are transmitted more often than cooler ones. Figure 10.2(b) gives an example of a broadcast schedule that is the time-multiplexed combination of three logical channels, each containing a different number of pages. Time-division multiplexing is potentially more flexible than frequency multiplexing in that it allows for a finer bandwidth partition. In particular, a logical channel can contain only one page, which results in a fine per-page transmission schedule. When the broadcast is scheduled on a per-page basis, pages are broadcast on the same physical channel with frequency determined by their popularity. A family of scheduling algorithms for time-multiplexed broadcast assumes that the data source has estimates of the stationary probabilities with which clients need pages. Estimations of these probabilities would be collected by the document selection module as part of the classification of page popularity. The objective of scheduling is to minimize the expected waiting time to download a page. The square-root law asserts that page i should be scheduled with frequency proportional to $\sqrt{p_i}$ [14], where p_i is the stationary probability that i is requested by clients. The square-root law results in transmission frequencies that are in general not integral, and thus the law can be interpreted as a linear relaxation of an integer programming problem [14]. The problem of finding an optimal integral cyclic schedule is NP-hard if pages have a broadcast cost [14], but not MAX-SNP-hard [50], and it can be solved in polynomial time if the server broadcasts only two pages [12]. A simple and practical 2-approximation algorithm is expressed by the *MAD* (Mean Aggregate Delay) rule [77, 14]. The MAD algorithm maintains a value s_i associated with each page i. The quantity s_i is the number of broadcast ticks since the last time page i was broadcast. The MAD algorithm broadcasts a page i with the maximum value of $(s_i+1)^2 p_i$. In particular, when all p_i's are equal, MAD generates a flat broadcast. MAD is a 2-approximation of the optimal schedule, and guarantees a cyclical schedule. Extensions include algorithms for the cases when broadcast pages have different sizes [49, 73] and when client objectives are described by polynomial utility functions [13].

The stationary access probabilities p_i do not express dependencies between data items. Consider the following elementary example. Page B is an embedded image in page A (and only in page A). Page A is not accessed very frequently, but when it is accessed, page B is almost certainly accessed as well. In this scenario, the stationary access probability p_B of page B is small, but the value of p_B is not fully expressive of the true access pattern to B. The problem of exploiting page access dependencies can be modeled as a graph-optimization problem, for which we have devised an $\widetilde{O}(\sqrt{n})$-approximation algorithm [60]. We have also proposed a sequence of simpler heuristics that exploit page access dependencies. We measured the resulting client-perceived delay on multiple Web server traces, and observed a speed-up over previous methods ranging from 8% to 91% [59].

10.4.4 Multicast Pull Scheduling

In multicast pull, clients make explicit requests for resources, and the server multicasts its responses. If several clients ask the same resource at approximately the same time, the server can aggregate those requests and multicast the corresponding resource only once. The multicast pull scheduling component resolves contention among client request for the use of the warm multicast channel and establishes the order in which pages are sent over that channel. The multicast pull scheduling component operates only at the server. There are many reasonable objective functions to measure the performance of a server, but by far the mostly commonly studied measure is average user-perceived latency, or equivalently average flow/response time, which measures how long the average request waits to be satisfied. In traditional unicast pull dissemination, it is well known that the algorithm Shortest Remaining Processing Time (SRPT) optimizes average user-perceived latency.

The most obvious reason that the situation is trickier for the server in multicast pull data dissemination, than for unicast pull data dissemination, since the server needs to balance the conflicting demands of servicing shorter files and of serving more popular files. To see that the situation can even be more insidious, consider the case that all of the documents are of unit size. Then the obvious algorithm is Most Requests First (MRF), which always transmits the document with the most outstanding requests. In fact, one might even initially suspect that MRF is optimal in the case of unit-sized documents. However, it was shown in [47] that MRF can perform arbitrarily badly compared to optimal. The basic idea of this lower bound is that it can be a bad idea to immediately broadcast the most popular document if more requests will arrive immediately after this broadcast. Thus, one sees that the multicast pull server has to also be concerned about how to best aggregate requests over time.

The standard redress in situations where a limited algorithm cannot produce optimal solutions is to seek algorithms that produce solutions with bounded relative error (i.e., competitive analysis). However, it was shown in [47, 30] that no algorithm can achieve even bounded relative error. As is commonly the case, the input distributions for which it is impossible to guarantee bounded relative error are inputs where the system load is near peak capacity. This makes intuitive sense as a server at near peak capacity has insufficient residual processing power to recover from even small mistakes in scheduling. In [46], we suggested that one should seek algorithms that are guaranteed to perform well, unless the system is practically starved. For example, we showed in [46] that the algorithms, Shortest Elapsed Time First, and the Multi-level Feedback Queue used by Unix and Windows NT, have bounded relative error if the server load is bounded an arbitrarily small amount away from peak capacity.

Our method to analyze scheduling problems has come to be known as *resource augmentation analysis* and has been widely adopted as an analysis technique for a wide array of online and offline scheduling problems. In our context, an *s-speed c-approximation* online algorithm A has the property that for all inputs the average user-perceived latency of the schedule that A produces with a speed s server, denoted by A_s, is at most c times Opt_c, the optimal average user-perceived latency for a speed 1 server. The notation and terminology are from [66]. Intuitively, Opt_1 is approximately equal to $\Theta(Opt_{1+\epsilon})$ unless the system load is near peak capacity. So, one way to interpret an s-speed c-approximation algorithm is that is it is $O(c)$-competitive if $Opt_1 = \Theta(Opt_s)$, or alternatively, if the system load is at most $1/s$.

We have shown in [30] that the algorithm Equi-partition, which broadcasts each file at a rate proportional to the number of outstanding requests for the file, is an $O(1)$-speed $O(1)$-approximation algorithm, that is, the average user-perceived latencies produced by Equi-partition have bounded relative error if the server load is below some constant threshold. This work also highlights the surprisingly close relationship between multicast pull scheduling and scheduling computational jobs, with varying speed-up curves, on a multiprocessor. Another way of viewing Equi-partition is that the bandwidth is distributed evenly between all requests. Hence, Equi-partition can be implemented by servicing the requests in a round-robin fashion.

Researchers, including us, have also considered the special case in which the data items are all of approximately the same size [7, 47]. This would be an appropriate job environment for a name server communicating IP address, or any server where all data items are small. It seems that the right algorithm in this setting is Longest Wait First (LWF), which was proposed in [29]. The algorithm LWF maintains a counter for each data item that is the sum over all unsatisfied requests for that page, of the elapsed time since that request. The algorithm LWF then always broadcasts the page with highest counter. LWF can be implemented in logarithmic time per broadcast using the data structure given in [48]. We show in [31] that LWF is an $O(1)$-speed $O(1)$-approximation algorithm.

An open research problem is to seek an algorithm that has bounded relative error when the load is arbitrarily close to the peak capacity of the server. That is, we seek an $(1 + \epsilon)$-speed $O(1)$-competitive algorithm. Such an algorithm is called *almost fully scalable* [71].

Another interesting dimension to the multicast push scheduling is the exploitation of the semantics of data and applications to develop highly efficient, application-specific algorithms. In Section 10.6, we briefly discuss two such algorithms developed in the context of decision-support applications.

10.4.5 Multicast Indexing

In multicast data dissemination, users monitor the multicast/broadcast channel and retrieve documents as they arrive on the channel. This kind of access is sequential, as it is in tape drives. On the other hand, the middleware combines one multicast push channel and one multicast pull channel, and so it must support effective tuning into multiple multicast channels. To achieve effective tuning, the client needs some form of directory information to be broadcasted along with data/documents, making the broadcast self-descriptive. This directory identifies the data items on the broadcast by some key value, or URL, and gives the time step of the actual broadcast. Further, such a directory not only facilitates effective search across multiple channels but it also supports an energy-efficient way to access data. The power consumption is a key issue for both handheld and mobile devices, given their dependency on small batteries, and also for all other computer products, given the negative effects of heat. Heat adversely effects the reliability of the digital circuits and increases costs for cooling, especially in servers [64]. In order to access the desired data, a client has to be in *active mode*, waiting for the data to appear on the multicast. New architectures are capable of switching from *active mode* to *doze mode*, which requires much less energy.

The objective of the multicast indexing component is to introduce a page directory in the multicast push channel for the purposes described above. The component is invoked after scheduling and feeds directly into the TAL to multicast the hot data set and the corresponding index. Multicast indexing has components both at the server and on the clients. In multicast push data dissemination, several implementations of multicast indexing have been proposed to encode a directory structure. These include incorporating hashing in broadcasts [44], using signature techniques [53], and broadcasting *index* information along with data [43, 26, 25, 81, 41]. Of these, broadcast indexing is the simplest and most effective in terms of space utilization. The efficiency of accessing data on a multicast can be characterized by two parameters.

- *Tuning Time:* The amount of time spent by a user in active mode (listening to channel).
- *Access Time:* The total time that elapses from the moment a client requests data identified by ordering key, to the time the client reads that data on channel.

Ideally, we would like to reduce both tuning time and access time. However, it is generally not possible to simultaneously optimize both tuning time and access time. Optimizing the tuning time requires additional information to be broadcast. On the other hand, the best access time is achieved when only data are broadcast and without any indexing. Clearly, this is the worst case for tuning time.

We have developed a new indexing scheme, called *Constant-size I-node Distributed Indexing* (CI) [6], that performs much better with respect to tuning time and access time for broadcast sizes in practical applications. This new scheme minimizes the amount of coding required for constructing an index in order to correctly locate the required data on the broadcast, thus decreasing the size of the index and, consequently, access time as well. Our detailed simulation results indicate that CI, which is a variant of the previously best performing *Distributed Indexing* (DI) [43, 81], outperforms it for broadcast sizes of 12,000 or fewer data items, reducing access time up to 25%, tuning time by 15% and saving energy up to 40%. Our experimental results on 1 to 5 channels also reveal that there is a trade-off between the various existing indexing schemes in terms of tuning and access time and that the performance of different schemes is dependent on the size of the broadcast [5].

Besides optimizing existing schemes, one open research goal is to develop a mixed-adaptive indexing scheme that is essentially optimal over all broadcast sizes.

10.4.6 Data Consistency and Currency

As multicast-based data dissemination continues to evolve, more and more sophisticated client applications will require reading current and consistent data despite updates at the server. For this reason, several protocols have been recently proposed [74, 15, 67, 68, 54, 28] with the goal of achieving consistency and currency in broadcast environments beyond local cache consistency. The objective of the data consistency components is to implement such protocols for consistency and currency. All these protocols assume that the server is stateless and does not therefore maintain any client-specific control information. To get semantic and temporal-related information, clients do not contact the server directly, instead concurrency control information, such as invalidation reports, is broadcast along with the data. This is in line with the multicast push paradigm to enhance the server scalability to millions of clients. When the scheduling algorithm selects a page to be multicast, there can be, in general, different versions of that page that can be disseminated. The two obvious choices are as follows.

- *Immediate-value broadcast:* The value that is placed on the broadcast channel at time t for an item x is the most recent value of x (that is the value of x produced by all transactions committed at the server by t).
- *Periodic-update broadcast:* Updates at the server are not reflected on the broadcast content immediately, but at the beginning of intervals called *broadcast currency intervals* or *bc-intervals* for short. In particular, the value of item x that the server places on the broadcast at time t is the value of x produced by all transactions committed at the server by the beginning of the current bc-interval. Note that this may not be the value of x at the server at the time x is placed in the broadcast medium if in the interim x has been updated by the server.

In the case of periodic-update broadcast, often the bc-interval is selected to coincide with the broadcast cycle so that the value broadcast for each item during the cycle is the value of the item at the server at the beginning of the cycle. In this case, clients reading all their data, for example, components of a complex document, within a bc-interval are ensured to be both consistent and current with respect to the beginning of the broadcast cycle. However, different components that are read from different broadcast cycles might not be mutually consistent even if current, and hence when they are combined by the clients, the resulting document may be one that had never existed in the server. The same problem exists also in the case of immediate-value broadcast – consider immediate-value broadcast as a periodic-update broadcast with a bc-interval of zero duration.

Our investigation is directed toward the development of a general framework for correctness in broadcast-based data dissemination environments [69]. We have introduced the notion of the *currency interval* of an item in the readset of a transaction as the time interval during which the value of the item is valid. On the basis of the currency intervals of the items in the readset, we developed the notion of temporal spread of the readset and two notions of currency (snapshot and oldest-value) through which we characterize the temporal coherency of a transaction. Further, in order to better combine currency and consistency notions, we have developed a taxonomy of the existing consistency guarantees [18, 34, 82] and show that there are three versions for each definition of consistency. The first (Ci) is the strongest one and requires serializability of *all* read-only client transactions with server transactions. This means that there is a global serialization order including all read-only client transactions and (a subset of) server transactions. The second version (Ci-S) requires serializability of some *subset* of client read-only transactions with the server transactions. This subset may for example consist of all transactions at a given client site. The last version (Ci-I) requires serializability of each read-only client transaction *individually* with the server transactions.

An open research problem is to investigate efficient ways to disseminate any control information. For example, consistency information can utilize the broadcast indexing structures. Another open research issue is to expand our theoretical framework to include the case of a cache being maintained at the clients. The goal is to develop a model that will provide the necessary tools for arguing about the temporal and semantic coherency provided by the various protocols to client transactions. In addition, it will provide the basis for new protocols to be advanced.

10.4.7 Client Cache

Clients have the option of caching the multicast pages locally. As a result, clients can find requested pages either in their local cache or by listening to the server multicast. The client-caching module provides buffer space and a cache management functionality to the client. The module is the main entry point of the client application into the middleware: a page is always requested to the cache module first, and it is passed on to the remaining client middleware components only in the case of a cache miss. The implementation of the client cache module takes into account the two primary factors of *consistency* and *replacement*. If the client caches a page, that page could contain stale values or its usage can lead to an inconsistent view of the original data set. Thus, appropriate strategies are needed for currency and consistency, which were discussed in Section 10.4.6. Moreover, if the client has a cache of limited size, it must perform *page placement and replacement*, that is, it must dynamically decide which pages are stored in the cache and which pages should be removed from the cache. Traditional policies for cache replacement include, for example, LRU (Least-Recently Used) and similar policies originally developed in the context of virtual memory systems [78]. The gist of LRU is that past accesses should predict future accesses, and so it should incur few page misses. In multicast push environments, caching aims at reducing the time spent waiting during the broadcast. By contrast, minimizing (say) the number of page faults in isolation might not bring any performance improvement. Consequently, a policy such as LRU can lead to suboptimal performance in multicast push environments. In the context of Web caching, a number of algorithms, such as Landlord/Greedy-Dual-Size [22, 83], take into account both the number of cache misses and the time spent to download a Web object. Such policies can be suboptimal in a multicast push environment for two reasons. First, Web caching policies are based on average estimates of resource download times, whereas in the multicast environment, the download time varies periodically depending on the server schedule. Second, *prefetching* can sometimes be executed without clients waiting on the broadcast [3], and can thus be used to complement a pure replacement strategy.

If stationary access probabilities exist and are known, several policies strike a balance between page popularity and access times [4, 3, 79]. For example, the PT algorithm maintains two values for each broadcast page i: p_i, the probability that page i will be requested, and t_i, the waiting time needed to load i once the current request has been satisfied. Then, PT keeps in the cache the set of pages with the largest values of $p_i t_i$ [3]. In certain applications, such as Web resource dissemination, hot pages have nearly the same stationary access probability [10] or such probabilities do not fully capture the underlying access process [57], so that it is desirable to have page replacement and prefetching algorithms that work with no probabilistic assumptions. Algorithms of this type are said to be *online algorithms* and are almost always evaluated through *competitive analysis* (Section 10.4.4, [19]). We have devised an algorithm (the *Gray algorithm*) that is online and is provably optimal in terms of competitive ratios [51]. We have subjected the Gray algorithm to extensive empirical investigation on both synthetic and Web traces, where it outperformed previous online strategies [58].

10.5 Building Block Integration

10.5.1 Integration

The interaction of the middleware building blocks should be carefully tuned to achieve optimal performance. Multicast data management components have often been designed, analyzed, and empirically evaluated in isolation. Important exceptions include [58], which addresses the interaction of caching and scheduling, and [40], which focuses on wireless environments. The integration of building blocks is still riddled with many additional open research issues. First, the scheduling of the multicast push determines the frequency and order in which data items are broadcast, and so it affects the appropriateness of an indexing scheme. Analogously, cache replacement strategies interact with consistency policies at the client side. Another issue arises from the interaction of dynamic document selection with scheduling and indexing.

Another type of interaction exists between the middleware and the underlying transport. For example, certain scheduling, indexing, and caching algorithms assume that the broadcast data is reliably delivered to all recipients in the order in which it is broadcast. Packet drops can potentially cause performance degradation, even when open-loop reliability solutions are adopted. Certain protocols, such as IP multicast, are intrinsically unreliable. Furthermore, packet losses can occur as a side effect of congestion control algorithms: intermediate components, such as router in router-assisted congestion control [62] or intermediate hosts in end-to-end multicast and in multiple-group sender-controlled congestion control [38] deliberately drop packets to avoid overrunning the downlink. Another issue arises when a layered multicast scheme is chosen (Section 10.3.2). Certain scheduling and indexing algorithms are optimized for the case when all receiving clients obtain the multicast pages in the same order. However, layering can change the order in which clients receive the multicast pages depending on how many layers have been subscribed to by clients. For example, a client that has subscribed several layers receives broadcast pages in a shuffled sequence as compared to a client that subscribed to one multicast layer [16].

However, indexing relies on the existence of a certain fixed ordering of the broadcast cycle that is common to all clients. Thus, layering can lead to suboptimal performance of indexing and scheduling. Analogously, scheduling establishes an ordering of the broadcast pages, which can be scrambled by a subsequent layered scheme. The interaction between push scheduling and multicast congestion control is discussed next in more detail.

10.5.2 Scheduling for Layered Multicast

Layered multicast leads to a different version of the multicast push scheduling problem in which the contents are scheduled on multiple channels and the bandwidth on channel l is given by the rates r_l. Unlike previous multichannel models (e.g., [14]), all contents are multicast on layer 0 (to guarantee that all receivers can eventually get all the contents) and receivers do not necessarily listen to all channels. Because of these added restrictions, it can easily be shown that the previously known lower-bounds on the average waiting time hold also for the layered multicast setting. Furthermore, previous algorithms for the multichannel problem can be made to run within one layer, which, in conjunction with the previous lower bound, proves that every c-approximation algorithm for the multichannel problem becomes a $2c$-approximation algorithm for the layered multicast problem. In particular, the algorithm of [50] gives a $(2+\epsilon)$-approximation algorithm that runs in polynomial time for any fixed $\epsilon > 0$ [56]. The main open questions in this area are to find better approximation algorithms and to empirically verify their performance.

10.6 Application: Real-Time Outbreak and Disease Surveillance

Section 10.2 describes the use of a data dissemination middleware for scalable Web contents delivery. This section discusses an additional application: the RODS (Real-Time Outbreak and Disease Surveillance) system. RODS is a healthcare alert system developed by the Center for Biomedical Informatics at the University of Pittsburgh [1]. The RODS system has been deployed since 1999 in Western Pennsylvania and since December 2001 in Utah for the Winter Olympic Games [35]. The core of RODS is the health-system-resident component (HSRC) whose function is data merging, data regularization, privacy protection, and communication with the regional system. Typically, users or applications request consolidated and summarized information as in OLAP (on-line analytical processing) business applications. For example, queries often involve joins over several large tables to perform a statistical analysis, for example, computing daily percentage of patients with a particular prodrome in a region for one month period. Also, currently RODS displays spatial-temporal plots of patients presenting with seven key prodromes through a Web interface.

RODS can use a data dissemination middleware to support the collection and monitoring of the large volume of data needed for the assessment of disease outbreaks, as well as the dissemination of critical information to a large number of health officials when outbreaks of diseases are detected.

We have implemented a prototype similar to RODS that utilizes the middleware to disseminate data for decision making, especially during emergencies and periods of crisis. In this prototype, the schema for the database was adapted from the RODS Laboratory. The database, implemented on one of the department's Oracle servers, was populated with randomly distributed data that would mimic the type of data that is collected in the real system. In particular, cases of disease and prodromes populate the database. In the prototype, the data was restricted to each of the 67 counties in Pennsylvania.

The server-side modules were written in Java 2 using the JDBC to Oracle conduits. The application server queries the database for the counts of each distinct disease and prodrome grouped by different attributes, for example by county, and passes them to the middleware. The middleware then transmits these counts in XML messages on the appropriate multicast channel.

The client-side modules were written in Java 2 as well. The client application accepts user input via a graphical user interface that displays the map of Pennsylvania broken down by county. When a user clicks on a county, the counts of disease and prodromes in that county are requested via the middleware (see Figure 10.3). If the requested data is not locally cached, then it is fetched from the multicast channel. When these counts are returned by the middleware, they are displayed on a separate window in a new table.

In the special context of RODS, we have investigated multicast push scheduling schemes that exploit the semantics of the OLAP data. More specifically, every requested document is a summary table, and summary tables have the interesting property that one summary table can be derived from one or more other summary tables. This means that a table requested by a client may *subsume* the table requested by another client. We proposed two new, heuristic scheduling algorithms that use this property to both maximize the aggregated data sharing between clients and reduce the broadcast length compared to the already existing techniques. The first algorithm, called *Summary Tables On-Demand Broadcast Scheduler* STOBS, is based on the *RxW* algorithm [2] and the second one, called *Subsumption-Based Scheduler* (SBS), is based on the *Longest Total Stretch First* (LTSF) algorithm [2]. Further, they differ on the used criterion for aggregate requests. Otherwise, both STOBS and SBS are non-preemptive and consider the varying sizes of the summary tables. The effectiveness of the algorithms with respect to access time and fairness was evaluated using simulation. Both SBS and STOBS are applicable in the context of any OLAP environment, both wired and wireless.

10.7 Conclusions

Multicast is a promising method to solve the scalability problem in the Internet. In this chapter, we have discussed a middleware that unifies and extends support for data management in multicast data dissemination. The chapter describes the overall objectives and architecture of the middleware, the function and interaction of its components, and the implementation of each component. Particular emphasis has gone into improving the performance of the middleware through the adoption of state-of-the-art algorithms, and research issues in this area have been surveyed.

The chapter has also described two applications of the middleware that ensure the scalability of bulk data dissemination. The first application is a *scalable Web server*, which exploits the middleware hybrid dissemination scheme to improve the performance of a Web hot spot. The second application is the *Real-Time Outbreak and Disease Surveillance* (RODS) system. RODS is a scalable healthcare alert system that disseminates information about prodromes and diseases over a wide geographical area. It can be used to scalable and efficiently alert health operator of disease outbreaks so that appropriate responses can be pursued in real-time.

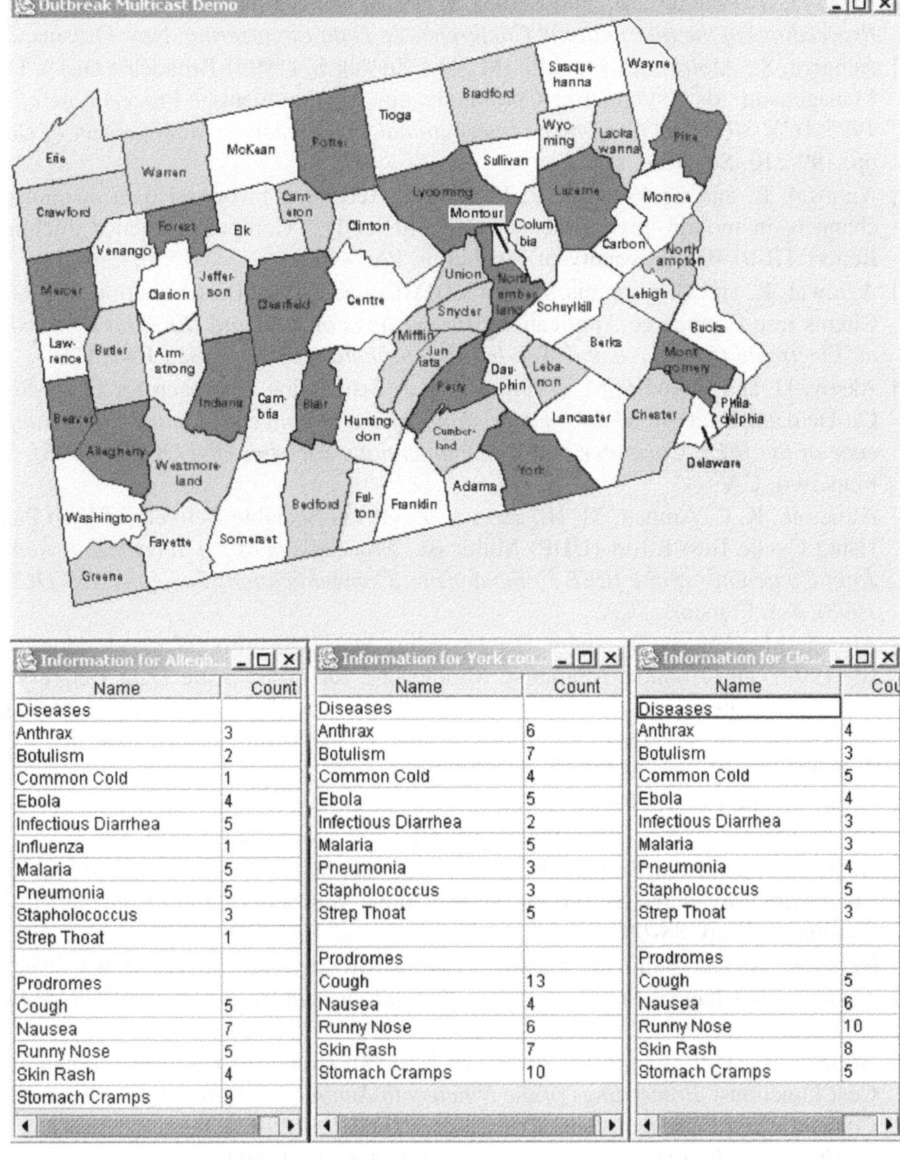

Name	Count
Diseases	
Anthrax	3
Botulism	2
Common Cold	1
Ebola	4
Infectious Diarrhea	5
Influenza	1
Malaria	5
Pneumonia	5
Stapholococcus	3
Strep Thoat	1
Prodromes	
Cough	5
Nausea	7
Runny Nose	5
Skin Rash	4
Stomach Cramps	9

Name	Count
Diseases	
Anthrax	6
Botulism	7
Common Cold	4
Ebola	5
Infectious Diarrhea	2
Malaria	5
Pneumonia	3
Stapholococcus	3
Strep Thoat	5
Prodromes	
Cough	13
Nausea	4
Runny Nose	6
Skin Rash	7
Stomach Cramps	10

Name	Cou
Diseases	
Anthrax	4
Botulism	3
Common Cold	5
Ebola	4
Infectious Diarrhea	3
Malaria	3
Pneumonia	4
Stapholococcus	5
Strep Thoat	3
Prodromes	
Cough	5
Nausea	6
Runny Nose	10
Skin Rash	8
Stomach Cramps	5

Figure 10.3 A snapshot of the emulated RODS systems

Bibliography

[1] http://www.health.pitt.edu/rods.

[2] Acharya, S. and Muthukrishnan, S. (1998) Scheduling On-Demand Broadcasts: New Metrics and Algorithms. *Proceedings of Fourth Annual ACM/IEEE International Conference on Mobile Computing and Networking*, Dallas, TX.

[3] Acharya, S., Franklin, M., and Zdonik S. (1996) Prefetching from a Broadcast Disk. *Proceedings of the International Conference on Data Engineering*, New Orleans, LA.

[4] Acharya, S., Alonso, R., Franklin, M., and Zdonik S. (1995) Broadcast Disks: Data Management for Asymmetric Communication Environments. *Proceedings of the 1995 ACM SIGMOD Conference International Conference on Management of Data*, pp. 199-210, San Jose, CA.

[5] Agrawal, R. and Chrysanthis, P. K. (2001) Accessing broadcast data on multiple channels in mobile environments more efficiently. Computer Science Technical Report TR-01-05, University of Pittsburgh, PA.

[6] Agrawal, R. and Chrysanthis, P. K. (2001) Efficient Data Dissemination to Mobile Clients in e-Commerce Applications. *Proceedings of the Third IEEE Int'l Workshop on Electronic Commerce and Web-based Information Systems*, San Jose, CA.

[7] Aksoy, D. and Franklin, M. (1998) RxW: A Scheduling Approach for Large-Scale On-Demand Data Broadcast. *Proceedings of the Seventeenth Annual Joint Conference of the IEEE Computer and Communications Societies (INFOCOM 1998)*, San Francisco, CA.

[8] Almeroth, K. C. Ammar, M. H., and Fei, Z. (1998) Scalable Delivery of Web Pages Using Cyclic Best-Effort (UDP) Multicast. *Proceedings of the Seventeenth Annual Joint Conference of the IEEE Computer and Communications Societies (INFOCOM 1998)*, San Francisco, CA.

[9] Altinel, M., Aksoy, D., Baby, T., Franklin, M. J., Shapiro, W., and Zdonik, S. B. (1999) Dbis-toolkit: Adaptable Middleware for Large Scale Data Delivery, in *SIGMOD 1999, Proceedings ACM SIGMOD International Conference on Management of Data (eds A. Delis, C. Faloutsos, and S. Ghandeharizadeh), June 1-3, 1999*, Philadelphia, PA, pp. 544–546; ACM Press.

[10] Arlitt, M. and Jin, T. (1999) Workload characterization of the 1998 World Cup Web site. Technical Report HPL-1999-35R1, HP Laboratories, Palo Alto, CA.

[11] Azar, Y., Feder, M., Lubetzky, E., Rajwan, D., and Shulman N. (2001) The Multicast Bandwidth Advantage in Serving a Web Site. *Networked Group Communication*, London, UK, pp. 88–99.

[12] Bar-Noy, A. and Shilo, Y. (2000) Optimal Broadcasting of Two Files Over an Asymmetric Channel. *Journal of Parallel and Distributed Computing*, **60**(4), 474–493.

[13] Bar-Noy, A., Patt-Shamir, B., and Ziper, I. (2000) Broadcast Disks with Polynomial Cost Functions. *Proceedings of the Nineteenth Annual Joint Conference of the IEEE Computer and Communications Societies (INFOCOM 2000)*, Tel Aviv, Israel.

[14] Bar-Noy, A., Bhatia, R., Naor, J., and Schieber, B. (1998) Minimizing Service and Operation Costs of Periodic Scheduling. *Proceedings of the Ninth ACM-SIAM Symposium on Discrete Algorithms*, pp. 11–20, San Francisco, CA.

[15] Barbará, D. (1997) Certification Reports: Supporting Transactions in Wireless Systems. *Proceedings of the IEEE International Conference on Distributed Computing Systems*, June 1997, Baltimore, MD.

[16] Battacharyya, S., Kurose, J. F., Towsley, D., and Nagarajan, R. (1998) Efficient Rate Controlled Bulk Data Transfer Using Multiple Multicast Groups. *Proceedings of the Seventeenth Annual Joint Conference of the IEEE Computer and Communications Societies (INFOCOM 1998)*, San Francisco, CA.

[17] Beaver, J., Pruhs, K., Chrysanthis, P., and Liberatore, V. (2003) The Multicast pull advantage. Computer science technical report, CSD TR-03-104, University of Pittsburgh.

[18] Bober, P. M. and Carey, M. J. (1992) Multiversion Query Locking. *Proceedings of the 1992 SIGMOD Conference*, pp. 497–510, May 1992, Madison, WI.

[19] Borodin, A. and El-Yaniv, R. (1998) *Online Computation and Competitive Analysis*, Cambridge University Press, New York.

[20] Bowen, T. G., Gopal, G., Herman, G., Hickey, T., Lee, K. C., Mansfield, W. H., Raitz, J., and Weinrib, A. (1992) The Datacycle Architecture. *Communications of the ACM*, **35**(12), 71–81.

[21] Byers, J. W., Luby, M., Mitzenmacher, M., and Rege, A. (1998) A Digital Fountain Approach to Reliable Distribution of Bulk Data. *Proceedings of SIGCOMM*, Vancouver, British Columbia, Canada.

[22] Cao, P. and Irani, S. (1997) Cost-Aware WWW Proxy Caching Algorithms. *Proceedings of the USENIX Symposium on Internet Technologies and Systems*, pp. 193–206, Monterey, CA.

[23] Yang-hua C., Rao, S. G., and Zhang, H. (2000) A Case for End System Multicast. *Proceedings ACM SIGMETRICS '2000*, pp. 1–12, Santa Clara, CA.

[24] Dabek, F., Brunskill, E., Kaashoek, F. M., Karger, D., Morris, R., Stoica, I., and Balakrishnan H. (2001) Building Peer-to-Peer Systems With Chord, a Distributed Lookup Service. *Proceedings of the 8th Workshop on Hot Topics in Operating Systems (HotOS-VIII)*, Elmau/Oberbayern, Germany.

[25] Datta, A., VanderMeer, D. E., Celik, A., and Kumar, V. (1999) Broadcast Protocols to Support Efficient Retrieval from Databases by Mobile Users. *ACM TODS*, 24(1), 1–79.

[26] Datta, A., Celik, A., Kim, J., VanderMeer, D., and Kumar, V. (1997) Adaptive Broadcast Protocols to Support Efficient and Energy Conserving Retrieval from Databases in Mobile Computing Environments. *Proceedings of the IEEE Int'l Conference on Data Engineering*, pp. 124–133, March 1997, Birmingham, UK.

[27] Deering, S. (1988) Multicast Routing in Internetworks and Extended Lans. *Proceedings of SIGCOMM*, pp. 55–64, Vancouver, British Columbia.

[28] Deolasee, P., Katkar, A., Panchbudhe, A., Ramamritham, K., and Shenoy, P. (2001) Dissemination of Dynamic Data. *Proceedings of the ACM SIGMOD International Conference on Management of Data*, ACM Press, Santa Barbara, CA, p. 599, Santa Barbara, CA.

[29] Dykeman, H., Ammar, M., and Wong, J. (1986) Scheduling Algorithms for Videotext Under Broadcast Delivery. *Proceedings of the IEEE International Conference on Communications*, Miami, FL.

[30] Edmonds, J. and Pruhs, K. (2003) Multicast Pull Scheduling: When Fairness is Fine. *Algorithmica*, **36**, 315–330.

[31] Edmonds, J., and Pruhs, K. (2004) A maiden analysis of Longest Wait First. *Proceedings of the ACM/SIAM Symposium on Discrete Algorithms*, pp. 818–827, New Orleans, Louisiana.

[32] Floyd, S., Jacobson, V., and McCanne, S. (1995) A Reliable Multicast Framework for Light-Weight Sessions and Application Level Framing. *Proceedings of ACM SIGCOMM*, pp. 342–356, Cambridge, MA.

[33] Franklin, M. and Zdonik, S. (1997) A Framework for Scalable Dissemination-Based Systems. *Proceedings of the 1997 ACM SIGPLAN Conference on Object-Oriented Programming Systems, Languages & Applications*, pp. 94-105, Atlanta, GA.

[34] Garcia-Molina, H. and Wiederhold, G. (1982) Read-Only Transactions in a Distributed Database. *ACM TODS*, **7**(2), 209–234.

[35] Gesteland, P. H., Gardner, R. M., Fu-Chiang T., Espino, J. U., Rolfs, R. T., James, B. C., Chapman, W. W., Moore, A. W., and Wagner, M. M. (2003) *Automated Syndromic Surveillance for the 2002 Winter Olympics*, August 2003.

[36] Gifford, D. K. (1990) Polychannel Systems for Mass Digital Communications. *Communications of the ACM*, **33**(2), 141–151.

[37] Handley, M. and Crowcroft, J. (1997) Network Text Editor (nte) a Scalable Shared Text Editor for Mbone. *Proceedings of ACM SIGCOMM*, pp. 197-208, Cannes, France.

[38] Handley, M., Floyd, S., Whetten, B., Kermode, R., Vicisano, L., and Luby, M. (2000) The Reliable Multicast Design Space for Bulk Data Transfer. RFC 2887.

[39] Holbrook, H. W. and Cheriton, D. R. (1999) IP Multicast Channels: EXPRESS Support for Large-Scale Single-Source Applications. *Proceedings of SIGCOMM*, Cambridge, MA.

[40] Hu, Q., Lee, W. -C., and Lee, D. L. (1999) Performance Evaluation of a Wireless Hierarchical Data Dissemination System. *Proceedings of MobiCom*, Seattle, WA.

[41] Hu, Q., Lee, W. -C., and Lee, D. L. (2000) Power Conservative Multi-Attribute Queries on Data Broadcast. *Proceedings of the 16th International Conference on Data Engineering*, pp. 157–166, February 2000, San Diego, CA.

[42] Imielinski, T. and Badrinath, B. (1994) Mobile Wireless Computing: Challenges in Data Management. *Communications of the ACM*, **37**(10), 18–28.

[43] Imielinski, T., Viswanathan, S., and Badrinath, B.R. (1994) Energy Efficient Indexing on Air. *Proceedings of the SIGMOD Conference*, pp. 25–36, Minneapolis, MN.

[44] Imielinski, T., Viswanathan, S., and Badrinath, B. R. (1994) Power Efficient Filtering of Data on Air. *Proceedings of the International Conference on Extending Database Technology*, Cambridge, UK.

[45] Jannotti, J., Gifford, D. K., Johnson, K. L., Fran s Kaashoek, M., and O'Toole, J. W. Jr. (2000) Overcast: Reliable Multicasting with an Overlay Network. *OSDI*, pp. 197–212, San Diego, CA.

[46] Kalyanasundaram, B. and Pruhs, K. (2000) Speed is as Powerful as Clairvoyance. *Journal of the ACM*, **47**(4), 617-643.

[47] Kalyanasundaram, B., Pruhs, K., and Velauthapillai, M. (2000) Scheduling Broadcasts in Wireless Networks. *European Symposium on Algorithms (ESA)*, Saarbruecken, Germany.

[48] Kaplan, H., Tarjan, R., and Tsiotsiouliklis, K. (2001) Faster Kinetic Heaps and Their Use in Broadcast Scheduling. *Proceedings of the ACM/SIAM Symposium on Discrete Algorithms*, Washington, DC.

[49] Kenyon, C. and Schabanel, N. (1999) The Data Broadcast Problem with Non-Uniform Transmission Times. *Proceedings of the Tenth ACM-SIAM Symposium on Discrete Algorithms*, pp. 547–556, Baltimore, MD.

[50] Kenyon, C., Schabanel, N., and Young, N. (2000) Polynomial-Time Approximation Scheme for Data Broadcast. *Proceedings of the Thirty-Second ACM Symposium on the Theory of Computing*, Portland, OR.

[51] Khanna, S. and Liberatore, V. (2000) On Broadcast Disk Paging. *SIAM Journal on Computing*, **29**(5), 1683–1702.

[52] Khanna, S. and Zhou, S. (1998) On Indexed Data Broadcast. *Proceedings of the Thirtieth ACM Symposium on the Theory of Computing*, pp. 463–472, Dallas, TX.

[53] Lee, W. C. and Lee, D. L. (1999) Signature Caching Techniques for Information Filtering in Mobile Environments. *Wireless Networks*, **5**(1), pp. 57–67.

[54] Lee, V. C. S., Son, S. H., and Lam, K. (1999) On the Performance of Transaction Processing in Broadcast Environments. *Proceedings of the International Conference on Mobile Data Access (MDA'99)*, January 1999, Hong Kong.

[55] Leong, H. V. and Si, A. (1995) Data Broadcasting Strategies Over Multiple Unreliable Wireless Channels. *Proceedings of the ACM International Conference on Information and Knowledge Management*, Baltimore, MD.

[56] Li, W., Zhang, W., and Liberatore, V. Dissemination Scheduling in Layered Multicast Environments. (In preparation).

[57] Liberatore, V. (1999) Empirical Investigation of the Markov Reference Model. *Proceedings of the Tenth ACM-SIAM Symposium on Discrete Algorithms*, pp. 653–662, Baltimore, MD.

[58] Liberatore, V. (2000) Caching and Scheduling for Broadcast Disk Systems. *Proceedings of the 2nd Workshop on Algorithm Engineering and Experiments (ALENEX 00)*, pp. 15–28, San Francisco, CA.

[59] Liberatore, V. (2001) Broadcast Scheduling for Set Requests. *DIMACS Workshop on Resource Management and Scheduling in Next Generation Networks*, Piscataway, NJ.

[60] Liberatore, V. (2002) Circular Arrangements. *29-th International Colloquium on Automata, Languages, and Programming (ICALP), LNCS 2380*, pp. 1054–1065, Malaga, Spain.

[61] Liebeherr, J. and Sethi, B. S. (1998) A Scalable Control Topology for Multicast Communications. *Proceedings of IEEE Infocom*, San Francisco, CA.

[62] Luby, M., Vicisano, L., and Speakman, T. (1999) Heterogeneous Multicast Congestion Control Based on Router Packet Filtering, RMT Working Group.

[63] McCanne, S., Jacobson, V., and Vetterli, M. (1996) Receiver-Driven Layered Multicast. *Proceedings of ACM SIGCOMM*, pp. 117-130, Palo Alto, CA.

[64] Mudge, T. (2001) Power: A First Class Design Constraint. *Computer*, **34**(4), 52–57.

[65] Peterson, L. L. and Davie, B. S. (2000) *Computer Networks*, Morgan Kaufmann, New York.

[66] Phillips, C., Stein, C., Torng, E., and Wein, J. (1997) Optimal Time-Critical Scheduling Via Resource Augmentation. *Proceedings of the ACM Symposium on Theory of Computing*, El Paso, TX.

[67] Pitoura, E. and Chrysanthis, P. K. (1999) Scalable Processing of Read-Only Transactions in Broadcast Push. *Proceedings of the 19th IEEE International Conference on Distributed Computing Systems*, pp. 432–441, June 1999, Vienna, Austria.

[68] Pitoura, E. and Chrysanthis, P. K. (1999) Exploiting Versions for Handling Updates in Broadcast Disks. *Proceedings of the 25th International Conference on Very Large Data Bases*, pp. 114–125, September 1999, Edinburgh, Scotland.

[69] Pitoura, E., Chrysanthis, P. K., and Ramamritham, K. (2003) Characterizing the Temporal and Semantic Coherency of Broadcast-Based Data Dissemination. *Proceedings of the International Conference on Database Theory*, pp. 410–424, January 2003, Siena, Italy.

[70] Prabhakara, K., Hua, K. A., and Oh, J. (2000) Multi-Level Multi-Channel Air Cache Design for Broadcasting in a Mobile Environment. *Proceedings of the IEEE International Conference on Data Engineering*, pp. 167–176, March 2000, San Diego, CA.

[71] Pruhs, K., Sgall, J., and Torng, E. (2004) Online scheduling. *Handbook on Scheduling*, CRC Press, (to appear) 2004.

[72] Rosenzweig, P., Kadansky, M., and Hanna, S. (1998) The Java Reliable Multicast Service: A Reliable Multicast Library. Technical Report SMLI TR-98-68, Sun Microsystems, Palo Alto, CA.

[73] Schabanel, N. (2000) The Data Broadcast Problem with Preemption. *LNCS 1770 Proceedings of the 17th International Symposium on Theoretical Aspects of Computer Science (STACS 2000)*, pp. 181–192, Lille, France.

[74] Shanmugasundaram, J., Nithrakashyap, A., Sivasankaran, R., and Ramamritham, K. (1999) Efficient Concurrency Control for Broadcast Environments. *ACM SIGMOD International Conference on Management of Data*, Sydney, Australia.

[75] Sigel, E. (1980) *Videotext: The Coming Revolution in Home/Office Information Retrieval*, Knowledge Industry Publications, White Plains, New York.

[76] Stathatos, K., Roussopoulos, N., and Baras, J. S. (1997) Adaptive Data Broadcast in Hybrid Networks. *Proceedings 23rd International Conference on Very Large DataBases*, pp. 326–335, Athens, Greece.

[77] Su, C. J. and Tassiulas, L. (1997) Broadcast Scheduling for Information Distribution. *Proceedings of the Sixteenth Annual Joint Conference of the IEEE Computer and Communications Societies (INFOCOM 1997)*, Kobe, Japan.

[78] Tanenbaum, A. S. (1992) *Modern Operating Systems*, Prentice Hall, Englewood Cliffs, NJ.

[79] Tassiulas, L. and Su, C. J. (1997) Optimal Memory Management Strategies for a Mobile User in a Broadcast Data Delivery System. *IEEE Journal on Selected Areas in Communications*, **15**(7), 1226–1238.

[80] Vicisano, L., Rizzo, L., and Crowcroft, J. (1998) TCP-Like Congestion Control for Layered Multicast Data Transfer. *Proceedings of the Seventeenth Annual Joint Conference of the IEEE Computer and Communications Societies (INFOCOM 1998)*, San Francisco, CA.

[81] Viswanathan, S., Imielinski, T., and Badrinath, B. R. (1997) Data On Air: Organization and Access. *IEEE Transactions on Knowledge and Data Engineering*, **9**(3), 353–372.

[82] Weihl, W. E. (1987) Distributed Version Management for Read-Only Actions. *ACM Transactions on Software Engineering*, **13**(1), 56–64.

[83] Young, N. (1998) On-Line File Caching. *Proceedings of the Ninth Annual ACM-SIAM Symposium on Discrete Algorithms*, pp. 82–91, San Francisco, CA.

11

Principles of Mobile Computing Middleware

Cecilia Mascolo, Licia Capra, Wolfgang Emmerich
University College, London.

11.1 Introduction

The popularity of wireless devices, such as laptop computers, mobile phones, personal digital assistants, smart cards, digital cameras, and so on, is rapidly increasing. Their computing capabilities are growing quickly, while their size is shrinking, allowing many of them to become more and more part of everyday life. These devices can be connected to wireless networks of increasing bandwidth, and software development kits are available that can be used by third parties to develop applications. The combined use of these technologies on personal devices enables people to access their personal information as well as public resources anytime and anywhere.

Applications on these types of devices, however, introduce challenging problems. Devices face temporary and unannounced loss of network connectivity when they move; they are usually engaged in rather short connection sessions; they need to discover other hosts in an *ad hoc* manner; they are likely to have scarce resources, such as low battery power, slow CPU, and little memory; they are required to react to frequent changes in the environment, such as change of location or context conditions, variability of network bandwidth, that will remain by orders of magnitude lower than in fixed networks.

When developing distributed applications, designers do not have to explicitly deal with problems related to distribution, such as heterogeneity, scalability, resource-sharing, and fault-tolerance. *Middleware* developed upon network operating systems provides application designers with a higher level of abstraction, hiding the complexity introduced by distribution. Existing middleware technologies, such as transaction-oriented, message-oriented, or object-oriented middleware have been built trying to hide distribution as much as possible, so that the system appears as a single integrated computing facility. In other words, distribution becomes *transparent*.

Middleware for Communications. Edited by Qusay H. Mahmoud
 ISBN 0-470-86206-8

These technologies have been designed and are successfully used for stationary distributed systems. However, as it will become clearer in the following, some of the requirements introduced by mobility cannot be fulfilled by these existing traditional middleware. First, the interaction primitives, such as distributed transactions, object requests, or remote procedure calls, assume a stable, high-bandwidth and constant connection between components. Furthermore, synchronous point-to-point communication supported by object-oriented middleware systems, such as CORBA, requires a rendezvous between the client asking for a service, and the server delivering that service. In mobile systems, on the contrary, unreachability is not exceptional and the connection may be unstable. Moreover, it is quite likely that client and server hosts are not connected at the same time, because of voluntary disconnections (e.g., to save battery power) or forced disconnection (e.g., loss of network coverage). Disconnection is treated as an occasional fault by many traditional middleware; techniques for data-sharing and replication that have been successfully adopted in traditional systems might not, therefore, be suitable, and new methodologies need to be explored.

Moreover, mobility introduces higher degrees of heterogeneity and dynamicity than traditional distributed systems. Mobile hosts might have to support different communication protocols, according to the wireless links they are exploiting; lookup operations are more elaborate than in distributed systems, because of location variability. Middleware reconfiguration becomes essential in order to adapt to highly varying context conditions.

Finally, traditional middleware systems have been designed targeting devices with almost no resource limitations, especially in terms of battery power. On the contrary, even considering the improvements in the development of these technologies, resources of mobile devices will always be, by orders of magnitude, more constrained.

The aim of this chapter is to give an overview of how the requirements usually associated to distributed systems are affected by physical mobility issues. We discuss how traditional middleware, and middleware built targeting mobile systems can fulfill these requirements. We provide a framework and a classification of the most relevant literature in this area, highlighting goals that have been attained and goals that need to be pursued.

The chapter is structured as follows: Section 11.2 describes the main characteristics of mobile systems and highlights the many extents to which they differ from fixed distributed systems. Section 11.3 presents a reference model for middleware systems based on the requirements that need to be fulfilled; Sections 11.4 to 11.8 contain a detailed and comparative review of existing middleware for mobile systems, based on the model given. For every requirement category, we describe its main characteristics, illustrate some examples of solutions proposed to date, and highlight their strengths and limitations. Section 11.9 contains discussion and future directions of research in the area of middleware for mobile computing.

11.2 Mobile Distributed Systems

In this section, we introduce a framework that we will use to highlight the similarities, but more importantly, the differences between fixed distributed systems and mobile systems.

11.2.1 Characterization of Distributed Systems

A distributed system consists of a collection of components distributed over various computers (also called *hosts*) connected via a computer network. This definition of distributed system applies to both traditional and mobile systems. To understand the differences existing between the two, we now investigate three concepts hidden in the previous definition: the concept of *device*, of *network connection*, and of *execution context*.

Type of device.
As a first basic distinction, devices in a fixed distributed system are stationary or *fixed*, while a mobile distributed system has at least some physically *mobile* devices. This is a key point: fixed devices vary from home PCs to Unix workstations to mainframes; mobile devices vary from personal digital assistants to mobile phones, digital cameras, and smart cards. While the former are generally powerful machines, with large amounts of memory and very fast processors, the latter have limited capabilities such as slow CPU speed, little memory, low battery power, and small screen size.

Type of network connection.
Fixed hosts are often *permanently* connected to the network through continuous high-bandwidth links. Disconnections are either explicitly performed for administrative reasons or are caused by unpredictable failures. These failures are treated as exceptions to the normal behavior of the system. Such assumptions do not hold for mobile devices that connect to the Internet via wireless links. The performance of wireless networks (i.e., GSM, GPRS networks, satellite links, WaveLAN, HiperLAN, Bluetooth) may vary depending on the protocols and technologies being used; reasonable bandwidth may be achieved, for instance, if the hosts are within reach of a few (hundreds) meters from their base station, and if they are few in number in the same base station cell. In some of the technologies, all different hosts in a cell share the bandwidth (i.e., if they grow in number, the quality of service rapidly drops). Moreover, if a device moves to an area with no coverage or with high interference, bandwidth may suddenly drop to zero, and the connection may be lost. Unpredictable disconnections cannot be considered as an exception any longer, but they rather become part of normal wireless communication. Either because of failures or because of explicit disconnections, the network connection of mobile distributed systems is typically *intermittent*.

Type of execution context.
With context, we mean everything that can influence the behavior of an application; this includes resources internal to the device, such as amount of memory or screen size, and external resources, such as bandwidth, quality of the network connection, location, or hosts (or services) in the proximity. In a fixed distributed environment, context is more or less *static*: bandwidth is high and stable, location almost never changes, hosts can be added, deleted, or moved, but the frequency at which this happens is by orders of magnitude lower than in mobile settings. Services may change as well, but the discovery of available services is easily performed by forcing service providers to register with a well-known location service such as LDAP or DNS. Context is extremely *dynamic*

in mobile systems. Hosts may come and leave generally much more rapidly. Service lookup is more complex in the mobile scenario, especially when the fixed infrastructure is completely missing, as for *ad hoc* systems (Section 11.2.4). Broadcasting is the usual way of implementing service advertisement; however, this has to be carefully engineered in order to save the limited resources available (e.g., sending and receiving is power consuming), and to avoid flooding the network with messages. Location is no longer fixed: the size of wireless devices has shrunk to the point that most of them can be carried in a pocket and moved around easily. Depending on location and mobility, bandwidth and quality of the network connection may vary greatly.

11.2.2 Traditional Distributed Systems

According to the framework previously described, traditional distributed systems are a collection of fixed hosts, permanently connected to the network via high-bandwidth and stable links, executing in a static environment. The distributed components running on these hosts need to interact with each other in order, for example, to exchange data or to access services. To facilitate interactions, the following requirements have to be guaranteed [6, 9]:

- *Fault tolerance:* The ability to recover from faults without halting the whole system. Faults happen because of hardware or software failures (e.g., software errors, aging hardware, etc.), and distributed components must continue to operate even if other components they rely on have failed.
- *Openness:* The possibility to extend and modify the system easily, for example, to respond to changed functional requirements. Any real distributed system will evolve during its lifetime. The system needs to have a stable architecture so that new components can be easily integrated while preserving previous investments.
- *Heterogeneity:* It calls for integration of components written using different programming languages, running on different operating systems, executing on different hardware platforms. In a distributed system, heterogeneity is almost unavoidable, as different components may require different implementation technologies.
- *Scalability:* The ability to accommodate a higher load at some time in the future. The load can be measured using many different parameters, such as, for instance, the maximum number of concurrent users, the number of transactions executed in a time unit, and the data volume that has to be handled.
- *Resource-sharing:* In a distributed system, hardware and software resources (e.g., a printer, a database, etc.) are shared among the different users of the system; some form of access control of the shared resources is necessary in order to grant access to authorized users of the system only.

11.2.3 Mobile Nomadic Systems

Nomadic systems can be considered a compromise between totally fixed and totally mobile systems. They are based on a core of fixed routers, switches, and hosts; at the periphery of this fixed network, base stations with wireless communication capabilities control message traffic to and from dynamic configurations of mobile hosts.

Let us consider the requirements isolated for distributed systems and see how the characteristics of nomadic systems (i.e., mobile devices, dynamic context, permanent connectivity) influence them.

- *Fault tolerance:* In most of the existing nomadic systems, disconnection is still treated as an occasional fault, and the general pattern of communication used by most of the applications relies on a continuous connection. Nomadic applications that require support for disconnected operations exist, because of, for example, frequent bandwidth drops or high connection costs. In these cases, devices should be able to rely on cached data, in order to perform some off-line computation.
- *Openness:* Mobile nomadic systems are characterized by rapidly varying execution context, both in terms of available resources and of devices and services in reach while moving. Adaptation and dynamic reconfiguration are called for, in order to react to context changes, and to interact with newly discovered services.
- *Heterogeneity:* The issue of heterogeneity is even more serious than in traditional distributed systems. Heterogeneity of hardware platforms, of programming languages, and of operating systems still exists. However, mobility seems to push the need for heterogeneity handling further. In a nomadic setting, for example, it is likely for a device to be relocated into an environment with different conditions.
- *Scalability:* The ability to accommodate an increasing number of services and hosts, while maintaining a reasonable quality of service, is still an issue in nomadic systems. Services are often residing on the core network; the main concern in terms of scalability is the ability to deal with the limitations of the wireless links at the periphery of the network.
- *Resource-sharing:* As in traditional systems, the services and resources of the network are shared. This implies that some monitoring on the access needs to be guaranteed, both in terms of authentication and concurrency. Having wireless links implies further weakness in terms of security. Concurrency in traditional systems is usually solved using transaction mechanisms; however, these are quite heavy and usually rely on a permanent and stable connection.

11.2.4 Mobile Ad Hoc Systems

Mobile *ad hoc* (or simply *ad hoc*) distributed systems consist of a set of mobile hosts, connected to the network through wireless links. They differ from traditional and nomadic distributed systems in that they have no fixed infrastructure: mobile hosts can isolate themselves completely, and groups may evolve independently. Connectivity may be asymmetric or symmetric depending, for instance, on the radio frequency of the transmission used by the hosts. Radio connectivity is, by default, not transitive. However, *ad hoc* routing protocols have been defined [20] in order to overcome this limitation and allow routing of packets through mobile hosts. Pure *ad hoc* networks have encountered so far limited applications that range from small *ad hoc* groups to share information in meetings for a short time, to military applications on battlefields, and discovery or emergency networks in disaster areas.

The requirements discussed for distributed and nomadic systems still hold in this kind of networks, even though they are further complicated.

- *Fault tolerance:* Replication usually permits applications to cope with fault tolerance in common distributed and, to some extent, even nomadic system. However, in *ad hoc* networks, the lack of infrastructure imposes that no hosts acting as servers can be assumed. The distinction between the role of servers and clients in mobile *ad hoc* networks is in fact blurred.
- *Heterogeneity:* Heterogeneity issues for *ad hoc* networks are much the same as the ones we have described for nomadic networks. Because of the lack of a fixed infrastructure, it is more likely that hosts equipped with different network adaptors meet and are willing to communicate.
- *Openness:* The fact that no servers are available on a core network implies that new behaviors cannot be downloaded from a specific and known location, but need to be retrieved, if at all existing. Moreover, because of higher dynamicity than in fixed or nomadic networks, the chances that a host will come into contact with a service never met before, and for which it has no specific interaction behavior, or that it will encounter a new environment, for which new functionalities are needed, increase.
- *Scalability:* Coordination of many hosts in a completely unstructured environment is very challenging. Decentralized solutions for service discovery and advertisement need to be put in place and exploited in order to obtain the best results.
- *Resource-sharing:* Again, the lack of structure implies that security is much more difficult to obtain. The existence of specific locations to register authentication data of hosts cannot be assumed; links are wireless, allowing for more risks of eavesdropping. Concurrency can be controlled using transactions; however, as in nomadic systems, this becomes more difficult because of mobility, scarce resources, and connectivity. Moreover, while in nomadic systems core hosts could register the transaction session while a host is handing off a cell, in *ad hoc* networks no such infrastructure can be assumed, and a host can just go out of range in the middle of a transaction.

In between nomadic and *ad hoc* types of network, there is a large range of other network solutions, which adopt aspects of both. We believe these heterogeneous networks, where fixed components interact with *ad hoc* areas, and where different connectivity technologies are used, are going to be the networks of the future.

11.3 Middleware Systems: A Reference Model

Building distributed applications, either mobile or fixed, on top of the network layer is extremely tedious and error-prone. Application developers have to explicitly deal with all the requirements listed in the previous section, such as heterogeneity and fault tolerance, and this complicates considerably the development and maintenance of an application. Given the novelties of mobile systems and the lack of adequate development support, some of the developed systems for mobile environments adopted the radical approach of not having a middleware but rather rely on the application to handle all the services and deal with the requirements, often using a context-aware approach that allows adaptation to changing context [5]. Sun provides J2ME (Java Micro Edition), which is a basic JVM and development package targeting mobile devices. Microsoft provides the .Net Compact Framework, which also has support for XML data and web services connectivity. However, this approach is not sufficient, as it completely relies on application designers to address most of the requirements middleware should provide, and, as we have said, these requirements are difficult to meet and even further complicated by mobility.

During the past years, middleware technologies for distributed systems have been built and successfully used in industry. However, as we argued before, although object-oriented middleware have been very successful in fixed environments, these systems might not be suitable in a mobile setting. Researchers have both adapted traditional middleware for use in mobile setting and have been, and are, actively working to design new middleware targeted to the mobile setting (Sections 11.4 to 11.8).

Before entering into the details of specific systems and solutions, we provide a general overview of middleware for fixed and mobile distributed systems based on this model.

11.3.1 Middleware for Fixed Distributed Systems

Middleware for fixed systems can be mainly described as resource-consuming systems that hide most of the details of distribution from application designers. Considering the reference model introduced before, middleware for fixed distributed systems can be further characterized as follows:

- *Fault tolerance:* Disconnection of hosts is considered infrequent and limited in time. Middleware treats disconnection as an exception and transparently recovers from the fault. Connectivity among the hosts is considered stable and synchronous communication paradigms are mostly used, even if some middleware offer support for asynchronous communication. Replication techniques are used in order to cope with faults and to recover data. Given the stability of the connection among the hosts, and the quite reliable knowledge of the topology of the network, servers for replication of data are usually established. This has implications on both scalability, as replicas distribute access load among different nodes, and on resource-sharing, as changes to the copies need to be synchronized.
- *Heterogeneity:* Different technologies are used to implement servers, and different hardware components are deployed. Middleware for fixed distributed systems hides these differences through a set of primitives for communication that are independent from the underlying software and hardware.
- *Openness:* Systems need to adapt to changes and to allow for new functionalities to be introduced. Middleware for distributed systems fulfills this requirement by defining specific interfaces that allow dynamic discovery of new functionalities, so that components can easily be integrated.
- *Scalability:* The ability to accommodate an increasing number of users and services is critical in current distributed systems. Middleware tackles scalability issues mainly by replication of services and through hierarchical structuring of (replicated) lookup services. Transparent relocation of services is also used for scalability purposes.
- *Resource-sharing:* Transaction and security services are often a crucial part of existing middleware. Transaction servers are used to ensure the integrity of the data in a distributed setting, while authentication mechanisms are put in place to ensure that only trusted users and hosts are accessing the services.

11.3.2 Middleware for Mobile Nomadic and Ad hoc Systems

As we have seen, nomadic systems share some characteristics with traditional distributed systems. However, the introduction of mobility imposes new constraints to the middleware that can be developed for these systems. Mobile *ad hoc* systems are characterized by a complete lack of infrastructure. Mobility in this setting complicates matters further. The development of middleware for these systems is at a very early stage. With regard to the reference model, the following hold:

- *Fault tolerance:* Disconnections are frequent in nomadic and, especially, in *ad hoc* systems; moreover, the connection is often unstable and/or expensive. The synchronous mechanisms for communication adopted in traditional middleware seem to fail in delivering the right service. Asynchronous communication seems to fit better in mobile scenarios. Mixed solutions that adopt synchronous communication integrated with caching and buffering techniques also have some space in the mobile middleware arena. The use of data replication techniques and caching assumes a crucial importance as the ability to cope with disconnections (both faulty and volunteer) is much more essential than in traditional systems. Off-line operations should be allowed, and data should be available on the device even if not connected to the server. Different techniques have been devised to allow replication and consistency of copies, and they will be described in the following sections.
- *Heterogeneity:* Mobility causes devices to encounter highly dynamic context and heterogeneous environment. Different behaviors are required to allow devices to interact in these situations; however, because of resource limitations, only middleware with small footprints should be used. In order to cope with this discrepancy, reconfigurable middleware systems based on a compositional approach have been developed.
- *Openness:* The fact that the hosts are mobile means that they are very likely to change context and meet new services. Adaptability of the middleware to new conditions and new context is essential and may imply the use of techniques to do application-dependent adaptation.
- *Scalability:* As in traditional systems, the ability to accommodate new hosts and services is essential. In nomadic settings, services may be replicated on the core network in a strategic way, so as to offer better quality of service to devices that are clustered in some areas. In order to offer good quality of service in crowded environments, some middleware have devised a leasing system, where resources are temporarily assigned to "users" and then withdrawn and reassigned, allowing a better resource allocation scheme that fits highly dynamic environments. In *ad hoc* networks, service discovery should be very decentralized, as no central discovery service can be put in place.
- *Resource-sharing:* Security in mobile systems is still a very immature field. Nomadic system security relies on authentication happening on core network servers, like in traditional systems. Mobile *ad hoc* network security is more difficult to achieve, again due to the lack of fixed infrastructure, and the existing middleware do not offer much support in this sense. Transactions in nomadic environments are supported. However, different protocols than the ones used in traditional systems are used, in order to take into account mobility aspects. In *ad hoc* settings, the absence of a core network does not allow complex transactions to take place. Research in how to obtain

resource integrity while having concurrent access is focusing on replication and weak consistency techniques.

Some middleware specifically targeting the needs of mobile computing have been recently devised [23]; assumptions such as scarce resources, and fluctuating connectivity have been made in order to reach lightweight solutions. Some of the approaches, however, only target a subset of the requirements listed before.

11.4 Fault Tolerance

Fault-tolerance has been essential in the adoption of middleware for traditional distributed systems. However, fault-tolerance in mobile systems needs to be handled slightly differently. In particular, two aspects of fault-tolerance have been investigated by middleware researchers: connectivity and data-sharing. We treat the two issues separately and show examples of mobile middleware that offer support for them.

11.4.1 Connectivity

Given the potential instability of wireless links, the possible low bandwidth or high costs, the frequent disconnections and the mobility involved, middleware for mobile systems should be able to provide support for fault-tolerance in terms of connectivity. Some adaptation of traditional middleware to mobile systems have been attempted, allowing for a more flexible way of treating connectivity faults, on the basis of caching and buffering.

Traditional middleware applied in mobile computing.
Object-oriented middleware has been adapted to mobile settings, mainly to make mobile devices interoperable with existing fixed networks in a nomadic setting. The main challenge in this direction is in terms of software size and protocol suitability. IIOP (i.e., the Internet Inter-ORB Protocol) is the essential part of CORBA that is needed to allow communication among devices. IIOP defines the minimum protocol necessary to transfer invocations between ORBs. IIOP has been successfully ported to mobile setting and used as a minimal ORB for mobile devices. In ALICE [11], handhelds with Windows CE and GSM network adaptors have been used to provide support for client-server architectures in nomadic environments. An adaptation of IIOP specifically for mobile (i.e., LW-IOP, Light-weight Inter-Orb Protocol) has been devised in the DOLMEN project: unsent data is cached and an acknowledgment scheme is used to face wireless medium unreliability. Actual names of machines are translated dynamically through a name server, which maintains up-to-date information of the host location. In [22], CORBA and IIOP are used together with the WAP (Wireless Access Protocol) stack in order to facilitate the use of CORBA services on a fixed network to mobile devices connected through WAP and a gateway. IIOP is used to achieve message exchange.

In general, the synchronous connectivity paradigm typical of the systems discussed above assumes a permanent connection that cannot be taken for granted in many mobile computing scenarios. These systems therefore usually target nomadic settings, where handoffs allow mobile devices to roam while being connected. Only minimal support for disconnection is introduced.

Semi-asynchronous communication paradigms have also been investigated. RPC-based middleware has been enhanced with queueing delaying or buffering capabilities in order to cope with intermittent connections. Examples of these behaviors are Rover [14] and Mobile DCE [27]. Totally asynchronous paradigms have been adapted too. An implementation of the message-oriented middleware JMS (Java Messaging Server has recently been released [28]. It supports both point-to-point and publish/subscribe communication models, that is, a device can either communicate with a single other host (through its queue) or register for a topic and be notified of all the messages sent to that topic. We believe that the use of publish/subscribe and message-oriented systems will be taken further as they offer an asynchronous communication mechanism that naturally fits into the mobile setting, providing support for disconnected operations.

Implementations of publish/subscribe systems, which still target fixed distributed systems, such as CORBA Component Model, begin to appear on the market [13]. It will probably not take long before this begins to be applied to nomadic settings.

Tuple space-based middleware.
The characteristics of wireless communication media favor a decoupled and opportunistic style of communication: decoupled in the sense that computation proceeds even in presence of disconnections, and opportunistic as it exploits connectivity whenever it becomes available. Some attempts based on events [32], or queues (Rover or Mobile JMS) have been devised. Another asynchronous and decoupled communication paradigm has also been isolated as effective in mobile settings: although not initially designed for this purpose (their origins go back to Linda [10], a coordination language for concurrent programming), tuple space systems have been shown to provide many useful facilities for communication in wireless settings. In Linda, a tuple space is a globally shared, associatively addressed memory space used by processes to communicate. It acts as a repository (in particular, a multiset) of data structures called *tuples* that can be seen as vector of typed values. Tuples constitute the basic elements of a tuple space systems; they are created by a process and placed in the tuple space using a *write* primitive, and they can be accessed concurrently by several processes using *read* and *take* primitives, both of which are blocking (even if nonblocking versions can be provided). Tuples are anonymous, thus their selection takes place through pattern-matching on the tuple contents. Communication is decoupled in both time and space: senders and receivers do not need to be available at the same time, because tuples have their own life span, independent of the process that generated them, and mutual knowledge of their location is not necessary for data exchange, as the tuple space looks like a globally shared data space, regardless of machine or platform boundaries.

These forms of decoupling assume enormous importance in a mobile setting, where the parties involved in communication change dynamically because of their migration or connectivity patterns. However, a traditional tuple space implementation is not enough. There are basic questions that need to be answered: how is the globally shared data space presented to mobile hosts? How is it made persistent? The solutions developed to date basically differ, depending on the answers they give to the above questions.

We now review a tuple-space middleware that have been devised for mobile computing applications: Lime [17]. Others exist such as T-Spaces [33] or JavaSpaces. L2imbo [7] is a tuple space system where quality-of-service–related aspects are added.

Lime.

In Lime [17], the shift from a fixed context to a dynamically changing one is accomplished by breaking up the Linda tuple space into many tuple spaces, each permanently associated to a mobile unit, and by introducing rules for transient sharing of the individual tuple spaces based on connectivity.

Each mobile unit has access to an *interface tuple space* (ITS) that is permanently and exclusively attached to that unit and transferred along with it when movement occurs. Each ITS contains tuples that the unit wishes to share with others and it represents the only context accessible to the unit when it is alone. Access to the ITS takes place using conventional Linda primitives, whose semantics is basically unaffected. However, the content of the ITS (i.e., the set of tuples that can be accessed through the ITS) is dynamically recomputed in such a way that it looks like the result of the merging of the ITSs of other mobile units currently connected. Upon arrival of a new mobile unit, the content perceived by each mobile unit through its ITS is recomputed, taking the content of the new mobile unit into account. This operation is called *engagement* of tuple spaces; the opposite operation, performed on departure of a mobile unit, is called *disengagement*. The tuple space that can be accessed through the ITS of a mobile unit is therefore shared by construction and transient because its content changes according to the movement of mobile units.

11.4.2 Data-sharing

Mobility complicates the way data and services are shared among the hosts. In particular, the limitations of the resources on the device and the unstable connectivity patterns make this task more difficult. Support for disconnected operations and data-sharing have been regarded as key points by middleware such as Coda [26], Bayou [29], and Xmiddle [16]. They try to maximize availability of data and tolerance for disconnections, giving users access to replicas; they differ in the way they ensure that replicas move towards eventual consistency, that is, in the mechanisms they provide to detect and resolve conflicts that naturally arise in mobile systems. Despite a proliferation of different proprietary data synchronization protocols for mobile devices, we still lack a single synchronization standard, as most of these protocols are implemented only on a subset of devices and are able to access a small set of networked data. This represents a limitation for both end users, application developers, service providers, and device manufacturers.

Xmiddle.

Xmiddle [16] allows mobile hosts to share data when they are connected, or replicate the data and perform operations on them off-line when they are disconnected; reconciliation of data takes place once the host reconnects. Unlike tuple space–based systems, which store data in flat unstructured tuples, Xmiddle allows each device to store its data in a tree structure (represented as XML files). Trees allow sophisticated manipulations because of the different node levels, hierarchy among the nodes, and the relationships among the different elements, which could be defined.

When hosts get in touch with each other, they need to be able to interact. Xmiddle allows communication through sharing of trees. On each host, a set of possible access points for the private tree is defined; they essentially address branches of the tree that can be modified and read by peers. The size of these branches can vary from a single node to a complete tree; unlike systems such as Coda, where entire collections of files have to be replicated, the unit of replication can be easily tuned to accommodate different needs.

In order to share data, a host needs to explicitly *link* to another host's tree. As long as two hosts are connected, they can share and modify the information on each other's linked data trees. When disconnections occur, both explicit (e.g., to save battery power or to perform changes in isolation from other hosts) and implicit (e.g., due to movement of a host into an out of reach area), the disconnected hosts retain replicas of the trees they were sharing while connected, and continue to be able to access and modify the data.

Xmiddle addresses *ad hoc* networks. No assumption is made about the existence of more powerful and trusted hosts, which should play the role of servers and on which a collection of data should be replicated in full.

11.5 Heterogeneity

The ability to allow integration of different hardware and software servers and clients is important in traditional systems and becomes essential in mobile systems. In the recent years, we have seen a proliferation of different small and mobile hosts with all sorts of network connectivity capabilities, displays, and resource constraints. All these hosts need to be networked and to communicate with each other. Overcoming heterogeneity comes with costs in terms of computational load, which conflicts with the scarce availability of resources of mobile hosts. The IIOP CORBA implementation that has been ported on mobile (discussed in Section 11.4) allows integration of different software servers and clients. However, research went beyond this and there have been some attempts towards solving heterogeneity issues at a higher level, by allowing static and dynamic reconfiguration of the middleware components on the mobile host. As it may have been noticed, heterogeneity issues are strictly related to openness issues. The use of reconfiguration techniques for dynamic adaptation is in fact at the basis of both.

Universal interoperable core.

UIC (Universally Interoperable Core) [30] is at the basis of the implementation of Gaia (Section 11.5); it is a minimal reflective middleware that targets mobile hosts. UIC is composed of pluggable set of components that allow developers to specialize the middleware targeting different devices and environments, thus solving heterogeneity issues. The configuration can also be automatically updated both at compile and runtime. Personalities (i.e., configurations) can be defined to have a client-side, server-side or both behaviors. Personalities can also define with which server type to interact (i.e., CORBA or Java RMI): single personalities allow the interaction with only one type, while multipersonalities allow interaction with more than one type. In the case of multipersonalities, the middleware dynamically chooses the right interaction paradigm. The size of the core goes, for instance, from 16 KB for a client-side CORBA personality running on a Palm OS device to 37 KB for a client/server CORBA personality running on a Windows CE device.

11.6 Openness

In mobile systems, openness mainly refers to the ability of the system to dynamically adapt to context changes. Adaptation allows, for example, to optimize the system behavior on the basis of current resource availability; to choose the protocol suite (e.g., communication protocol, service discovery protocol, etc.) that better targets the current environment; to easily integrate new functionalities and behaviors into the systems, and so on.

In traditional distributed systems, the principle of *reflection* has been exploited to introduce more openness and flexibility into middleware platforms, as presented in Chapter 2. In a mobile setting, the need for adaptation is pressing, because of the high dynamicity of context. This need is coupled with resource limitations on portable devices that forbid the deployment of complex and heavy-weight middleware platforms. Reflection may help solving these needs: a middleware core with only a minimal set of functionalities can be installed on a device, and then, through reflection, the system can be reconfigured dynamically to adapt to context changes.

Reflection is not the only principle investigated towards the solution of this problem. Researchers in context-aware computing have studied and developed systems that collect context information and adapt to changes, even without exploiting the principle of reflection. User's context includes, but is not limited to

- location, with varying accuracy depending on the positioning system used;
- relative location, such as proximity to printers and databases;
- device characteristics, such as processing power and input devices;
- physical environment, such as noise level and bandwidth;
- user's activity, such as driving a car or sitting in a lecture theater.

In particular, location has attracted a lot of attention and many examples exist of applications that exploit location information to offer travelers directional guidance, such as the Shopping Assistant [3] and CyberGuide [15]; to find out neighboring devices and the services they provide, to send advertisements depending on user's location, or to send messages to anyone in a specific area. Most of these systems interact directly with the underlying network OS to extract location information, process it, and present it in a convenient format to the user. One of their major limitations concerns the fact that they do not cope with heterogeneity of coordinate information, and therefore different versions have to be released that are able to interact with specific sensor technologies, such as the Global Positioning System (GPS) outdoors, and infrared and radio frequency indoors.

To enhance the development of location-based services and applications, and reduce their development cycle, middleware systems have been built that integrate different positioning technologies by providing a common interface to the different positioning systems. Examples include Oracle iASWE [18], and many others are coming out. We will review some of these systems when discussing heterogeneity issues.

Odyssey.

The mostly application transparent approach adopted by Coda has been improved introducing context-awareness and application-dependent behaviors in Odyssey [25], and allowing the use of these approaches in mobile computing settings. Odyssey assumes that applications reside on mobile clients but access or update data stored on remote, more capable and trustworthy servers; once again, the nomadic scenario is targeted.

Odyssey proposes a collaborative model of adaptation. The operating system, as the arbiter of shared resources, is in the best position to determine resource availability; however, the application is the only entity that can properly adapt to given context conditions and must be allowed to specify adaptation policies. This collaborative model is called *application-aware adaptation*.

Although better suited to the mobile environment than its predecessor Coda, Odyssey suffers from some limitations: the data that can be moved across mobile hosts (i.e., a collection of files) may be too coarse-grained in a mobile setting, where hosts have limited amount of memory, and connection is often expensive and of low quality.

Gaia.

[4] shares the idea of offering the ability to change the behavior of the middleware and of the application on the basis of the knowledge about the changing context; they differ in the way they achieve this goal. Gaia converts physical spaces and the ubiquitous computing devices they contain into active spaces, that is, programmable computing systems with well-defined behavior and explicitly defined functionality.

11.7 Scalability

Scalability issues are related to the ability of accommodating large numbers of hosts, both in terms of, for instance, quality of service and discovery of services. The fact that mobile links are unstable, expensive and sometimes limited in terms of bandwidth implies that quality of service rapidly decreases when the number of hosts involved increases. There have been some attempts to use middleware in order to solve scalability issues related to quality of service and we will give some examples. Discovery of services when hosts are mobile is also challenging, especially in completely *ad hoc* scenarios where no centralized discovery service can be used. Some of the middleware already presented, and some that we are going to introduce, offer a solution to these issues.

11.7.1 Discovery

In traditional middleware systems, service discovery is provided using fixed name services, which every host knows the existence of. The more the network becomes dynamic, the more difficult service and host discovery becomes. Already in distributed peer-to-peer network [19], service discovery is more complex as hosts join and leave the overlay network very frequently. In nomadic systems, service discovery is still very similar to service discovery in traditional systems, where a fixed infrastructure containing all the information and the services is present. However, in terms of more *ad hoc* or mixed systems, where services can be run on roaming hosts, discovery may become very complex and/or expensive.

Most of the *ad hoc* systems encountered till now have their own discovery service. Lime and Xmiddle use a completely *ad hoc* strategy where hosts continuously monitor their environment to check who is available and what they are offering. A trade-off between power and bandwidth consumption (i.e., broadcast) and discovery needs to be evaluated. Recently, some work on Lime for service advertisement and discovery has been devised [12]. Standard service discovery frameworks have appeared in the recent years: UPnP [31], Jini [2], and Salutation [24]. UPnP stands for Universal Plug and Play and it is an open standard for transparently connecting appliances and services, which is adopted by the Microsoft operating systems. UPnP can work with different protocols such as TCP, SOAP, HTTP. Salutation is a general framework for service discovery, which is platform- and OS-independent. Jini is Java-based and dependent on the Java Virtual Machine. The purpose of these frameworks is to allow groups of hosts and software components to federate into a single, dynamic distributed system, enabling dynamic discovery of services inside the network federation.

Jini and JMatos.

Jini [2] is a distributed system middleware based on the idea of federating groups of users and resources required by those users. Its main goal is to turn the network into a flexible, easily administered framework on which resources (both hardware devices and software programs) and services can be found, added, and deleted by humans and computational clients. The most important concept within the Jini architecture is the service. A service is an entity that can be used by a person, a program, or another service. Members of a Jini system federate in order to share access to services. Services can be found and resolved using a lookup service that maps interfaces indicating the functionality provided by a service to sets of objects that implement that service. The lookup service acts as the central marketplace for offering and finding services by members of the federation. A service is added to a lookup service by a pair of protocols called *discovery* and *join*: the new service provider locates an appropriate lookup service by using the first protocol, and then it joins it, using the second one. A distributed security model is put in place in order to give access to resources only to authorized users.

Jini assumes the existence of a fixed infrastructure, which provides mechanisms for hosts, services, and users to join and detach from a network in an easy, natural, often automatic, manner. It relies on the existence of a network of reasonable speed connecting Jini technology-enabled hosts. However, the large footprint of Jini (3 Mbytes), mainly due to the use of Java RMI, prevents the use of Jini on smaller devices such as iPAQs or PDAs. In this direction, Psinaptic JMatos [21] has been developed, complying with the Jini Specification. JMatos does not rely on Java RMI for messaging and has a footprint of just 100 KB.

11.7.2 Quality of Service

In the existing examples of use of traditional middleware in the mobile setting, the focus is on the provision of services from a backbone network to a set of mobile hosts: the main concerns in this scenario are connectivity and message exchange. In case of a less structured network, or in case services must be provided by mobile hosts, traditional middleware paradigms seem to be less suitable and a new set of strategies needs to be used. The importance of monitoring the condition of the environment, and adaptation to application needs, maybe through communication of context information to the upper layers, becomes vital to achieve reasonable quality of service.

Given the highly dynamic environment and the scarce resources, quality-of-service provision presents higher challenges in mobile computing. Nevertheless, researchers have devised a number of interesting approaches to quality-of-service provision to mobile hosts [5]. Most of the time, the hosts are considered terminal nodes and the clients of the service provision, and the network connectivity is assumed fluctuating but almost continuous (like in GSM settings).

Mobiware.
Probably the most significant example of quality-of-service–oriented middleware is Mobiware [1], which uses CORBA, IIOP, and Java to allow service quality adaptation in mobile setting. In Mobiware, mobile hosts are seen as terminal nodes of the network and the main operations and services are developed on a core programmable network of routers and switches. Mobile hosts are connected to access points and can roam from one access point to another.

Mobiware mostly assumes a service provision scenario where mobile hosts are roaming but permanently connected, with fluctuating bandwidth. Even in the case of the *ad hoc* broadband link, the host is supposed to receive the service provision from the core network through the cellular links first and then some *ad hoc* hops. In more extreme scenarios, where links are all *ad hoc*, these assumptions cannot be made, and different middleware technologies need to be applied. One of the strengths of Mobiware is the adaptation component to customize quality-of-service results.

11.8 Resource-sharing

Resource-sharing is a very delicate and important aspect of mobile computing middleware. It involves two main issues that we are now going to tackle: transactions and security. Traditional middleware have successfully dealt with these issues for fixed distributed systems; however, those solutions seem to be cumbersome and inappropriate in a mobile scenario. Some new solutions for these issues have been proposed, especially in the area of nomadic system, where the core network helps in reusing traditional middleware solutions; however, the state of the art is very poor in terms of *ad hoc* systems.

11.8.1 Transactions

We already introduced techniques for data-sharing and consistency in Section 11.4. These techniques can, for some aspects, be considered as resource-sharing techniques. However, given the tight link to connectivity, we introduced them in that section. Those techniques are probably the best that can be achieved in mobile *ad hoc* network where no infrastructure can support transactions. More elaborate techniques for resource-sharing can be developed in nomadic systems where transaction servers can be located on the core network. For example, T-Spaces allows transactions to be managed on the server to guarantee consistent access to tuples. Protocols for dealing with transaction on nomadic systems have been developed. However, there is little novelty with respect to the traditional middleware transaction solutions. We now illustrate a couple of the most original examples; the ideas behind these approaches are quite similar to the ones presented while describing

Bayou and Coda. Jini (Section 11.7) also offers leases for resource-sharing, and Lime (Section 11.4) has an approach to transactions based on engagement and disengagement of tuple spaces. However, the following middleware is an example of a system more focused on the idea of transactions.

Kangaroo.
The main idea behind Kangaroo [8] is that distributed transactions necessary for consistent resource-sharing in mobile nomadic environments are subject to faults due to disconnections. In order to deal with these disconnections, Kangaroo models hopping transactions where the state of the started transaction is recorded at the base station where the host is connected. After disconnection, the host, having recorded the last base station that offered connectivity, can resume the transaction using the log at the base station. Base stations contain agents that manage the hopping transactions. Kangaroo offers two modes of operation: compensating and split. In compensating mode, the failure of any subtransaction at a base station causes the failure of the chain of all the transactions. In split mode, which is the default mode, the failure of a subtransaction at a base station does not cause the failure of the subtransactions committed after the previous hops. Kangaroo targets nomadic settings, where it offers interesting performance improvement, dealing with disconnections. However, given the fact that it completely relies on the existence of server-side databases, Kangaroo does not seem to be the appropriate solution for *ad hoc* systems.

11.8.2 Security

Security requirements for mobile networks are similar to those for traditional systems: authentication, confidentiality, integrity, nonrepudiation, access control, and availability. However, the mobility aspect introduced and the lack of infrastructure in the case of *ad hoc* networks increase the number of challenges for the implementation and fulfillment of the above requirements. In nomadic systems, techniques already used in traditional systems can be applied, as the existence of servers where authentication data can be stored is assumed. However, given the involvement of wireless links, this has to be done in combination with some network-level security (e.g., WEP, message encryption) to avoid eavesdropping. The limitation in terms of resources (e.g., battery power, processing power) also limits the kinds of encryption that can be used as this is power draining. In case of *ad hoc* networks, the absence of any infrastructure makes things more challenging. Servers for authentication cannot be used, hosts should trust each other on the basis of simple and maybe application-specific information. Security requirement in mobile environments is highly related to some of the other requirements we discussed in this chapter. Most importantly, a host can be "attacked" by jamming his wireless connections to others. This problem is related to scalability and quality-of-service issues described in Section 11.7.

Middleware for nomadic systems including the fulfillment of security requirements are many. They all adopt the server-based authentication strategy adopted in traditional systems. In terms of *ad hoc* networks, we have not seen very mature solutions put forward in this direction. All the middleware we have presented rely on trust of the hosts involved in the applications.

11.9 Conclusions

Mobile computing middleware research still faces many challenges that might not be solvable adapting traditional middleware techniques. We believe the future mobile networks will be heterogeneous in the sense that many different hosts will be available on the market, with possibly different operating systems and user interfaces. The network connectivity will also be heterogeneous even if an effort towards complete coverage through different connection technologies will be made. For these reasons, mobile computing middleware will have to adapt and be customizable in these different dimensions, both at start-up time (i.e., in case of adaptation to different operating systems) and at runtime (i.e., in case of adaptation to different connection technologies).

We also believe application-dependent information could play an important role in the adaptation of the behavior of the middleware and in the trade-off between scarce resource availability and efficient service provision. In this direction, the effort of presentation of the information to the application, and the gathering of application-dependent policies, is an important presentation layer issue that should be integrated in any mobile computing middleware.

Discovery of existing services is a key point in mobile systems, where the dynamicity of the system is, by orders of magnitude, higher than in traditional distributed systems. Recently, interesting research advances in peer-to-peer systems have focused on discovery issues that might be applicable, at least partially, to mobile settings. However, considerations on the variability of the connection, of the load and of the resources might be different for mobile scenarios. Furthermore, the integration of quality-of-service consideration into the service advertisement and discovery might enable some optimization in the service provision.

Another direction of research concerns security. Portable devices are particularly exposed to security attacks as it is so easy to connect to a wireless link. Dynamic customization techniques seem to worsen the situation. Reflection is a technique for accessing protected internal data structures and it could cause security problems if malicious programs break the protection mechanism and use the reflective capability to disclose, modify, or delete data. Security is a major issue for any mobile computing application and therefore proper measures need to be included in the design of any mobile middleware system.

Bibliography

[1] Angin, O., Campbell, A., Kounavis, M., and Liao, R. (1998) The Mobiware Toolkit: Programmable Support for Adaptive Mobile Networking. *Personal Communications Magazine*, Special Issue on Adapting to Network and Client Variability, August 1998; IEEE Computer Society Press, **5**(4), 32–43.

[2] Arnold, K., O'Sullivan, B., Scheifler, R. W., Waldo, J., and Wollrath, A. (1999) *The Jini[tm] Specification*, Addison-Wesley, Boston, MA.

[3] Asthana, A. and Krzyzanowski, M. C. P. (1994) An Indoor Wireless System for Personalized Shopping Assistance. *Proceedings of IEEE Workshop on Mobile Computing Systems and Applications*, Santa Cruz, CA, December 1994, pp. 69–74; IEEE Computer Society Press.

[4] Cerqueira, R., Hess, C. K., Romn, M., and Campbell, R. H. (2001) Gaia: A Development Infrastructure for Active Spaces. *Workshop on Application Models and Programming Tools for Ubiquitous Computing (held in conjunction with the UBICOMP 2001)*, September 2001, Atlanta, GA.

[5] Chalmers, D. and Sloman, M. (1999) A Survey of Quality of Service in Mobile Computing Environments. *IEEE Communications Surveys Second Quarter*, **2**(2), 2–10.

[6] Coulouris, G., Dollimore, J., and Kindberg, T. (2001) *Distributed Systems – Concepts and Design*, 3rd edition, Addison-Wesley, Boston, MA.

[7] Davies, N., Friday, A., Wade, S., and Blair, G. (1998) L2imbo: A Distributed Systems Platform for Mobile Computing. *ACM Mobile Networks and Applications*. Special Issue on Protocols and Software Paradigms of Mobile Networks, **3**(2).

[8] Dunham, M. H., Helal, A., and Balakrishnan, S. (1997) A Mobile Transaction Model That Captures Both the Data and Movement Behavior. *ACM/Baltzer Journal on Special Topics in Mobile Networks and Applications (MONET)*, **2**, 149–162.

[9] Emmerich, W. (2000) *Engineering Distributed Objects*, John Wiley & Sons, Chichester, UK.

[10] Gelernter, D. (1985) Generative Communication in Linda. *ACM Transactions on Programming Languages and Systems*, **7**(1), 80–112.

[11] Haahr, M., Cunningham, R., and Cahill, V. (1999) Supporting CORBA Applications in a Mobile Environment (ALICE). *5th International Conference on Mobile Computing and Networking (MobiCom)*, August 1999; ACM Press, Seattle, WA.

[12] Handorean, R. and Roman, G.-C. (2002) Service Provision in Ad-hoc Networks, in *Coordination*, Volume 2315 of *Lecture Notes in Computer Science* (eds F. Arbab and C. L. Talcott), pp. 207–219; Springer York, UK.

[13] iCMG. (2002) K2: CORBA Component Server, http://www.componentworld.nu/.

[14] Joseph, A. D., Tauber, J. A., and Kaashoek, M. F. (1997) Mobile Computing with the Rover Toolkit. *IEEE Transactions on Computers*, **46**(3).

[15] Long, S., Kooper, R., Abowd, G., and Atkenson, C. (1996) Rapid Prototyping of Mobile Context-Aware Applications: The Cyberguide Case Study. *Proceedings of the Second Annual International Conference on Mobile Computing and Networking*, White Plains, New York, November 1996, pp. 97–107; ACM Press.

[16] Mascolo, C., Capra, L., Zachariadis, S., and Emmerich, W. (2002) XMIDDLE: A Data- Sharing Middleware for Mobile Computing. *International Journal on Personal and Wireless Communications*, **21**(1).

[17] Murphy, A. L., Picco, G. P., and Roman, G.-C. (2001) Lime: A Middleware for Physical and Logical Mobility. *Proceedings of the 21st International Conference on Distributed Computing Systems (ICDCS-21)*, May 2001, Phoenix, AZ, pp. 524–533.

[18] Oracle Technology Network (2000) Oracle9i Application Server Wireless, http://technet.oracle.com/products/iaswe/content.html.

[19] Oram, A. (2001) *Peer-to-Peer: Harnessing the Power of Disruptive Technologies*, O'Reilly.

[20] Perkins, C. (2001) *Ad-hoc Networking*, Addison-Wesley, Boston, MA.

[21] Psinaptic (2001) JMatos, http://www.psinaptic.com/.

[22] Reinstorf, T., Ruggaber, R., Seitz, J., and Zitterbart, M. (2001) A WAP-based Session Layer Supporting Distributed Applications in Nomadic Environments. *International Conference on Middleware*, November 2001, pp. 56–76; Springer, Heidelberg, Germany.

[23] Roman, G.-C., Murphy, A. L., and Picco, G. P. (2000) Software Engineering for Mobility: A Roadmap. *The Future of Software Engineering – 22nd International Conference on Software Engineering (ICSE2000)*, May 2000, pp. 243–258; ACM Press, Limerick, Ireland.

[24] Salutation Consortium (1999) Salutation, http://www.salutation.org/.

[25] Satyanarayanan, M. (1996) Mobile Information Access. *IEEE Personal Communications*, **3**(1), 26–33.

[26] Satyanarayanan, M., Kistler, J., Kumar, P., Okasaki, M., Siegel, E., and Steere, D. (1990) Coda: A Highly Available File System for a Distributed Workstation Environment. *IEEE Transactions on Computers*, **39**(4), 447–459.

[27] Schill, A., Bellmann, B., Bohmak, W., and Kummel, S. (1995) System support for mobile distributed applications. *Proceedings of 2nd International Workshop on Services in Distributed and Networked Environments (SDNE)*, pp. 124-131; IEEE Computer Society Press, Whistler, British Columbia.

[28] Softwired (2002) iBus Mobile, http://www.softwired-inc.com/products/mobile/mobile.html. Sun JavaSpaces, 1998.

[29] Terry, D., Theimer, M., Petersen, K., Demers, A., Spreitzer, M., and Hauser, C. (1995) Managing update conflicts in Bayou, a weakly connected replicated storage system. *Proceedings of the 15th ACM Symposium on Operating Systems Principles (SOSP-15)*, Copper Mountain Resort, CO. pp. 172-183.

[30] Ubi-core (2001). Universally Interoperable Core. http://www.ubi-core.com.

[31] UPnP Forum (1998) Universal Plug and Play, http://www.upnp.org/.

[32] Welling, G. and Badrinath, B. (1998) An Architecture for Exporting Environment Awareness to Mobile Computing. *IEEE Transactions on Software Engineering*, **24**(5), 391–400.

[33] Wyckoff, P., McLaughry, S. W., Lehman, T. J., and Ford, D. A. (1998) T Spaces. *IBM Systems Journal*, **37**(3), 454–474.

12

Application of Middleware Technologies to Mobile Enterprise Information Services

Guijun Wang, Alice Chen, Surya Sripada, Changzhou Wang
Boeing Phantom Works

12.1 Introduction

Mobile devices have been increasingly powerful, ubiquitously available, and network-connected [9]. There are a variety of mobile devices including handheld devices, tablet computers, and mobile phones. At the same time, wireless infrastructure provides ample bandwidth, wide coverage, and seamless integration with wired networks for communications. Mobile devices and networks offer the potential for providing enterprise information services to users anywhere anytime. For example, an engineer from factory floor can enter data from where they are collected, access product information at their convenience, and collaborate with experts to resolve problems on the spot.

The capabilities of providing information services to users anywhere anytime have tremendous value to enterprise workforces and customers. Business processes can be improved, workforces can be more productive, quality of services to customers can be more satisfactory, and value-added services can be easily created and offered. To achieve such capabilities, middleware platforms and services play a crucial role in integrating end-user devices, heterogeneous networks (wireless and wired), enterprise business processes, enterprise information systems, and enterprise data services [11].

Middleware platforms and services such as J2EE and .Net provide the common software services and runtime management functions to applications. Component-based application servers and integration servers based on these platforms are integral parts of enterprise

Middleware for Communications. Edited by Qusay H. Mahmoud
 ISBN 0-470-86206-8

application integration solutions. For example, IBM's Websphere, BEA's Weblogic, and Oracle's 9iAS are some of the J2EE-based products for application servers and integration servers. Components implementing business service logic are deployed in application servers and components implementing business process coordination logic are deployed in integration servers. Application servers and integration servers, as middleware platforms, integrate with backend enterprise data services. Enterprise data services host business data, which are accessed and updated by business components in application or integration servers. Common middleware services including security, transaction, and messaging provide common mechanisms for secure and flexible interactions between distributed system components.

Extending Enterprise Information Services (EIS) to mobile devices (mobile EIS) faces a number of challenges including the variation of devices, unpredictable connectivity, diverse enterprise information systems, and different types of data contents. Devices such as handheld devices, tablet computers, and mobile phones have different display dimensions, different communication capacities, and different computing resources. Data contents must be tailored to the profiles of each type of devices. Furthermore, mobile devices tend to join and leave the network frequently. Yet jobs submitted from the devices and on application servers or integration servers must continue their courses of executions. Execution results must be retained until devices are connected to the network and delivered when users are authenticated.

Similarly, mobile EIS must integrate with diverse enterprise information systems ranging from email service, directory service, and timekeeping service, to resource planning service, product support service, and engineering service. Mobile EIS requires an integrated and effective architecture to bring these diverse services to mobile users. The different types of data contents represent yet another dimension of the complexity for mobile EIS. Simple text, email with attachment, images, engineering drawings, and voice messages are some of the types of data contents. These contents must be adapted to the profiles of mobile devices and the preferences of users [10].

To meet with these challenges in mobile EIS, our approach focuses on integrated mobile service architecture and reliable and flexible middleware services for enterprise systems integration. The architecture is service-based. It is capable of meeting the challenges and integrating with traditional enterprise systems in a flexible, extensible manner. The middleware services are based on the open platform J2EE. They support the requirements for integrating mobile devices with enterprise information systems.

In this chapter, we present necessary technologies, wireless and middleware in particular, for mobile EIS. We describe our approach for delivering enterprise information services to mobile users. The approach consists of an integrated mobile service architecture and reliable and flexible middleware services. We describe the architecture and discuss its design features and benefits for mobile EIS. Middleware services consist of reliable messaging based on the Java Message Service (JMS) [14], a set of services based on J2EE features such as Security and JavaServer Pages, and XML-based data templates for flexible content representation and tailoring [15]. We then present in detail on how to represent and tailoring contents for mobile devices and how to integrate with traditional enterprise systems. Finally, we outline some challenges and future directions in mobile EIS.

12.2 Wireless Technologies

The emergence of wireless networking technologies makes it possible to provide information services to enterprise workforces and customers anywhere anytime. To enable mobile EIS, wireless technologies must integrate with enterprise networks. To develop a successful architecture and software solution to mobile EIS, we need to understand wireless technologies including their state-of-the-arts, integration issues in enterprise networking, and constraints.

Networks are essential to modern enterprises, where distributed systems must be integrated from communications to applications. These networks may include WAN, MAN, and/or LAN from different service providers. In addition, it is not uncommon that an enterprise has stovepipes of network technologies for different applications. These networks mostly are non-TCP/IP based such as SNA, DECnet and X.25, and so on. In the last decade, in order to reduce the network operating cost, enterprises have been consolidating their networks into two if not into one network. The goal is to migrate all their networks to a TCP/IP based network regardless in the WAN/MAN/LAN environments. With the availability and maturity of wireless LAN technologies such as IEEE 802.11b/a/g and the coming of the 4G-based WAN and the 802.16-based or 802.20-based MAN technologies, the wireless access to the existing wired enterprise network infrastructure is inevitable. Figure 12.1 is an example of how wireless networks can be hierarchically integrated into an enterprise Intranet.

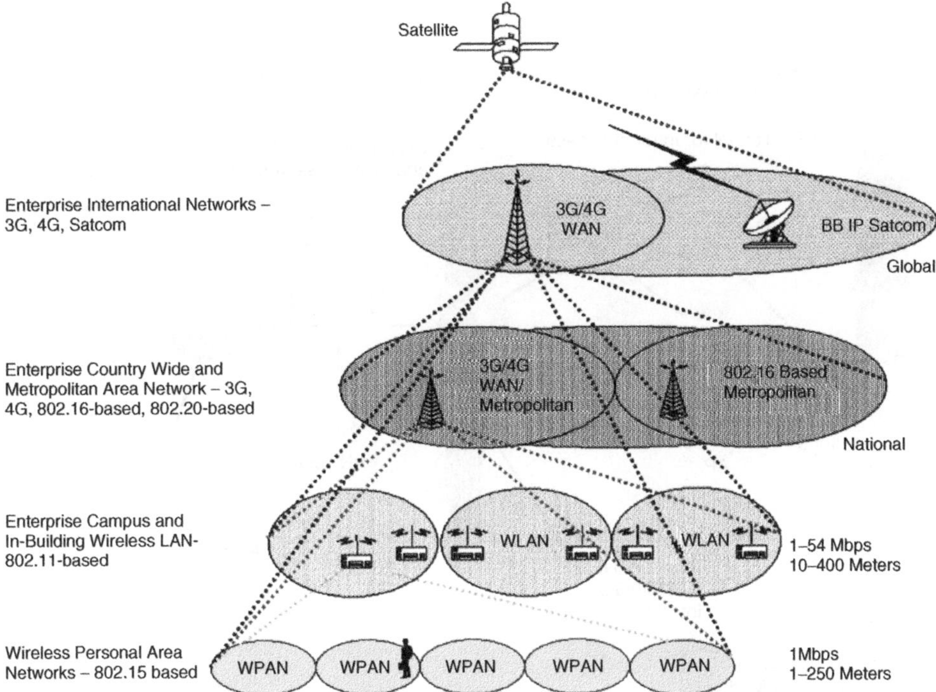

Figure 12.1 Integration of wireless networks to enterprise intranet

Technically, there are still several issues with wireless technology deployment in the enterprises today. The most important issue to enterprises is the wireless security. Since the wireless network media is air, the frames transmitted over the air need to be secured. However, the existing 802.11 wireless encryption protocol WEP (Wireless Encryption Protocol) has some flaws and hence is not adequate for deployment. In addition, since wireless LAN access points are cheap and very portable, any enterprise visitor can plug an AP (access point) into an existing Intranet LAN drop for unauthorized network access. Therefore, wireless enterprise communications need to not only confirm to the enterprise's wired network security policy but also be treated as traffics from the public Internet. While the standard 802.11x (i.e., manual authentication at switch port via RADIUS) and 802.11i (i.e., centralized authentication and dynamic key distribution) based WPA (WLAN Protected Access) products are still evolving, to meet the wireless applications requirements, enterprises have to deploy various short-term solutions in order to provide two levels of security that are normally required by enterprise security policy.

Figure 12.2 gives an example that illustrates how a wireless device may access an enterprise network. Typically, at an enterprise security perimeter, there are Radius and IPSec VPN servers and/or reverse proxies that provide the authentication, authorization, and accounting (AAA) processing, encryption, and VPN tunnel addresses binding services. On a user's wireless device, there is an SSL or IPSec client installed and a wireless Network Interface Card (NIC). To obtain wireless link services from a WLAN (e.g., 802.11a/b/g) AP, the wireless device's NIC driver will have to be configured that it is the same as the network id (called SSID) of the servicing WLAN AP and its 128-bit encryption key.

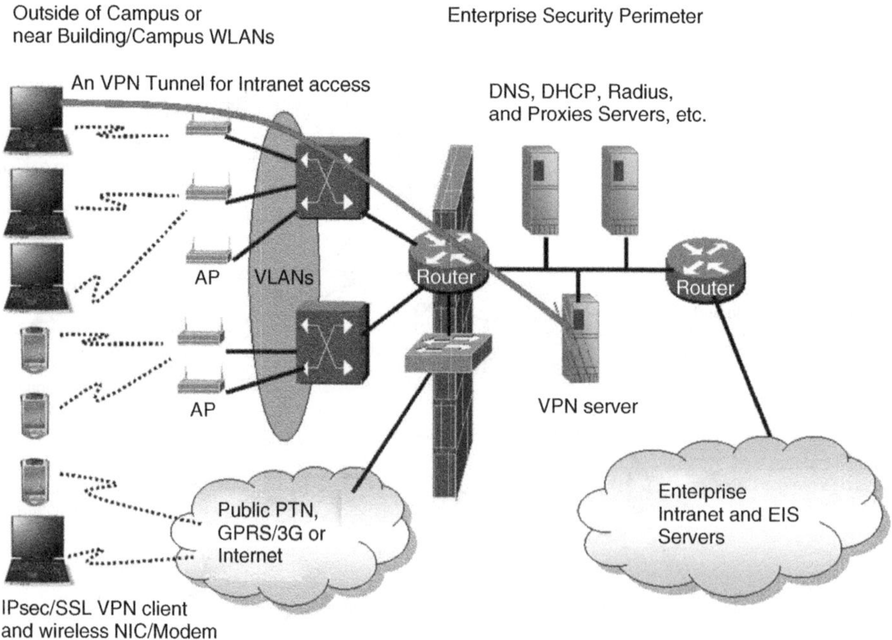

Figure 12.2 Access to enterprise networks from wireless devices

Let us assume the wireless device has an IP address assigned either by hard coded, Network Address Translation (NAT) or Dynamic Host Configuration Protocol (DHCP). To prevent wireless traffics that bypass the security perimeter or firewall to access the enterprise Intranet directly, the switch ports that the wireless APs are connecting to are either configured on some virtual LANs (VLANs) or controlled and managed by the switches, a security appliance, or a centralized AP manager. Hence, only certain preconfigured and authorized IP subnet addresses are allowed to access the enterprise security perimeter while the rest traffics will be refused for services. In order to pass the enterprise security perimeter to access the enterprise Intranet, every user wireless device has to establish a SSL or IPSec VPN tunnel with the VPN server first. To authenticate an authorized user, every mobile user is also issued with a hard or soft smart card that can generate a one-time-password when the user's key is entered. If the one-time-password passes the AAA validation by the Radius server, the VPN server will then establish a tunnel that binds between the VPN client and server.

Once the mobile user is connected to the network, there may be more user logons required, depending on the type of network the application server is on and the application security requirements. Hence, a centralized single sign-on requirement (e.g., via XML signatures and SAML –Security Assertions Markup Language protocol) is also on the horizon.

In addition to the wireless security issue, among the other known enterprise wireless network deployment concerns are the wireless QoS, IP mobility, application persistence, wireless LAN/WAN integration, and the constraint of mobile devices issues. The solutions of these issues do not warrant the wireless network connectivity. Therefore, a solution is to push the connectivity problem to the upper layer. A messaging-based asynchronous mode operations can be best served in a wireless environment.

Mobile IP intends to address the mobile roaming issue when users traverse through different IP subnets that are connected by different routers. The latency between the inter subnet APs handover and the roaming DHCP client to get a new routable IP address update and the Domain Name Service (DNS) update may cause a running application to drop its connection. This is especially true with applications that involve with the X client and media player. Mobile IP and IPv6 are intended to provide the solution to do the packets rerouting from the home agent. However, their infrastructure deployments in enterprises are very few today. In addition, because both Mobile IP and IPv6 use a client–agent model, the cost of the installation of mobile IP client on mobile devices may prohibit the deployment of Mobile IP and IPv6 in enterprise if the Mobile IP client were not preloaded on the mobile device from manufactory. There is an industry trend to use the SIP protocol in the future. For now, a short-term fix to this problem is the usage of VLANs. Even though VLAN configuration and management mechanisms are getting mature, this approach is still not scalable for big enterprise. There are other vendor approaches such as using network appliances or DHCP proxies to assign NAT IP addresses and manage the mobile devices address table at the proxy. The biggest concern of these vendor-specific approaches is the product maturity and the viability of these vendors in the long run.

To address the application persistency issue, both the Internet Engineering Task Force (IETF) and the Open Group are working on the solutions. As far as the wireless LAN/WAN integration issues, there are very few products that put the GPRS/3G/802.11-based technologies on the network access card for mobile device and because the wireless LAN technologies are still evolving, we do not expect a mature integration solution in the next couple of years.

Even if all these issues are addressed, mobile devices still have basic constraints such as bandwidth, display screen size, and power shortage. For instance, an iPAQ with an improved 802.11 CF NIC card, even with the power management on, can last only for three to four hours, which is not enough for an eight-hour work shift.

What we can conclude is that in a mobile user environment, the network connectivity between a mobile device and its application server or peer can never be assumed. How shall we deal with these mobility-related network issues? On the basis of the OSI reference model, it is reasonable to push some of the services to the layer above if a service is not provided below, that is, to push the mobile network issues to upper layers such as transport, session, presentation, and application layers. Middleware fits into the session layer. A message-oriented asynchronous mode middleware is very suitable for mobile network environment, especially when mobile users come and go and come back sometime after.

While the networks are converging into one, to reduce the cost of ownership of applications, enterprises are also streaming their process by integrating or reducing the number of applications. Hence, the use of middleware is proliferating. Web technologies also provide a common ground for presentation layer consolidation. To provide reliable, flexible, and adaptive mobile services, various presentation and application front-ends or gateways may be introduced for the purposes of integration. We believe architecture design and integration middleware are two key elements for successful mobile EIS. We will elaborate on design issues and middleware solutions in the following sections.

12.3 Middleware Technologies for Enterprise Application Integrations

Middleware is a layer of software between applications and Operating Systems (OS). Its main purpose is to provide common abstractions and services for applications to use system resources and communicate with others in distributed systems. It consists of a development time framework and runtime services for distributed applications. For example, applications can use mechanisms in middleware to publish messages, and other applications can use similar mechanisms to receive messages. For object-oriented applications, objects in one application can make method invocations on objects in another application distributed across networks. Middleware runtime services manage object lifecycles, dispatch requests to objects, and make distributed communications transparent to applications.

A middleware development time framework includes design patterns, interfaces, and hooks for applications to access runtime services and communicate with others in distributed environments. It also often provides common services such as Naming, Persistence, Transaction, and Security for applications. Runtime services include lifecycle management, resource management, request dispatch, and message distribution. Figure 12.3 shows the role of middleware in abstracting platform and network specifics from applications and facilitating information exchanges between them.

Enterprise application integration (EAI) consolidates enterprise applications, removes application stovepipes, and connects applications to applications and applications to data in order to support enterprise business processes. Middleware plays an essential role in EAI. Middleware paradigms for EAI include:

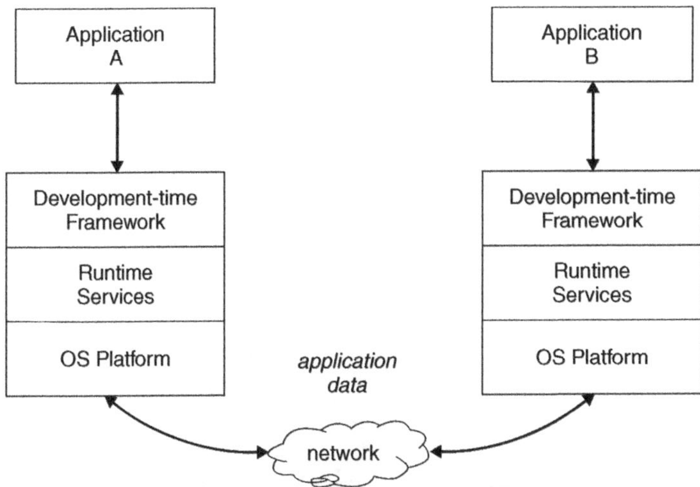

Figure 12.3 Middleware plays a critical role in distributed application integration. Reproduced by permission of IBM, in: OpenCard Framework 1.2 Programmer's Guide, 4 ed. (1999) OPENCARD CONSORTIUM

- Transaction Processing Monitors, which provide tools and services for applications to access databases.
- Remote Procedure Calls (RPC), which enable one application to call a remote application the same way as calling a local application. Protocols such as Simple Object Access Protocol (SOAP) for web services [1] are examples of this middleware paradigm.
- Message-Oriented Middleware (MOM), which enables message exchanges in a data-centric manner for distributed applications. Messages encapsulate business data and any meta-data-like properties associated with it. MOM manages message collections and distributions for distributed applications. Senders and receivers of messages are decoupled, which affords a flexible EAI architecture.
- Object Request Broker (ORB), which supports interactions between distributed object-oriented applications. A good example of ORB middleware is OMG's Common Object Request Broker Architecture (CORBA).
- Component-Based Application Server, which provides a component-based development framework and a set of runtime services based on a component platform such as J2EE [2] and .Net [3]. It integrates essential features of middleware services from paradigms such as RPC, ORB, and MOM in a single component-based middleware. For example, the J2EE platform supports both synchronous method calls as in an ORB and asynchronous messaging as in a MOM between distributed components.

Component-based application servers such as those based on J2EE and .Net provide comprehensive component-based development [4] frameworks and runtime services for EAI. A typical EAI architecture consists of web servers in the front end, component-based application servers in the middle, and data servers in the backend. Figure 12.4 shows the roles of web server, component-based application server, and data server in EAI.

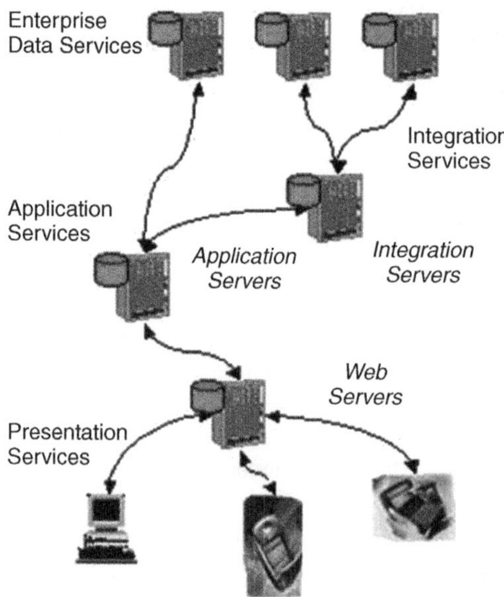

Figure 12.4 EAI divides services in multiple tiers on different types of servers

The J2EE platform provides a comprehensive middleware framework and runtime system for developing and deploying business applications on web servers and application servers. For example, features such as Java Server Page (JSP) and Servlet together with XML-based data representation support flexible information presentations from web servers to users. Components that implement business logic are deployed in application servers and connected to enterprise data servers. Major J2EE-based application server providers include IBM, Oracle, BEA Systems, and IONA.

To support message-oriented interactions between distributed applications, J2EE includes the Java Message Service (JMS) specification. JMS is based on the MOM paradigm. MOM plays an important role to some EAI cases, particularly if MOM is used in conjunction with other types of middleware to provide total integration for enterprise systems. Major advantages of MOM for EAI include loose coupling, data-centric, relatively simple interfaces, and flexibility. Additional message reliabilities can be achieved via reliable MOM architecture and persistence services. The advantages of JMS-based MOM are its open architecture with standard interfaces and multivendor supports, and its seamless integration with other J2EE capabilities.

Challenges for mobile EIS include the variation of devices, unpredictable connectivity, diverse enterprise information systems, and different types of data contents. Furthermore, the problem of information collection, access, and dissemination via mobile devices and wireless networks must be considered within a total EAI context. A MOM based on JMS provides a good solution to mobile EIS because of its advantages for EAI. On the other hand, middleware is only part of the solution to the mobile EIS problem. An effective and integrated architecture for mobile EIS in the context of integration with diverse enterprise systems is another critical part of the solution.

12.4 An Integrated Architecture for Mobile Enterprise Information Services

12.4.1 Enterprise Requirements

Because of the complexity and diversity of mobile devices and enterprise systems, an integrated architecture is required to provide enterprise information services to users anywhere anytime. Enterprise information types include business data collected on the spot (e.g., factory floor and point of sale), routine office data (e.g., email, voice messages, and time keeping), product data (e.g., bill of material, parts, and configuration), engineering data (e.g., design manuals, drawing, and operation manuals), and service data (e.g., maps, maintenance aids, and sales). For example, an engineer who is testing a product on the factory floor should be able to collect the test data on the spot using a mobile device and send the data to enterprise systems for further processing. Similarly, he/she should be able to access the product data and engineering data on the spot from the mobile device whenever he/she needs it. An alternative approach using paper forms and walking around the factory floor would be error-prone, time-consuming, and ineffective.

Enterprise requirements for such an integrated mobile EIS architecture include:

- Available: Provide enterprise information services to users anywhere anytime via mobile networks and devices.
- Flexible: Integrate multiple types of mobile devices with diverse enterprise information systems.
- Deployable: Require no or minimal changes to existing enterprise information systems.
- Adaptable: Automatically adjust services and tailor data content according to user preference and device profiles.
- Secure: Support enterprise authentication and authorization to establish a security context for accessing any enterprise information systems.
- Reliable: Messages between mobile devices and backend data servers must be stored persistently until successful delivery or any predetermined threshold is reached (e.g., deleted after a fixed number of attempts or certain amount of time has expired).
- Dependable: The architecture must be able to reconfigure itself dynamically when certain hosts become overloaded or fail.
- Scalable: The architecture must be able to handle a large number of service requests concurrently.
- Extensible: The architecture should support the addition of new services without any changes to existing services.

12.4.2 Design Considerations and Our Approach

Important issues in the design of our mobile EIS architecture can be divided into the following categories:

- Front-end issues: How to deal with variations of devices.
- Wireless and wired communication protocols: How to deal with variations of protocols and entry points to secure enterprise systems.

- Backend enterprise systems: How to deal with variations of enterprise systems and how to simplify the addition or removal of backend enterprise systems from mobile EIS architecture.
- Mobile service platform: How to have a single dependable, scalable, and extensible platform for mobile EIS, which integrates front-end devices to backend enterprise information systems.
- Middleware for integration: How to best utilizing middleware for reliable interactions between mobile devices and the mobile service platform.
- Data content: How to access enterprise data, represent it, and tailor it to device profiles.

Our approach is an integrated mobile service platform, which consists of a number of services including profile management for users and devices, configurable gateways for connecting devices to the platform via various protocols, configurable proxies for integrating with backend enterprise systems, JMS-based messaging service for reliable interactions between mobile devices and the platform, and XML-based data representation and transformation for tailoring data content for mobile devices.

12.4.3 An Integrated Mobile EIS Architecture

The enterprise requirements and design considerations state concerns in different categories: mobile devices, connectivity to enterprise networks via wired or wireless network protocols, middleware for information exchanges and integrations, connectivity to enterprise information services, data contents, and quality concerns including security, extensibility, and dependability. We take a separation-of-concerns approach for architecture design.

To deal with device profiling, access protocols, secure login, and content presentation, we use gateways hosted on web servers. A gateway is an entry point for a device using a particular protocol (e.g., HTTP) to log on to enterprise systems. To address security concerns, we use secure network infrastructure features such as IPSec, VPN, and HTTPS to ensure communication securely. A user will get authentication and authorization from enterprise security services at login time. A gateway is responsible for presenting data content to connected devices according to device profiles. A profile of a device is an XML-based property file that contains information about the device's capability including color, dimension, and resolution. From an implementation point of view, a gateway is implemented as a Java Servlet in our prototype.

To be able to scale to large number of concurrent users, provide dependable services, and integrate with backend enterprise systems email service, production service, and engineering service, a mobile EIS application server is used to host critical integration functions. These functions include interfaces with backend enterprise systems, access methods and protocols, and content transcoding. Each type of backend enterprise systems has one such set of functions, aggregated in one entity called an *Infolet.* Each Infolet uses a standard XML-based template, but customizes on specifics. The template includes attributes such as host name, port number, protocol type, access interfaces and parameters, transcoding services if any, and security credentials. The mobile EIS application server provides a standard execution engine for executing Infolets to interact with

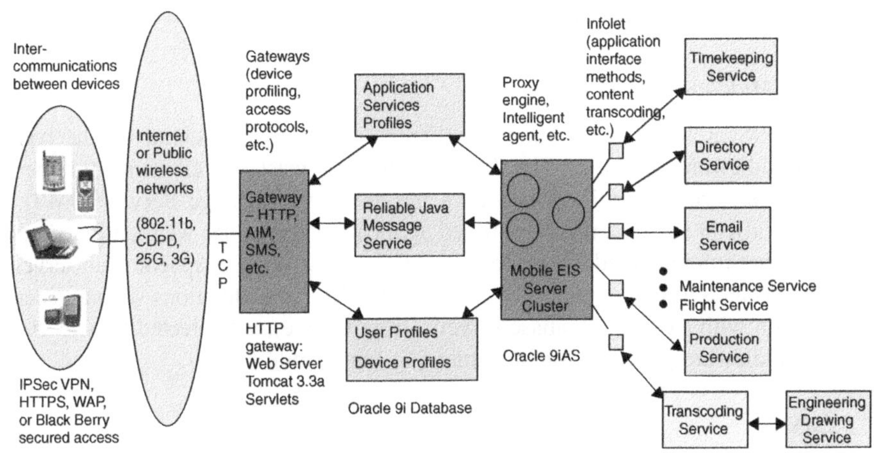

Figure 12.5 An integrated mobile EIS architecture

backend enterprise systems. Several of the servers can form a server cluster for dependability.

To support extensibility, reliability, and characteristics of unpredictable mobile connections, gateways on web servers and Infolets in mobile EIS application servers are connected with a JMS-based middleware, which provides reliable message queues for request and response messages between them. A gateway sends user's request messages to a request message queue and retrieves response messages from a response message queue. Similarly, an Infolet retrieves request messages from a request queue and sends response messages to a response message queue. Messages can be made persistent by storing in a database.

Figure 12.5 illustrates the architecture of our mobile EIS. It is adapted from the iMobile architecture in collaboration with AT&T Labs-Research [5, 6]. It is a service-based architecture, where the services are loosely coupled and have well-defined interfaces.

12.4.4 Deployment and Operation

Our architecture does not require mobile devices to install sophisticated software in order to access mobile EIS. A web browser or any other native interface software on mobile devices is sufficient to connect to the gateways on web servers. For example, pocket PC such as iPAQ can connect to a gateway referenced by a URL using a web browser. Once connected to a gateway, the iPAQ client can log in and view available services. Service requests will be received by the gateway, which wrap the requests in messages to JMS queues. Furthermore, device profiles are stored in a database on the mobile EIS platform. Infolets take request messages from a request queue, access information from enterprise information services backend, transform the data content from a response based on the device profiles, and finally wrap the data content in a response message and send it to a reply queue. Gateways take messages from reply queues, examine message headers to identify clients, and present response data content to rthe clients.

12.5 J2EE-Based Middleware in Mobile EIS

12.5.1 J2EE Middleware Platform

The mobile EIS architecture as shown in Figure 12.5 is implemented on the basis of the J2EE middleware platform. The J2EE platform consists of elements for the web servers such as Java Server Page and Java Servlet, Java Message Service (JMS), and Enterprise JavaBean (EJB) component-based application server. As we mentioned earlier, the gateways are implemented as Servlets on web servers. We use Apache web server and Tomcat 3.3a as its Servlet container. Because of practical considerations such as licensing and integration with existing database servers (from Oracle), we selected Oracle 9iAS as our J2EE middleware platform for implementing our mobile EIS.

12.5.2 JMS

JMS is a critical piece of middleware for decoupling Gateway web servers and Infolet application servers. It is essential for dealing with the characteristics of mobile devices such as unpredictable connectivity and variations in capability. It also supports quality attributes such as flexibility, extensibility, and reliability.

12.5.2.1 JMS Styles of Messaging

JMS supports two styles of messaging: Point-to-Point (PTP) and Publish-and-Subscribe (Pub/Sub). The PTP style allows a sender (producer) to send messages to a receiver (consumer) via a message queue. The queue retains all messages until the messages are consumed or expired. A message is consumed if the queue receives a success acknowledgment from a receiver, in which case, the message will be purged from the queue. A message is expired when current time is later than its expiration time as specified in the message header. An expiration time of 0 indicates the message will never expire. The P2P style is suitable for situations where every message from a sender must be consumed by a receiver. It features the reliability of messaging. Figure 12.6 illustrates the PTP style of JMS.

The Pub/Sub style allows publishers to publish messages to a topic and subscribers to subscribe to messages of a topic. A Topic roughly corresponds to the Queue in the P2P style. But there could be multiple publishers sending messages to the same topic and multiple subscribers receiving messages from the same topic. A subscriber can receive messages published only after its subscription to a topic is established. Furthermore, using the durable subscription type, a subscriber is able to receive messages published while it is inactive. The JMS system takes care of distributing the messages arriving from a topic's

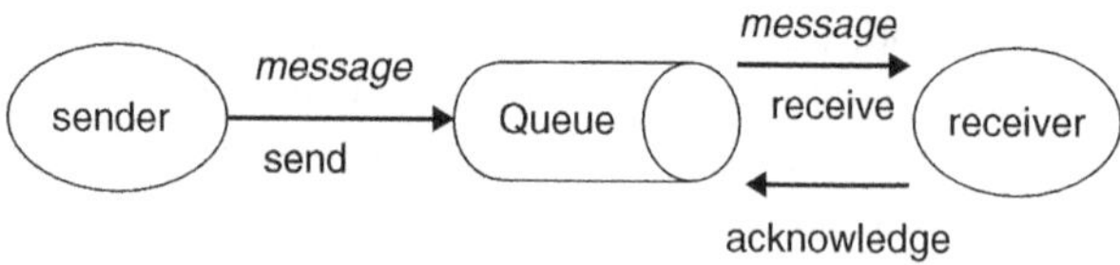

Figure 12.6 PTP style of messaging in JMS

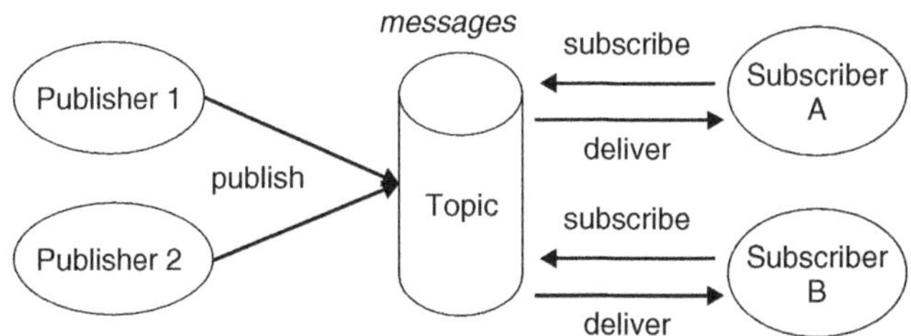

Figure 12.7 Pub/Sub style of messaging in JMS

multiple publishers to its multiple subscribers. Topics retain messages only as long as it takes to distribute them to current subscribers. The Pub/Sub style is suited for situations where multiple publishers send messages to the same topic and messages are intended for multiple consumers. Figure 12.7 illustrates the Pub/Sub style of messaging in JMS.

12.5.2.2 JMS Programming Concepts

JMS provides a small number of concepts for programmers to build a JMS application, either as a producer of messages or a consumer of messages or both. These concepts include:

- *Administered objects:* Connection Factories and Destinations. A connection factory is an object responsible for creating a specific type of connections, either queue-based or topic-based connections, depending on the style of messaging. A destination is the object that a producer sends messages to or a consumer receives messages from. In other words, a destination is either a queue or a topic. JMS specifies the interfaces for the two classes of objects. Concrete implementations of them are often provided by JMS products and are configured at the system start-up time. In an application, these objects can be obtained by a lookup service, for example, using code like this: Context ctx = new InitialContext(); QueueConnectionFactory queueConnectionFactory = (QueueConnectionFactory)ctx.lookup("QueueConnectionFactory")
- *Connections.* A connection encapsulates the distributed communications (via TCP socket, for example) between a client (producer or consumer) and a messaging server (the JMS provider). It creates and manages sessions, each of which is a thread of execution context for producers and consumers who want to exchange messages.
- *Sessions.* A session is a single-threaded context for managing the reception and dissemination of messages. It is used to create message producers, message consumers, and messages. A session is also capable of providing a transactional context with which a set of messages is grouped into an atomic unit of work.
- *Message Producers.* A message producer sends messages to a destination.
- *Message Consumers.* A message consumer receives messages from a destination.
- *Messages.* Messages are the data objects exchanged in messaging.

Figure 12.8 shows how all these concepts fit together in a JMS application.

JMS defines interfaces for each of the concepts. The following table summarizes the general concepts and their specifics in the PTP and Pub/Sub styles.

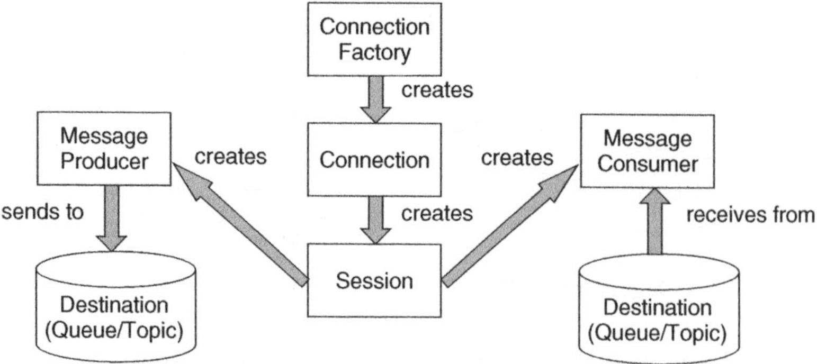

Figure 12.8 Relationships of JMS programming concepts

General JMS	PTP	Pub/Sub
ConnectionFactory	QueueConnectionFactory	TopicConnectionFactory
Destination	Queue	Topic
Connection	QueueConnection	TopicConnection
Session	QueueSession	TopicSession
MessageProducer	QueueSender	TopicPublisher
MessageConsumer	QueueReceiver, QueueBrowser	TopicSubscriber

12.5.2.3 JMS Messages

JMS messages are data objects that can be produced and consumed by applications. A Message is composed of the following parts:

- Header – All messages support the same set of header fields. A message's header fields are used to identify and route messages. Each field has set/get methods. For example, every message has a unique identifier, represented in the header field JMSMessageID. Another header field JMSDestination represents the queue or the topic to which the message is sent. A complete list of header fields is shown in the table below.
- Properties – Provides a built-in facility for adding optional properties to a message. Properties could be JMS predefined properties, application-specific, or provider-specific.
- Body – JMS provides six types of message body. JMS specifies interfaces for creating messages of each type and for filling in their contents. The six types are:

  - StreamMessage: A message whose body contains a stream of Java primitive values. It is filled and read sequentially.
  - MapMessage: A message whose body contains a set of name-value pairs where names are Strings and values are Java primitive types.
  - TextMessage: A message whose body contains a java.lang.String.
  - ObjectMessage: A message that contains a Serializable Java object.
  - BytesMessage: A message that contains a stream of uninterpreted bytes.
  - Message: A message that is composed of header fields and properties only. This message type is useful when a message body is not required.

Header field set by	Meaning	Header field set by
JMSDestination	The destination to which the message is being sent	send or publish method
JMSDeliveryMode	The mode by which the message should be delivered	send or publish method
JMSPriority	A ten-level priority value	send or publish method
JMSTimestamp	The time when a message is being handed off to a message server	send or publish method
JMSReplyTo	A Destination where a reply to the message should be sent	Client
JMSRedelivered	Boolean value indicating whether it is redelivering a message	JMS provider (server)
JMSExpiration	The time at which the message should expire	send or publish method
JMSMessageID	A String uniquely identifies each message	send or publish method
JMSCorrelationID	An identifier to link one message (e.g., a response message) with another (e.g., a request message)	Client
JMSType	A type identifier	Client

12.5.3 JMS in Our Mobile EIS

The JMS provider in our mobile EIS is Oracle9iAS. Because of the reliability requirement, we use only the PTP style of messaging. PTP mandates that every message from a sender must be consumed by a receiver. We create two types of queues: one for request messages, and another for response messages. A response message ties to its original request message using the JMSCorrelationID header field. The value of this field uniquely identifies the client (a user of a wireless device), which includes user name, session id, IP address, and port number.

At runtime, a gateway is a sender of messages to the request queue. An Infolet is a receiver of messages from the request queue. Similarly, the Infolet is a sender of response messages to the response queue, while the gateway is a receiver of messages from the response queue. A gateway is active when a client is connected to the gateway web server. Naturally, it only produces or consumes messages when the client is connected. Because messages are guaranteed to stay in the queue until it is consumed, we effectively deal with the challenge of unpredictable connectivity of mobile devices. Because gateways (on web servers) are decoupled from Infolets (on application servers), request messages can continuously be processed independent of client's connectivity situations.

Figure 12.9 shows how we use JMS queues, producers, and consumers for reliable, extensible, and scalable services in our mobile EIS. A message is retrieved by a mobile EIS server from the request queue and dispatched to an Infolet for processing. Messages are either the type of TextMessage, which consists of XML strings, or BytesMessage, which consists of raw data.

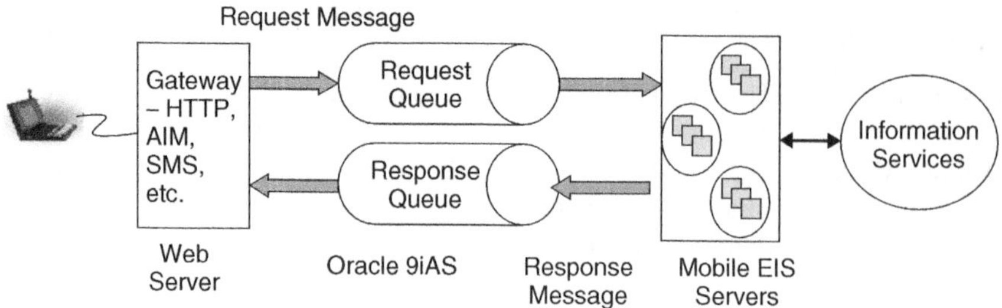

Figure 12.9 Our mobile EIS uses message queues for reliable point-to-point messaging

12.6 Data Representation and Presentation in Mobile Enterprise Information Services

Enterprise Information Services store data in many different formats and locations, ranging from database systems to directory servers to web pages. On the other hand, enterprise information users often use different mobile devices with diverse computation and presentation capabilities ranging from laptops to PDAs to cell phones. To enable the information flow between information systems and devices, the data stored in information systems must be adapted and transformed into appropriate content and format for each type of mobile devices [12, 13].

The adaptation and transformation mechanism shall be efficient. Ideally, supporting a new device shall not require changing of existing services, and adding a new service shall not require considering capacities of each and every supported device. This requires the separation of data contents provided by the services from the data presentations on different devices, and a common format to encode data contents for unifying the interface between applications and mobile devices. We used two abstractions, Infolets and gateways, to achieve the separation half of the requirement. Infolets are applications aforementioned to access data sources and transform data sources into XML-based messages. Gateways are responsible for extracting information from messages and presenting it to devices. Infolets and gateways share access to the device profiles in a database as shown in Figure 12.5 (the architecture figure).

The eXtensible Markup Language (XML) technology is another half of the solution for this requirement. The XML document format has been widely used in industry to encode, store, and transfer various types of data. In addition, the companion XSLT specification [16] provides industrial-strength template-based transformation mechanism on XML documents. Hence, Infolets access data sources and generate the data content in XML format, which are then adapted and transformed into suitable format using XSLT templates for various devices.

Some of the information services also provide multimedia content such as engineering drawings and technical manuals, which often requires special adaptation engines. Such an engine may have its own template or rule specification language to facilitate customizable adaptation. For example, an image adaptation engine may transform various vector or raster images into thumbnail JPEG images with specific dimension and color requirements.

Different types of devices may require different XSLT templates or transforming rules to meet the specific presentation capacities of the devices. Usually, a device profile is compiled for each type of devices, and the system selects the appropriate template or rule (and possibly passes some parameters to it) according to the matched device profile. A device profile may describe:

1. The acceptable presentation format, such as WML, cHTML, HDML, XHTML Basics, HTML, or simply plain text.
2. Screen dimension, resolution, and color or gray-scale depth.
3. Language support, including character set, encoding, and font support.
4. Sound and image output capacity.
5. User input capacity, such as keyboard, pen, and voice.
6. Embedded scripting language as well as downloadable and executable software support.
7. Encryption and compression support.
8. Operating system and device model.
9. Network connectivity and bandwidth, which is often dynamic as the same device may be connected using different networks at different time, and the network characteristics may change over time.

As an example, in our implementation, users may navigate engineering drawings on iPAQ Pocket PCs. The information about a set of drawings is retrieved from the engineering database and composed into XML. See Listing 6.1 for an example. The XML document is transformed into the simple HTML suitable for iPAQ. See Listing 6.2 for an example. The transformation is done using the XSLT shown in Listing 6.3. Users may choose one drawing from the list in Listing 6.2.

For simplicity, the XSLT transform in Listing 6.3 generates simple HTML documents suitable for Pocket IE. This approach is not desirable when new devices with enhanced or different capacities are added, since the XSLT templates need to be modified or new XSLT templates need to be created for each application like this. A better solution is to use a common XML schema with generic representation directives, such as the Mobile Form defined in Microsoft Mobile Internet Toolkit, so that all applications can generate XML documents in this schema and an adaptation server then transforms the XML documents for individual devices.

```
<?xml version="1.0"?>
<plist>
 <array>
 <string>wdm/diag1.gif</string>
 <string>wdm/diag2.gif</string>
 <string>wdm/diag3.gif</string>
 </array>
</plist>
```

Listing 6.1 The Input XML Example

```
<html><head><title>Wiring Diagrams</title></head><body>
<h1>Available Wiring Diagrams</h1>
<a href="Gate?is=BWD&amp;dg=diag1">diag1</a><br/>
```

```
<a href="Gate?is=BWD&amp;dg=diag2">diag2</a><br/>
<a href="Gate?is=BWD&amp;dg=diag3">diag3</a><br/>
</body></html>
```

Listing 6.2 The Output HTML Example

```
<xsl:template match="/plist/array">
 <html><head><title>Wiring Diagrams</title></head><body>
 <h1>Available Wiring Diagrams</h1>
 <xsl:apply-templates select="string" />
 </body></html>
 </html>
 </xsl:template>
<xsl:template match="string">
<xsl:variable name="diagram" select="substring-after(text(),'/')"/>
 <a href="Gate?is=BWD\&amp;dg=\{\$diagram\}">
 <xsl:value-of select="\$diagram" />
 </a> <br/>
</xsl:template>
```

Listing 6.3 The XSLT Template (fragment)

The drawing itself is too large to fit in Pocket PC 320 × 240 screen. We use a customized image adaptation engine to break the drawing into small pieces with navigation links to their neighboring pieces. Figure 12.10a illustrates a portion of an example drawing. Figure 12.10b and Figure 12.10c illustrate two pieces on the right top portion of the

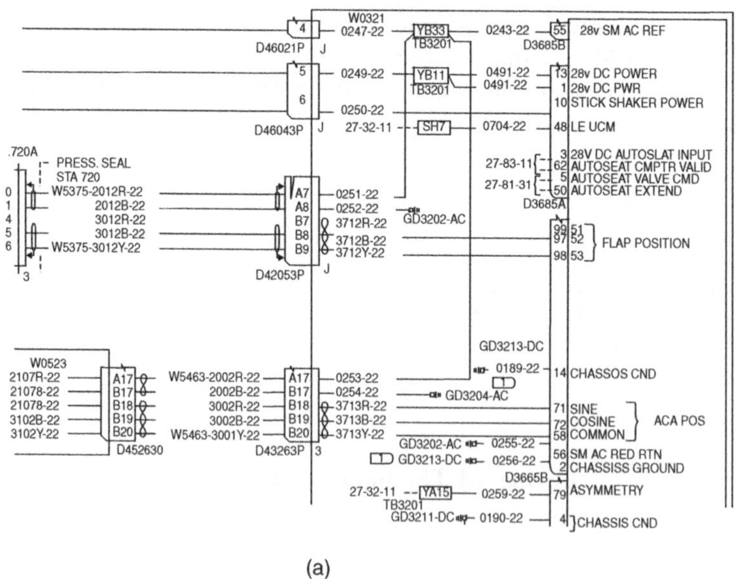

(a)

Figure 12.10a Example of a drawing and its transformed pieces 10b and 10c

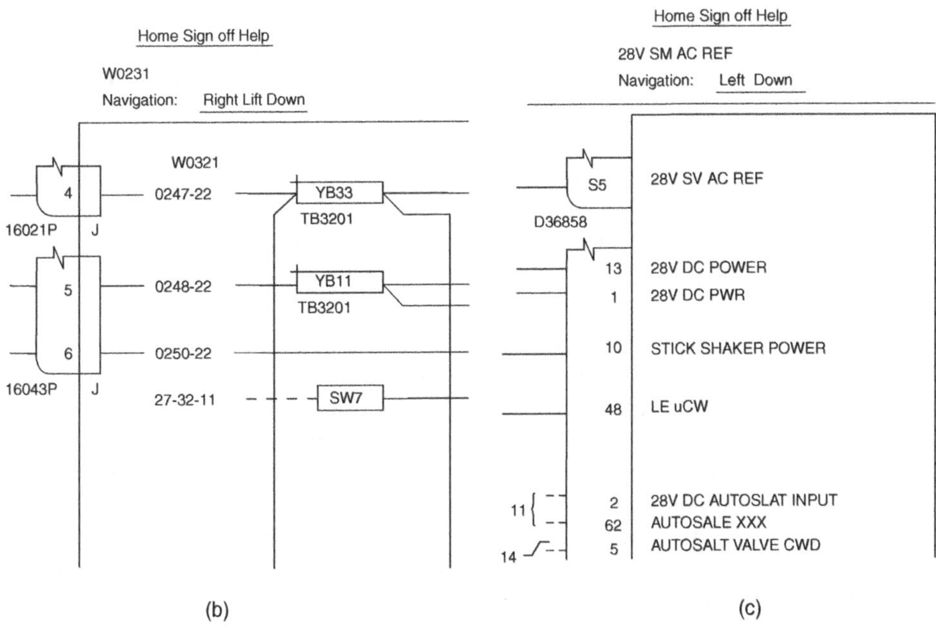

Figure 12.10(b and c) Transformed pieces

drawing on a Pocket IE window. The bold title (e.g. W0321) is obtained from the text annotation associated with the original drawing.

Legacy information services usually provide information in formats other than XML. To enable mobile access to this information, one approach is to develop adaptor or wrapper for each legacy system to generate information in XML format. Alternatively, a general adaptation engine may be used to adapt information directly to the target device, for example, to transcode large HTML pages into a set of small HTML pages suitable for PDAs. In our current implementation, we use the first approach as transcoding functions in Infolets.

12.7 Challenges and Future Directions

Mobile enterprise information services (mobile EIS) differ from consumer-oriented mobile services in several aspects. These differences must be recognized in developing and selecting middleware technologies for mobile EIS. Enterprise has well-established business processes and procedures. Enterprise systems have security services in place especially at the entry points from public Internet to enterprise internal networks. Consumer-oriented mobile services, on the other hand, have relatively simple services and loosely defined processes and security procedures. Since mobile EIS tie to business objectives including reduced time, increased revenue, and improved product quality, such services require dependability, efficiency, extensibility, and simplicity in architecture design and system deployment. For example, software deployed on mobile devices should be simple and as much standard-based as possible (e.g., a web browser common to all). Consumer-oriented mobile services could have any proprietary software on consumer devices. Furthermore, mobile EIS have relative stable and well-managed set of services. Consumer-oriented mobile services, on the other hand, tend to show more dynamic and *ad hoc* nature.

We believe our architecture design and middleware solution for mobile EIS have adequately met those challenges we outlined from the beginning. Those challenges include the variation of devices, unpredictable connectivity, diverse enterprise information systems, and different types of data contents. Yet major issues remain in the following areas.

- Security issues: In our current system, a user receives authentication and authorization at the entry points to the enterprise, namely, the web servers in our mobile EIS platform. But some of the backend systems, the engineering drawing service, for example, require additional logins and additional security mechanisms such as encryption. Ideally, a user is required to login at a single point and receive a security context good for any of the enterprise services. This is hard because communications between wireless devices and enterprise systems could go through public wireless networks and enterprise internal wired networks. In addition, the diversity of enterprise information systems from timekeeping, payroll, email, to production also makes it hard.
- Content tailoring to device profiles: It may not be that difficult to tailor data content to devices of different physical constraints. But it is very difficult to tailor it to the user's context (e.g., location) and tasks (what he/she wants/needs to do). To support such type of tailoring, mobile EIS must be context-aware and orient content toward user tasks. This capability requires explicit context information and user task modeling.
- Integration issues: The diversity of enterprise information systems presents a diverse set of interfaces to communicate with these systems. Our current mobile EIS platform uses Infolets as proxies to communicate with these systems. Ideally, our platform should use a uniform interface to communicate with any of such systems and dynamically discover and integrate with services from them. In this regard, industry standards in Web Services [1] are likely to play a significant role to achieve this capability.
- Quality of Service (QoS) issues: As mobile services are extended to more users and information is more dynamically produced, disseminated, and consumed, effective QoS management is essential to deliver the most relevant and time critical information to users. QoS management is a set of functions and mechanisms to optimize resource utilization to differentiate information processing on the basis of QoS characteristics such as priority, criticality, and timeliness. Our current implementation of mobile EIS uses a Point-to-Point style of messaging. Load balancing is the only QoS management function supported. As information becomes more dynamic, our mobile EIS must support the Pub/Sub style of messaging and a complete set of QoS management functions.
- Efficiency issues: As each round of sending and receiving messages between mobile devices and enterprise systems goes through multiple types of networks and several hosts, the amount of information in the messages should be large. In other words, it would not be efficient if it takes several rounds of messages to complete a unit of work. There are two types of solutions to the efficiency issues. One is content caching, where data that a user may need is stored closer to the user. For example, a directory search could yield a long list of results. The results could be cached in a web server,

which is the closest host to the user, so that results do not have to be fetched from the directory server deep in the enterprise as the user demands to see more results. Another is defining coarse-grained messages so that one message carries out a unit of work. For example, a flight-planning task is a unit of work that requires aggregating information from multiple services including airport service, weather service, flight manual service, and map service. It would not be efficient if request messages have to be sent to each of these services separately from the user.

Advanced applications such as Aircraft Maintenance also require the support of business processes requiring efficient handling of "stream queries" and "migrating transactions" [7]. Similarly, other applications require the support for other kinds of "stream" and "continuous" queries [8].

Our future research directions are to improve our mobile EIS platform to address these issues.

12.8 Summary and Conclusions

In this chapter, we have presented the architecture and its middleware solution for mobile enterprise information services (mobile EIS). We discussed the wireless network technologies and middleware technologies related to mobile EIS. Enterprise application integration (EAI) is a driving requirement for integrating mobile devices with enterprise information systems. Component-based application servers such as those based on J2EE or .Net are essential middleware platforms to EAI.

In the context of EAI, we outlined the enterprise requirements, design considerations, and approach to our architecture design for mobile EIS. Key features and their benefits are described. The usage of Java Message Service (JMS) in our mobile EIS is described in detail. In particular, we described the features and programming models of JMS and the rationale and usage of Point-to-Point style of messaging in our current implementation. Content tailoring to devices of different capabilities is critical to present information to users according to physical constraints of devices. We presented our solution, which uses XML-based content representation and XSLT templates. Finally, we discussed open issues and our future directions in designing and developing mobile EIS.

Middleware for mobile EIS must support business objectives and enterprise requirements. While the J2EE-based middleware platform provides adequate features for developing software solutions to mobile EIS, architecture design is the key to meet those challenges we outlined from the beginning. Those challenges include the variation of devices, unpredictable connectivity, diverse enterprise information systems, and different types of data contents. Furthermore, enterprise requirements on security, extensibility, simplicity, dependability, and scalability are important design constraints. Our approach is an integrated service-based architecture. We selected Oracle9iAS as our provider of J2EE-based middleware platform for implementing our architecture.

Components in our architecture are designed on the basis of the separation of concerns principle and have well-defined roles, responsibilities, and quality attributes. We fully utilized J2EE-based middleware features for integrating these components and integrating our mobile service platform with the rest of the enterprise systems. In particular, we used gateways as an abstraction to handle various types of devices connecting to our mobile EIS platform via various network protocols. We used Infolets as an abstraction to integrate with diverse types of backend enterprise information systems. Gateways and Infolets are decoupled by reliable message queues on the basis of JMS. We believe JMS-based messaging system provides mature and adequate functions to our mobile EIS.

Our solution to mobile EIS including the integrated architecture and application of middleware technologies seamlessly integrates mobile clients with enterprise systems. The solution allows us to decouple issues of interacting with diverse types of mobile devices from issues of integrating with a variety of enterprise systems (e.g., ERP, email, and employee services). It satisfies our enterprise requirements for designing and developing mobile EIS discussed in the chapter. Among the challenges of mobile EIS, our solution handles the variation of devices, unpredictable connectivity, and diverse enterprise information systems well. Its ability to tailor different types of data contents to a variety of devices (another challenge) is adequate for the case of transforming engineering drawings for presentations to devices of different sizes and resolutions. How good the presentations are, that is, the usability of the presentations from the end user's perspective is yet to be studied. Content tailoring to device profiles is one of the future directions we are pursuing. Other future directions include Security, Integration, QoS, and Efficiency.

12.9 Acknowledgment

We thank Yih-Farn R (Robin) Chen, Huale Huang, Rittwik Jana, Serban Jora, Radhakrishnan Muthumanickam of AT&T Labs - Research for their collaboration in developing the Mobile EIS architecture and implementation described in this paper.

Bibliography

[1] Newcomer, E. (2002) *Understanding Web Services: XML, WSDL, SOAP, and UDDI*, Pearson Education.

[2] Sun Microsystems, Java 2 Platform, Enterprise Edition (J2EE), http://java.sun.com/j2ee/, May 2003.

[3] Ferguson, D. and Kurlander, D. (2001) *Mobile.NET*, APress.

[4] Szyperski, C. (2002) *Component Software: Beyond Object-Oriented Programming*, Addison-Wesley Longman.

[5] Chen, Y., Huang, H., Jana, R., Jim, T., Hiltunen, M., Muthumanickam, R., John, S., Jora, S., and Wei, B. (2003) iMobile EE – An Enterprise Mobile Service Platform. *ACM Journal on Wireless Networks*, **9**(4), 283–297, 2003.

[6] Chen, Y., Huang, H., Jana, R., John, S., Jora, S., Reibman, A., and Wei, B. (2002) Personalized Multimedia Services Using a Mobile Service Platform. Proceedings of WCNC 2003, Orlando, FL, March 2002.

[7] Sripada, S. M. and Moeller, P. (1994) Query Processing and Transaction Management in Mobile Environments, in *Proceedings of MOBIDATA NSF Workshop on Mobile and Wireless Information Systems* (eds T. Imielinski and H. F. Korth), Rutgers University, New Brunswick, NJ.

[8] Arasu, A., Babcock, B., Babu, S.,Datar, M., Ito, K., Motwani, R., Nishizawa, I., Srivastava, U., Thomas, D., Varma, R., and Widom, J. (2003) STREAM: The Stanford Stream Data Manager. *IEEE Data Engineering Bulletin*, **26**(1), 19–26.

[9] Mahmoud, Q. H. and Vasiu, L. (2002) Accessing and Using Internet Services from Java-enabled Handheld Wireless Devices: A Mediator-based Approach. *Proceedings of the 4th International Conference on Enterprise Information Systems*, Ciudad Real, Spain, April 2002, pp. 1048–1053.

[10] Mahmoud, Q. H. (2001) MobiAgent: An Agent-based Approach to Wireless Information Systems, in *Agent-Oriented Information Systems* (eds G. Wagner, K. Karlapalem, and Y. Yu. E. Lesperance), iCue Publishing, Berlin, Germany.

[11] Mascolo, C., Capra, L., and Emmerich, W. (2002) Mobile Computing Middleware, in *Networking 2002 Tutorial Papers* (eds E. Gregori, G. Anatasi, and S. Basagni), LNCS 2497, Springer-Verlag, Berlin, Germany.

[12] Hwang, Y., Seo, E., and Kim, J. (2002) WebAlchemist: A Structure-Aware Web Transcoding System for Mobile Devices. *Proceedings of Mobile Search Workshop*, Honolulu, Hawaii, May 2002.

[13] Phan, T., Zorpas, G., and Bagrodia, R. (2002) An Extensible and Scalable Content Adaptation Pipeline Architecture to Support Heterogeneous Clients. *Proceedings of the 22nd International Conference on Distributed Computing Systems*, July 2002, Vienna, Austria.

[14] Sun Microsystems, Java Message Service (JMS), http://java.sun.com/products/jms/, May 2003.

[15] W3C, *Extensible Markup Language (XML) 1.0*, Second Edition, W3C Recommendation, http://www.w3.org/TR/REC-xml, May 2003.

[16] W3C, *XSL Transformation*, W3C Recommendation, http://www.w3.org/TR/xslt, May 2003.

13

Middleware for Location-based Services: Design and Implementation Issues

Peter Langendörfer, Oliver Maye, Zoya Dyka, Roland Sorge, Rita Winkler, Rolp Kraemer
IHP, Im Technologiepark

13.1 Introduction

The implementation of distributed applications is a tedious and error-prone task. In order to simplify it and provide suitable platforms, a lot of research was done during the last 10 to 15 years. These platforms provide an abstraction from details such as semantics of RPCs, marshaling, and service discovery. CORBA [1], DCOM [2], and Java RMI [3] represent relatively mature and well-understood technologies. All these approaches target more or less all-potential applications.

A special class of applications are location-aware services. In order to provide a benefit to its users, these services require information about the current position of the user, about the profile of the user, and the capabilities of their mobile device. When it comes to the implementation of such a service, the application programmer has to deal with around a dozen different mobile terminals, nearly a handful of different positioning systems, and three or four different protocols for wireless communication. In order to simplify the development of location-aware services, specialized platforms that provide the following components are needed:

- Profile handling
- Location handling
- Personalization
- Event handling.

Middleware for Communications. Edited by Qusay H. Mahmoud
 ISBN 0-470-86206-8

In addition, a concept that allows the setting up of an infrastructure that runs platform components is needed. The design of the components and of the infrastructure concept influences significantly the efficiency and reliability of applications running on top of the platform. Thus, all design decisions have to be made carefully, for example, whether to use replication to improve performance and how to ensure consistency in these cases.

In this chapter, we present our own platform called PLASMA (Platform Supporting Mobile Applications). It consists of components for location and event handling, as well as a component for profile handling. The event-handling component provides its own event-definition language that can be used to realize push services or to define user-related events. This central component has a very good performance. Each of its instances can handle up to 1000 events per second. In order to ensure good performance of PLASMA, replication and filtering mechanisms are used to minimize internal communication and to minimize the effort in event evaluations.

This chapter is structured as follows. We start with a detailed view of PLASMAs architecture: infrastructure features, platform components of the client and the server side. Then we present the concepts behind selected platform components. Performance measurements and guidelines for deployment of an efficient PLASMA infrastructure are provided in Section V. Thereafter, we report on lessons learned from our initial approach and from the implementation to the current state. We will conclude with a short summary of the major results and an outlook on further research steps.

13.2 Related Work

In the literature, several approaches for the design of a service platform have been reported.

In [4], a platform called MASE was designed to support mobile users. It supports the development of applications such as city guides, mobile multimedia services, and so on. Features such as tracking of a mobile user and the finding of locally available services are not supported by this platform.

The AROUND project [5] proposes a discovery mechanism for location-based services along with an interesting proximity model that introduces spatial relations such as "is included in" or "is adjacent to".

An important aspect of a location-aware service platform is keeping track of mobile objects in a scalable way [6]. One approach for locating objects in a worldwide distributed system has been worked out within the Globe project [7]. The concept is based on a worldwide distributed search tree that adapts dynamically to an object's migration pattern to optimize lookups and updates. The main problem with this approach is the root of the tree, which has to maintain $10^9 - 10^{12}$ objects [8]. Another approach [9, 10] uses several centralized registers in which each object can be found. But to be able to find an object inside this register, it must be updated in proportion to the movement of the user. This leads to a large burden on the communication in the fixed area network, which should be avoided when several million objects must be observed.

Leonhardi [11] proposes a generic and scalable location service as one of a couple of infrastructure components supporting the development of mobile applications. A thorough analysis of various protocols for updating a mobile's location information can be found in [12].

A context and location-aware approach for support of tourist information system is presented in [13]. The interesting aspects here are that they rely fully on Web technology, that is, UDDI [14], SOAP [15], and WSDL [16], and that the context-aware content is compiled on the fly when the request arrives. Performance issues are not addressed in [13].

In [17], the roles of a proxy in location-based services is investigated. Consequences of placing a proxy between the client and a server-based infrastructure in terms of protection of personal information are elaborated. Furthermore, the authors introduce a privacy model to personal location information.

13.3 Architecture

The major design goals of PLASMA are adaptability, scalability, and openness. PLASMA should be equally suitable for small hot spots such as shopping malls or train stations as well as worldwide deployment. An example for the latter case could be if a certain airline wants to equip all its lounges with PLASMA.

PLASMA consists of a set of platform components called *engines* (which provide the basic functionality such as managing the location information of mobile clients, etc.) and a set of so-called PLASMA servers (which provide management services inside the infrastructure).

In order to achieve scalability and adaptability, we developed an architecture that allows to replicate and distribute PLASMA servers over several server machines or to combine them on a single server machine, respectively. In order to achieve openness, we realized a so-called mediator, which is used to communicate with clients that do not run PLASMA. The PLASMA architecture, its servers, and management are described in Section 13.3.1. The basic mechanisms needed to implement location-aware services are provided by seven platform components that are presented in Section 13.3.2.

13.3.1 Infrastructure

PLASMA may be deployed ina hierarchical structure and it allows replicating the infrastructure servers. The tree structure allows to split geographical regions with high density of mobile devices into several new geographical regions, so that the load per infrastructure node is reduced. On one hand, this approach provides simple means to do efficient load sharing. On the other hand, in a worldwide deployment scenario, a single tree may become a bottleneck, since its root has to keep track of up to worldwide 10^9 to 10^{12} objects. In order to avoid this drawback, PLASMA may be deployed with several independent trees. With this approach, it would be impossible for two mobile devices to find each other if they are located in different trees. In order to support worldwide search facilities, we introduced the object register that stores for each object the information in which tree it is currently active. This concept is very similar to those of the home agent of mobile IP [18] or the home location register of Global System for Mobile Communications (GSM) [19].

PLASMA allows to introduce several hierarchical levels between the root of the tree and its leaves. It is even possible to have no intermediate levels. This approach was also introduced in order to allow easy clustering of geographical regions when load sharing should be used. Our experiences show, however, that flat structures should be preferred at

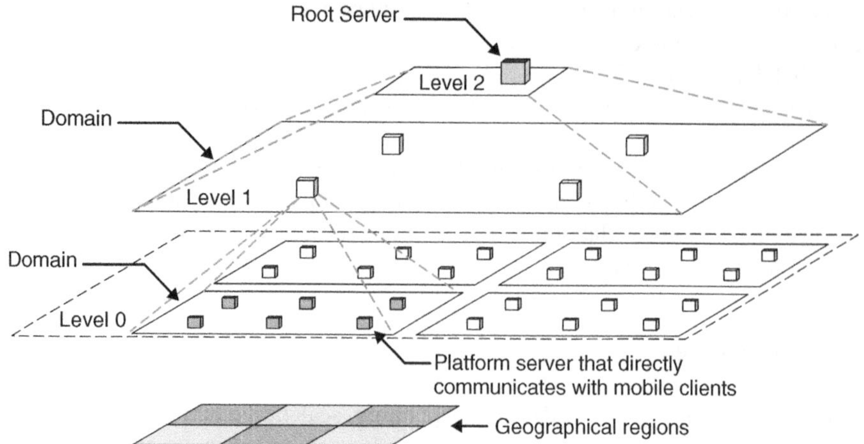

Figure 13.1 Hierarchical structure of PLASMA's infrastructure

least in local deployments (see Section V.B). To summarize, PLASMA provides a whole set of parameters that can be used to achieve good scalability.

One example for a hierarchical structure is shown in Figure 13.1.

The platform servers in the lowest hierarchy level (i.e., level 0 in Figure 13.1) are also referred to as *Leaf Servers*. They directly communicate with mobile clients. Each of these platform servers is responsible for a particular geographical region. So, clients inside a certain geographical region communicate with the server that is responsible for that region. The geographical regions can have any shape, but they must be disjunctive and must completely cover the area in which the platform services should be available. The size of the geographical regions can be adapted in such a way that the entire workload is homogeneously distributed between all platform servers. For hot spots such as city centers, convention domes, and so on, with many mobile users, small geographical regions are suitable, whereas for rural areas large regions are adequate.

A platform server in a higher hierarchy level is called *Domain Server*. It is the communication partner for a certain number of servers in the next lower level. Thus, the geographical region of a server in level 1 is the combination of the geographical regions of its child nodes in level 0. This concept spans a pyramid of platform servers, each responsible for one geographical region and child node servers, respectively.

The top of this pyramid is referred to as *Root Server*. The entire platform can consist of more than one root server, that is, it is built up out of more than one hierarchical structure. The number of root servers as well as the number of hierarchy levels in a pyramid, the number of platform servers in a level, and the size of geographical regions can be adapted according to the requirements of a particular deployment scenario.

In order to store user-specific data that is relevant to the middleware, profiles are assigned to each mobile user. A profile contains information on the user's terminal as well as privacy-related settings, such as a flag indicating whether the user's position is to be exposed to other users. More than that, possibly defined auras and events are all stored in the profile. These user profiles are managed by so-called *Profile Servers* . Each profile is stored persistently at exactly one Profile Server, making this server the authoritative original source for that profile. Of course, a single Profile Server may store several, even thousands of profiles. One of the tasks of PLASMA is to provide multiple copies of the same profile at different locations (i. e., to different platform servers) in a consistent way (see Section IV.B).

Deployment scenarios may use several profile servers or even several PLASMA trees. In order to support an easy and efficient way to identify the correct profile server for a given mobile client, we use another platform entity called *Object Register*. It forms, in some sense, the administrative top of the platform. The object register provides a mapping of an object's id to the address of the object's profile server. Moreover, also the reference of the current root server can be requested from the object register. This data is important when searching for the location of a certain object. The "correct" object register can be identified from only knowing an object's id, as the contact address of the register is part of each object's id. Thus, the object register can be used to connect several PLASMA trees.

Since our platform is based on Java, any client that wants to use the service offered by the platform should have a Java virtual machine running on it. Very thin terminals such as mobile phones or other devices, which do not have a Java virtual machine (JVM), cannot use the service. The need to use Java as programming language and the need to run a JVM on the mobile device significantly reduces the number of potential applications and PLASMA clients, respectively. In order to enlarge the number of potential clients and applications, we decided to introduce a so-called *Mediator*, which realizes a gateway between the Java and the non-Java world. We selected Simple Object Access Protocol (SOAP) as communication protocol between the Java and the non-Java world. The benefit of this approach is that SOAP can be used to realize remote procedure calls between any two programming languages. Therefore, we need only one kind of mediator instead of one per programming language. The major PLASMA APIs are described in Web Services Description Language (WSDL) [16] so that interfaces for different programming languages can be generated automatically. Basically, PLASMA can be used as a huge WebService [20] that is invoked by sending SOAP requests to the Mediator. The latter is really implemented as a web service, which exposes the PLASMA API to the non-Java world. It translates SOAP calls into the PLASMA internal protocol. It is connecting to and interacting with PLASMA servers like a single Java-Client (see Figure 13.2).

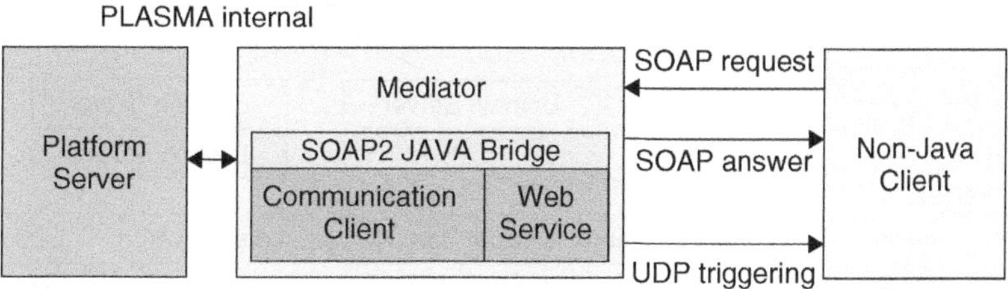

Figure 13.2 For the support of non-Java clients, the mediator translates the protocol for the platform communication into SOAP calls and vice versa. PLASMA events are sent as UDP packets to the non-Java clients

The major part of the methods provided by the mediator works in a synchronous way, that is, the calling instance on the mobile blocks until the result is received. The benefit of this approach is that no resources are wasted for polling in order to check whether the result is delivered. But this concept does not work for asynchronous scenarios in which PLASMA sends a message to the mobile client, for example, if one of the events defined by the client has become true. On a Java client running the PLASMA proxies, PLASMA directly calls internal functions of the client. This is not possible with a non-Java Client that is realized as a WebService Client. One possible solution is to run a SOAP listener and parser on a non-Java Client in order to be capable to receive messages from the infrastructure. But this would increase the load on the mobile client drastically. In order to reduce this overhead, we decided to use non-SOAP solution for the messages that are sent from the infrastructure to the client. The major part of these messages, for example, information about a landmark, is useful only for a very short time, that is, the loss of such a message does not cause much harm. Thus, a reliable transport protocol such as TCP is not really needed. Therefore, we decided to minimize the communication overhead and to use User Datagram Protocol (UDP) to send events from PLASMA via the mediator to the client. If the client has registered for PLASMA events, it opens a UDP socket and listens on a predefined port. The event can then be handled by the application running on the mobile client in a suitable way.

Since PLASMA provides a WSDL API, application programmers are not bound to any programming language or operating system running on a mobile client.

13.3.1.1 Platform Management

PLASMA allows to configure which of the above-mentioned servers are clustered together on a single machine as well as on which machine they are installed. Thus, it is possible to run all servers on a single machine (see Figure 13.3a) in order to provide PLASMA functionality in a small hot spot. If a large area is to be covered PLASMA, servers may be structured as a tree and the PLASMA servers may be distributed over the whole tree. This allows solutions such as a single object register for two or more physically separated hot spots (see Figure 13.3b). This flexibility opens up a new problem.

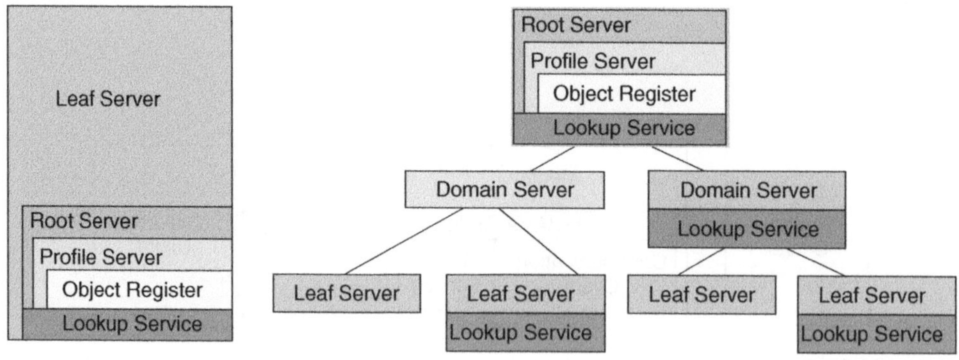

Figure 13.3 Two sample deployment configurations of the PLASMA platform; a) single server b) hierarchically structured

In order to provide the platform functions in a correct and consistent way, the platform servers need to have certain items of knowledge about the infrastructure:

- For searching mobile clients as well as for the handover mechanism, the platform server needs to know all its neighbors, that is, leaf servers with geographical regions that are adjacent to its own region.
- For the search mechanism, it needs to know the platform server in the next hierarchy level, which is responsible for the domain to which the platform server itself belongs.
- In order to create a new object, it needs to know at least one object register.
- To store and retrieve object profiles, it needs to know at least one profile server.

Since the PLASMA infrastructure can be deployed in many different configurations, the above-mentioned information has to be provided dynamically, that is, the platform servers need a way to retrieve all necessary information independent of the actual setting. Therefore, we decided to use Jini lookup services [21] to provide and maintain this information. For each domain, one logical lookup service exists. Each lookup service contains the following information:

- All PLASMA servers in the domain of the Jini lookup service, including the geographical region belonging to each server.
- The addresses of object registers available in this domain.
- The addresses of data bases available in this domain.
- Whether the domain is a search domain or not.
- How to contact a lookup service of the next hierarchy level upwards.
- How to contact a lookup service of the next hierarchy level downwards, which is responsible for a particular domain.

So, every platform server can access the information necessary to contact other PLASMA servers, such as profile servers from the lookup services. The lookup service in the top hierarchy knows all root servers with the corresponding lookup services in the domains directly below the root servers.

An automatic reconfiguration of PLASMA is supported by the fact that all platform servers register appropriate lookup services, requesting to be notified if changes in the platform structure occur.

13.3.2 Platform Components

In order to provide useful location-aware applications, it is necessary that the current position of a user can be queried by the application. This means a suitable positioning system has to be part of the infrastructure and must be able to submit its information to PLASMA. This information has to be processed in a suitable manner. In addition, potential users require the possibility to define areas of interest (which we call *aura*) as well as the means to define events (such as "please notify me when a color printer is less than 5 m away"). In order to support large-scale deployment, a handover mechanism is needed. Thus, the platform consists of the following components:

- sighting proxy
- database engine

- event engine
- auras & objects engine
- communication engine
- lookup engine
- profile engine
- handover engine
- security & privacy engine.

All these components are used on the infrastructure side. Java clients only need to implement the communication engine as well as proxies of the Location Management API, Lookup service API, and Profile Database API. Non-Java clients do not need platform-specific components at all. The capability to process SOAP requests is sufficient. Thus, applications can use well-defined APIs provided by the components of the platform.

13.3.2.1 Platform Components of the Infrastructure Side

In this section, we present all platform components that have to be deployed on an infrastructure side. Figure 13.4 shows all components of the platform that are realized in the infrastructure.

- *Sighting proxy* : The sighting proxy receives the position information of registered clients. It converts the position information and the client ID provided by the sighting system into geographical coordinates and object ID that are required by the platform.

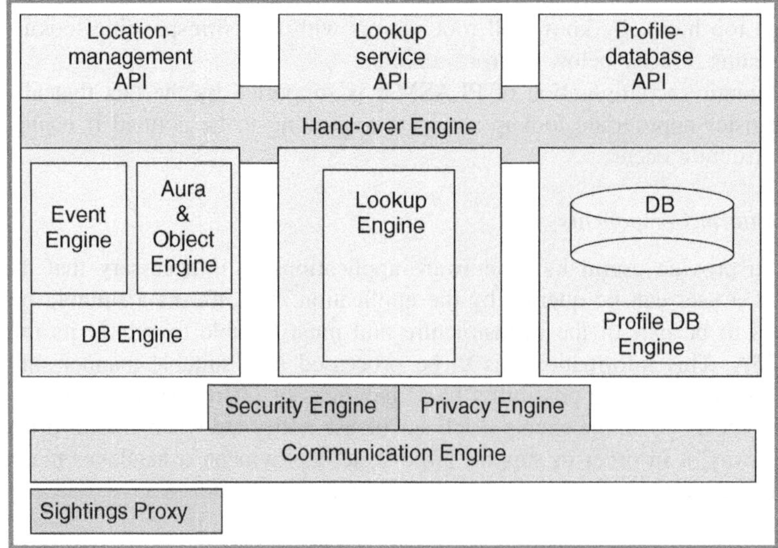

Figure 13.4 PLASMA components and their structure inside a platform server.

Then, sighting information is forwarded to the database engine. The position information may be delivered by any positioning system, for example, by GPS in outdoor scenarios or by IR beacons in indoor scenarios. The sighting proxy resides only on platform servers of the lowest hierarchy level, since only these servers communicate with mobile clients and receive sightings from an underlying sighting system.

- The *database engine* (DB) stores the sighting information and provides it to other platform components and to the application upon request. More than this, significant sightings, for example, an object's first sighting ever or a sighting that is relevant for further event processing, are automatically forwarded to the event engine. This filtering dramatically reduces the communication effort inside a platform unit.
- The *auras & objects engine* handles an aura, which is the spatial scope of interest for an object. For instance, a user can inquire all objects that are currently located in a particular aura, or all objects that have auras overlapping with a particular aura. To perform the queries of users, the engine for auras and objects uses rendering algorithms. Moreover, with the concepts of auras and objects, events such as "object A entered aura X" can be defined. Thus, the purpose of the auras & objects engine is to calculate which objects are located in which aura and to determine which auras intersect each other.
- The *event engine* is responsible for interpreting and monitoring user-specified events. It has to process all sighting notifications coming from the DB engine in order to observe and possibly fire user events. For example, to find out, if two persons are near each other, it deploys the auras & objects engine that is able to determine if one position is within a certain aura of another one. Our event-definition language as well as the event-evaluation process is explained in detail in the selected components section.
- The *communication engine* handles target references, which are used for the communication between clients and platform servers and for platform-internal communication. In fact, a target reference is made up of the machine's IP address and the number of the port used by PLASMA. The communication engine is also responsible for an initial communication establishment for newly appearing devices, for example, when powered on.
- *Lookup engine*: The platform supports queries on the Jini lookup system. Users can register their own services with this lookup system or can just use services offered by other devices. Since the registration of services with a Jini lookup service is done via the platform, locations and auras can be defined for the services. The platform search mechanisms can be used to find services. The lookup engine at the server side handles the queries and sends appropriate objects to the client. Moreover, the lookup engine also supports the platform management, for example, by finding different platform components such as next neighbor platform servers, object register, profile server, and so on, to organize the communication within the platform.
- The *profile engine* is responsible for the support of personalization. Each mobile terminal can provide a profile to the platform that contains some common, platform-related properties of the terminal and the user. For example, capabilities of the terminal, as well as user preferences such as security-related information could be specified in this profile. This profile is stored in the profile database. The primary copy of the profile is managed on the profile server responsible for a certain object. In order to speed up applications, local copies of profiles are needed. Our replication and consistency model is explained in the following section.

- *Handover engine:* When a mobile object leaves a geographical region and enters a new one, the handover engine transfers all object-related data (communication references, auras, profile, registered events) from the old leaf server to the new one.
- *The security & privacy engine* provides means to protect user data against misuse. Its negotiation mechanisms enable the mobile device to select a cipher mechanism for the communication with the infrastructure [22]. The selected algorithms are used to encrypt the messages exchanged with the infrastructure as well as to provide digital signatures. Thus, eavesdropping can be avoided. Data stored in the infrastructure are protected via access rights that can be changed only by the users themselves. That is, they can authorize selected services or users to access their data, for example, to allow these to localize them.

13.3.2.2 Platform Components on the Client Side

We distinguish between Java-enabled mobile clients (JEMOC) and SOAP clients (SOC). When a new object is created for the service platform, the CreateObject function creates a standard profile for this object. The type entry of the standard profile is set to JEMOC, and for SOCs, this is changed into SOC.

JEMOCs unit implements a proxy for the platform functions, which are available on the server side. Requests are sent to and received from, respectively, the leaf server responsible for the client's current location. SOCs even do not implement Java Proxies. Applications on SOCs communicate directly with the mediator, which is responsible for the client's current position.

Both client types are active with respect to the sighting system by reporting their current position to the sighting proxy.

13.4 Concepts of Selected Components

13.4.1 Event Engine and Auras and Objects Engine

The event engine supports the definition and evaluation of distributed events. It takes into account the current position of all objects involved in the event. This holds true also for the auras defined for those events. This evaluation scheme may interfere with the privacy needs of the relevant users. The users are in full control over their data and thus they might have decided to be invisible, that is, nobody may see their current position. From the viewpoint of the event engine, this means that the current position of the user is unknown. In order to allow a proper definition and evaluation of events, we defined an event-definition language based on tri-state logic.

13.4.1.1 Defining and Processing User Events

Any user may register with the platform to be notified at certain events, say, when he/she or another user reaches a certain location or a certain object. As outlined below, this kind of interaction between the platform and the client utilizes the implemented event management.

Often, a running mobile client generates sightings, consisting of information on its identity and its current location. That location information must be reported to the platform,

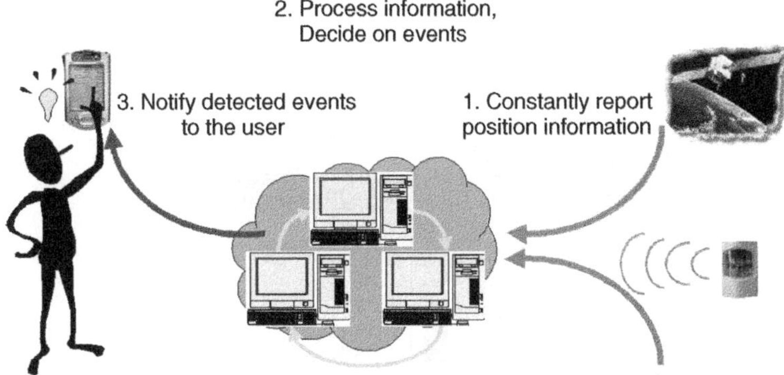

Figure 13.5 Data flow between positioning systems, the PLASMA infrastructure, and mobile clients

processed appropriately. It may be combined with other position data, and as a result, the platform decides whether a certain user event has become true. If so, it fires a notification to the end user. Figure 13.5 illustrates the event processing.

The location information emitted by the user's device is transmitted to the sighting proxy of the leaf server that is responsible for the geographical region covering the user's position. The whole processing of the position information is event-driven (see Figure 13.6). The platform only acts on whatever action triggers it. This could be the movement of an object, for example.

If an event is defined on a certain object, this means that the user wants to be notified when the object is inside a certain area. In this case, the event engine, which is the module designed to manage user events, registers with the underlying DB engine for the movement of the object. When the object moves, its sighting is received by the

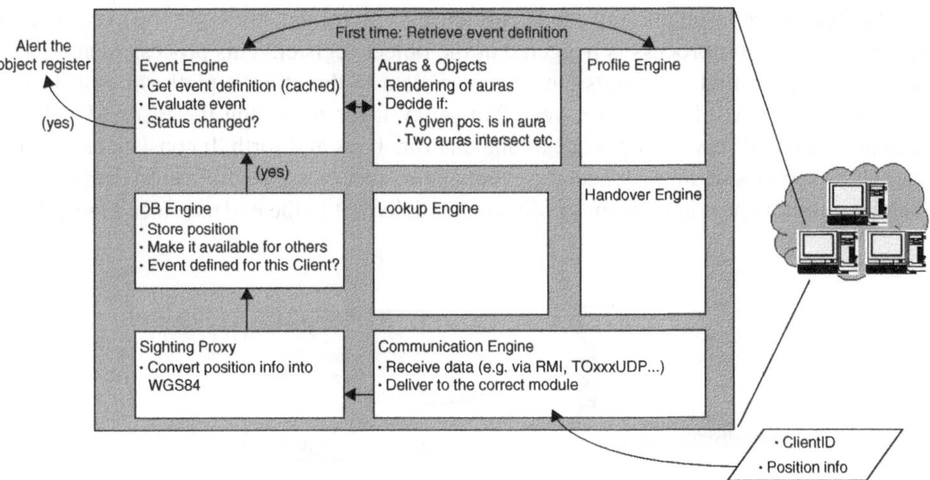

Figure 13.6 Processing of sightings inside a leaf platform server

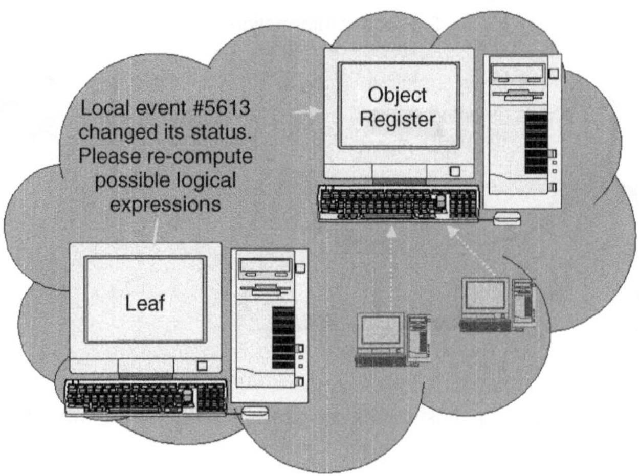

Figure 13.7 Signaling the new status of atomic events to the overall management unit at the object register

communication engine and forwarded to the sighting proxy. Here, the data are converted as necessary and forwarded to the DB engine. The DB engine checks whether the event engine should be notified and, if yes, triggers the event engine to check the corresponding event condition. If the condition is met, the notification process will be triggered. In this way, the DB engine filters sightings that are not related to an event.

A user may have registered to be notified upon a combination of several atomic events, for example, "if person A meets B AND C is at D." To handle this, in the next step the new status of an atomic event is signaled to the object register's event engine. Here, the manager for the overall event resides. This manager is responsible for combining the atomic events according to the definition and deciding whether the whole event has become true. This process is depicted in Figure 13.7.

If the logical recomputation of the event-definition term indicates that the event is to be fired, the notification process is triggered at the object register. This process identifies the end user to be notified and sends the necessary data to the client's platform unit. In fact, the client-side PLASMA unit is a very thin proxy layer providing the PLASMA API to the application and forwarding requests and answers back and forth. It consists of the same modules as server-side unit. Hence, the event engine receives the notification data and calls a handling routine provided by the application to deliver it to the end user (see Figure 13.8).

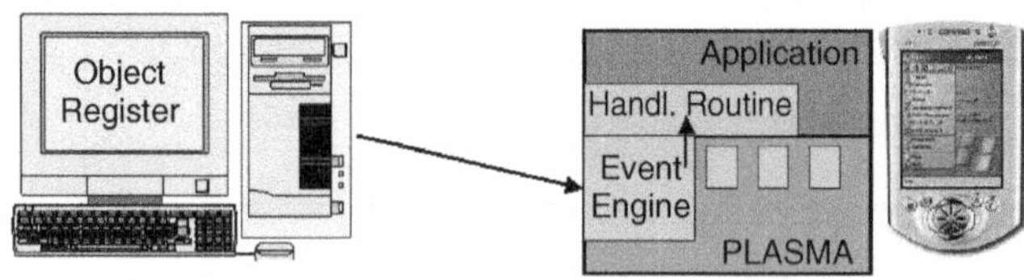

Figure 13.8 Notification of the application on the mobile client about an event that has become true

Event-driven processing is not only used for user defined events but also for other platform operations. In general, components of the platform register with other components to be notified when for the component relevant data, (aura, profile, etc.) have changed. We use event-driven processing to save processing time and reduce efforts in communication, thus keeping the workload of a platform server as low as possible.

13.4.1.2 Aura –The Spatial Scope of Interest

The minimal information that is needed to realize a location-aware service is the current position of the user. It can be determined by positioning systems (indoor and outdoor) or by using the cell identifier in which the mobile has been sighted. Using this information, location-aware applications such as eGraffiti can be realized [23]. When it comes to the design of more sophisticated location-aware services, additional information about the surrounding of the mobile and the geographical scope of interest is required.

This kind of information can be provided on a physical basis as well as on a logical basis. The latter is used in the approach described in [5]. The major drawback of this approach is that areas of interest cannot be specified in a simple way and that they have to be adapted to each new environment. Therefore, we introduced the concept of auras and objects. An aura is a spatial scope of interest. It can have any number and kind of dimensions and defines a space of interest for an object (see Figure 13.9). Since it takes only the distance to the current position into account, it can be used in any new environment without requiring adaptation. Each object can have several auras, which are totally independent from each other. In our model, an object can be a physical entity like a printer or logical object. Logical objects are not predefined but can be specified by the application designers. A service, for example, which is only available in a specific room, is a logical object that has an aura that is identical to this room.

For instance, a logical object could be an information service for a building, fair booth, and so on. The aura of this information service could be identical to the physical dimension of the building or fair booth. A mobile client could request all objects whose auras overlap with the mobile client's aura.

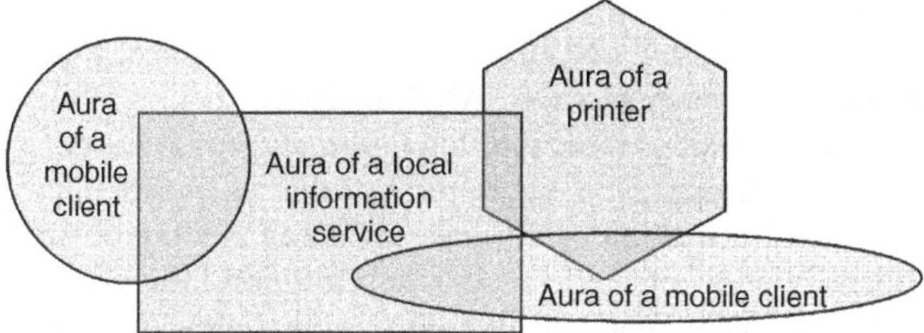

Figure 13.9 Samples of auras with different shapes and different owners. Reproduced by permission of Springer, from: "A Scalable Location Aware Service Platform for Mobile Applications based on Java RMI" Drogehorn, Olaf, Singh-Kurbel, Kirti, Franz, Markus et al. taken from Linnhoff-Popien, Claudia, Hegering, Heinz-Gerd (eds.) Trends in Distributed Systems. Towards a Universal Service Market, Third International IFIP/GI Working Conference, USM 2000, Munich, Germany. Proceedings. 2000 Pitt3-540-41024-Keller4 (Lecture Notes in Computer Science 1890), pp. 296–301.

On the basis of this approach, it is the task of rendering algorithms to check whether an object or an aura crosses another aura or another object. Which properties these logical objects can have is not specified inside the service platform but in the definition of each application.

With this concept, we are in the position to create a large range of innovative applications. For example, the users can define their own events such as "please inform me if my colleague is around." Within our approach, they can define what "around" means, perhaps each position that is less than 5 m away is considered to be within their aura, and each time one of their colleagues approaches them, they will receive the corresponding event. On the basis of this information, several useful actions can be triggered, such as automatic identification of a free meeting room, ordering some coffee, and so on.

13.4.1.3 Event-Definition Language

In order to enable end users as well as application programmers to efficiently define events, we have defined our own event-definition language. The users can describe the event upon which they want to be notified by means of that language. The difficult part is to find an event-definition language that is easy to learn from a user's perspective and which has a sufficient expressiveness. In natural language, events such as *"A meets B"* or *"A is near B"* can be expressed easily. But for automated evaluation, they have to be formalized with respect to what *"meets"* and *"is near"* really means. Thus, in a first step, we mapped these event definitions to the more formal expression: *"The position of A is inside aura No. X of B"*. Such events can be expressed using the following syntax:

`{A – ObjectID} ENTERS {AuraNum} OF {B – ObjectID}` or
`{A – ObjectID} LEAVES {AuraNum} OF {B – ObjectID}` respectively.

In order to realize these events, the two language elements "ENTERS" and "LEAVES" have been defined. An event using "ENTERS" becomes true if and only if A was previously outside the given aura of B and is now inside, whereas "LEAVES" fires when A was previously inside B's aura and is now out.

Each event can be prefixed by a NOT operator to revert the meaning. Two events can be combined by the logical operators OR, AND, or XOR.

The state of events is undefined if the position information of at least one of the involved objects is not accessible. This can be due to the fact that the position information of that object is not available because the users have switched off the device, or since their privacy policy does not allow others to access their position information. To handle such situations correctly, a tri-state logic was introduced into the event-evaluation algorithms. Therefore, the result of an evaluation can either be TRUE, FALSE, or UNKNOWN. Consequently, the semantic of the logical operators mentioned above were redefined accordingly. Table 1 to Table 3 show the definition of the tri-state logic for the operations OR, AND, and XOR.

Table 13.1 Tri-state logic definition for logical OR

OR		B		
		TRUE	FALSE	UNKNOWN
	TRUE	TRUE	TRUE	TRUE
A	FALSE	TRUE	FALSE	UNKNOWN
	UNKNOWN	TRUE	UNKNOWN	UNKNOWN

Table 13.2 Tri-state logic definition for logical AND

AND		B		
		TRUE	FALSE	UNKNOWN
	TRUE	TRUE	FALSE	UNKNOWN
A	FALSE	FALSE	FALSE	FALSE
	UNKNOWN	UNKNOWN	FALSE	UNKNOWN

Table 13.3 Tri-state logic definition for logical XOR

XOR		B		
		TRUE	FALSE	UNKNOWN
	TRUE	FALSE	TRUE	UNKNOWN
A	FALSE	TRUE	FALSE	UNKNOWN
	UNKNOWN	UNKNOWN	UNKNOWN	UNKNOWN

The state UNKNOWN requires a special handling of the NOT operator, since NOT UNKNOWN definitely may not result in TRUE or FALSE. Therefore, NOT UNKNOWN is defined to be UNKNOWN.

13.4.2 Profile Servers and Profile Database

Profile servers form the interface between other platform servers and the profile database. Their responsibility is to assure nonconflicting access (retrieval, modification, creation, and deletion of profiles) to the profile database as well as to notify platform servers of profile updates. The latter implies interaction with the database engine in order to determine the platform server where a certain object can be found. The profile server must keep track of all objects requesting update notifications.

In addition to the actual profile information, the profile database contains a timestamp that indicates when the profile was last modified. This timestamp serves as a basis to decide if a client initiating a profile modification has an up-to-date version of the profile and is thus allowed to modify the profile. Every time a server requests a profile, the timestamp will be returned as well.

A profile server accesses its profile database via the interface provided by the database engine (which in turn encapsulates a JDBC driver). In order to identify the profile server hosting the database for an object's profile, the object register contains a reference to this profile server for each object.

13.4.2.1 Basic Mechanisms

Object profiles are stored on one or more profile servers. These servers can reside on the same node as the platform server or object register. The information indicating which profile server has the profile of a given object is stored in the object register entry of that object. The object register managing a certain object can be identified by the object ID, which consists of the IP address of the object register and a unique identifier of the mobile device.

An object's profile is frequently accessed by the leaf platform server where the object currently resides. The profile is needed to perform various algorithms (event handling, handover, etc.). To reduce communication between leaf servers and profile servers, a copy of each object's profile is kept at the current platform server. The platform server requests this replica from the profile server when the object is sighted for the first time.

In the simplest case, the replica can be read in a one-time read operation, also called "read once" or "dirty read". With this, the platform server gets an up-to-date copy of the requested profile, which is valid at the time of the retrieval, but no further guarantees can be made. However, other objects can modify a profile meanwhile, for example, in order to register events. In order to guarantee validity of a cache copy over a period of time, there must be a way to propagate profile updates. If a modified profile is written to the profile server, all platform servers that hold a copy of this profile must be informed. In order to receive updates of a certain profile, a platform server has to register at the corresponding profile server. This will guarantee that the platform server will get an updated version of this profile pushed as soon as whenever it is modified (see Figure 13.10). The mechanism applied here to detect and prevent potential conflicts is similar to the one called "callback promise" in the Andrew file system [25].

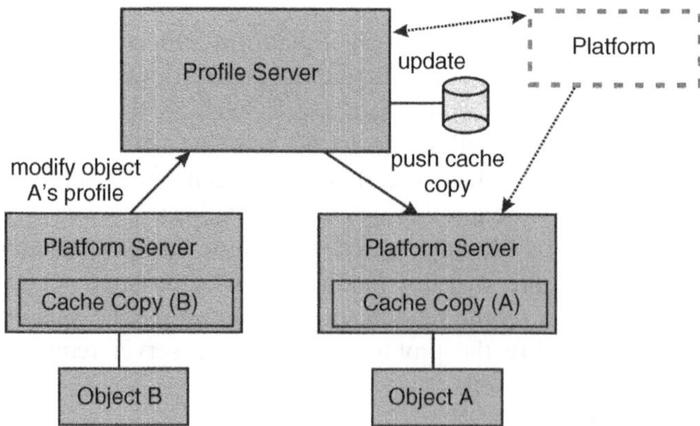

Figure 13.10 A platform server can register at the profile server to be notified when a certain profile is modified. Then, a copy of the profile is sent to the platform server after each update

If the platform server is no longer in charge for a certain profile, it is supposed to de-register at the profile server using the "close cache copy" function.

In order to modify a profile, a transaction-based write operation is offered by the profile server. A server that wants to alter a profile has to provide the time stamp of the last write access it knows about. This is compared with the timestamp of the last write operation that the profile server has confirmed. If the timestamp of the profile server equals the other one, the write operation is executed. In effect, we use a very simple form of the gossip architecture to ensure consistency [26].

The design of the notification service is based on the assumption that handovers happen much more frequently than modifications of a user's profile. Therefore, the profile server does not store a reference to the *platform server* that must be notified of changes, but a reference to the *object* to which the profile belongs. An object and its current platform server can be located at anytime by using the search algorithm with the object's ID.

13.5 Measurements

In this section, we present the measurements of the event rate of different PLASMA configurations. In order to evaluate whether our approach is applicable to real-world scenarios with a significant number of mobile clients, we measured the event-notification rate. The event rate is the most useful performance parameter since most of the PLASMA components are involved in the event-evaluation process (see sect. III). Thus, it is one the most complex parameters. In addition, the event mechanism is one of the most useful features of PLASMA. This holds true for application programmers as well as for end users.

13.5.1 Testbed Settings

13.5.1.1 Network and Computer Parameters

In our testbed, we used up to three PCs and a notebook that functioned as the measurement client. The PCs were connected through a wired 100 Mbit Ethernet network. The client (notebook) was connected through a wired 100 Mbit Ethernet network or an 11 Mbit WLAN network. The configuration of the PCs and the notebook are given in Table 4.

13.5.1.2 Deployment Scenarios

In order to obtain the performance and scalability parameters, we investigated the following configurations of PLASMA:

Table 13.4 Hardware and software configuration of testbed computers

PC's name	Type of processor	Operating system	RAM
PC1 –Root PC2 –Leaf1 PC3 –Leaf2	Intel ® Pentium ® 4 2.53 GHz	Microsoft Windows2000 Version 5.00.2195	256 MB
Client Notebook	Intel ® Pentium ® III Mobile 1.2 GHz	Microsoft Windows2000 Version 5.00.2195	256 MB

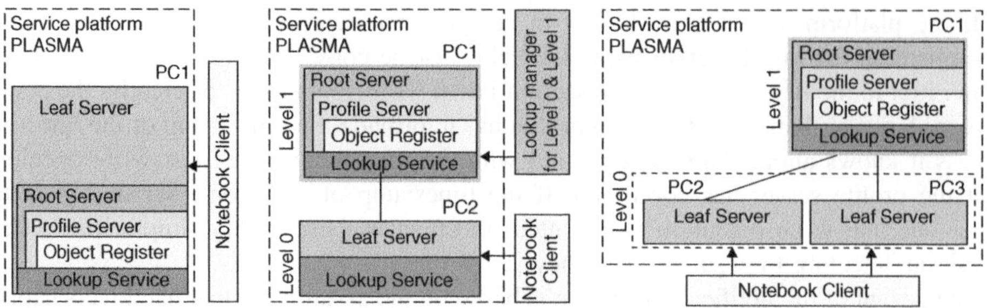

Figure 13.11 Various PLASMA platform trees: (a) The trivial tree–A single server (b) One root and one leaf (c) One root, two leafs

Single-Server Scenario: Here, all PLASMA servers (Lookup Server, Root Server, *Profile* Server), the Object Register, and the Lookup Service are installed on a single machine. Thus, the Leaf server provides all PLASMA features (see part (a) of Figure 13.11).

One-Leaf Server: In contrast to the single-server scenario, PLASMA was now configured in two hierarchical levels. On the leaf level, a simple leaf server for managing and communicating with the mobile client was installed. A Lookup Service was also running at the leaf level. All PLASMA servers and the Object Register were installed on a single machine at the root level. All Lookup Services in this scenario were managed by a Lookup Manager, which ran on the Root Server (see part (b) of Figure 13.11).

Two-Leaf Servers: Here, we had the same configuration as in the one-leaf server scenario, except that we now have two-leaf servers. In addition, the leaf server has no own Lookup services running. Instead, they use the Lookup service of their root node (see part (c) of Figure 13.11).

13.5.1.3 Event Definition and Position Changes

On the notebook, there was one client with a PLASMA application running. For this client, two events were defined. Event One was "client_2 enters the aura of client_1" and the second event was "client_2 leaves the aura of client_1." In our measurement scenario, client_1 was the owner and the only addressee of these events. In addition, we generated sightings (position information of mobile clients that is reported to PLASMA) on the notebook. In order to get a considerable workload, we generated 70 sightings every 10 or 100 ms, respectively. To ensure that the measurements are reproducible, we used a slightly modified version of the random walk model. All position updates stem from the random walk model, except those of client_1 and client_2. Client_1 is not moving at all, whereas the position of client_2 alternates between two predefined locations, one inside the aura of client_1 and one outside.

13.5.2 Results

In order to measure the event rate we sent periodical sightings, of which 10% would trigger an event to the sighting proxy. We did this ten times for each scenario and logged all triggered events in intervals of 120 s. The values given in the following tables show the average of the results of these measurements.

Table 13.5 Average values for events per burst and events per second, when the communication is done via 100 Mbit wired network

Time sec	Events/Burst	Events/sec	Comments
120	6.96	623	70 Sightings and 7 Events per burst were generated every 10 ms
120	14.90	961	160 Sightings including 16 Events were generated every 10 ms

Table 13.6 Average values for events per burst and events per second, when the communication is done via a 11-Mbit wireless link with differing additional amount of traffic

Time sec	Events/Burst	Events/sec	Comments
120	3.5	130	Traffic over the Access Point unknown
120	2.6	91	Additional Traffic over the Access Point
120	4.8	170	Without additional Traffic over the Access Point

13.5.2.1 Single-Server Scenario

The results we measured in the single-server scenario show that PLASMA can handle up to 960 events per second (see Table 5). This is the upper limit for the events PLASMA can manage if it has to cope with burst of 160 sightings and 16 events per burst.

In order to determine whether the measured performance limits the number of mobile clients that can be managed by a leaf server, we repeated the experiment. But this time, we sent the sightings via a wireless link with 11 Mbps. The result was that the event rate dropped to 120 events per second. The event rate decreased even further to 70 if non-PLASMA traffic goes over the wireless link (see Table 6). This indicates clearly that the wireless link and not the performance of PLASMA limits the number of clients that can be managed by a single PLASMA server. A single PLASMA server can manage up to 10 access points if 120 events are triggered within the area covered by each of these access points.

13.5.2.2 One- and Two-Leaf Server Scenario

The performance of PLASMA decreased dramatically in the one- and the two-leaf server scenario. Here, only four events could be processed per second (see Table 7). The reduction of the event rate is caused by the overhead that results from the fact that the events are evaluated remotely (see Section VI.A). The performance difference between the single-server and the two other scenarios indicates clearly that scalability cannot be achieved by introducing a hierarchy. A flat structure in which the profiles and parts of the object

Table 13.7 Average values for events per burst and events per second, in the one- and two-leaf server scenarios as well as in the single-server scenario when the communication is done via a 100-Mbit wired network; 70 Sightings including 7 events per burst sent every 100 ms;

Scenario	Time sec	Events/Burst	Events/sec
One-leaf server:	120	0.480	**4.80**
Two-leaf server	120	0.483	**4.83**
Single Server	120	7.000	**69.00**

Table 13.8 Average values for the time needed to gather and marshal all necessary information on the old leaf server and to de-marshal and scatter this information on the new leaf server. Measurement was repeated 100 times

Time to prepare data for handover on the old leaf server	Time to extract data on the new leaf server after handover
31 ms	**31 ms**

register entries of the mobile clients are replicated on the leaf servers that are responsible for the mobiles will lead to better performance, since it is possible to evaluate the majority of the events locally.

If the evaluation of events is executed on the leaf servers, large geographical regions can be split up into several smaller ones. The time needed to perform a handover between two PLASMA leaf server influences the number of events that can be processed from these servers. In order to verify this, we measured the time it takes to gather and marshal all information that has to be transferred from the old leaf server to the new one. Table 8 shows our measurement results. The whole handover takes about 100 ms –excluding the connection reestablishment. Both leaf servers involved in the handover are busy with the data handling for about 30 s. This reduces the number of events that can be processed by a factor of 0.03 per handover. We are focusing on indoor scenarios, which means that the mobile devices are moving relatively slowly. If we assume a walking speed of 5000 m/h, the mobiles are moving about 1.4 m/s. Since a single leaf server can handle up to ten access points, it can cover an area about 1600 sq. m if we assume that its signal covers 40 m times 40 m indoor. Thus, handovers will be seldom necessary.

13.5.3 Conclusions

The measurements presented here show clearly that PLASMA can handle nearly 1000 events per second. In addition, the single-server scenario indicates that the wireless communication link is the bottleneck in the event processing. Thus, PLASMA should be deployed in such a way that each leaf server is responsible for no more than ten access points. This limits the number of potential events to 1000, which can be handled by a PLASMA leaf server.

Splitting large geographical regions into smaller ones in order to reduce the workload of the leaf server has to be done very carefully. The measurements indicate that a hierarchically structured infrastructure introduces a significant overhead in the PLASMA internal communication. This influences an optimal deployment in two ways. First, the number of hierarchy levels of the infrastructure should be kept small. Second, event evaluation should be executed on the leaf server that is responsible for the relevant mobile client. Thereby, the communication overhead is minimized. Normally, the majority of events defined for a mobile client will be related to its vicinity. This means that they can be evaluated by the responsible leaf server without additional communication. For sure, there can be events that change their state since something happens remotely, say, a colleague enters the airport. The processing of those events may take a second or two. This does not cause any harm since the user's reaction will also last several minutes, for example, meeting the colleague at the bar normally means both have to walk to the bar.

To summarize, PLASMA shows a good performance for medium-sized geographical regions. The possibility to structure PLASMA hierarchically allows to connect geographical regions in an effective way. It allows to search for other mobile clients not only in the vicinity of the mobile client but also remotely. If the event evaluation can be done on the leaf servers, the hierarchical structure has very little influence on the performance. In addition, the execution of the handover mechanisms has nearly no influence on the overall performance.

13.6 Realization of PLASMA

In this section, we will briefly report on our original design decisions and on the experiences we gained, as well as on the implementation of the platform together with a demonstrator.

13.6.1 Design Decisions and Lessons Learned

In the design phase, we chose RMI instead of CORBA because of its reduced overhead. Furthermore, there is currently no CORBA version available that can run on handheld devices. But JAVA virtual machines are available in different flavors for mobile devices, such as EPOC, WinCE, or PalmOS. Most of these cover the set of functions required to support complete RMI communication based on standard sockets. The often-mentioned bottleneck of using an RMI registry is left out in our scenario, because we use the remote references of other platform functions directly. Because of performance problems, we had to rethink this decision. Using RMI only for communication turned out to be not a good choice. The overhead in terms of full object serialization/de-serialization introduced relatively high costs subject to the platform performance and in some case even constituted a performance bottleneck. For example, one of the objects transmitted took about 4 kB after serialization. But the information that was really "important" was not more than 20 bytes! This means that more than 99% of the traffic was just overhead. As a corrective action, the protocol was downgraded to plain TCP socket communication, which offers sufficient functionality and significantly improved the platform performance. As the interaction between PLASMA entities relies only on simple data transfer, the benefits of a remote method invocation (RMI) scheme do not apply and hence, its costs can be avoided. Now, in general every interaction between any two platform entities, such as a request or a response, involves data that fits into one TCP packet.

In a first implementation, we tried to achieve a very high level of transparency, in the sense of location transparency of all platform components. This meant that the algorithms were scenario insensitive, so that on the platform level, the algorithms had the same flow regardless of whether they ran on a single server . So, all data that was exchanged between two PLASMA components was prepared to be sent via the network before transmission, even when both components were located in the same JVM. This was due to the fact the all knowledge about the actual deployment was hidden from the platform. The decision whether the receiving PLASMA component was remote was taken by underlying protocol layers. This resulted in an event-processing rate of 2 events per second in the single-server scenario. With that performance, PLASMA would have never satisfied real-world performance requirements. In order to improve the performance, we introduced a wrapper that knows which components are locally available and which are not. This now avoids the communication overhead. The event-processing rate increased by two orders of magnitude for single-server scenarios. But as soon as the event evaluation has to be executed on a remote server, the performance dropped down to the original one because of the communication overhead. Intelligent replication schemes such as replicating the object registers at the leaf servers can avoid this communication overhead, so that the event-processing rate of the single-server scenario can be achieved in nearly all deployment scenarios.

13.6.2 Implementation Notes

The platform described here was implemented in 100% pure Java. For the server side, SUN's JDK 1.2.2 was used as the standard to develop, test, and evaluate the result. Nevertheless, experiments with different implementations (Borland), compatible JDK's (1.3.1, 1.4.0), alternative operating system (Linux), and varying hardware (Pentium II, III, IV) impressively demonstrated the great flexibility introduced by Java.

At the client side, the implementation standard was JDK 1.1.8 or the corresponding level of PersonalJava. Again, the platform together with the demo application on top of it ran in several VMs (Insignia's Jeode, NSICom's Creme) and on different hardware platforms, such as SH3, MIPS, StrongARM, and XSCALE. In the end, especially the client-side platform unit benefited from the decision to use Java as the programming language, as PLASMA can now address this manifold variety of mobile computing devices out there that is capable of running a JVM.

In order to determine the current position of the mobile device, a positioning system has to be integrated. In our demonstrator, the position is determined by IR beacons, which periodically send out an integer. This number is received by the IR port of the mobile device. The access to the mobile's IR port was partly implemented as native code. From our point of view, this does not violate the 100% pure Java restriction for the platform, since the positioning support module is not part of the platform itself. Instead, it is specific to the positioning system and to the hardware deployed.

Our demonstrator consists of an IR-based positioning system, some logical objects such as printing services, and of our so-called "local pilot," which is realized in Java on top of PLASMA. The local pilot includes a basic guiding and service discovery functionality.

It is also capable to display the multimedia content that is available in front of posters presenting the scientific work of our institute.

13.7 Conclusions

We have presented the design and implementation of our location-aware platform PLASMA. We discussed all the so-called PLASMA servers, such as root, leaf, and profile server, that have to installed on the infrastructure side. We also described the concept of the PLASMA mediator, which is used to integrate non-Java clients into the PLASMA infrastructure using SOAP and WSDL. We also described all the PLASMA components and their functionality in general in order to provide a satisfying overview of PLASMA's features.

We discussed in detail the design issues of those components that are related to PLASMA's event-definition and evaluation process. In particular, we briefly introduced our event-definition language, including our tri-state logic for event evaluation and the concept of auras to describe logical and/or geographical regions of interest. In addition, we presented the communication from and to profile and database servers, as well as the applied replication and consistency mechanisms.

The measurements shown here prove that PLASMA has a very good performance when it is deployed in a single-server scenario. But they also indicate that the performance of PLASMA drops drastically if a hierarchical structure is deployed and events have to be evaluated remotely. Our analysis of these results showed that this performance problem can be solved if object register entries and the profiles of the mobile clients are replicated on the leaf servers.

We reported on our experiences with implementing PLASMA in pure Java. It really provides what it promises: very high portability. In the Sections 13.6.1 and 13.6.2, we also provided some details about speeding up our first PLASMA version.

Currently, we are working on privacy and security aspects, which we consider to be crucial when it comes to users' acceptance. The most significant parameters of the security features are processing speed and energy consumption on the client side. In order to increase the former and decrease the latter, we decided to go for a hardware/software codesign approach. In order to increase privacy, we are working towards a privacy negotiation scheme [27, 28].

Bibliography

[1] Siegel, J. (2000) *CORBA 3 Fundamentals and Programming*, 2nd edition, John Wiley & Sons, New York, ISBN: 0-471-29518-3.

[2] Microsoft Corporation (1996) DCOM Technical Overview, http://msdn.microsoft.com/library/default.asp?url =/library/en-us/dndcom/html/msdn_dcomtec.asp, last viewed September 24, 2003.

[3] Pitt, E. and McNiff, K. (2001) *java.rmi: The Guide to Remote Method Invocation*, Addison-Wesley, Boston, MA.

[4] Keller, B., Park, A. S., Meggers, J., Forsgern, G., Kovacs, E., and Rosinus, M. (1998) UMTS: A Middleware Architecture and Mobile API Approach. *IEEE Personal Communications*, **5**(2), 32–38.

[5] José, R., Moreira, A., Meneses, F., and Coulson, G. (2001) An Open Architecture for Developing Mobile Location-Based Applications over the Internet. *Proceedings of the 6th IEEE Symposium on Computers and Communications*, Hammamet, Tunisia, July 3–5; IEEE Computer Society, Los Alamitos, ISBN 0-7695-1177-5.

[6] Black, A. and Artsy, Y. (1990) Implementing Location Independent Invocation. *IEEE Transactions on Parallel and Distributed Systems*, **1**(1), 107–119.

[7] van Steen, M., Tanenbaum, A. S., Kuz, I., and Sips, H. J. (1999) A Scalable Middleware Solution for Advanced Wide-Area Web Services. *Distributed Systems Engineering*, **6**(1), 34–42.

[8] van Steen, M., Hauck, F. J., Homburg, P., and Tanenbaum, A. S. (1998) Locating Objects in Wide-Area Systems. *IEEE Communication Magazine*, **36**(1), 104–109.

[9] Hohl, F., Kubach, U., Leonhardi, A., Schwehm, M., and Rothermel, K. (1999) Nexus: An Open Global Infrastructure for Spatial-Aware Applications. *Proceedings of the Fifth International Conference on Mobile Computing and Networking (MobiCom 99)*, Seattle, Washington, DC, August 15–19; ACM Press, New York.

[10] Leonhardi, A. and Kubach, U. (1999) An Architecture for a Distributed Universal Location Service. *Proceedings of the European Wireless '99 Conference*, Munich, Germany, October 6–8; ITG Fachbericht, VDE Verlag, Berlin, pp. 351–355.

[11] Leonhardi, A. (2003) *Architektur eines verteilten skalierbaren Lokationsdienstes*, Dissertation, Mensch & Buch Verlag, Berlin, ISBN: 3-89820-537-1.

[12] Leonhardi, A. and Rothermel, K (2001). A Comparison of Protocols for Updating Location Information Baltzer. *Cluster Computing Journal*, **4**(4), 355–367.

[13] Pashtan, A., Blattler, R., Heusser, A., and Scheuermann, P.(2003) CATIS: A Context-Aware Tourist information System. *Proceedings of the 4th International Workshop of Mobile Computing*, Rostock, Germany, June 17–18.

[14] UDDI Technical White Paper, UDDI.org, September 2000, http://www.uddi.org/pubs/Iru_UDDI_Technical_White_Paper.pdf, last viewed September 24, 2003.

[15] Simple Object Access Protocol (SOAP) 1.1, World Wide Web Consortium (W3C), http://www.w3.org/TR/SOAP/, last viewed September 24, 2003.

[16] Web Services Description Language (WSDL) 1.1, World Wide Web Consortium (W3C), http://www.w3.org/TR/wsdl, last viewed September 24, 2003.

[17] Escudero, A. and Maguire, G. Q. (2002) Role(s) of a Proxy in Location Based Services. *PIMRC 2002, 13th IEEE International Symposium on Personal, Indoor and Mobile Radio Communications*, Lisboa, Portugal, September 16–18.

[18] Perkins, C. (1997) *Mobile IP Design, Principles and Practice*, Addison-Wesley, Boston, MA.

[19] ETSI (1991) General Description of a GSM PLMN, GSM Recommendations.

[20] Web Services Architecture, World Wide Web Consortium (W3C), http://www.w3.org/TR/ws-arch/, last viewed September 24, 2003.

[21] Edwards, W. K. (1999) Core Jini, Prentice Hall, Munich, ISBN: 3-8272-9592-0.

[22] Langendörfer, P., Dyka, Z., Maye, O., and Kraemer, R. (2002) A Low Power Security Architecture for Mobile Commerce. *Proceedings of the 5th IEEE CAS Workshop on Wireless Communications and Networking*, Pasadena, CA, September 5–6.

[23] Burrell, J. and Gay, G. K. (2002) E-graffiti: Evaluating Real-World Use of a Context-Aware System. *Interacting with Computers* (Special Issue on Universal Usability), **14**(4), 301–312.

[24] Olaf, D., Kirti, S.-K., Markus, F., Roland, S., Rita, W., Klaus, D. (2000) A Scalable Location Aware Service Platform for Mobile Applications based on Java RMI, in *Trends in Distributed Systems: Towards a Universal Service Market* (eds C. Linnhoff-Popien and H.-G. Hegering), Proceedings, Lecture Notes in Computer Science 1890, Springer, ISBN 3-540-41024-4, pp. 296–301; *Third International IFIP/GI Working Conference, USM 2000*, Munich, Germany, September 12–14.

[25] Coulouris, G., Dollimore, J., and Kindberg, T. (2001) *Distributed Systems: Concepts and Design*, Chapter 8, Addison Wesley, Boston, MA.

[26] Ladin, R., Liskov, B., Shrira, L., and Ghemawat, S. (1992) Providing High Availability Using Lazy Replication. *ACM Transactions on Computer Systems*, **10**(4), 360–391.

[27] Langendörfer, P. and Kraemer, R. (2002) Towards User Defined Privacy in location-aware Platforms. *Proceedings of the 3rd international Conference on Internet computing*, Las Vegas, June 24–27; CSREA Press, Atlanta.

[28] Bennicke, M. and Langendörfer, P. (2003) Towards Automatic Negotiation of Privacy Contracts for Internet Services. *Proceedings of the 11th IEEE Conference on Networks (ICON 2003)*, Sydney, Australia, September 28–30, October 1; IEEE Society Press, Sydney.

14

QoS-Enabled Middleware for MPEG Video Streaming

Karl R.P.H. Leung[1], Joseph Kee-Yin Ng[2], Calvin Kin-Cheung Hui[3]

[1]*Hong Kong Institute of Vocational Education (Tsing Yi),*
[2,3]*Hong Kong Baptist University*

14.1 Introduction

Online video–streaming service has become a popular service in the open network. Video streaming is the concurrent processes of sending video images from servers to clients over a network and playing back of these video images at the clients' video players. Online video–streaming service system is an online, firm real-time, multiusers system that operates over an uncontrollable, performance unstable and nonpredicatable open network, such as the Internet. MPEG has become the *de facto* standard for video encoding [8, 18]. One of the challenges of the MPEG video–streaming systems is to provide good quality of services (QoS) to the clients. There are many software, such as the Real Player, for online MPEG video playback. Because of the characteristics of the open network, it is ineffective, if not impossible, to tune the QoS of online video streaming by tuning the network performance. There are many studies in the topic of enhancing online video streaming [1–7, 9–10, 15, 27, 29–6, 33, 36]. The approaches of all these studies can be classified into the following three categories.

1. Replacing the existing video-encoding data format by new encoding algorithms and data structures that are more suitable for dealing with video images in open networks.
2. Replacing existing popular protocols by new protocols and new mechanisms for transmitting video images over the open networks.
3. The video is prepared in several versions, each of different resolution or frame sizes, that is, in different QoS levels. These versions are playback concurrently in different

Middleware for Communications. Edited by Qusay H. Mahmoud
 ISBN 0-470-86206-8

streams. According to the network performance, the server selects the stream that QoS fits the client most.

The main disadvantage of the former two approaches is the need of new video players or communication systems. Practically, these approaches are not viable. The main disadvantage of the last approach requires huge resources, including processing and bandwidth. Hence, none of these approaches is popular.

We consider middleware as a software system that is placed between servers and clients in order to provide value-added services to the client-server system. There are studies on QoS-enabled multimedia middleware [24, 15, 27, 31, 6]. Most of the middleware are designed for the implementation of these approaches. They tune the QoS with reference to the observed bandwidth at the server side. However, the throughput between the server and each individual client on the open network can be different. This approach cannot provide a customized QoS tuning for each individual client. Furthermore, many of these studies do not take the variations of different multimedia encoding format into account. Hence, the QoS tuning scheme may not fit the characteristics of MPEG video encoding format well. Moreover, none of these middleware can support well the approach of providing multiple QoS level streaming with a single playback stream.

In order to allow users playing back the video with their favorite video player and to allow streaming services to be provided over the open network, we tackle the problem by making use of a distributed QoS-enabled middleware. This middleware is placed between the video servers and client video players. It consists of a serverware and some clientware. The serverware is placed at the server side, while each client is installed with a clientware. Since the throughput between the server and each individual client can be different, the best possible QoS of video playback for clients are different among each other. Furthermore, when network is congested, some frames have to be dropped. However, the drop of QoS can be mapped to some skipped MPEG frame patterns. Hence, the serverware need not transmit all MPEG frames to the clients but only some selected frames according to the corresponding best possible QoS of the clients on the fly. These are the value-added services to be provided by the distributed middleware.

Quality of Services is a key success factor for a video-streaming system. An appropriate metric for measuring the QoS is of ultimate importance. Two common metrics measure the QoS for video by the rate of transmission of bytes or frames. We call the former metric *QoS-Byte* and the latter *QoS-Frame*. These two metrics assume that QoS is directly and linearly proportional to the amount of bytes or frames being played. However, MPEG consists of three different kinds of frames. These frames are of different sizes and have different impacts on the quality of the video. This is a kind of Group Frame Scheme (GFS) of variable bit rate data format that does not satisfy the assumptions of QoS-Byte and QoS-Frames. Consequently, both QoS-Byte and QoS-Frame are not the best metrics for measuring the QoS of MPEG video over open networks. We have proposed a QoS-Index that measures QoS playback with reference to human perspective [22]. Human perspective is the ultimate judgment of the playback quality. QoS-Index, hence, is a better metric for measuring MPEG playback quality. With reference to the QoS-Index, we designed a metric, QoS-GFS, which is more suitable for real-time QoS tuning in the online MPEG video–streaming system.

14.2 Related Works

14.2.1 Overview of MPEG

MPEG was originally designed for storing video and audio on digital media. MPEG compression is found also suitable for transmitting video frames over computer network and then has become a *de facto* standard video encoding format. The basic idea of this compression scheme is, first, predict motion from frame to frame in the temporal direction. Then, discrete cosine transforms (DCT) is applied to organize the redundancy in the spatial directions. The MPEG coding scheme is a kind of *group frame scheme* (GFS). In MPEG standard, there are three types of frames. The intra-frames or I-frames are simply frames coded as still images. They can be played without reference to another other frames. The second kind of frames are the predicted-frames or P-frames. P-frames are constructed from the most recently reconstructed I- or P-frames. The third kind of frame is the bi-directional-frames or B-frames. These are predicted from the nearest two I- or P-frames, one from the past and one in the future.

In general, the frame size of an I-frame is larger than that of a P-frame; and the frame size of a P-frame is larger than that of a B-frame. Since a B-frame needs two subsequent P-frames or I-frames to reconstruct the current B-frame, I-frames and P-frames have to be sent before their two dependent B-frames. Figure 14.1 shows the relationship between the original frame sequence and the transmission sequence for a typical MPEG video stream. Except for the very first I-frame, the transmission sequence follows the pattern of "PBBPBBPBBIBB" and repeats itself.

In an ideal environment, all video frames are received in the correct order and on time such that the decoder can reconstruct and playback the video properly. However, in the real world, owing to network congestion or system faults, some video frames may be lost

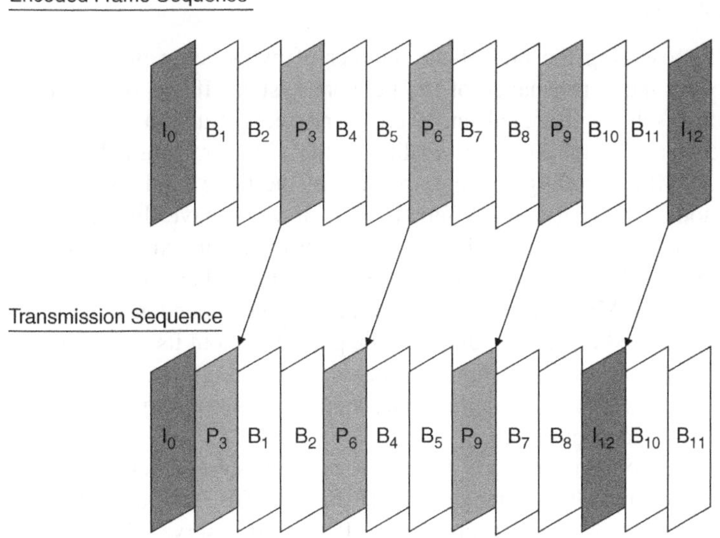

Figure 14.1 Encoded frame and transmission sequence of an MPEG stream

or delayed. If the lost frame is a B-frame, the decoder can simply ignore it. However, if the lost frame is a P-frame, the decoder will not be able to reconstruct up to four B-frames that are dependent on it. If the lost frame is an I-frame, the entire *group of pictures* (GOP) is unable to be reconstructed. It is easy to see that the decreasing order of significance of MPEG frames is I-Frame, P-Frame, and B-Frame.

14.2.2 Quality of Services

Quality of Services (QoS) of video playback usually is commonly measured by the processed number of frames or bytes. We call the former approach *QoS-Frame* and the latter approach *QoS-Byte*. Human perspective is the key attribute of measuring the quality of video playback. However, QoS-Byte and QoS-Frame do not take human perspective into consideration. In order to obtain better measurement, we developed a QoS-Index [22] that measures MPEG video playback with reference to human perspective and frames being playback. In the rest of this section, QoS-Byte, QoS-Frame, and QoS-Index are briefly reviewed.

14.2.2.1 QoS-Frame

QoS-Frame is the number of frames that is received or is being utilized by the client. Counting the number of frames received on time or being utilized by the client is the only attribute for measuring the quality of video playback. However, MPEG is a kind of group frame scheme (GFS) that consists of different kinds of frames where different kind of frame is serving different purposes in the playback. Hence, the impact on the QoS when losing a frame is depending on the kind of frames being lost. Consequently, QoS-Frame is not an appropriate metric for measuring the QoS of group frame scheme encoded video playback, such as MPEG.

14.2.2.2 QoS-Byte

QoS-Byte is measuring the rate of the number of bytes being processed on time. QoS-Byte can reflect the performance of the network system. If the size of all the frames of a video-coding scheme is equal and there is no reconstruction dependency among the frames, QoS-byte can measure the quality of the video transmitted properly. However, in the case of MPEG coding scheme, the size of the three kinds of frames are different from one another. Furthermore, different kinds of video have different frame size ratio. For instance, with reference to the videos we use in our experiment, the frame ratio I: P: B of Action Video (RACING01.MPG) is 4:4:1, whereas Talking Head Video is 13:6:1 (NEWS0001.MPG). Furthermore, because of the dependence features of B-frames, a completely received B-frame is unable to be played without its depending P- or I-frames. This means that more bytes or frames received by the client may not lead to better QoS. Hence, QoS-Byte is also not an appropriate metric for measuring the QoS of group frame scheme encoded video playback.

Furthermore, QoS-Byte and QoS-Frame are not consistent with each other. For instance, the size of a 12-frame GOP of pattern "IBBPBBPBBPBB" for our Talking Head Video is about $13 + 3 \times 6 + 9 \times 1 = 42$ units in size. If the Video server skipped 4 B-Frames, the QoS-Byte is dropped by $4/42 = 9.5\%$ but the QoS-Frame is dropped by $4/12 = 33.33\%$.

Hence, we need a more appropriate metric for measuring the QoS of group frame scheme–encoded video playback.

14.2.2.3 QoS-Index

Human perspective is the key factor in measuring the quality of video playback. We have seen that both QoS-Byte and QoS-Frame have not incorporated human perspective in the measurement. We have proposed a QoS-Index that measures the playback of MPEG frames with human perspective being incorporated [22].

QoS-Index is a value that shows the relations between the number of frames displayed, the distribution of skipped frame, and the maximum interval in a display pattern or a GoP. The number of displayed frames is measured by QoS-Frame. The maximum interval in a display pattern is some well-defined value of different display patterns. The distribution of skipped frames are measured by the variance of the interframe distance and the variance of the variance of the interframe distance. And to facilitate better comparison the QoS-Index is a normalized value of the performance metrics with respect to a GoP. For detail on how the QoS-Index is constructed, please refer to [22].

14.2.3 Video Distribution

There are many commercially available streaming video systems, namely, RealPlayer & RealSystem G2 from Real Network [28], VDOLive Player from VDOnet [34], InterVU Player [11] & NetShow from Microsoft [19], and Streamworks from Xing Technology [32]. However, the QoS of these systems mainly described as best effort system with their performance highly depending on the available bandwidth and on the availability of huge buffers at the client side. There is also no firm real-time QoS tuning and effective feedback mechanism implemented in these systems. Moreover, many of these systems have their own proprietary video formats that are not compatible with each other.

There are studies on enhancing VBR transmission scheme. One of the approaches is enhancing the traditional scheduling algorithms for better real-time transmission of multimedia data [9, 13, 14, 16]. Ott et al. [25] and Lam et al. [17] studied another approach by smoothing the schemes for VBR video transmission.

Some researchers, such as Ismail, Lambadaris, and Devetsikiotis [12], Izquierdo and Reeves [13], and Krunz, Sass, and Hughes [16], study the MPEG video characterization and modeling. These studies help the approach, such as the one reported by Pancha and El [26], of studying variable bandwidth allocation schemes to obtain better performance in transmission by referencing to the MPEG-1 video coding standard in the design.

Network congestion is the most notable problem in open network. There are many studies [1–7, 9, 10] tackling this problem by designing new transmission schemes for video distribution or designing new distributed video systems over the open network. QoS has been a concern in some studies. Bolot and Turletti proposed a rate control mechanism for transmitting packet video over the Internet [4]. Reibman et al. [14, 29] and Reininger et al [30] made use of an adaptive congestion control scheme to study the problem of transmitting VBR video over ATM networks. Hasegawa and Kato implemented a video system with congestion control on the basis of a two-level rate control [10]. All these QoS-based studies are referenced to the network QoS metrics that cannot truly reflect the QoS perceived by human beings.

All these studies have some enhancements in providing better QoS of online MPEG video services over the open network. However, new video systems or new communication protocols are required in order to implement these results. These are not practically viable. This is because the cost of converting the huge amount of MPEG productions to the new data format would be costly. It is also difficult to change the commonly installed communication protocols all over the world.

The problem of online MPEG video streaming is a conglomerate of problems due to the variable bit rate (VBR) characteristics of MPEG and the unstable, uncontrollable, and unpredictable nature of the open network. Providing better quality of online video streaming has been a hot research topic. However, most of these studies only address the topic by tackling one or the other of the subproblems.

A viable approach of enhancing the QoS of online video distribution is to use the *de facto* video coding standard with the popular communication protocols. Our group has been studying this topic for some time. We have suggested ways to make use of a feedback mechanism to maintain the QoS of the distributed video system [21, 35, 36]. Furthermore, we noticed that video quality cannot be accurately measured by traditional QoS metrics that are solely based on byte- or frame-processing rate. We proposed a QoS-Index of MPEG video that is based on human perspective [22]. There are few research like ours that study providing human perspective QoS in online MPEG video streaming with existing MPEG video system on common open network. We are going to report our study of tuning the QoS of online MPEG video streaming by means of a distributed middleware between the video servers and video clients.

14.3 Requirements for QoS-enabled MPEG-Streaming Middleware

14.3.1 Environment Analysis

Streaming MPEG video in the open network is a kind of client-server system. The server system can consist of one or more MPEG video servers providing services to the clients concurrently. We make the environment further stringent by assuming that the server system can send frames to only one client at a time. Clients play the MPEG videos with their favorite video players such as the Windows media player. Video in MPEG-1 encoding format are transmitted from the video servers to clients through open networks, such as the Internet, the HTTP protocol. Figure 14.2 gives an idea of the environment of the problem.

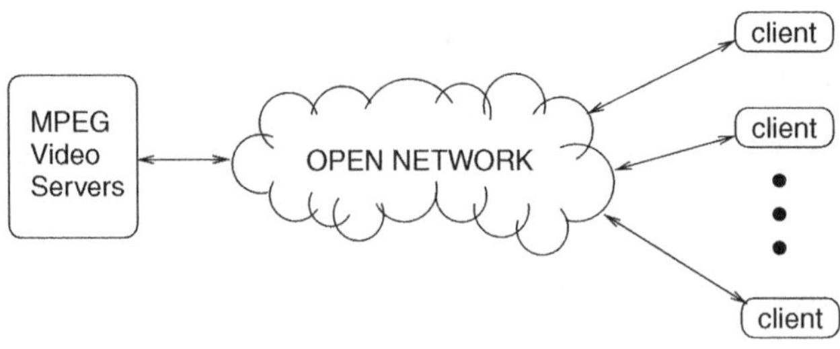

Figure 14.2 An overview of the problem

The performance of the open network is nonpredicatable, unstable, uneven, and uncontrollable. Different clients may communicate with the server at different rate. Neither the server nor the clients can influence the performance of the open networks. Hence, it is impossible to tune the QoS by regulating the performance of the open network. Hence, the only components that can be used to tune the QoS are the servers and the clients.

14.3.2 Requirements for the Middleware

On the basis of the characteristics of this client-server system, we identified the following requirements for the middleware providing online MPEG video services.

1. Play the MPEG videos with the best possible QoS at the client sides.
2. Users can use their favorite video player to watch the playback.
3. The QoS tuning scheme should be communication protocols–independent.
4. The scheme should support multiusers.
5. Minimize the number of frames that arrived late and have to be discarded by the clients.

14.4 QoS Facilities

All video-streaming systems have a transmission scheme for streaming contents to clients. All QoS-enabled systems measure QoS with reference to some QoS metrics and possess a scheme to tune the QoS with reference to these metrics. In this section, we discuss a QoS metric, QoS-GFS, in Section 14.4.1; a four-sessions transmission scheme in Section 14.4.2; and a QoS tuning scheme in Section 14.4.3. These are the facilities being adopted in the distributed MPEG video–streaming middleware.

14.4.1 QoS-GFS

We have seen that QoS-Frame and QoS-Byte are inappropriate for measuring the QoS of VBR video streaming, and QoS-Index is a better metric for measuring the MPEG video playback quality. We develop the GFS-based metric, QoS-GFS, for the reference of our MPEG video–streaming middleware. The QoS-GFS is founded on the MPEG characteristics that I-Frames have more impact on the QoS than P-Frames, and P-Frames have more impact on the QoS than B-Frames. It references to the QoS-Index and hence human perspective is incorporated in the metric. QoS-GFS is a simple metric that is more convenient to be implemented in video-streaming system.

We assume that the pattern of group of pictures of MPEG video is IBBPBBPBBPBB. We found out that for a given number of missing frames in a GOP, viewers find the QoS is better if the missing frames are of symmetric patterns [20, 23]. Furthermore, skipping an 'I' frame is equivalent to skipping the whole GoP and skipping 'P' frames have severe impact on the playback quality. Hence, in tuning the playback QoS, skipping 'B' frames would be a better approach. Following this approach, we define five grades of skipping frames as shown in Table 14.1. The best grade is 5 and the lowest grade is 1. Grade 5 is given when there is no skipped frame. Then with every skipping of two B-frames, the grade is lowered by one. Then Grade 1 is given when all B-frames are skipped.

Table 14.1 QoS-GFS levels

QoS Level	Transmitted Frame(s)	Skipped Frame(s)	GoP Pattern
5	12	0	IBBPBBPBBPBB
4	10	2	IBBPB.PBBPB.
3	8	4	IB.PB.PB.PB.
2	6	6	IB.P..PB.P..
1	4	8	I..P..P..P..

14.4.2 Transmission Scheme

Following the MPEG GFS properties and unstable bandwidth properties of the open network, we designed an MPEG frames transmission scheme such that MPEG video servers can stream MPEG frames with reference to the significance of the frame types. This transmission scheme also supports the server to provide streaming services to multiple clients and allows clients to join in at different times.

We adopted the time-sharing approach for providing services to multiple clients. Time-sharing is a simple and effective approach to provide services to all clients concurrently. Within a given period of time, all the clients have to be served once. We call this period a *round* and the duration of a round is denoted by T_{round}, which is in ms. The duration of a T_{round} is determined by the server's system administrator. For example, if a round is defined to be 100 ms and the frame rate of the video is 30 frames per second, then three frames are sent to each client in each round.

A round can consist of up to four types of sessions, namely, Session 0, Session 1, Session 2, and Session 3 where each of these sessions are serving different purposes in providing the best allowed QoS to each individual clients.

14.4.2.1 Video Transmission Sessions

An MPEG video clip is structured as a sequence of groups of pictures. It is desirable to align the round with a group of frames. We can conceptually divide the video sequence into segments. Each segment contains all the frames of the groups to be transmitted in a round. For instance, Figure 14.3 shows a GOP being divided into four segments and being transmitted in four rounds. Each segment contains either an I-frame or a P-frame, plus two B-frames.

In MPEG video playback, I- and P-frames are more significant than B-frames. Consequently, we send I-frames and P-frames before B-frames. We call the session of sending

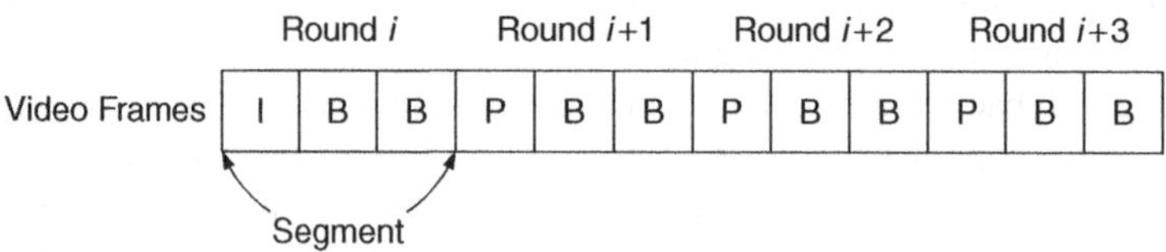

Figure 14.3 A GOP of MPEG video to be transmitted in 4 rounds

Table 14.2 Summary of transmission scheme

Transmission	Real-time		Non real-time	
Session	Session 0	Session 1	Session 2	Session 3
Description	I- or P-frame only. Drop the P-frames when the cycle reaches the limit.	B-frame only. Drop the frames when the cycle reaches the limit.	For initial buffering under transmission control	For advance buffering under transmission control

I-frame or P-frames Session 0 and the session of sending B-frames Session 1. In case there is time left in a round after Session 0 & Session 1, the video server will consider using this duration for sending either start-up frames to entertain new requests or sending extra frames to the clients. We call the duration of the former action Session 2 and the latter one Session 3.

The time taken up by Session i, $i = 0, 1, 2, 3$, is denoted by T_i where $i = 0, 1, 2, 3$ respectively. When there is time left in a round after a session, but the time left is estimated to be insufficient for the session that followed, the time left will be discarded and next round will start immediately.

It is easy to notice that the frames being transmitted in Session 0 and Session 1 are subject to real-time requirement, that is, the frames have to be received by the clients before the deadline; otherwise these frames will be discarded. On the other hand, Session 2 and Session 3 are not subjected to real-time requirement. While Table 14.2 gives a summary of the four sessions, detailed descriptions of these four sessions are described as follows.

Session 0.

A round starts with Session 0. We call the duration consumed by Session 0 T_0. Clients are being served in turn before the T_{round} is over. Clients are served in turn. If all the clients cannot be served within T_{round}, if I-frames are being sent in this round, the round duration is extended until all clients have received their I-frames. If P-frames are being sent in this round, the unsent P-frames are discarded. Since without the I-frames no image can be reconstructed from the P-frames and B-frames being sent afterward, I-frames have to be sent even though the round is delayed.

Session 1.

After Session 0, if T_0 is smaller than T_{round}, there is time left for Session 1. B-frames are sent to the clients according to the service QoS level determined for each of the clients in Session 1. The duration taken up by Session 1 is denoted by T_1. Again, these B-frames are sent to clients in turn. If T_{round} is reached and Session 1 has not been finished, the unsent B-frames are discarded.

Session 2.

After Session 1, if there is time left in a round, that is, $T_0 + T_1 < T_{round}$, and if there are requests for video services, and if these requests can be entertained, start-up frames will be sent to these clients in Session 2. We denote the time taken up by Session 2 T_2. Whether a new request is going to be entertained is decided with reference to the video servers' loading, network performance and the QoS of the of clients being served.

Session 3.
If there is time left in a round after Session 0 and Session 1, that is, $T_0 + T_1 < T_{round}$, video servers can also consider sending extra frames to the clients during this period if there is free space in the clients' buffer. We call this session Session 3. We call the duration consumed by Session 3 T_3. This is a session for increasing the transmission rate when bandwidth is available. The decision of having Session 3 is made upon the allowed QoS levels of the clients, the network performance, and the time left in the T_{round}. Since the bandwidth between the server and each client can be different, the throughput between the servers and every individual client can be different too. Consequently, not all clients can enjoy Session 3. Then, clients cannot be served in turn, but are served by selection according to their throughput independently. Clients are selected one at a time for sending extra frames in this session.

Example.
We illustrate the session control with an example as shown in Figure 14.4. Originally, there are three clients being served by the server. There is time left after Session 1 in the round, that is, $T_0 + T_1 < T_{round}$, the server decides to send start-up frames to Client 4. The throughput of the system is so good that there is time left in the round after Session 2, that is, $T_0 + T_1 + T_2 < T_{round}$, the server can send extra frames to clients. Since the throughput of Client 1 is not so good, extra frames are sent to Client 2, Client 3, and Client 4 only. It should be noticed that although Client 4 is newly joined, its performance is also taken into consideration in Session 3. Furthermore, since clients are selected one

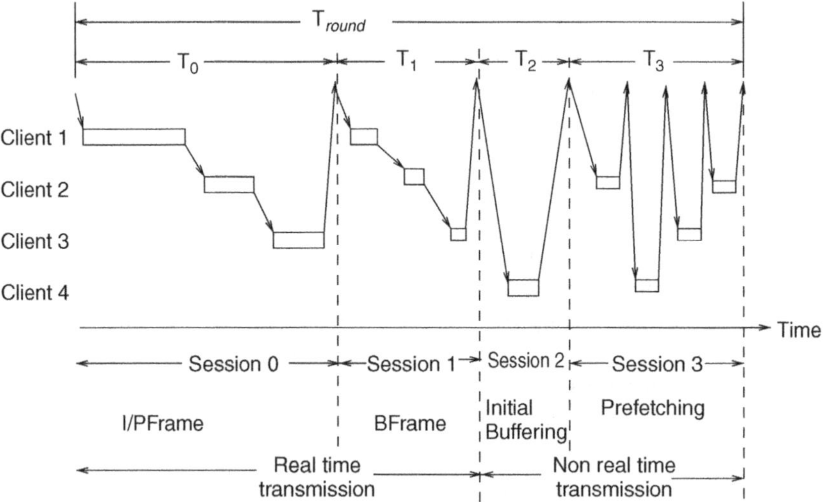

Figure 14.4 A round with four sessions

at a time in this session, the sequence of clients being served are not in the order as in Session 0 and Session 1.

14.4.3 QoS Tuning Scheme

With the QoS-GFS levels, we designed a feedback-control QoS Tuning Scheme. This scheme consists of ClientQoS, ServerQoS, and SystemQoS. Each client determines its own QoS, the ClientQoS, and feedback to the server. The server also determines the ServerQoS, which is the QoS in serving each individual client. The SystemQoS of a client is the minimum of the ClientQoS and ServerQoS for this client.

ClientQoS.
ClientQoS is determined by the amount of frames buffered at the client. Following the grading of QoS-GFS, there are five levels of QoS, where five is the best level and one is the worst level. Each of these levels is mapped to the amount of frames left in the receiving buffer of the client. The smaller the amount of MPEG frames left in the client's buffer, the lower the QoS level. These amounts of QoS level are determined by the video service provider with reference to the average frame size of the MPEG video being streamed.

ServerQoS.
The serverware determine the ServerQoS for each individual client. This reflects the performance of sending MPEG frames to that client. If the server can complete the sessions of sending I-frames, P-frames or B-frames, that is, Session 0 or Session 1, earlier than the end of the round, the Server QoS for this client is raised by one. On the other hand, if the session cannot complete before the end of the round, the ServerQoS of this client is dropped by one.

14.5 Architecture

This middleware is placed between the video servers on the server side and the HTTP MPEG players on the client sides as shown in Figure 14.5. This middleware consists of two main components, the serverware on the server side and the clientware at the client sides. Serverware and clientware is of one-to-many relation, as the server is serving a number of clients concurrently. The middleware uses the QoS Control Scheme and QoS-GFS to tune the QoS for each individual clients. It also uses the transmission scheme for transmitting frames to clients.

Serverware and clientware are synchronized by feedback-control mechanism. Clientware feedback their ClientQoS and the amount of frames waiting to be played back to the serverware. Referencing to these information, the serverware tune the QoS for each individual client on the fly.

14.5.1 Clientware

A clientware is installed at each individual client. Clientware are responsible for the following functions.

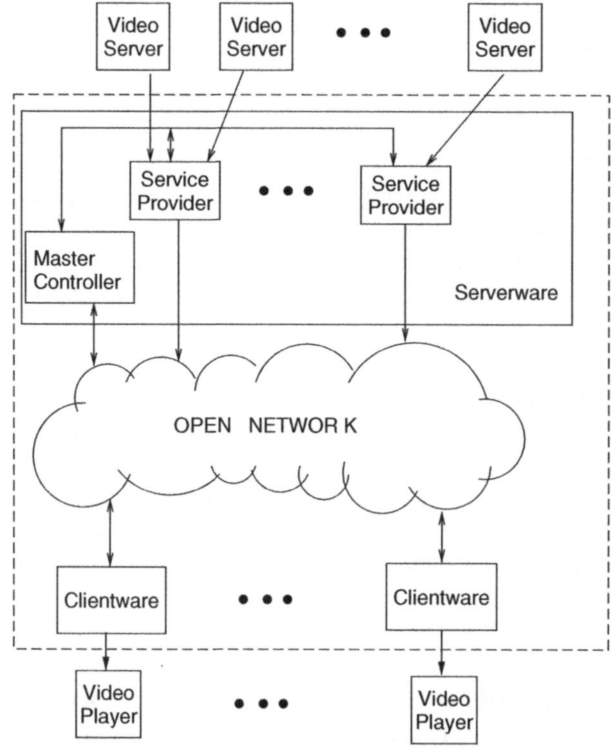

Figure 14.5 Distributed MPEG-streaming middleware architecture

- To receive frames from serverware
- To provide temporary storage for the frames
- To feed frames including the substitute frames to the video player
- To detect the buffer status and QoS level and feedback these information to the serverware.

These functions are performed by the components FramesReceiver, FrameBase, PlaybackPort, FrameBufferLevelDetector, QoSDetector, and Logging. The architecture of the clientware is shown in Figure 14.6.

FramesReceiver.

FramesReceiver consists of three components, namely, TCPModule, UDPModule, and PacketHandler. The TCPModule and UDPModule are responsible for receiving packets from the serverware, depending on using TCP or UDP protocols. The PacketHandler is responsible for removing the header and tailer of packets and pass it to the FrameBase.

FrameBase.

FrameBase consists of buffer for storing the received packets. Because of packet size limit, a frame may have to be split into several packets and transmitted to the clientware

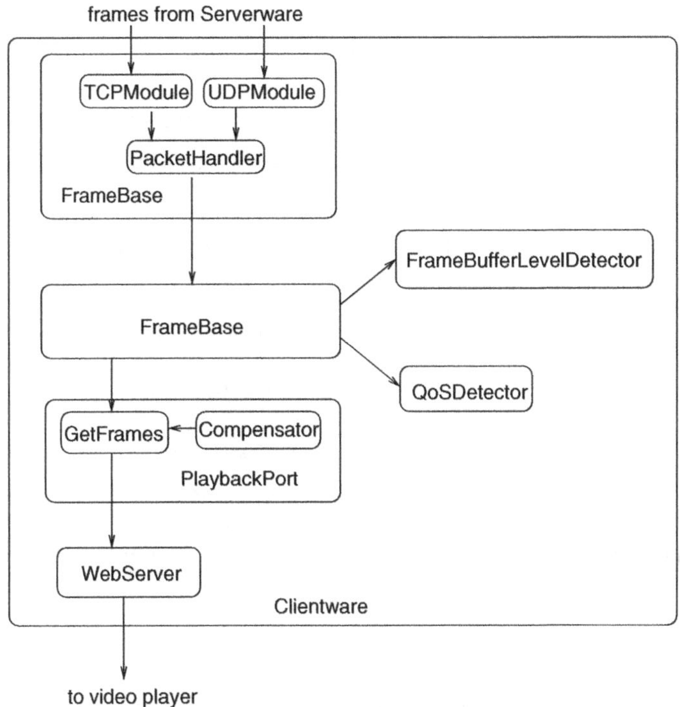

Figure 14.6 Clientware architecture

separately. FrameBase stores all these packets in the buffer and puts them back into frames when all the packets of the frames are received. Frames may not arrive at the client side in the proper sequence; it is the FrameBase that puts frames in sequence. FrameBase is also responsible for feeding frames to the PlaybackPort for video playback.

PlaybackPort.

The PlaybackPort serves two functions. The GetFrames component retrieves frames from the FrameBase. It is also the GetFrames that is responsible for the timing and sequencing of sending out frames. If the retrieved frame is a corrupted one or is missing or late due to network delay, the Compensator of PlaybackPort replaces the frame by a null P-Frame, that is, a P-Frame with null movement vector. These frames are then fed to the WebServer for streaming to the MPEG video player.

WebServer.

The local video player communicates with the WebServer by taking the WebServer as a local host. In connecting to the WebServer, the URL of the video server is passed to the WebServer in the form of parameter. The WebServer then connects with the serverware of the video server site and performs the housekeeping process such as user authentication and video selection. Afterward, the QoS-controlled MPEG video–streaming service takes place.

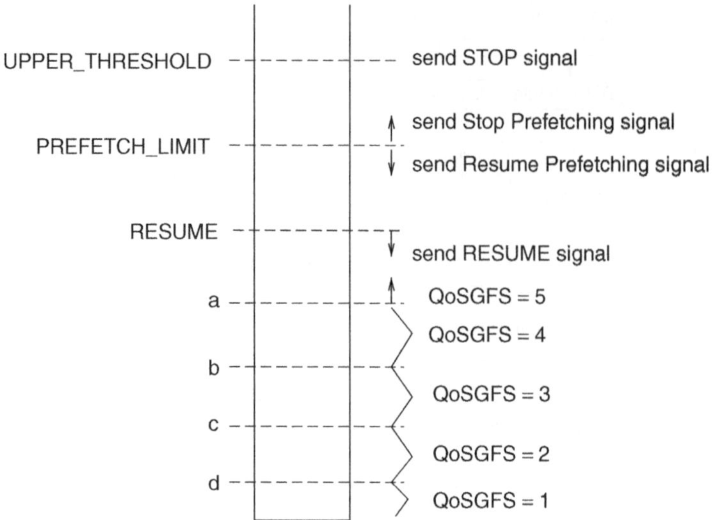

Figure 14.7 The marks for FrameBufferLevelDetector & QoSDetector

FrameBufferLevelDetector.

The FrameBufferLevelDetector detects the level of buffer in byte. A Stop Prefetching signal is sent to the serverware, requesting the serverware to stop sending extra frames to this client when the buffer level raise through PREFETCH_LIMIT mark. When the buffer level drops through the PREFETCH_LIMIT mark, it sends the Resume Prefetching to the serverware, informing the serverware to resume sending extra frames to it when the bandwidth is possible. When buffer level raises through the UPPER_THRESHOLD mark, it sends a Stop signal to the serverware, requesting the serverware to stop sending any frame to it because the buffer is full. When the buffer level drops through the RESUME mark, a Resume signal is sent to the serverware, requesting to resume sending frames to it. The marks are shown in Figure 14.7.

QoSDetector.

The QoSDetector continuously detects the buffer level in terms of amount of frames. This is to determine the ClientQoS and feedback the ClientQoS to the serverware, that is, this is the implementation of the ClientQoS of the QoS Tuning Scheme. According to the QoS-GFS, there are five levels of QoS and the best QoS level is level five. Four QoS level marks, namely, a, b, c and d, are labeled at the buffer, each marking different buffer level as shown in Figure 14.7. These marks, usually, are put below the RESUME mark. The QoSClient above mark a is five. When the buffer drops through a QoS level mark, the QoS level is reduced by one. For example, when the buffer level drops through mark a, the QoS level is four. When the buffer level drops through mark b, the QoS level is three. On the other hand, when the buffer level rises through a QoS level mark, the QoS level is raised by one. Hence, when the buffer level rises through mark b, the QoS level is raised from three to four.

14.5.2 Serverware

The serverware has the following responsibilities.

- To determine the best QoS of MPEG video playback that can be supported by each individual client on the fly.
- To transmit MPEG frames to clients according to their best QoS that can be supported.

The first responsibility is performed by the component MasterController, while the second one is performed by the components ServiceProviders. There is only one MasterController, but they can have many ServiceProviders. Furthermore, each ServiceProvider can serve a number of clients. Figure 14.8 shows the architecture of the serverware with a MasterController and one ServiceProvider.

MasterController.
MasterController is the key component of the middleware. The responsibilities of MasterController are

1. to regulate the streaming of frames to all the clients; and
2. to determine the QoS of each individual client on the fly.

SessionController.
SessionController manages the video-streaming services to be provided to clients. These include deciding whether new requests from clients are entertained, session operation, and determining the duration of sessions, that is, it is the controller in the implementation of the transmission scheme. SessionController manages the session operation by sending the session type to the StreamSelector of the ServiceProvider for execution.

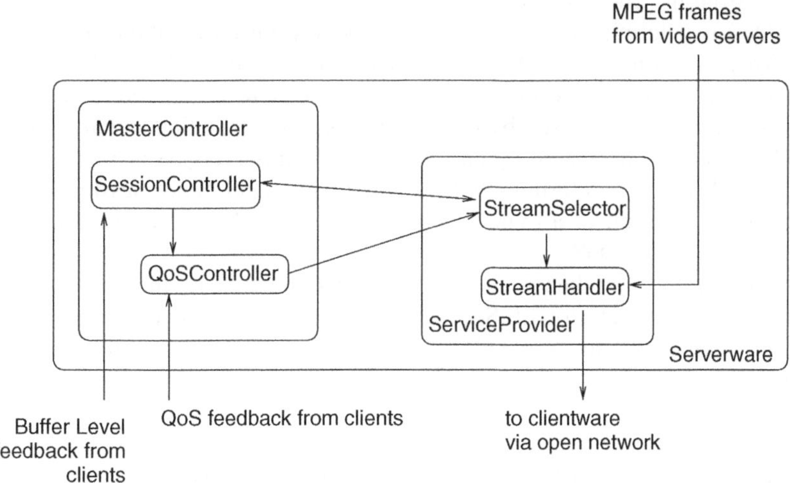

Figure 14.8 Architecture of a Serverware with a MasterController and one ServiceProvider

When a client requests for streaming service, SessionController determines whether this request can be entertained by referencing to the throughput of the whole system. Throughput is indicated by the QoS level of all the clients being served at that moment. If the QoS of most of the serving clients are low, the new request will not be entertained. Otherwise, some Session 2 is to be scheduled for sending the start-up frames.

If there is time left in a round and if there are new clients waiting to be served, SessionController will decide whether using the time left as Session 2 for sending start-up frames to the new client or using it as Session 3. If there is no new client waiting to be served, SessionController will make the time left as Session 3. If the time left is used as Session 3, SessionController will look for clients with the spare bandwidth for sending extra frames.

SessionController also monitors the feedbacks from clients. These feedbacks include Stop, Resume, Stop Prefetching and Resume Prefetching. When a Stop signal is received from a client, this means that the buffer at that client is almost full. SessionController will not arrange sending any frame to this client until the Resume signal is received from this client. When a Stop Prefetching signal is received from a client, it means that the buffer level of this client is very high. SessionController will not arrange sending extra frames to this client even when there is time left in a round for Session 3 until the Resume Prefetching signal is received.

SessionController also helps in determining the server QoS. It informs the QoS Controller the start time of a session. When the session completion feedback signal from the StreamHandler of ServiceProvider is received, SessionController forwards this signal to QoSController. QoSController can then determine the server QoS of this client with reference to the timing of these signals.

QoSController.

The main function of QoSController is to determine the QoS of each individual client. It implements the SystemQoS and the ServerQoS of the QoS Tuning Scheme. The SystemQoS of a client is the minimum of ClientQoS and ServerQoS of that client. ClientQoS is received from the client's feedback.

ServerQoS is determined with reference to the session information from SessionController. When a session starts, QoSController is informed. If the session is a Session 0 or a Session 1 and a session completion signal is received before the duration is expired, ServerQoS is increased by one. If the session completion signal is late or missing, the ServerQoS is decreased by one. The maximum ServerQoS is 5 and the minimum ServerQoS is 1, that is, conform with the scale of ClientQoS. Since Session 2 and Session 3 are using the time left of a round, the performance of these two session types has no influence on the ServerQoS.

The SystemQoS of all clients are determined continuously by the QoSController. Furthermore, the StreamSelector of ServiceProvider is being updated continuously. This arrangement is to allow the ServiceProvider transmitting frames to clients with the best allowed QoS.

ServiceProvider.

ServiceProvider consists of a StreamSelector and a StreamHandlers. The functions of ServiceProvider is to transmit frames to the clients in the assigned QoS. A ServiceProvider can be installed on a machine different from the SessionController. Furthermore, a system can have more than one ServiceProvider.

StreamSelector.
The main purpose of StreamSelector is managing frame transmission to clients. Information from SessionController includes the session type and clients to be served. If the session type is Session 0 or Session 1, clients are served in a round-robin manner. StreamSelector instructs the StreamHandler to transmit frames to the clients with the serving QoS, one by one. If the session type is Session 2, StreamSelector instructs the SteamHandler to set up a new stream for the new client and start sending the start-up frames to the client. If the session type is Session 3, StreamSelector instructs the StreamHandler to transmit frames to each of the selected clients with the serving QoS, one at a time.

StreamHandler.
StreamHandlers are the first-line components for performing QoS-enabled MPEG streaming. It gets the MPEG frames from video servers and stores them stream by stream. When a client is to be served, according to the serving QoS, frames may be discarded by the StreamHandler, if necessary. Since frame size may be larger than the packet size for transmission to the client, frames are fragmented to fit the packets. StreamHandler wraps the packets with either TCP or UDP transmission protocol tags, depending on the protocol being used between the server and the client. Then these packets are sent to the client.

14.6 Experiments

14.6.1 System Setup

We have developed the distributed QoS-enabled MPEG-streaming middleware for studying its viability. The system configuration of the experimental system and the testing environment is as follows:

- Servers and clients: Windows 2000 Server (for server), Windows 2000 professional (for clients)
- Video Servers: Pentium II 800 MHz PC
- Clients: Pentium II 766 MHz PC
- Network: Ethernet (10 Mbps) for both video servers and clients
- Video Stream: 3 videos (action type - RACING01.mpg, news type – news.mpg and cartoon type – anim0001.mpg) with video clips resolution of 352 288 at 25 fps
- GOP pattern: IBBPBBPBBPBB
- Average bandwidth for the MPEG with no dropped frame: 1150 kbps
- Number of clients: varies between 1 to 10
- Upper threshold (UPPER_THRESHOLD) of the client buffer (in Mbyte): 0.6
- Startup latency (START_LATENCY) (in number of frames): 51
- TRIGGER_LEVEL (in number of frames): 40
- Prefetch buffering level(in number of frames) is equal to STARTUP_LATENCY
- Fixed PREFETCHING_LIMIT (ADVBUFFER_LIMIT) to 1

- The default value of trigger level for client QoS (in number of frames): a = 22, b = 16, c = 7, and d = 3.

14.6.2 Experiment Results

We validate the effectiveness of the middleware by comparing the overall throughput of the system with the middleware and without the middleware. We measure the throughput by means of QoS-Index. With our previous study [20, 23], the MPEG video playback quality also depends on the video content or the type of video. Three types of video, namely, action, talking head, and cartoon, have been identified. Action videos, such as car racing videos, are the videos with relatively big changes in the images. Talking head videos, such as news broadcast, are the videos with relatively small changes in the images. Cartoons are the video, whose image contents, in general, are drawings.

In our study, we validate the effectiveness of the middleware with all these three kinds of video. Figure 14.9 show the QoS-Index of action video with and without middleware. and Figure 14.10 and Figure 14.11 show the QoS-Frame and QoS-Byte of action video with and without middleware respectively. The QoS-Index, QoS-Frame, and QoS-Byte of with and without middleware of talking head type video are shown in Figure 14.12, Figure 14.13, and Figure 14.14 respectively. Figure 14.15, Figure 14.16 and Figure 14.17, respectively, show the QoS-Index, QoS-Frame, and QoS-Byte for cartoon videos with and without middleware.

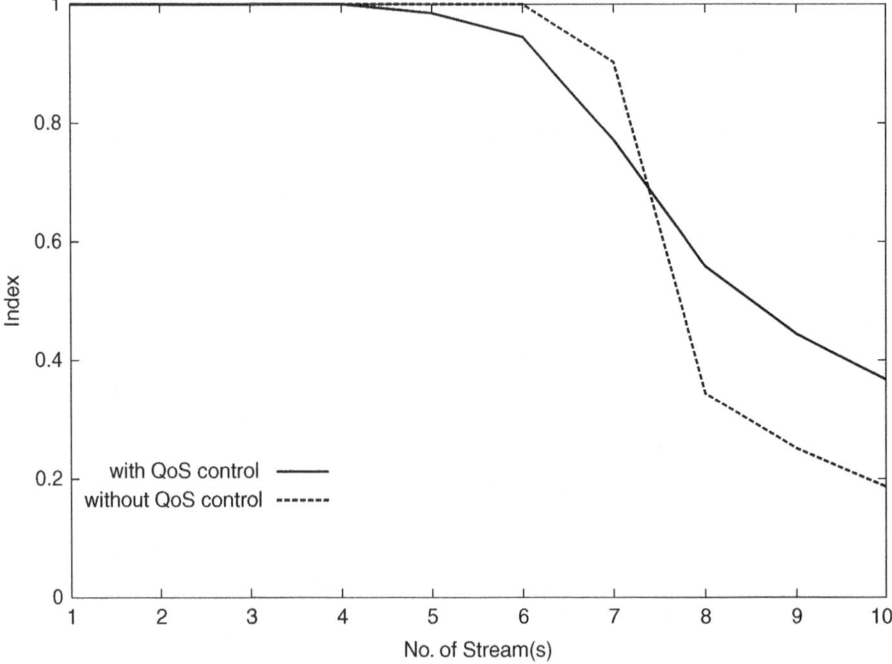

Figure 14.9 QoS-index of system with/without QoS control for action video

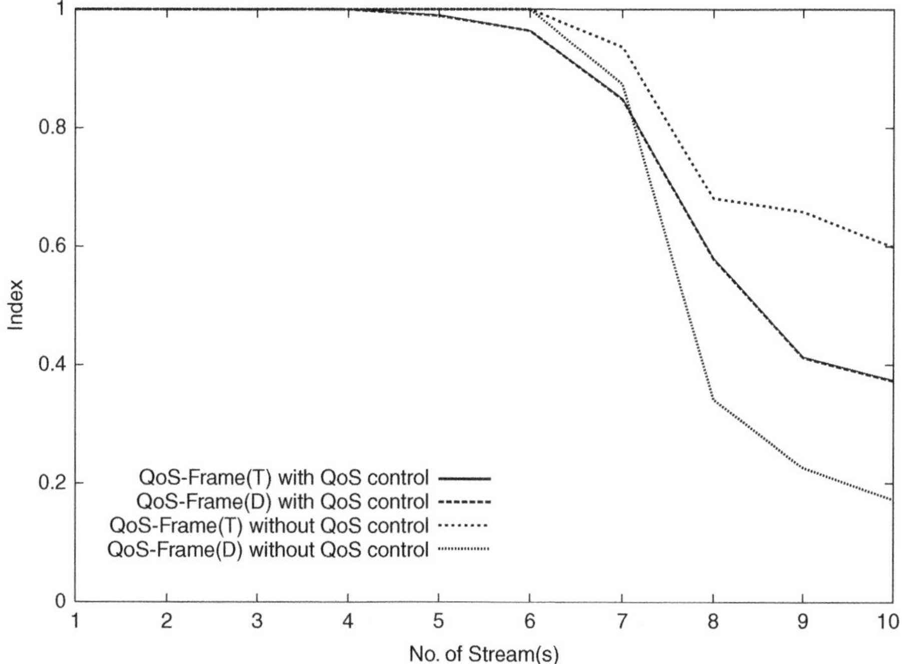

Figure 14.10 QoS-Frame of system with/without QoS control for action video

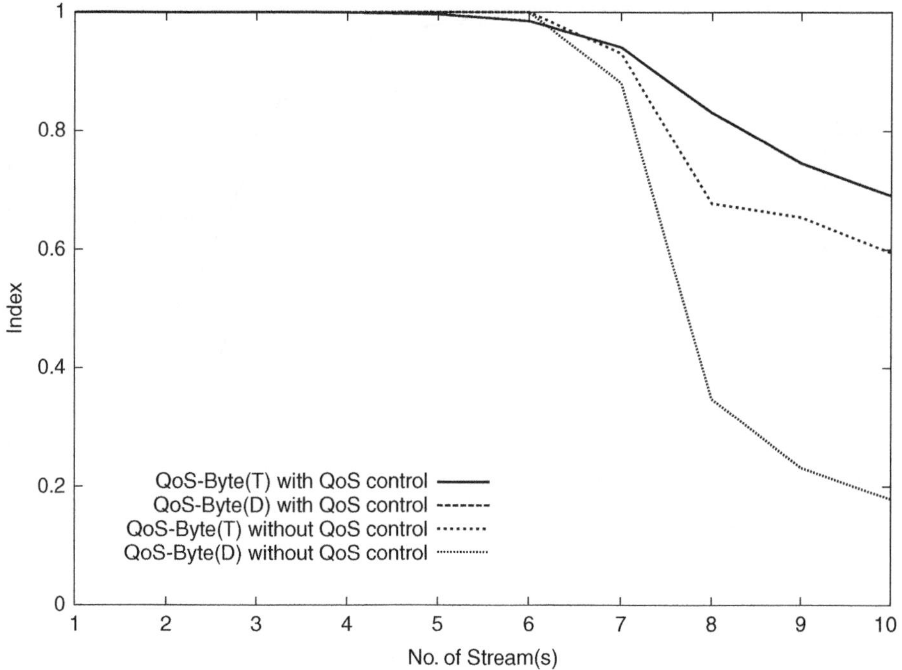

Figure 14.11 QoS-Byte of system with/without QoS control for action video

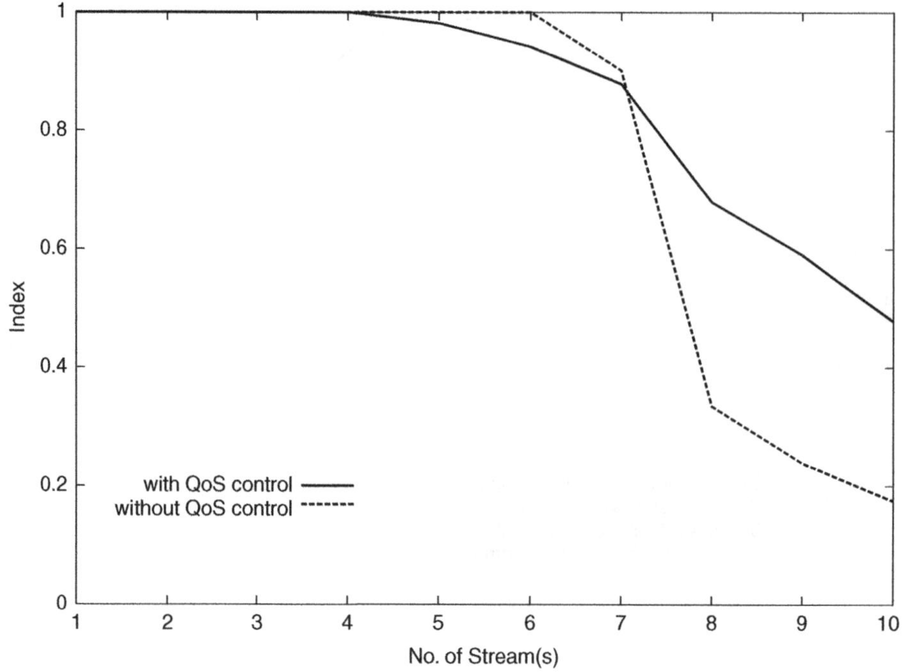

Figure 14.12 Talking head: QoS-index w/wo QoS control

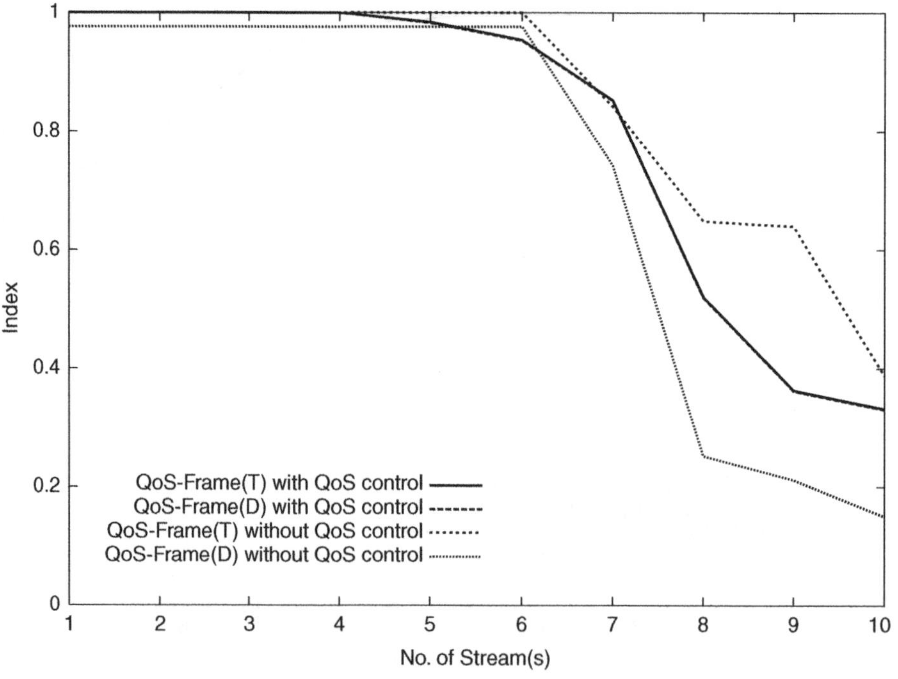

Figure 14.13 Talking Head: QoS-Frame w/wo QoS control

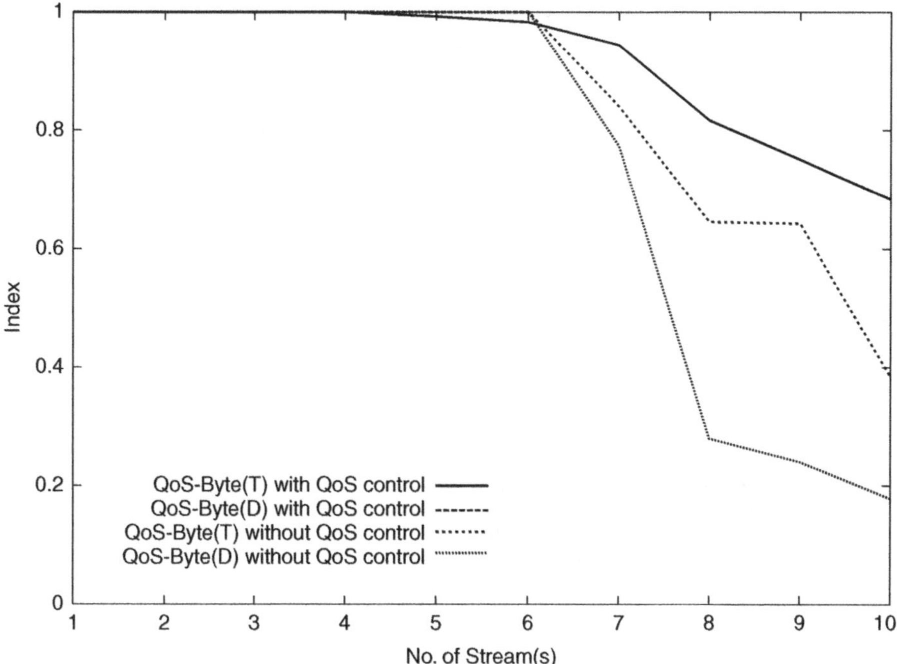

Figure 14.14 Talking head: QoS-Byte w/wo QoS control

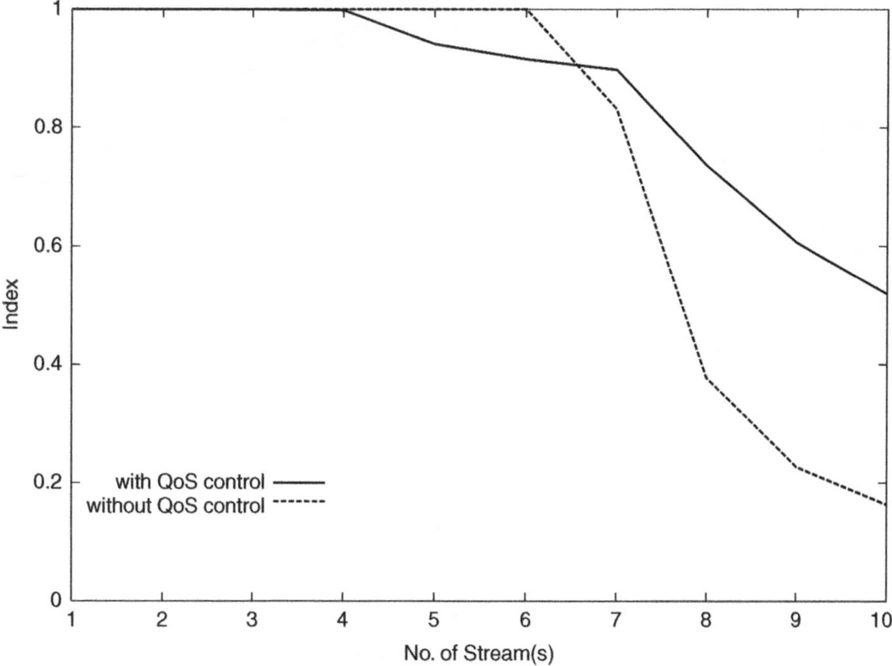

Figure 14.15 Cartoon: QoS-Index w/wo QoS control

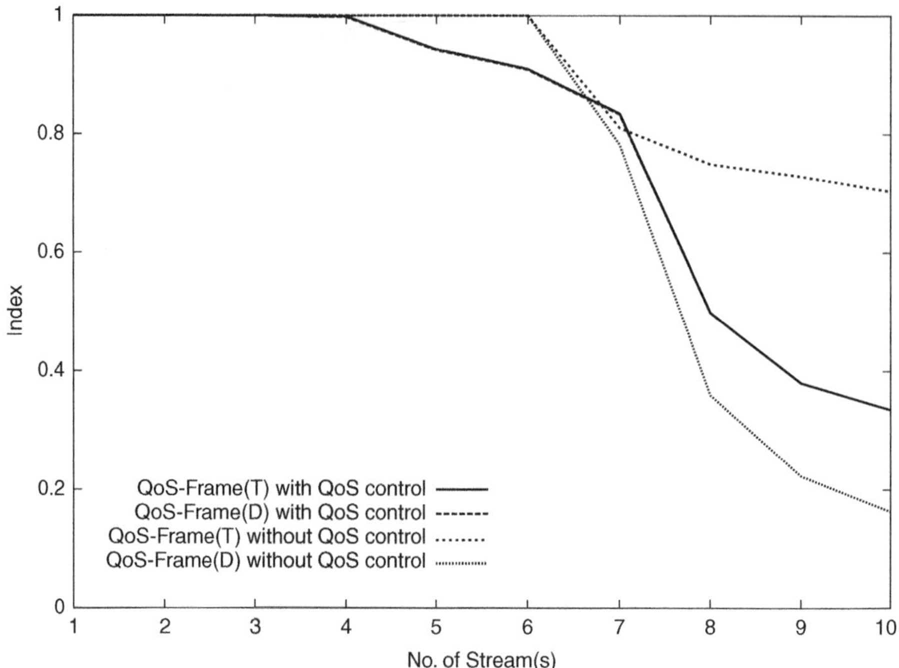

Figure 14.16 Cartoon: QoS-frame w/wo QoS control

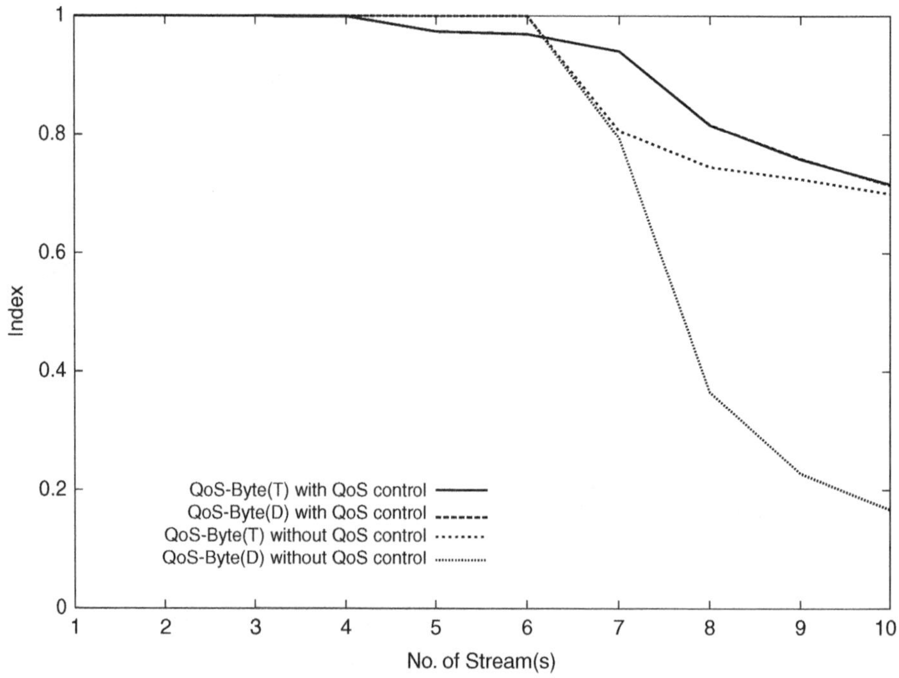

Figure 14.17 Cartoon: QoS-byte w/wo QoS control

We find that the effectiveness of the middleware on all these three types of videos is consistent with each other. Hence, we do not separate our discussion on the effectiveness according to the types of video.

14.6.3 QoS in QoS-Index

This study validates the effectiveness of the middleware by measuring the QoS perceived by the viewers at the clients. The QoS of using middleware when dealing with seven or more streams out-perform the QoS without using the middleware by around 50%. Because of the overhead of the middleware, the QoS of using middleware is not as good as without middleware by around 10% when dealing with three to six streams. When dealing with three streams or less, the QoS are the same. This means when there is sufficient bandwidth, whether there are middleware makes no difference.

14.6.4 QoS-Frame and QoS-Byte

In all cases, with the middleware, the QoS-Frame of transmission is equal to the QoS-Frame of Display. This implies that all the frames being transmitted are being played back. No frame is discarded by the clients. Again in all cases, without the middleware, starting from dealing with six streams, the QoS-Byte and QoS-Frame of transmission is larger than that of playback. Furthermore, the more clients the server is serving, the greater the difference in the QoS of transmission and playback. This implies that some frames are discarded by the clients, that is, the bandwidth has not been effectively utilized because of the transmission of discarded frames.

14.7 Discussions

Usage of the middleware.

The middleware provides transparent QoS control for MPEG video streaming. On the server side, the video servers communicate with the serverware as if they were communicating with the clients directly. The WebServer of clientware is acting as a local web server or a proxy. Hence, the video player can access the clientware simply in the way of accessing local host together with the destination of the video server site. Then the clientware will connect to the serverware of the video server site and perform the requested housekeeping process and MPEG video streaming. Hence, the video players communicate with the clientware as if they were communicating with the video servers. Furthermore, QoS control developers can use their QoS metric, QoS tuning scheme, and transmission scheme by replacing the FrameBufferLevelDetector and QoSDetector at the clientware and MasterController at the serverware.

To conclude, this distributed middleware is providing transparent QoS control MPEG-streaming service to regular users and also provides a platform for QoS control developers.

Effectiveness of middleware.

With our experiment results in Section 14.6.2, this middleware can effectively enhance the QoS to be provided to individual clients. Furthermore, our middleware also effectively reduces the amount of frames being discarded by the clients. This is a reduction in the waste of bandwidth and hence the bandwidth is effectively used in the video streaming.

Independent of playback device.
This architecture is designed for video streaming with the HTTP protocol. Clients can use any HTTP protocol MPEG player to enjoy the streaming. This is because a Web Server is included in the architecture, which streams MPEG frames to the video player with the HTTP protocol. With this design, all users can enjoy the enhanced streaming quality with their favorite MPEG players.

Independent of open network.
The open network is very complex, performance unstable and nonpredicatable, and uncontrollable. However, network performance is one of the key factors affecting the video streaming quality. Our middleware is a design that allows the adoption of QoS control scheme, bypassing the difficulties in controlling the network. It copes with the unstable network performance and adjusts the frames being sent to the clients with a feedback-control mechanism.

Customize QoS control.
Server can provide video streaming different QoS to different clients. QoS of each client are customized in real time individually. This provides the most fit QoS to each individual client. Our design considers the throughput of the server and each client individually and independently. This makes all the clients to enjoy the best quality of video streaming with their allowed throughput among themselves and the server.

Novel middleware design I: Distributed middleware.
Most middleware is a single system being installed at the server to provide value-added services to the clients. Few of these middleware can provide feedbacks of each individual client, and hence the servers do not have good information for providing customized services to each individual client. Our design extends the notion of middleware to client-server. It is a client-server system, which is placed between the server and the clients. This design allows the server to provide customized services to each individual client.

Novel middleware design II: Dynamic number of components.
The total number of components of this middleware is the number of clients being served plus the number of ServiceProviders set up at the serverware plus one. There is no predefined number of clients and not all clients are being served at the same time. Hence, the number of components of this middleware design is dynamic.

14.8 Acknowledgment

The work reported in this chapter was supported in part by the RGC Earmarked Research Grant under HKBU2074/01E.

14.9 Conclusion & Future Works

There are few video systems that can provide the support for streaming MPEG video over an uncontrolled network while still providing some guarantee on the Quality of Services (QoS) of the video being displayed. Throughout this paper, we propose a new distributed MPEG video streaming middleware that allows clients using their favorite HTTP MPEG video players to enjoy the video-streaming services over the open network. This middleware is a distributed system that is placed between the video server and the clients to provide the streaming service. This distributed middleware consists of a serverware and some clientware and is working with a feedback-control mechanism. The serverware is located at the video server side while each client is installed with a clientware. The serverware and clientware work together to tune the QoS of each individual client by means of a QoS-tuning scheme. MPEG frames are transmitted from serverware to clientware by a transmission scheme. Our experiment shows that this middleware can effectively enhance the QoS of MPEG video playback and increase the number of clients to be served at the same time with reasonable QoS. For demonstration purpose, interested users can download our experimental system at `http://www.comp.hkbu.edu.hk/~jng/QMPEGv2`.

Bibliography

[1] Baiceanu, V., Cowan, C., McNamme, D., Pu, C., and Walpole, J. (1996) Multiple Applications Require Adaptive CPU Scheduling. Proceedings of the Workshop on Resource Allocation Problems in Multimedia Systems, Washington DC, December 1–2, 1996.

[2] Bolosky, W. J., Barrera, J. S., III, Draves, R. P., Fitzgerald, R. P., Gibson, G. A., Jones, M. B., Levi, S. P., Myhrvold, N. P., and Rashid, R. F. *Proceedings of the Sixth International Workshop on Network and Operating System Support for Digital Audio and Video (NOSSDAV '96)*, April 23–26; Shonan Village International Conference Center, IEEE Computer Society, Zushi, Japan.

[3] Bolosky, W., Fitzgerald, R., and Douceru, J. (1997) Distributed Schedule Management in the Tiger Video Fileserver. Proceedings of SOSP'97, SaintMalo, France, October 1997; Also available at http://www.research.microsoft.com/rearch/os/bolosky/sosp/cdrom.html.

[4] Bolot, J. and Turletti, T. (1994) A Rate Control Mechanism for Packet Video In the Internet. Proceedings of INFOCOM'94, Toronto, Canada, June 1994.

[5] Cen, S., Pu, C., and Staehli, R. (1995) A Distributed Real-time MPEG video Audio Player. Proceedings of the 5th International Workshop on NOSSDAV'95, April 18–21, Durham, New Hampshire, April 18–22 1995.

[6] Chorus Systems (1996) Requirements for a Real-time ORB, ReTINA, Technical Report RT/RT-96-08.

[7] Cowan, C., Cen, S., Walpole, J., and Pu, C. Adaptive Methods for Distributed Video Presentation. ACM Computing Sureys, 27(4), 580–583.

[8] Gall, D. L. (1991) Mpeg: A Video Compression Standard for Multimedia Applications. Communications of the ACM, 34(4), 46–58.

[9] Gringeri, S., Khasnabish, B., Lewis, A., Shuib, K., Egorov, R., and Basch, B. (1998) Transmission of MPEG-2 Video Streams Over ATM. IEEE Multimedia, **5**(1), 58–71.

[10] Hasegawa, T., Hasegawa, T., and Kato, T. (1999) Implementation and Evaluation of Video Transfer System over Internet with Congestion Control Based on Two Level Rate Control. Proceedings of RTCSA'99, New World Renaissance Hotel, Hong Kong, China, pp. 141–148.

[11] InterVU. Intervu. www.intervu.com, September 1999.

[12] Ismail, M. R., Lambadaris, I., Devetsikiotis, M., and Kaye, A. R. (1995) Modeling Prioritized MPEG Video Using TES and Frame Spreading Strategy for Transmission in ATM Networks. Proceedings of IEEE INFOCOM'95, Boston, MA, April 1995, pp. 762–770.

[13] Izquierdo, R. and Reeves, D. (1995) Statistical Characterization of MPEG VBR Video at the SLICE Layer. Proceedings of SPIE Multimedia Computing and Networking, Volume 2417, San Diego, CA, September 1995.

[14] Kanakia, H., Mishra, P., and Reibman, A. (1993) An adaptive Congestion Control Scheme for Real-Time Packet Video Transport. Proceedings of ACM SIGCOMM '93, San Francisco, CA, September 13–17 1993, pp. 20–31.

[15] Kath, O., Stoinski, F., Takita, W., and Tsuchiya, Y. (2001) Middleware Platform Support for Multimedia Content Provision. Proceedings of EURESCOM Summit 2001: 3G Technologies and Applications, Heidelberg, Germany.

[16] Krunz, M., Sass, R., and Hughes, H. (1995) Statistical Characteristics and Multiplexing of MPEG Streams. Proceedings of IEEE INFOCOM'95, Boston, MA, April 1995.

[17] Lam, S., Chow, S., and Yau, D. (1994) In Algorithm for Lossless Smoothing of MPEG Video. Proceedings of ACM SIGCOMM 94, London, UK, pp. 281–293.

[18] Mitchell, J. L., Pennebaker, W. B., Fogg, C. E., and Legall, D. J. MPEG Video Compression Standard, pp. 1–49.

[19] NetShow from microsoft, Available at http://www.microsoft.com.

[20] Ng, J. and Lee, V. (2000) Performance Evaluation of Transmission Schemes for Real-Time Traffic in a High-Speed Timed-Token MAC Network. Journal of Systems and Software, 54(1), 41–60.

[21] Ng, J., Wai, H. K., Xiong, S. H., and Du, X. W. (1998) A Distributed MPEG video Player System with Feedback and QoS Control. Proceedings of the Fifth International Conference on Real-Time Computing Systems and Applications (RTCSA'98), Hiroshima, JAPAN, October 1998, pp. 91–100.

[22] Ng, J., Leung, K., Wong, W., Lee, V., and Hui, C. (2002) A Scheme on Measuring MPEG Video QoS with Human Perspective. Proceedings of the *8th* International Conference on Real-Time Computing Systems and Applications (RTCSA 2002), Tokyo, Japan, March 18–20, pp. 233–241.

[23] Ng, J. (1996) A Study on Transmitting MPEG-I Video over a FDDI Network. Proceedings of the Third International Workshop on Real-Time Computing Systems and Applications (RTCSA'96), November 1996, pp. 10–17; IEEE Computer Society Press, Seoul, Korea.

[24] OMG (1997) Control and Management of A/V Streams. OMG Document, telecom/97-05-07 edition.

[25] Ott, T., Lakshman, T., and Tabatabai, A. (1992) Scheme for Smoothing Delay-sensitive Traffic Offered to ATM Networks. Proceedings of IEEE INFOCOM'92, Florence, Italy, pp. 776–765.

[26] Pancha, P. and Zarki, M. E. (1993) Bandwidth Requirement of Variable Bit Rate MPEG sources in ATM Networks. Proceedings of IEEE INFOCOM'93, Hotel Nikko, San Francisco, CA, pp. 902–909.

[27] Plagemann, T., Eliassen, F., Hafskjold, B., Kristensen, T., MacDonald, R., and Rafaelsen, H. (2000) Flexible and Extensible QoS Management for Adaptable Middleware. Proceedings of International Workshop on Protocols for Multimedia Systems (PROMS 2000), Cracow, Poland, October 2000.

[28] RealNetworks, Realplayer and realsystems g2, www.realaudio.com.

[29] Reibman, R. and Berger, A. (1992) On VBR video Teleconferencing over ATM Networks. Proceedings of IEEE GLOBECOM'92, Orlando, FL, pp. 314–319.

[30] Reininger, R., Raychaudhuri, D., Melamed, B., Sengupta, B., and Hill, J. (1993) Statistical Multiplexing of VBR MPEG Compressed Video on ATM Networks. Proceedings of IEEE INFOCOM'93, Hotel Nikko, San Francisco, CA, pp. 919–926.

[31] Siqueira, F. and Cahill, V. (2000) Quartz: A QoS Architecture for Open Systems. 20th International Conference on Distributed Computing Systems (ICDCS'00), Taipei, Taiwan, April 2000.

[32] Tech, X. Streamworks, www.xingtech.com, September 1999.

[33] Tokuda, H. (1994) Operating System Support for Continuous Media Applications, ACM Press, Addison-Wesley Publishing Company, pp. 201–220.

[34] VDOnet. Vdolive player, www.clubvdo.net.

[35] Wai, H. K. and Ng, J. (1998) The Design and Implementation of a Distributed MPEG Video System. Proceedings of the First HK ACM Postgraduate Research Day, October 1998, pp. 101–107.

[36] Wai, H. K. (1999) Priority feed back mechanism with quality of service control for MPEG video system. MSc Thesis, Department of Computer Science, Hong Kong Baptist University.

15

Middleware for Smart Cards

Harald Vogt[1], Michael Rohs[1], Roger Kilian-Kehr[2]

[1]*Swiss Federal Institute of Technology (ETH Zurich)*
[2]*SAP Corporate Research*

15.1 Introduction

Smart cards are credit card–sized plastic cards with an integrated microcontroller chip. This chip is protected against physical and logical tampering, thus unauthorized access to internal data structures is virtually impossible. This makes a smart card an excellent device for storing secret cryptographic keys and other sensitive data. In practice, smart cards are used for applications such as digitally signing documents, ticketing, controlling access to desktop computers, and authenticating the users of mobile phone networks. Sometimes, smart card functionality is provided by other appliances such as USB "crypto tokens," rings, or the GSM SIM (subscriber identity module). Although these devices look differently, they technically just differ in the interface to the host they connect to, thus we will treat them as smart cards as well.

In order to work together with a host computer, smart cards require an additional device that provides an electrical interface for data exchange, a so-called smart card reader. Nowadays they can be found built into an increasing number of desktop computers. This makes smart card services available to the full range of PC and Web/Internet applications and lets them play a major role in payment schemes and for access control on the Internet.

The need for middleware and system support became immediately clear when smart cards started to gain ground on the common computing platforms. Application developers cannot be expected to deal with the large number of different, manufacturer-dependent interfaces that are offered by smart card readers and smart card services. Therefore, smart card middleware is necessary for mediating between these services and application software. In such a system, it should be possible to replace one type of smart card with another one offering similar features, without affecting the application level. The major requirements to smart card middleware are therefore encapsulation of communication specifics, interoperability, and system integration.

Middleware for Communications. Edited by Qusay H. Mahmoud
 ISBN 0-470-86206-8

Standardization plays a major role in the smart card business. The lower system and communication layers have been standardized for a long time. The ISO 7816 standard [1] was established around 1989 and specifies not only the form factor and the electrical properties of the communication interface but also a basic file system and card holder verification (the well-known PIN), among other features. Ever since, it was extended to embrace new requirements of smart card vendors. For the integration with host computers, PC/SC [2] is the *de facto* industry standard since it was adopted by the major desktop operating systems.

In this chapter, we present an overview of the existing approaches to smart card middleware. Some of them were inspired by the development of powerful microprocessors for smart cards that made it possible to run a Java interpreter. Generally, the availability of the Java programming language for smart cards increased their popularity and made lots of new applications possible. As we will see later in the chapter, this also made it easier to integrate them into distributed computing environments. This shows that, with the right kind of middleware, the power of smart cards can be effectively utilized.

15.2 ISO 7816

We start by briefly introducing the most basic communication principles of ISO-compliant smart cards. The electronic signals and transmission protocols of smart cards are standardized in ISO 7816 [1]. This standard consists of several parts, which define the characteristics of smart cards on various layers. It comprises physical characteristics, like the position of the contacts on the card, electrical characteristics, transmission protocols between card and card reader, instruction sets of frequently used card commands, and more. ISO 7816 does not cover the interface between the card reader and the PC.

15.2.1 Communication between Card and Card Reader

When a card is inserted into the card reader, the card is powered up and the RST (reset) line is activated. The card responds with an "answer to reset" (ATR), which gives basic information about the card type, its electrical characteristics, communication conventions, and its manufacturer. An interesting portion of the ATR are the "historical characters" or "historicals." The historicals are not prescribed by any standard. They mostly contain information about the smart card, its operating system, version number, and so on, in a proprietary format. In addition to the ATR, ISO 7816-4 [3] defines an ATR file that contains further information about the ATR. The ATR and the ATR file play an important role for the integration of smart cards into middleware systems. The actual communication between card and card reader takes place according to the protocol parameters specified in the ATR.

15.2.1.1 Communication Protocols

The ATR contains information about the communication protocols the card supports. The protocols used most often are called $T = 0$ and $T = 1$. $T = 0$ is an asynchronous character-oriented half-duplex protocol that is relatively easy to implement. $T = 1$ is a block-oriented half-duplex protocol. It is more complex than $T = 0$ and ensures the reliable delivery of data blocks.

15.2.1.2 Application Layer

On the application layer, requests and responses between card and card reader are exchanged in so-called application protocol data units (APDUs). Command APDUs are requests from the card reader to the card, while response APDUs contain the corresponding answers of the smart card. APDUs are defined independently of the communication protocol ($T = 0$ or $T = 1$).

Command APDUs.
A command APDU specifies a request to the card. It consists of a header of fixed length and an optional body of variable length. The structure of command and response APDUs is shown in figure 15.1. The CLA byte specifies the instruction class, which also defines the requested card application (e.g., 0xA0 for GSM). The INS byte is the operation code of the instruction to execute. P1 and P2 are parameters for the operation. The body of a command APDU has a variable structure, depending on whether data bytes are sent in the command APDU, expected in the response APDU, or both. The Lc byte specifies the number of data bytes, the Le byte the number of bytes expected in the response APDU.

Response APDUs.
They consist of an optional body, containing the application data of the answer corresponding to the request, and a mandatory trailer, consisting of a status word that indicates the correct execution of the command or the error reason.

15.3 Data Structures on Smart Cards

ISO 7816-4 specifies a file system as one possibility to organize the data on the card's EEPROM. All mutable data is stored in files of various types. Access to sensitive files can be bound to security conditions, like the authentication of the user to the card. Smart card files have an inner structure: they can be byte- or record-oriented. In addition to the core file system itself, ISO 7816-4 defines an instruction set to access and administer the files in a well-defined manner. The ISO 7816-3 file system is organized hierarchically: There are directory files containing other files, and elementary files, containing application data.

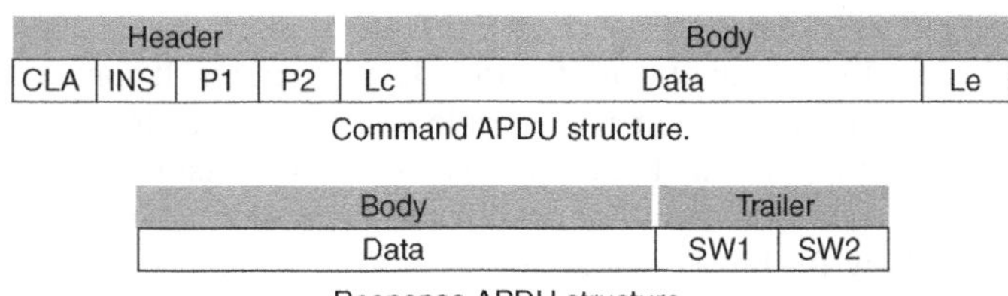

Figure 15.1 Basic structure of command and response APDUs

Files consist of a header for administrative data – such as file type, file size, and access conditions – as well as a body for the actual data. On many cards, the file hierarchy is fixed. In this case, files cannot be created or deleted, but only the contents of existing files can be changed.

Smart card files are named by 2-byte *file identifiers (FIDs)* and addressed by path names consisting of the concatenation of several FIDs. Before a file can be accessed, it has to be selected using a special SELECT FILE command APDU. Most file operations implicitly operate on the currently selected file.

Dedicated files (DFs) are directory files that group files of the same application or category. In addition to FIDs, DFs are addressed via *application identifiers (AIDs)*, which are 5 to 16 bytes long. An AID is used to select the card application that is contained in the identified DF. JavaCard applets, which are described below, are also selected using AIDs.

15.3.1 Command Sets

ISO 7816-4 defines an extensive set of so-called "basic interindustry commands." These commands are not targeted to specific applications, but are generally applicable to a wide array of different needs, which is in contrast to the application-specific commands that are defined, for example, for GSM SIMs [4]. The goal of specifying these commands is to unify access to smart card services. Unfortunately, this goal has not been fully achieved, because some smart cards do not completely implement all commands or command options, while others provide vendor-specific commands for the same purpose. It could also be argued that it is not useful to attempt a standardization of this kind at this relatively low level of abstraction. The basic interindustry commands can be classified as follows:

- **File selection commands:**
  SELECT FILE
- **Read and write commands (transparent EFs):**
  READ/WRITE/UPDATE/ERASE BINARY
- **Read and write commands (record-oriented EFs):**
  READ/WRITE/UPDATE/APPEND RECORD
- **Authentication commands:**
  VERIFY, INTERNAL/EXTERNAL AUTHENTICATE, GET CHALLENGE
- **Transmission-oriented commands:**
  GET RESPONSE, ENVELOPE, MANAGE CHANNEL

15.4 JavaCards

A JavaCard [5] basically is a smart card that runs Java programs. It contains the JavaCard Runtime Environment (JCRE) [6] that is capable to execute a restricted version of Java byte code. JCRE is standardized and hides the specifics of the smart card hardware. Binary compatibility was a major motivation for developing JavaCard: "The standards that define the Java platform allow for binary portability of Java programs across all Java platform implementations. This 'write once, run anywhere' quality of Java programs is perhaps the most significant feature of the platform" [7].

Java is a secure and robust language, since it is type safe; it disallows direct memory access, and its execution is controlled by an interpreter. It is known to many software developers, which further lowers the hurdle of smart card programming. It has to be said though, that the significant resource constraints on smart cards fundamentally change the "feel" of programming a smart card compared to standard Java programming on a PC. Programming a smart card in Java is, however, less error-prone and more efficient than programming it in the assembly language of the smart card's microcontroller. Other examples of high-level languages for smart cards are Windows for Smart cards [8], MultOS [9], and BasicCard [10].

15.4.1 Hardware Architecture

The JCRE, which includes the JavaCard virtual machine (VM) and a basic class library, is located in the ROM of the card. It takes on the role of the smart card operating system, controlling access to card resources such as memory and I/O. JavaCard applets and their persistent state are stored in the EEPROM. The RAM contains the runtime stack, the I/O buffer, and transient objects. To the external world, a JavaCard looks like an ordinary ISO 7816 compliant smart card, that is, it communicates by exchanging APDUs. This means that JavaCards work in situations where legacy ISO 7816 cards have been used before. The downside of this approach is that JavaCard developers have to deal with APDUs. For each application, the developer needs to specify an APDU protocol, and manually encode and decode APDUs.

JavaCard version 2.2 has introduced a middleware layer – called JavaCard RMI – which hides the APDU layer and therefore relieves the developer from manually developing an APDU protocol. Java Card RMI will be described further down.

15.4.2 Runtime Environment

Figure 15.2 shows the JavaCard software architecture. The lowest layer implements the communication protocols, such as $T = 0$ and $T = 1$, memory management, and cryptographic functionality. For performance reasons, this layer is partly realized in hardware and coded in the machine language of the smart card microprocessor. The central component of the architecture is the JavaCard virtual machine (VM), which provides an abstraction of the concrete hardware details, such as the microprocessor instruction set, and controls access to hardware components, such as I/O. It offers a runtime environment for JavaCard applets and isolates the individual applets from each other. The JavaCard API component provides access to the services of the JavaCard framework.

The actual functionality of the card is implemented as a set of JavaCard applets, each representing a different card service. In addition to applets, user libraries can be installed, which may be used by applets, but cannot be externally invoked.

The applet installer is an optional component of the JCRE that – if it exists – accepts APDUs containing authorization data, the code of the applet to install, and linkage information. The JCRE specification describes applet installation and deletion in general terms only, in order to leave freedom of implementation details to smart card manufacturers. This led to various proprietary installer implementations, which is in contrast to the idea of a standardized open runtime environment. One example architecture for secure installation of new applications on multiapplication cards is the Global Platform [11]. It

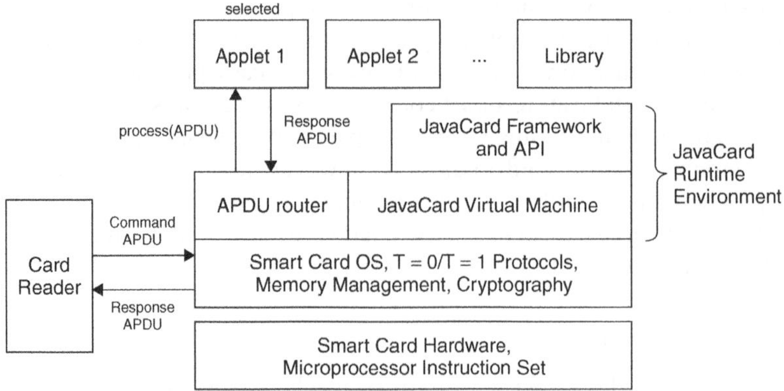

Figure 15.2 JavaCard software architecture

comprises mechanisms for loading and installing application code and for general card content management.

A complete implementation of the Java 2 VM [12] is not feasible with current hardware, in particular, if one considers that smart cards are low-cost mass products that are produced as cheaply as possible. To meet the constraints of smart card hardware, two strategies have been followed. Firstly, the VM itself was thinned out and secondly, the binary format of Java class files was simplified. Costly elements of the Java language such as dynamic class loading, the standard Java security architecture, garbage collection, and object cloning have been removed. The data types `long`, `char`, `float`, and `double` are not available, and support for `int` is optional. The 16-bit `short` is the standard data type for integer operations. JavaCard does not use the – relatively verbose and flexible – class file format, but the CAP file format, which contains the binary data of a whole package. Linkage information is outsourced to a separate EXP file.

Conceptually, the JavaCard runtime environment acts as a server: It continuously runs in an infinite loop waiting for command APDUs, routing them to the currently selected applet for processing, and sending back response APDUs. An applet is selected for processing by using a special SELECT FILE APDU that contains its application identifier (AID)[1]. For each APDU the JCRE receives, it first checks whether it is a SELECT FILE APDU. If so, the current applet is deselected and the new one is selected. Subsequently, each APDU is dispatched to the selected applet by calling its "process" method. The "process" method takes the command APDU as a parameter.

Each applet runs in its own context. JCRE itself runs in a special context with extended privileges. The border between two contexts is called a "software firewall." Objects are associated with their originating context and can be accessed only from within this context, with one exception: To provide a way for applets to interact, an object of another context can be accessed if it implements the `javacard.framework.Shareable` interface.

JCRE provides a simple transaction processing system. It guarantees atomic changes to a related set of persistent objects that are included in a transaction. When the transaction successfully commits, all of its operations have been executed. When it fails, intermediate

[1] It also contains a channel identifier, which is not discussed here.

states are rolled back to the state before the transaction. This is crucial for reliability and fault tolerance of security-critical operations. Changing single fields is atomic, as well as copying arrays with `Util.arrayCopy`. A set of methods controls the beginning, commit, and abort of a transaction.

The JavaCard 2.2 API is subdivided into seven packages:

- java.lang and java.io — JavaCard versions of some Java exceptions
- java.rmi — JavaCard version of the Remote interface
- javacard.framework — Base classes of the JavaCard framework
- javacard.framework.service — A framework for implementing an applet as a set of services
- javacard.security — Classes related to cryptography
- javacardx.crypto — Classes related to export-restricted cryptography.

The central classes of the API are *Applet*, *APDU*, and *JCSystem*:

JavaCard applets are extensions of `javacard.framework.Applet`. They must implement the *process* method, which is called by the JCRE to handle incoming command APDUs, and the *install* factory method to create the applet object instance. Calling *register* completes the successful instantiation of a new applet. The example in the next section illustrates how these methods are used.

The class `javacard.framework.APDU` encapsulates communication with the external world according to the ISO 7816-3 standard. It manages the I/O buffer and sets the data direction of the half-duplex communication. It works with the protocols $T = 0$ and $T = 1$. As we will see later, JavaCard RMI hides APDU handling from the implementer of card applications.

`javacard.framework.JCSystem` is used for accessing various resources of the card and services of the JCRE. It provides methods for controlling transactions, for interactions between applets, and for managing object persistence.

15.4.3 Developing JavaCard Applets

The JavaCard development kit is available for download from Sun's web site [13]. It includes a set of tools, a reference implementation for simulating smart cards, and documentation. A JavaCard applet is first compiled using a standard Java compiler, such as *javac*. The applet classes may only reference classes of the JavaCard API as well as any additional classes that are available on the card. The result is then converted into CAP and EXP files, containing the data for the applet package and linkage information, respectively. To perform the conversion process, the converter needs linkage information for the JCRE. To test the applet, it can be downloaded into a smart card simulator. The simulated card of the JavaCard reference implementation contains an installation applet that downloads the CAP file data in a series of APDUs.

In the following, we give a step-by-step example to illustrate how to develop an applet with JavaCard. The example applet is a prepaid card with an initial balance of 1000 money units. There are operations for debiting a certain amount and for querying the remaining balance. If the amount to be debited is negative or greater then the remaining amount, an exception is thrown.

The first step in developing a JavaCard applet is to define an APDU protocol. The set of command APDUs, expected response APDUs, and error conditions have to be specified. The APDU protocol is the externally visible interface of the smart card and therefore has to be described precisely. It should include the preconditions for applying each command, the coding of the APDUs and the individual parameters, the possible errors, and their associated status words. In our example, the APDU protocol looks as follows:

```
GET_BALANCE command APDU:
 CLA  INS  P1   P2   Le
 0x80 0x10 0x00 0x00 0x02

GET_BALANCE response APDU:
 BaHi BaLo SW1  SW2
 0xXX 0xXX 0x90 0x00
```

We assume signed two-byte monetary values. The first byte of the response contains the high-order byte of the balance, the second one contains the low-order byte. The status word 0x9000 (SW1, SW2) usually signals the successful execution of an operation.

```
DEBIT command APDU:
 CLA  INS  P1   P2   Lc   AmHi AmLo Pin1 Pin2 Pin3 Pin4 Le
 0x80 0x20 0x00 0x00 0x06 0xXX 0xXX 0xYY 0xYY 0xYY 0xYY 0x00
```

The data part of the command APDU contains the amount to be debited, high-order byte first, followed by a four-byte pin.

```
DEBIT response APDU (success):
 SW1 SW2
 0x90 0x00

DEBIT response APDU (bad argument):
 SW1  SW2
 0x60 0x00

DEBIT response APDU (PIN error):
 SW1  SW2
 0x60 0x01
```

After having specified the APDU protocol, the class structure and the individual methods of the applet classes need to be designed. The central component is an applet class (called `PrePaidApplet` in the example), which has to be derived from `javacard.framework.Applet`. Its *install* method needs to create an instance of the applet and register it with the JCRE. To work on incoming requests, the JCRE calls the applet's *process* method. It checks class and instruction bytes of the command APDUs and dispatches them to the proper method. Finally, AIDs need to be assigned to the applet and its package.

```
package prepaid;

import javacard.framework.ISO7816;
import javacard.framework.APDU;
import javacard.framework.ISOException;
```

```
import javacard.framework.OwnerPIN;
import javacard.framework.JCSystem;

public class PrePaidApplet extends javacard.framework.Applet {

    // code of CLA byte in the command APDU header
    public final static byte CLA = (byte)0x80;

    // codes of INS byte in the command APDU header
    public final static byte GET_BALANCE = (byte) 0x10;
    public final static byte DEBIT = (byte) 0x20;

    // status words
    public final static short SW_BAD_ARGUMENT = (short)0x6000;
    public final static short SW_PIN_ERROR = (short)0x6001;

    // state
    private short balance;
    private OwnerPIN pin;

    private final static byte[] pinData =
        { (byte)0x31, (byte)0x32, (byte)0x33, (byte)0x34 };

public PrePaidApplet() {
    balance = (short)1000; // initial balance
    pin = new OwnerPIN((byte)3, (byte)4);
    pin.update(pinData, (short)0, (byte)4);
    register(); // register the applet with JCRE
}

public static void install(byte[] aid, short s, byte b) {
    new PrePaidApplet();
}

public void process(APDU apdu) throws ISOException {
    if (selectingApplet()) return; // return if this is the selection APDU

    byte[] buffer = apdu.getBuffer();

    if (buffer[ISO7816.OFFSET_CLA] ! = CLA)
        ISOException.throwIt(ISO7816.SW_CLA_NOT_SUPPORTED);

    switch (buffer[ISO7816.OFFSET_INS]) {
        case DEBIT:
            debit(apdu);
            return;

        case GET_BALANCE:
            getBalance(apdu);
            return;

        default:
            ISOException.throwIt(ISO7816.SW_INS_NOT_SUPPORTED);
    }
}
```

```
public void debit(APDU apdu) {
    byte[] buffer = apdu.getBuffer();
    byte lc = buffer[ISO7816.OFFSET_LC];
    byte bytesRead = (byte)apdu.setIncomingAndReceive();

    if (lc ! = 6 || bytesRead ! = 6)
        ISOException.throwIt(ISO7816.SW_WRONG_LENGTH);

    if (!pin.check(buffer, (short)(ISO7816.OFFSET_CDATA + 2),
                           (byte)pinData.length))
        ISOException.throwIt(SW_PIN_ERROR);

    short amount = (short)
        ((buffer[ISO7816.OFFSET_CDATA] << 8) |
         (buffer[ISO7816.OFFSET_CDATA + 1] & 0xff));

   if (amount < 0 || amount > balance) {
       ISOException.throwIt(SW_BAD_ARGUMENT);
   }

   JCSystem.beginTransaction();
   balance -= amount;
   JCSystem.commitTransaction();
}

public void getBalance(APDU apdu) {
    byte[] buffer = apdu.getBuffer();
    short le = apdu.setOutgoing();
    if (le < 2) ISOException.throwIt(ISO7816.SW_WRONG_LENGTH);

    apdu.setOutgoingLength((short)2);
    buffer[0] = (byte)(balance >> 8);
    buffer[1] = (byte)balance;
    apdu.sendBytes((short)0, (short)2);
  }

}
```

The *debit* method in the example shows how transactions can be used in JavaCard applets. In this case, the transaction is not strictly necessary, because the JCRE guarantees the atomic update of an individual field.

This simple example shows that Java programming on a smart card fundamentally differs from programming Java on a full-fledged PC. The severe resource constraints have to be considered all the time. The lack of garbage collection means that all objects should be created at the time of applet construction and never dynamically during the processing of a request, except for transient arrays. JavaCard 2.2 allows for deleting objects upon request, but this feature should be used sparingly, because it is resource intensive. Writing to EEPROM, which is used for persistent storage on the card, is orders of magnitudes slower than writing to RAM. Even worse, an EEPROM cell can be overwritten only a limited number of times. Therefore, the number of EEPROM writing operations should be minimized.

15.5 PC/SC: Data Communications

In 1997, an industry consortium formed by smart card manufacturers and IT companies released the specification of a standard smart card interface for PC environments, called PC/SC [2]. The main components of PC/SC have been integrated into the Windows operating system, and most card readers come with device drivers that are compatible with PC/SC. The specification was also taken up by an open source project for implementation within the Linux operating system, called MUSCLE [14].

PC/SC addresses the following issues:

- The lack of a standardized interface to card readers, which has significantly hampered the deployment of applications in large numbers.
- The lack of a high-level programming interface to IC cards.
- It was anticipated that IC cards would be shared by multiple applications in the near future, thus a sharing mechanism was required.
- To be widely deployed, such a framework must work with (nearly) all existing smart card technologies while being extensible to embrace technological advancements.

The PC/SC specification describes an architecture for the integration of smart card readers and smart cards with applications. The main concepts of this architecture are the Resource Manager (RM) and the Service Provider (SP). The RM keeps track of installed card-reader devices and controls access to them. There is a single instance of the RM within a given system, and all accesses to smart card resources must pass through it. Thus, it can grant or deny applications access to smart card services, depending on access rights and concurrency constraints.

An SP represents a specific smart card type, providing a high-level interface to the functionality of that card type, for example, file system operations or cryptographic operations. In principle, an SP can represent more than one card type, and one card type can provide more than one SP. However, for each card type, a "primary" service provider is identified. A database (which is part of the RM) keeps track of the mapping from card types (identified by their ATR) to SPs.

The RM is essentially part of the operating system. Its state is visible only through its interface functions, which are prefixed by "`SCard...`". Before smart card resources can be addressed, an application has to establish a "context" with the RM through the `SCardEstablishContext` call:

```
SCARDCONTEXT hCardContext;
SCardEstablishContext(SCARD_SCOPE_USER, NULL, NULL, &hCardContext);
```

The resulting context handle is used in subsequent interactions with the RM. For example, when a new smart card type is introduced to the system. This step associates a card type with a service provider. It is necessary if the card type is to be recognized by the RM. After a card type is made known to the RM, the RM can notify applications when a smart card of that type is inserted into a card reader.

A card type is registered with the smart card database through the system call `SCard-IntroduceCardType`. A type is identified by an ATR string (as described above) and an associated byte mask that rules out insignificant and volatile parts of an ATR. Additionally, a user-recognizable name can be assigned to a card type, which can be used later when searching for inserted smart cards. A card type needs to be registered only once. Later, when any card with a matching ATR is inserted, applications can query the database for supported interfaces. This mechanism decouples applications from specific smart cards: from the application's point of view, it is important to know whether a smart card can be used through a certain interface, for example, providing user authentication. Details such as the command structure or the contents of the card's file system should be hidden from the application.

Note that there is a rather static model of smart card applications behind this architecture. It assumes that the functionality of a smart card does not change frequently, if at all. The mapping from smart card types to interfaces is fixed, and it is not possible for cards to advertise their capabilities dynamically. Thus, smart cards of the same type but of different configurations, such as Java cards with slightly different card applets, would have to be assigned different types. This requires that they respond with different ATRs and are introduced to the system separately, which might be hard to achieve in practice.

For the implementation of a service provider, it is necessary to be able to directly exchange APDUs with a smart card. The Resource Manager provides the `SCardTransmit` system call for this purpose. This function requires detailed knowledge of the command structure of the used smart card. It can also be used by applications, but then most of the interoperability capabilities of a PC/SC environment are lost. Within the MUSCLE implementation of PC/SC, which is still under development, this is currently the only means of interaction with smart cards, since service providers are not yet supported.

Some operations require the exchange of several APDUs that may not be interleaved by APDUs from other applications. For example, decrypting a large message requires that several APDUs, each containing a small data chunk, are sent to the card. An application that is not authorized by the user must not be allowed to intersperse its own data and thus gain access to confidential information. The RM grants temporary exclusive access for commands that are enclosed within a `SCardBeginTransaction/SCardEndTransaction` pair. The SCardEndTransaction call even allows resetting the card, which makes it necessary that the users authenticate themselves again for subsequent security-sensitive operations.

In summary, a PC/SC-compliant environment regulates access to smart card resources (card readers and smart cards) and provides applications with high-level interfaces to services offered by smart cards. A PC/SC implementation therefore acts as a middleware component for smart card applications. It supports the division of responsibility for the implementation of smart card–aware applications among three groups:

- The card-reader manufacturers provide compliant drivers for their devices;
- smart card vendors implement service providers and registration routines for their cards;
- application developers make use of basic Resource Manager functionality but mainly rely on the service providers.

Information about the use of smart cards in the Windows environment can be found in the MSDN library [15] in the Security/Authentication section.

15.6 OpenCard Framework

The OpenCard Framework (OCF) is a Java-based middleware for smart cards. OCF serves as an application platform, unifying access to a large variety of cards and terminals. In addition, it provides a standardized programming interface for application developers and developers of smart card services. In contrast to JavaCard, OCF focuses on the card-external part of smart card development.

OCF is realized as a set of *"jar"* files and has to be installed on the PC to which the terminal is connected. OCF runs in the VM of the application that uses OCF instead of as a separate OS service.

15.6.1 Architectural Concepts

OCF consists of two subsystems: one for card terminals and one for card services. Card terminals encapsulate physical card readers. Card services are used by applications to talk to the actual services located on the card. The architecture of OCF is depicted in Figure 15.3.

15.6.1.1 Card Terminals

The card terminal layer abstracts from the reader device actually employed, thus guaranteeing the independence of an application from a particular reader type. Card-reader

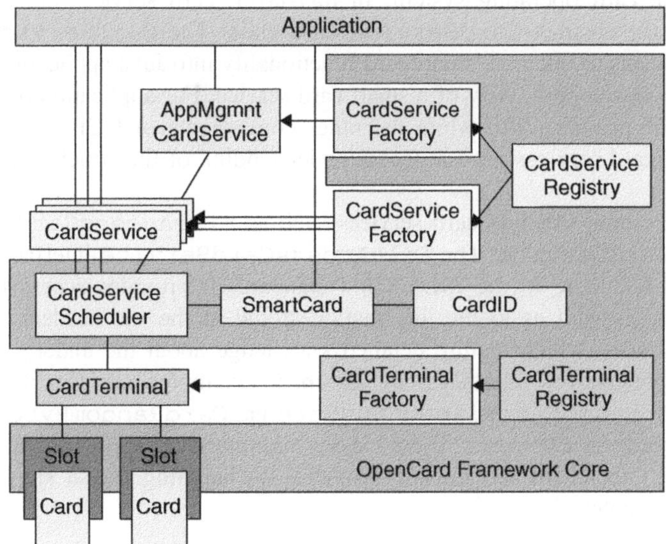

Figure 15.3 Architecture of the OpenCard framework

manufacturers who want to make their devices available for OCF need to provide OCF drivers, which are realized as implementations of the abstract classes `CardTerminal` and `CardTerminalFactory`.

Each card reader is represented by a `CardTerminal` instance. Its card slots are modeled as instances of class `Slot`. The ATRs of a card are encapsulated in `CardID` objects. The card terminals connected to a PC are usually statically configured, but can also be added dynamically.

Mutually exclusive access to a card in a card slot is guarded by `SlotChannel` objects. An object owning a `SlotChannel` instance has exclusive access to a smart card, until the slot channel is released again. The class `CardTerminal` provides methods for checking for card presence, getting slot channel objects, and for sending and receiving APDUs. For extended functionality of special card readers, such as number pads and finger print readers, OCF provides separate Java interfaces that can be implemented by the concrete `CardTerminal` implementations.

The singleton object `CardTerminalRegistry` records all terminals installed in the system. During start-up, the registry reads configuration data from one or more "opencard.properties" files. At runtime, additional terminals might be added. The registry creates `CardTerminal` objects by using `CardTerminalFactory` objects, whose class names are listed in the configuration file. These factories are responsible for instantiating the OCF drivers for a particular family of terminals, for example, those of a certain manufacturer. Upon card insertion and removal, card terminals generate events. Objects can register at the registry for those events.

15.6.1.2 Card Services

The card service layer defines mechanisms for accessing card-resident functionality. It provides high-level interfaces to the individual card services and abstracts from particularities of the card operating system. In this layer, card services are represented as extensions of the abstract `CardService` base class. The `CardService` abstraction provides the means to integrate smart card functionality into Java applications and defines the application developer's view of a smart card service. For application developers, card services are the primary API, while the other components of OCF are less important. Developers of card services need a deeper understanding of the structure of the OCF, of course.

OCF defines some standard card services such as `FileAccessCardService` and `SignatureCardService`. The `FileAccessCardService` interface encapsulates smart cards that adhere to the ISO 7816-4 standard. It provides access to binary and structured files, as well as to the information stored in the file headers. Around these low-level interfaces, which require detailed knowledge about the underlying ISO 7816-4 standard, higher-level interfaces are grouped. Examples are `CardFile`, `CardFileInputStream`, `CardFileOutputStream`, `CardRandomByteAccess`, and `CardRandomRecordAccess`. These classes and interfaces approximate the usual Java I/O semantics. Other card services are provided for handling digital signatures and for managing card applets.

15.6.2 Configuration

The terminals and card services available in the system are statically configured. The configuration data are stored in a so-called "opencard.properties" file, which looks as follows:

```
OpenCard.terminals = com.foo.opencard.fooCardTerminalFactory|fooName|fooType|
                     COM1
OpenCard.services  = com.abc.opencard.abcCardServiceFactory \
                     org.xyz.opencard.xyzCardServiceFactory
```

The entry `OpenCard.terminals` contains the fully qualified class name of the available readers, together with parameters for their instantiation. The entry is used to instantiate the terminals that will then be stored in the card terminal registry. The `OpenCard.services` entry lists the available card service factories. The factories are instantiated and stored in the card service registry.

15.6.3 Programming Model

The basic programming model of the OCF is illustrated with an example. It shows how to read the contents of the EF_{ICCID} file of a GSM SIM card with OCF. EF_{ICCID} contains a 10 byte identification number, has FID 0x2FE2, is located in the root directory (FID 0x3F00), and is of type "transparent." The code for accessing the card from within an application via OCF is shown below.

```
try {
    // initialize OCF
    SmartCard.start();

    // 1) waiting for a card
    CardRequest cr = new CardRequest(FileAccessCardService.class);
    SmartCard sc = SmartCard.waitForCard(cr);

    // 2) getting a reference to the card service
    FileAccessCardService facs = (FileAccessCardService)
        sc.getCardService(FileAccessCardService.class, true);

    // 3) using the card service
    byte[] data = facs.read(new CardFilePath(":2fe2"), 0, 10);

    // do something with the data...

    sc.close();
} catch (Exception e) {
    e.printStackTrace();
} finally {
    try {
        SmartCard.shutdown();
```

```
        } catch (Exception e) {
            e.printStackTrace();

        }

}
```

Before OCF can be used, it needs to be initialized by executing the SmartCard.start method. This triggers reading the configuration file, which contains information about the terminal and service factories. By calling createCardTerminals, the card factories are instructed to instantiate the individual CardTerminal objects, which are inserted into the card terminal registry. In a second step, the card service factories are inserted into the card service registry. The counterpart to SmartCard.start is SmartCard.shutdown, which ensures that any system resources allocated by OCF are released again.

After initializing OCF, interaction with the card takes place in three steps: First we wait until an appropriate card is inserted, then we get a reference to the card service, and eventually we use it. In the example, we are interested in a card for which a FileAccessCardService is applicable. This is specified with a card request object. The CardRequest class allows for a detailed description of the kind of card and service we are interested in. Possible parameters include the desired card service, the terminal in which the card has to be inserted, and the maximum waiting time.

The SmartCard.waitForCard method is called with the card request object as a parameter. It waits until the request is satisfied or the timeout is reached. Upon inserting a card, the request conditions are checked one by one. The CardServiceFactory objects are queried, whether they know the card type (method getCardType) and the requested service (method getClassFor). When such a card service factory is found, SmartCard.waitForCard returns with a reference to a SmartCard object. The desired card service is then instantiated by calling the getCardService method on that reference. Internally, the card service registry checks for a card service factory that is able to instantiate the card service (method getCardServiceInstance). In the example, the actual usage of the card can now be coded in a single line.

As an alternative to waiting for an appropriate card, OCF provides an event mechanism. Events are triggered upon insertion or removal of cards. Objects that are interested in these events have to implement the CTListener interface and have to register with the card terminal registry. If events are detected, the registry calls cardInserted and cardRemoved, respectively.

15.6.4 Summary

OCF is organized in a two-layer architecture – abstracting from card reader devices and providing high-level Java interfaces of smart card services. It uses an application-centric model in which the card acts as a passive service provider and applications specify which service is required. OCF is statically configured, with terminal and service factories stored in a configuration file. The approach does not allow for proactive card services that supply their functionality to the outside world. To enable this functionality, the card would have to be queried for its services, which would then have to be made available in some kind of lookup service.

OCF is designed for usage from within the same Java VM that runs the application. This prohibits the concurrent use of smart cards by different applications. If the reader is connected via the serial interface, for example, the serial port is blocked by the first application using it. Moreover, OCF's mechanisms for mutual exclusion are limited to a single VM.

15.7 JavaCard RMI

JavaCard RMI (JCRMI) [16–18] extends JavaCard with a subset of standard Java RMI. It provides a distributed object-oriented model for communicating with on-card objects. In this model, the card acts as a server that makes remote objects accessible for invocation by off-card client applications. Client applications invoke methods on the remote interfaces of these objects, as if they would be locally available.

JCRMI is built on top of the APDU-based messaging model. It completely hides the APDU layer by automatically encoding (marshaling) and decoding (unmarshaling) method invocations, request parameters, and result values. This relieves the implementer of the card applet as well as the developer of the client application from the tedious tasks of specifying the APDU protocol and coding the parameters manually. As the JavaCard example above shows, if APDUs are used directly, a significant portion of code is devoted to decode incoming and encode outgoing APDUs.

The idea of applying distributed object-oriented and RPC-like principles to smart cards was first implemented by Gemplus. They developed a system called "direct method invocation" (DMI), which was part of their GemXpresso rapid application development environment [19].

15.7.1 On-Card JCRMI

JCRMI covers on-card programming as well as the client side. The on-card part will be described first.

15.7.1.1 Remote Objects

In the same way as in standard RMI, objects that can be invoked remotely are specified in terms of one or more remote interfaces. Remote interfaces extend `java.rmi.Remote`. The methods of a remote interface must declare `java.rmi.RemoteException` in their throws clause. An example is shown below. It defines the remote interface for the JavaCard example given above. This interface has to be made available to the off-card client application.

```
package prepaidrmi;

import java.rmi.Remote;
import java.rmi.RemoteException;
import javacard.framework.UserException;

public interface PrePaid extends Remote {

    public static final short BAD_ARGUMENT = (short)0x6000;
    public static final short PIN_ERROR = (short)0x6001;
```

```
    public void debit(short amount, byte[] pin)
        throws RemoteException, UserException;

    public short getBalance()
        throws RemoteException;

}
```

The limitations of the JCRE and JCVM impose a number of constraints on the type of parameters and return values:

- Parameters can only be of primitive type (boolean, byte, short, and optionally int) or single-dimensional arrays of primitive types (boolean[], byte[], short[], and optionally int[]).
- Return values can be of any type supported as an argument type, type void, or any remote interface type.

Additionally, functional constraints exist. For example, remote references to on-card objects cannot be passed from the client VM to another VM.

15.7.1.2 Remote Object Implementation

After the remote interface has been designed, it needs to be implemented as a JavaCard class that will later execute on the card. In addition to implementing the remote interface `PrePaid`, class `PrePaidImpl` in the example below extends `javacard.framework.service.CardRemoteObject` and calls its default constructor upon creation. This ensures that the object is "exported" and therefore ready to accept incoming remote method calls. Only exported remote objects may be referenced from outside the card. As an alternative to extending `CardRemoteObject`, `PrePaidImpl` could have called `CardRemoteObject.export(this)` in its constructor. Note that the implementing class does not need to handle APDUs at all. They are completely hidden by the RMI layer, resulting in much cleaner code.

```
package prepaidrmi;

import javacard.framework.service.CardRemoteObject;
import javacard.framework.Util;
import javacard.framework.OwnerPIN;
import javacard.framework.UserException;
import java.rmi.RemoteException;

public class PrePaidImpl extends CardRemoteObject implements PrePaid {

    private short balance;
    private OwnerPIN pin;

    public PrePaidImpl(short initialBalance, byte[] pinData) {
        super(); // export this instance
        balance = initialBalance;
```

```
        pin = new OwnerPIN((byte)3, (byte)4);
        pin.update(pinData, (short)0, (byte)4);
    }

    public void debit(short amount, byte[] pinData)
        throws RemoteException, UserException
    {
        if (!pin.check(pinData, (short)0, (byte)pinData.length))
            UserException.throwIt(PIN_ERROR);

        if (amount < 0 || amount > balance) {
            UserException.throwIt(BAD_ARGUMENT);
        }

        balance -= amount;
    }

    public short getBalance() throws RemoteException {
        return balance;
    }
}
```

15.7.1.3 A JavaCard Applet Using JCRMI

As for non-JCRMI applets, the central class for applets providing remote objects is derived from `javacard.framework.Applet`. The only difference is that APDU objects that are passed in to the *process* method are not handled by the applet itself. Instead, a dispatcher service passes them to an `RMIService` object. The dispatcher is part of a service framework that routes incoming APDUs through a list of services. One after the other, they process the incoming – or later the outgoing – APDU. In the example below, the JCRMI service is the only service registered.

There are two types of APDUs defined in the JCRMI protocol: applet selection APDUs and method invocation APDUs. The JCRMI service handles both of them.

The constructor of the JCRMI service takes a reference to an initial remote object. This will be the first remote object that off-card client applications can talk to. In the example, it is also the only one. Upon applet selection with an appropriate selection command APDU, the JCRMI service returns the object ID and the fully qualified name of the initial reference in the response APDU. Depending on the format requested in the selection APDU, this might either be the fully qualified class name – `prepaidrmi.PrePaidImpl` in the example – or the fully qualified interface name, which would be `prepaidrmi.PrePaid` in the example. This information is used as a bootstrap for the off-card client application, allowing it to instantiate the appropriate stub class for talking to the card object.

Incoming method invocation APDUs contain the object ID of the invocation target, the ID of the method to invoke, and the marshaled parameters. The JCRMI service is responsible for unmarshaling the parameters and invoking the requested method of the referenced object. After the remote object implementation has done its work, the RMI service marshals the result value or the exception code, respectively, in the response APDU and sends it back.

```
package prepaidrmi;

import java.rmi.Remote;

import javacard.framework.APDU;
import javacard.framework.ISOException;
import javacard.framework.service.RMIService;
import javacard.framework.service.Dispatcher;

public class PrePaidApplet extends javacard.framework.Applet {

    private Dispatcher disp;
    private RMIService serv;
    private Remote prePaid;
    private final static byte[] pinData =
        { (byte)0x31, (byte)0x32, (byte)0x33, (byte)0x34 };

    public PrePaidApplet() {
        prePaid = new PrePaidImpl((short)1000, pinData);
        serv = new RMIService(prePaid);

        disp = new Dispatcher((short)1);
        disp.addService(serv, Dispatcher.PROCESS_COMMAND);

        register(); // register applet with JCRE
    }

    public static void install(byte[] aid, short s, byte b) {
        new PrePaidApplet();
    }

    public void process(APDU apdu) throws ISOException {
        disp.process(apdu);
    }

}
```

15.7.2 *Off-Card JCRMI*

Having described the card-resident part, we will now show how an off-card client application can use JavaCard objects remotely via standard RMI.

The client application typically runs on the computing device to which the card reader is connected. This could be a stationary PC running J2SE or a mobile phone running J2ME, for example. The client-side API is defined in package `com.sun.javacard.javax.smartcard.rmiclient`. It is designed to be independent from the Java platform and card access technology used. It provides the client application with a mechanism to initiate an RMI session with a JavaCard applet and to obtain an initial reference to the main remote object of an applet. In combination with the on-card service framework it allows the client to introduce various transport level policies, for example, transport level security. The main classes `JavaCardRMIConnect` and `CardObjectFactory`, as well as the `CardAccessor` interface will now be briefly introduced.

Interface `CardAccessor` declares method `byte[] exchangeAPDU(byte[] apdu)` to exchange data with the card. Implementations of this interface hide the actual card access mechanism employed. Sun's reference implementation, for example, runs on J2SE and uses OCF as the card access mechanism. Its implementation class is called `OCFCardAccessor`.

Custom policies, such as transport level security, can be realized by extending a Card-Accessor class and providing a special implementation of the `exchangeAPDU` method. The custom method first applies the policy to the bytes exchanged, such as encryption or adding a message authentication code (MAC), before it calls the `exchangeAPDU` method of the super class in the same way it decrypts responses or verifies their MACs. On the card, this is paralleled by the JavaCard service framework, as discussed above. First, a security service would get access to the incoming data, decrypting them or verifying the MAC, and would then hand them on to the next service in the chain, for example, the JCRMI service.

`JavaCardRMIConnect` objects are initialized with a `CardAccessor` object. A session with an applet is started with one of the `selectApplet` methods, which have the following signatures:

```
public byte[] selectApplet(byte[] appletAID)
public byte[] selectApplet(byte[] appletAID, CardObjectFactory cOF)
```

The second parameter specifies a `CardObjectFactory`, as explained below. Method `Remote getInitialReference()` is used to get the initial `java.rmi.Remote` reference of the initial object of an applet. The client usually casts this method to the actual object interface, which is interface `PrePaid` in the example.

Extensions of abstract class `CardObjectFactory` are responsible for instantiating client stubs on the client. `JavaCardRMIConnect` uses the card object factory passed to the `selectApplet` method or a default one. The default card object factory simply locates the stub class of the remote object implementation and instantiates it. The name of this class is returned in the remote reference descriptor that is sent in response to the select APDU – `prepaidrmi.PrePaidImpl` in the example. The corresponding stub name is `prepaidrmi.PrePaidImpl_Stub`. Of course, this requires that the stub was previously compiled using the standard RMI compiler "rmic"[2] and that it is reachable via the class path.

An alternative card object factory, called `JCCardProxyFactory`, does not require the availability of a stub class. It generates the stub object dynamically from the list

[2] "rmic" has to be called with the "-v1.2" option.

of remote interfaces that the remote class implements. This requires the dynamic proxy generation mechanism of JDK 1.3 or above.

15.7.2.1 Reference Implementation of the JavaCard RMI Client-Side

The reference implementation runs on J2SE and uses OCF for card access. This essentially means that the OCF library files "base-core.jar" and "base-opt.jar" need to be present in the client class path and that an "opencard.properties" file has to be available, which contains the configuration for JCRMI. The main classes of the reference implementation are `OCFCardAccessor` and `OCFCardAccessorFactory`. They reside in package `com.sun.javacard.ocfrmiclientimpl`. `OCFCardAccessor` extends the OCF class `CardService` and implements the JCRMI client-side interface `CardAccessor`. In its `exchangeAPDU` method, it simply passes the unmodified command APDU to the underlying OCF `CardTerminal` and returns the result APDU. Class `OCFCardAccessorFactory` is the factory of the `OCFCardAccessor` card service and has to be listed in the "opencard.properties" file.

15.7.2.2 Client RMI Example

We will now give a complete example of how a client application might access an on-card object using RMI. The example uses the OCF card accessor and is structured like the basic OCF example shown above.

```
package prepaidrmiclient;

import java.rmi.*;
import javacard.framework.*;
import com.sun.javacard.javax.smartcard.rmiclient.*;
import com.sun.javacard.ocfrmiclientimpl.*;
import opencard.core.service.*;
import prepaidrmi.PrePaid;

public class PrePaidClient {

    private static final byte[] PREPAID_AID =
    {(byte)0xa0, 0x00, 0x00, 0x00, 0x62, 0x03, 0x01, 0xc, 0x8, 0x01 };

     public static void main(String[] argv) throws RemoteException {

         try {
             // initialize OCF
             SmartCard.start();

             // wait for a smart card
             CardRequest cr = new CardRequest(OCFCardAccessor.class);
             SmartCard sc = SmartCard.waitForCard(cr);

             // obtain an OCFCardAccessor for Java Card RMI
             CardAccessor ca = (CardAccessor)
               sc.getCardService(OCFCardAccessor.class, true);
```

```
            // create a Java Card RMI instance
            JavaCardRMIConnect jcRMI = new JavaCardRMIConnect(ca);

            // select the Java Card applet
            jcRMI.selectApplet(PREPAID_AID);

            // alternative possibility to the previous line:
            // CardObjectFactory factory = new JCCardProxyFactory(ca);
            // jcRMI.selectApplet(PREPAID_AID, factory);

            // obtain the initial reference
            PrePaid pp = (PrePaid) jcRMI.getInitialReference();

            // get the balance
            short balance = pp.getBalance();
            System.out.println("Balance = " + balance);

            // debiting 15 money units
            byte[] pin = new byte[] {0x31, 0x32, 0x33, 0x34};
            pp.debit((short)15, pin);

            // get the new balance
            balance = pp.getBalance();
            System.out.println("Balance = " + balance);

        } catch(UserException e) {
            System.out.println("Reason code = 0x" +
              Integer.toHexString(0xffff & e.getReason()));
        } catch (Exception e) {
            System.out.println(e);
        } finally {
            try {
                SmartCard.shutdown();
            } catch (Exception e) {
                System.out.println(e);
            }
        }
    }
}
```

The program first waits for a card that can be accessed via an OCFCardAccessor. Then, a JavaCardRMIConnect (jcRMI) instance is created and initialized with an OCFCardAccessor. The jcRMI is used to select the applet by its AID and to obtain the reference for the applet's initial remote object. The functionality of the card is subsequently available via local method calls on the `PrePaid` proxy object. All APDU interactions are hidden by the RMI layer. In the example, the jcRMI implicitly uses the default card object factory, called `JCCardObjectFactory`, which requires a stub object. Alternatively, the `JCCardProxyFactory` could have been used, which dynamically creates a proxy object and does not need a stub. The required code is shown in the example, below the line reading `selectApplet(PREPAID_AID)`.

15.7.3 Summary

Viewing the card as a component in a distributed object-oriented system, such as RMI, has benefits for the JavaCard applet developer as well as for the client application developer. The APDU layer is completely hidden by the RMI middleware and tedious APDU protocol issues no longer need to be dealt with. In this model, the client application gets a reference to a local stub or proxy object and calls it like a local object. Although the range of types allowed for parameters and return values is limited, the available types are still sufficient for most JavaCard-based client-server interaction. Extending card accessor implementations opens the possibility to realize custom policies on the transport layer. On the card, JCRMI is accessed using the JavaCard service framework, which allows for flexible handling of APDUs in a chain of services.

15.8 PKCS #11 Security Tokens

For the storage of high-quality cryptographic keys and for the execution of cryptographic operations, such as digitally signing a document, humans are dependent on computers. Especially for the protection of cryptographic secrets – encryption and signature keys – secure storage is required, since the breaking of such a secret would have potentially hazardous implications.

Smart cards are security tokens that were especially designed for this purpose. However, a smart card is a device resulting from a design trade-off: protection against convenience. Plastic cards were in use before secure chips were available, and smart cards are simply the continuation of this design with additional features. When one abstracts from the physical shape and restrictions imposed by usability requirements, and concentrates on the functionality as a security device, the notion of a security token emerges, a physical object that serves for security purposes and can be attached to a host system such as a PC. There is a vast variety of ways to connect smart cards and other security tokens to a PC: serial interfaces such as RS-232 and USB with varying speed and connectors, Firewire, Infrared, the PCI bus, and so forth. Each of these interfaces requires different device drivers. Additionally, each vendor provides their own programming library.

In order to facilitate application development making use of cryptographic functionality, the "Public-Key Cryptography Standard" [20, 21] was conceived. Part 11 of this standard, called "cryptoki" [22], describes the interface to a generic security token. The description is intentionally very broad, comprising a large variety of data structures, algorithms, and operation modes. Thus, virtually any cryptographic device fits into part of this description and can be made accessible from PKCS #11-compatible software, assuming that the application is sufficiently flexible to adapt to the varying capabilities of different tokens.

PKCS #11 takes on an object-oriented view of a cryptographic token. A token is viewed as capable of storing three kinds of objects: application data, keys, and certificates. Objects are characterized by their attributes and can differ in lifetime and visibility. As an example, consider a symmetric key that is generated during a session of an application with a cryptographic token. During the session, the key is used to encrypt and decrypt data. Before the session is closed, the key can be stored persistently on the token. Alternatively, the key can simply be discarded.

There are a number of products available that come with interfaces that are compatible with the PKCS #11 standard. They include smart cards, USB crypto tokens, PCI extension cards, and other varieties. On desktop computers, they can be used to log into the system, sign email, or access restricted web sites. For example, the Netscape/Mozilla browser suite can employ PKCS #11 tokens [23]. Other products, such as the IBM 4758 cryptographic coprocessor, are employed in high-security server environments, for example, for banking applications.

15.9 Smart Cards as Distributed Objects

Employing the CORBA [24] middleware approach to implementing a distributed object system, it is possible to integrate smart cards as well. Smart card services are described just like other objects in the CORBA Interface Description Language (IDL). By using an adapted IDL compiler, such as described in [25], special objects for accessing the smart card are created. Basically, the invocations of operations on such objects have to be mapped to corresponding APDU exchanges. Special care has to be taken, however, regarding the security of the system. For example, it would be a bad idea to transmit a PIN in clear text from a user interface object to the smart card object for user authorization.

15.10 Smart Card Middleware for Mobile Environments

In the GSM and the future UMTS world, smart cards (as subscriber identity modules, or SIMs) are the fundamental building blocks for user authentication in mobile networks. Besides their primary role of authenticating subscribers of a mobile operator, they offer additional functionality that is addressed by smart card middleware. In this section, we concentrate on two such technologies: (a) the SIM Application Toolkit (SAT), which offers an application platform in GSM SIMs and which is already used in various operator-based services, forming the basis of the so-called "SAT Browser" technologies, and (b) the current activities in the Java community to specify interfaces between mobile handsets such as mobile (smart) phones and security modules such as SIMs.

15.10.1 SIM Application Toolkit

The "SIM Application Toolkit (SAT)" [26] is an interface implemented by GSM mobile phones offering among others the following services to the GSM SIM:

- *DisplayText(text}*: Displays the supplied text on the display of the mobile phone.
- *GetInput([title][,type])*: Displays an optional title text and queries the user for input. Several syntactic categories such as digits, hidden input, and so on, are supported. The text entered by the user is returned to the SIM.
- *SelectItem([title][,item]...)*: Displays an optional title and a number of items among which the user can choose. The number of the chosen item is returned to the SIM.
- *ProvideLocalInformation(type)*: Return localization and network measurement results, depending on the given type selector. In particular, it can be used to yield the *network cell identifier* and *location area* information enabling the rough localization of the user's current position.

- *SendShortMessage([title,]dest,payload):* Sends a short message with the given payload to the destination.

Central to the toolkit is the so-called *proactive command manager* that is responsible for managing a *proactive session*. Such a session can be initiated by a smart card–resident applet wishing, for example, to execute the toolkit command *SelectItem*. The applet invokes the appropriate method in the SIM API that in turn activates the proactive command manager that sends a response APDU to the mobile phone in the form of a status word (SW1, SW2) = (91, length). This response code indicates to the mobile station (MS) that the SIM wishes to start a proactive session. The MS then fetches the next command with the given length that contains the proactive command, which in our example contains the items the user has to select from. It then decodes the proactive command contained in the response APDU and in case of a *SelectItem* displays the menu items on the screen of the mobile phone. After the user has selected an item, the MS compiles a so-called *terminal response* APDU that contains, among other information, the index of the item the user has selected. This response is intercepted by the proactive command manager that in turn resumes the applet and passes the user's selection back to the SIM toolkit application.

Besides the commands listed above, the SIM toolkit further supports a number of mechanisms for registering timers that can be used to wake up an applet at regular intervals, registering for certain types of events such as the arrival of an SMS [27] or a change in the current network cell. The most important triggering mechanisms are the arrival of an SMS and the selection of the applet in the phone's SIM-specific menu.

Summing up, the SIM application toolkit allows to temporarily exchange the role of client (which is now the SIM) and server (which is the MS offering services to the SIM). It can be seen as a platform on top of which card-resident applications can be implemented that have access to an API that allows to perform user interaction and communication.

15.10.1.1 A Sample Application

Several European telecommunication operators run SAT-based services, for example, in the financial sector. Users can start a session with their preferred bank via selection from the SAT menu of their mobile phone. The card-resident banking applet is invoked and enters a SAT proactive session to query the user for authentication information such as a PIN using the proactive command *GetInput()*. By exchanging encrypted Short Messages (SMS) via some gateway within the operator's infrastructure, the toolkit applet can run appropriate challenge/response protocols with the bank (often 3DES based on [28]) to authenticate users and confidentially transfer financial information such as account balance or trigger financial transactions.

Thus, SIM toolkit is an application platform implemented by the SIM and the MS to offer smart card–based services for applications with high need for security.

15.10.1.2 SIM Toolkit Browser Technology

Although the SIM toolkit technology offers an interesting platform for certain application domains in the mobile world, it suffers from the general problem that for the realization of a particular service, an appropriate card-resident applet has to be written and installed on the subscriber's smart card.

SIM toolkit browsers such as the USIM Application Toolkit (USAT) Interpreter [29] try to overcome this limitation by providing a generic platform for the provision of such services by implementing a generic browser application that permanently resides inside the card. Services are implemented with the help of a network-resident gateway component that is in direct contact with the card-resident browser using appropriate communication channels such as SMS or USSD (unstructured supplementary service data).

In the above example of the financial service, the SAT browser would upon invocation dynamically load so-called *pages* that contain in a compact representation the encoding of the SIM toolkit commands it should execute. It then interprets these SAT commands by starting the necessary user interaction and sends the results of the user interaction back to the gateway that takes appropriate action. By exchanging messages between the SIM and the gateway, this process continues until the application is finished.

SAT browser technology essentially follows the approach of traditional Web browsers, where the server sends to the client a description of the data to render together with some interaction controls. Some additional primitives are added, for example, for securing the communication between the gateway, the SIM, and optional third parties such as banks. Commercially available gateways and browsers even go so far that the page descriptions are written in an encoding similar to the Wireless Markup Language (WML) that is then recoded in the gateway into a form suitable for the SAT browser.

Summing up, SAT browser technology directly builds upon the SIM toolkit by making it even simpler to use middleware for running security-sensitive applications and services in the context of SIMs and mobile networks.

15.10.2 J2ME Smart Card Middleware

Although SIM toolkit-based solutions potentially offer high security and with the advent of the USAT interpreters also high flexibility in the provision of services, they suffer from the inherent limitation in the communication bandwidth between a smart card and its outer world. It is simply not feasible to digitally sign large documents via a SAT interface that is bound to the length of APDUs in the exchange of information. Therefore, there is considerable interest in bridging the gap between mobile terminals and security tokens such as SIMs by means of appropriate *application-level* interfaces that enable a terminal-hosted application to exploit the security services of smart cards.

Java and, in particular, the Java 2 Micro Edition (J2ME) with the Connected (Limited) Device Configuration (CDC/CLDC) emerges as an interesting platform for mobile applications. On the other hand, additional card-resident security services such as the WAP Wireless Identity Module (WIM) enter the market. Typically, these are implemented as additional applications on a SIM that are independent from the card's GSM/UMTS core. Since the SIM itself is designed to be one of possibly many applications hosted by the Universal Integrated Circuit Card (UICC) platform, the WIM and possibly other security services could be easily hosted by future UICCs. However, there is currently no standard or interface defined that enables a J2ME application on a mobile terminal to make use of the security services potentially offered by the UICC/USIM.

The Security and Trust Services API specification defines an optional package for the J2ME platform and is currently being drafted by an expert group working on the Java Specification Request 177 [30]. The purpose of this effort is to define a collection of APIs that provide security and trust services to J2ME applications by integrating a "security element." Interaction with such an element provides the following benefits:

- Secure storage to protect sensitive data, such as the user's private keys, public key (root) certificates, service credentials, personal information, and so on.
- Cryptographic operations to support payment protocols, data integrity, and data confidentiality.
- Secure execution environment for custom and enabling security features.

J2ME applications would rely on these features to handle value-added services, such as user identification and authentication, banking, payment, loyalty applications, and so on. A security element can be provided in a variety of forms. For example, smart cards are a common implementation. They are widely deployed in wireless phones, such as SIM cards in GSM phones, UICC cards in 3G phones, RUIM cards in CDMA phones, and WIM applications in a SIM or UICC card in WAP-enabled phones. For example, in GSM networks, the network operator enters the network authentication data on the smart card, as well as the subscriber's personal information, such as the address book. When the subscriber inserts the smart card into a mobile handset, the handset is enabled to work on the operator's network.

15.11 JiniCard

Smart cards are typically used in a variety of places, because they are highly portable and are carried by their users wherever they may go. In each environment, a card and its services are present for a short term only and may appear and disappear without prior notice. Yet smart cards depend on their environment to be useful, as they lack an energy supply and user interface capabilities. These usage characteristics call for a middleware layer that provides an effortless integration into different environments without the requirement for any manual setup or configuration. As soon as a card is inserted into a card reader, the discovery of its services should happen spontaneously.

The JiniCard [31, 32] architecture integrates smart cards into Jini federations following the requirements outlined above. Jini's [33] objective is to provide simple mechanisms that enable devices to plug together to form an impromptu community – without any planning, installation, or human intervention. JiniCard makes card services accessible as Jini services. It supports any card that adheres to the ISO 7816-3 communications standard. The main idea of the JiniCard framework is to keep all functionality that is required to interact with a specific card service remotely on the net. It is loaded only when the respective card is actually inserted into the reader. The local environment is thus not forced to know the details of all kinds of cards it might interact with, but just their high-level Jini interfaces.

The JiniCard architecture consists of two layers. The lower layer provides the abstraction of the card reader as a Jini service, in order to make smart cards accessible as network components. The upper layer consists of a smart card exploration mechanism, which identifies the available services. It also provides a well-defined platform for the execution of card-external parts of card services, which are instantiated as the result of the exploration process.

The service exploration process is triggered by reading the ATR of a card. The ATR is mapped to a set of "card explorers" that are dynamically downloaded from the Internet, if they are not yet available locally. The exploration process is then delegated to the appropriate card explorers, which identify the card's services. The result is a set of `ServiceInfo` objects that each contain a code base URL for the service and various Jini-specific attributes. The services are then downloaded and instantiated within the runtime environment provided by the JiniCard framework, which also manages their registration in the Jini lookup service.

From the programmer's perspective, JiniCard provides an extensible framework [34] that helps to easily build the card-external parts of smart card applications. New card services can be deployed by uploading new card explorers to a well-known Web server and by providing corresponding card service implementations.

15.12 Smart Cards on the Internet

The use of Internet protocols for accessing smart card functionality was proposed for the integration of smart cards into commercial systems on the World Wide Web. The idea is that a Web user points his browser to the items he wants to buy and initiates the payment transaction through a click on a certain URL. This triggers a smart card session to start. The user is prompted to enter his PIN, which authorizes the smart card to carry out the payment. Many payment protocols involving smart cards have been proposed, but here we are interested in the issues that arise when a smart card is interfaced through an Internet protocol. The research projects described were inspired by the availability of JavaCard, which finally turns smart cards into programmable computing platforms.

15.12.1 A Browser Interface for Smart Cards

The first prototypical system that employed a Web browser to access a smart card was proposed by Barber [35]. A local HTTP server translates requests (in the form of URLs) into APDU exchanges with a smart card. The type of exchange depends on the path given in the URL. For example, some path triggers the upload of a new applet onto the card, while another path allows the user to directly state an APDU to be sent to the card. Yet another path provides a high-level interface to a specific card service, such as a payment transaction.

Parameters are given in the URL request and are translated into APDUs by the underlying component, called "card server." The card server can be configured such that it only accepts requests originating from the local machine, thus providing a simple mechanism for restricting access to the smart card to the local user. This approach is quite flexible, since it allows for arbitrary functionality within the card server. For example, it can prompt the user for confirmation, or interact with the keypad attached to the card reader. The card server, as described in the original paper, is itself based on the OpenCard Framework, which provides the basic smart card functionality.

15.12.2 Smart Cards as Mobile Web Servers

A different approach is based on the observation that every GSM mobile phone is essentially a portable smart card reader, carrying the SIM for customer authentication.

Thus, in many countries the basic infrastructure for carrying out secure business transactions is already in place. However, technical limitations of mobile phones as well as business interests of network operators are obstacles to making this infrastructure openly available. The WebSIM approach [36] tries to overcome these limitations by making a well-defined set of commands residing on the SIM available on the Internet.

The SIM is turned into a Web server through a proxy, which resides on a server on the Internet and accepts HTTP requests and turns them into messages in the SMS format. These messages are sent through a GSM gateway to the user's mobile phone where they are delivered to a JavaCard applet that resides on the SIM. Thus, the applet is able to react to HTTP requests. A result to such a request is returned also via SMS to the gateway and then delivered to the requestor as an HTTP response.

The SIM applet has a number of possibilities of executing requests. First, it can access the SIM's resources, such as the phone book entries that are stored in a smart card file. Thus, it is possible to update these entries through a Web browser interface without taking the SIM out of the mobile phone. Also, the applet can employ the SIM Application Toolkit to interact with the mobile phone, and thus with the user. For example, it is possible to show a list of items on the phone's display and ask the user to choose one of them. Another possibility is to use the SIM to create a digital signature on some data that are sent together with the request. Parameters like that, or an item list, are encoded within the request URL.

The address of a WebSIM is composed of a fixed Internet address, which is the name of the WebSIM proxy, and the telephone number of the mobile phone (which is actually the address of the SIM within the GSM network).

15.12.3 Internet Smart Cards

A similar, but more general approach, is the integration of smart cards at the TCP/IP level [37–39], which makes it possible to access smart cards directly from the Internet, for example, through a Web browser. This is achieved by implementing a simplified TCP/IP stack as a JavaCard applet on the smart card itself. This makes it possible to route TCP/IP packets across the card-reader interface to the card.

Webcard, for example, follows a simple protocol. It understands HTTP GET requests, followed by a URL pointing at a smart card file. If the URL path is a valid file name, the content of this file is returned as an HTTP response (the HTTP status line is supposed to be stored in the file together with the actual content). If the named file does not exist, the content of a default file is returned, indicating an error condition.

Impressively, as of the time of writing, a Webcard instance has been running for years and is still accessible on the Web through the URL http://smarty.citi.umich.edu/. A nice feature is the link that points to the source code of Webcard, which is stored on the smart card itself.

Making security-sensitive services available on the Internet requires effective access control mechanisms. Data exchange between an application and a smart card often requires confidentiality, for example, if the user PIN is presented to the smart card. Webcard does not support such security-sensitive operations. Instead, this was addressed by another project [38], where a secure link is established between a host application and the smart card before any sensitive data is exchanged. This makes it possible to access smart card services even if the connection between the host and the smart card cannot be trusted to be secure. Thus, applications such as SSH and Kerberos can employ the cryptographic functionality of a remote smart card through such a secure link.

15.13 Conclusion

Smart cards are not merely "dumb" (albeit secure) storage devices for cryptographic secrets, but have evolved to "real" computing platforms that are able to host complete and useful applications. This became possible through the progress in microprocessor technology and the availability of Java for smart cards. We hope it has become clear that making use of their advanced features requires sophisticated middleware concepts that are tailored to specific application areas. Many approaches are still evolving, especially in the mobile telecommunication world. Thus, although this chapter on middleware concludes, the story on smart card middleware continues.

Bibliography

[1] International Organization for Standardization (ISO) (1989) *International Standard ISO/IEC 7816: Identification Cards – Integrated Circuit(s) Cards with Contacts.*

[2] PC/SC Workgroup (1997) *Interoperability Specification for ICCs and Personal Computer Systems.*

[3] International Organization for Standardization (ISO) (1995) *International Standard ISO/IEC 7816-4: Identification Cards – Integrated Circuit(s) Cards with Contacts – Part 4: Interindustry Commands for Interchange.*

[4] European Telecommunications Standards Institute (ETSI) (1999) *Digital cellular Telecommunications System (Phase 2+); Specification of the Subscriber Identity Module – Mobile Equipment (SIM–ME) Interface*, GSM 11.11 version 8.1.0 Release 1999.

[5] JavaCard homepage, http://java.sun.com/products/javacard, April 2004.

[6] Sun Microsystems, Inc. (2002) *Java Card 2.2 Runtime Environment (JCRE) Specification.*

[7] Sun Microsystems, Inc. (2002) *Java Card 2.2 Virtual Machine Specification.*

[8] Windows for Smart Cards, http://www.microsoft.com/technet/security/prodtech/smrt card, April 2004.

[9] MultOS, http://www.multos.com, April 2004.

[10] BasicCard, http://www.basiccard.com, April 2004.

[11] Global Platform, http://www.globalplatform.org, April 2004.

[12] Java 2 Platform, Standard Edition (J2SE), http://java.sun.com/j2se, April 2004.

[13] http://java.sun.com/products/javacard/downloads, April 2004.

[14] *Movement for the Use of Smart Cards in a Linux Environment*, PC/SC Implementation for Linux, http://www.linuxnet.com, April 2004.

[15] Microsoft Developer Network, http://msdn.microsoft.com, April 2004.

[16] Sun Microsystems, Inc. (2002) *Java Card 2.2 Runtime Environment (JCRE) Specification*, Chapter 8, Remote Method Invocation Service, pp. 53-68.

[17] Sun Microsystems, Inc. (2002) *Java Card 2.2 RMI Client Application Programming Interface*, White paper.

[18] Sun Microsystems, Inc. (2002) *Java Card 2.2 Application Programming Notes*, Chapter 3, *Developing Java Card RMI Applications*, pp. 21–42.

[19] Vandewalle, J.-J. and Vétillard, E. (1998) Smart Card-Based Applications Using Java Card. *Proceedings of the 3rd Smart Card Research and Advanced Application Conference (CARDIS'98)*, Louvain-la-Neuve, Belgium, LNCS 1820, pp. 105–124; Springer-Verlag, Berlin, Germany.

[20] Public-Key Cryptography Standards, RSA Labs, http://www.rsasecurity.com/rsalabs/pkcs, April 2004.

[21] Kaliski, B. S. Jr. (1993) *An Overview of the PKCS Standards.* ftp://ftp.rsasecurity.com/pub/pkcs/doc/overview.doc, April 2004.

[22] RSA Security Inc. (1999) *PKCS #11 – Cryptographic Token Interface Standard*, Version 2.10.

[23] PKCS #11 Conformance Testing, http://www.mozilla.org/projects/security/pki/pkcs11, April 2004.

[24] Object Management Group, http://www.omg.org, April 2004.

[25] Chan, A. T. S., Tse, F., Cao, J., and Leong, H. V. (2002) Enabling Distributed CORBA Access to Smart Card Applications. *IEEE Internet Computing*, **6**(3), 27–36.

[26] European Telecommunications Standards Institute (ETSI) (1999) *Digital cellular telecommunications system (Phase 2+); Specification of the SIM Application Toolkit for the Subscriber Identity Module – Mobile Equipment (SIM–ME) interface*, GSM 11.14 version 8.1.0 Release 1999.

[27] European Telecommunications Standards Institute (ETSI) (2000) *Digital cellular Telecommunications System (Phase 2+); Technical Realization of the Short Message Service (SMS)*, GSM 03.40 version 7.4.0 Release 1998.

[28] European Telecommunications Standards Institute (ETSI) (1999) *Digital Cellular Telecommunications System (Phase 2+); Security Mechanisms for the SIM Application Toolkit; Stage 2*, GSM 03.48 version 8.1.0 Release 1999.

[29] 3rd Generation Partnership Project; Technical Specification Group Terminals (2002) *USIM Application Toolkit (USAT) Interpreter Architecture Description*, 3GPP TS 31.112 V5.2.0.

[30] Sun Microsystems, Inc. (2003) *Security and Trust Services API (SATSA) for Java Micro Edition*, Community Review Draft Version 1.0, Draft 0.9, Available at http://jcp.org/.

[31] Kehr, R., Rohs, M., and Vogt, H. (2000) Issues in Smartcard Middleware, in *Java on Smart Cards: Programming and Security* (eds I. Attali and T. Jensen), LNCS 2041, pp. 90–97, Springer-Verlag, Berlin, Germany.

[32] Kehr, R., Rohs, M., and Vogt, H. (2000) Mobile Code as an Enabling Technology for Service-oriented Smartcard Middleware. *Proceedings of the 2nd International Symposium on Distributed Objects and Applications (DOA'2000)*, Antwerp, Belgium; IEEE Computer Society, pp. 119–130.

[33] Waldo, J. (2000) *The Jini Specifications*, Second Edition, Addison-Wesley.

[34] JiniCard API, http://www.inf.ethz.ch/~rohs/JiniCard, April 2004.

[35] Barber, J. (1999) The Smart Card URL Programming Interface. *Gemplus Developer Conference*, Paris, France, 21–22 June, http://www.microexpert.com/smartcardurl.html, April 2004.

[36] Guthery, S., Kehr, R., and Posegga, J. (2000) How to turn a GSM SIM into a Web Server. *IFIP CARDIS 2000*.

[37] Rees, J. and Honeyman, P. (2000) Webcard: A Java Card Web Server. *IFIP CARDIS 2000*, CITI Tech Report 99-3.

[38] Itoi, N., Fukuzawa, T., and Honeyman, P. (2000) Secure Internet Smartcards. *Java Card Workshop*, Cannes, CITI Tech Report 00-6.

[39] Urien, P. (2000) Internet Card, a Smart Card as a True Internet Node. *Computer Communications*, **23**(17), 1655–1666.

16

Application-Oriented Middleware for E-Commerce

Jesús Martínez, Luis R. López, Pedro Merino
University of Malaga, Spain.

16.1 Introduction

Software applications for electronic commerce have constituted a strategic element for Internet business in recent years. Security for these applications is essential in order to obtain the confidence of users and companies involved, and special care must be taken with authentication, confidentiality, and integrity of data and transactions. However, guaranteeing practical security on Internet applications has been traditionally difficult, because of the unreliability of the environments in which those applications are executed, or the mobility requirements for users and servers that need to interchange sensitive data, like credit card numbers or data for bank transfers.

Smart cards have become key elements to accomplish security requirements for these applications (see Chapter 15). As flexible and tamper-proof-resistant computational devices, smart cards combine their storage capabilities along with the execution of secure applications inside them [7], also performing cryptographic operations with the aid of a crypto-coprocessor. These hardware devices, used within e-commerce applications, avoid the exposure of sensitive data, private keys, or cryptographic routines to external attackers that could compromise or manipulate the off-card application itself, when running over hostile environments. For example, smart cards are well suited for public key cryptosystems [3]. The use of a public/private key pair requires the use of such cards, because private keys need to be stored and used within a trustworthy environment. Note that it is not the case for a normal PC operating system.

It might seem that the need for the card to be physically connected to the host in which the application uses it comprises additional security enforcement. However, it limits the conception of new uses for these devices within distributed environments. Traditionally,

Middleware for Communications. Edited by Qusay H. Mahmoud
 ISBN 0-470-86206-8

smart cards have been used for access control applications, electronic purses, or digital identification cards, but they are not considered as first-class network resources even though they act as small personal data repositories, execute applications on demand, or perform digital signatures [19] on behalf of external applications.

When opening smart card services to distributed applications, special care must be taken with security in the design and implementation of usual middleware components such as stubs, servants, or interfaces.

This chapter introduces a solution for the integration of smart cards in a public transport ticketing service using middleware for ticket management, dealing with advanced mobility issues in both users and ticketing cards. The *public transport* concept should be considered more general than urban transport, implying long-distant journeys and a much higher cost per service. Therefore, customer identity or deferred payment details, which are irrelevant for an urban transport system, are considered mandatory in a public one. The design of the architecture must take into account this high service cost, providing strong security mechanisms to avoid forged or duplicated tickets, that could lead to severe economic losses to service providers.

We propose an architecture based on "digital tickets," which are easily managed by the electronic equipment involved. These tickets, stored on the smart card, offer some new exciting possibilities as transferability, tampering detection, or remote selling, which can add some value to customers and service providers. For this purpose, a new middleware API that remotely manages readers, cards, and transactions has been designed. A high-security environment is also provided using both the middleware security services available in the selected CORBA platform and the security mechanisms implemented as algorithms on the card.

The chapter is organized as follows: Section 2 describes the previous work in networked smart card applications and Section 3 describes the whole ticketing system. Section 4 reviews the security requirements for the middleware platform selected along with the new object library and interfaces designed. Section 5 briefly describes the characteristics of the prototype implemented for the transport ticketing system. We conclude with a summary and some suggestions for future work.

16.2 Previous Work on Networked Smart Card Applications

Recently, there has been a growing interest in designing new methods to obtain the resources and services provided by smart cards from network applications. This section provides an overview of the different proposals existing in recent literature, describing not only the solution adopted but also how they deal with security aspects.

Smart cards themselves constitute autonomous computer environments, so interactions between off-card and on-card applications could be seen as a kind of (soft) distributed computing. In this form, the broker used in traditional middleware platforms is substituted by serial protocols (T0, T1, TLP) between the off-card application and the reader that acts as a bridge to the smart card service, delivering APDUs as protocol units (or references to files, applications, and object methods). Further information related to this subject can be found in Chapter 15.

The smart card as a network resource was first proposed in [1]. The paper describes a method for storing and retrieving medical information for patients that is queried and updated later by doctors via Web browser interfaces. Mobility is accomplished by the definition of a Web server running on hosts where readers are connected. When the card is inserted, a dynamic server module (named JCWS: Java Card Web Servlet) accesses the medical information contained on it, and converts the data to suitable HTML tags that sends to the doctor Web browser. This scheme uses the client/server paradigm, and does not worry so much about security. Perhaps secure hypertext protocols such as *https* could improve the interchange of personal identification numbers or confidential patient's logs. Anyway, this approach cannot be considered a distributed proposal, because real mobility is not granted: Web browsers have to know explicitly where the location (URL) of the card in the network is for requesting the HTML information to the Web server. There is also the problem of guaranteeing that the server runs within a trustful environment (protected against manipulation), because the application that manages the card information also deals with sensitive information in clear text.

A valuable contribution focused on the design of a high-security scheme for networked applications that use smart cards was introduced in [10]. When using the card for digital signatures within a distrusted environment, there is no guarantee that a document (for example, a contract) signed by the card is identical to the one previously visualized by the user on the computer screen. Assuming that a Personal Digital Assistant (or PDA) increases security because of the negligible exposure of these devices to malicious software, the paper proposes a solution consisting of a combination of a PDA for visualizing the document that needs to be signed and a smart card reader, both connected to the network through a Jini federation. The overall system is called PCA: Personal Card Assistant. Using Jini, services are registered by servers. When a client requests any of them, it receives a serialized proxy that will manage the communication at the client side. Security is achieved when the PDA and the smart card are tightly coupled, using a public key cryptosystem. Therefore, devices know each other's public key. First, the user revises the document in the PDA screen, sending it to the reader along with the Card Holder Verification code. The complete message is previously ciphered (signed) with the PDA's private key and then ciphered again with the card's public key. On reception, the smart card only has to decipher the message, also confirming that it corresponds to the PDA. The last step consists of signing the document with the card's private key. Note that the PCA system does not increase security *per se*, since any PDA can be attacked, just like any network computer. However, PDAs are under the direct control of the person who wishes to apply a digital signature to a document, since in practice they are trustworthier for that person. It is interesting to note that a similar approach that substitutes the network and the Jini services with the infrared communication capabilities of PDAs has been proposed in [12]. This proposal assumes the need of multiple signatures for documents, as occurs with real contracts, when two or more entities are involved. Documents are transmitted using the infrared ports. The PDA itself is attached to the smart card reader by a serial port.

From the world of mobile phones, GSM networks may also be considered as distributed platforms for applications executed on the extended smart cards that have substituted traditional Subscriber Identity Modules (SIMs). These kinds of applications follow the SimToolKit recommendation [4], being able to control the phone terminal. For an introduction into the SIM Application Toolkit, see Chapter 15. When carriers provide SIMs with cryptographic capabilities, they can be employed in emerging mobile-commerce applications. A proposal that makes extensive use of the GSM platform and the new

SimToolKit cards to implement a complete ticketing system was introduced in [11]. This work also introduces delegation and payment scenarios.

The OrbCard framework [2] is the closest proposal to the one presented in this chapter. That work is devoted to expose the services of JavaCard applets (on-card applications) [20] to network clients through CORBA middleware. This process is fully automated with the aid of the OrbCard compiler that uses an IDL file as input. This file includes a description of the services available in the JavaCard applets, along with the APDUs responsible for invoking them. With this information, the compiler is able to generate Java classes that constitute the client stubs, an adapter that acts as a gateway between the smart card applet and remote client requests, and finally a mapper class that converts remote method invocations to the corresponding APDUs needed by the card. Using middleware enables dialogs with smart card services to clients, possibly written in different languages. Nevertheless, using Java as the native language for the adapters (those that communicate with readers and cards using a low-level driver) has proven not to be efficient enough in terms of speed. C or C++ based solutions seem to be more indicated when critical applications, like e-commerce ones, need synchronous responses within the middleware environment. Finally, security is not covered in this proposal. Instead, the authors give some suggestions for future work that include the use of public key cryptosystems.

16.3 A Public Transport Ticketing System for e-Commerce

As stated before, the design of an architecture for a public transport ticketing system must take into account the high service cost, providing strong security mechanisms to avoid forged or duplicated tickets, that could lead to severe economic losses to the company involved. Moreover, as it is not mandatory for people involved in an e-commerce transaction to share the same physical place, some additional measures must be taken into account to assure that people are who they claim to be. In order to accomplish this in a secure way, the use of tamper-proof hardware is required, because standard software solutions are vulnerable to hacking practices. This is absolutely essential when storing sensitive data, such as privileges or personal keys.

The architecture presented in this section is based on the concept of electronic ticketing (e-tickets). All the agents and scenarios are built around this piece of digital data. More precisely, this e-ticket is the proof of a successful transaction, and includes all the necessary mechanisms to prevent any alteration of its content. These electronic tickets will be safely stored in a smart card that suits the whole security design requirements.

16.3.1 The System Architecture

This section describes the system architecture, whose main components are depicted in Figure 16.1. We will first describe services available and next we will introduce the agents that take part in the system, together with their main functionality.

16.3.1.1 System Services

The ticketing system is designed to offer two main services: (a) ticket emission and validation (on board) and (b) user queries and transaction logs.

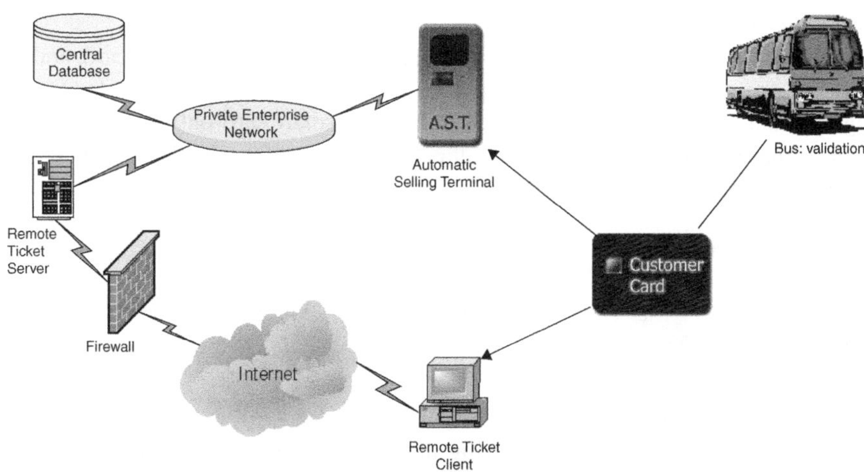

Figure 16.1 The public transport ticketing system architecture

Ticket Emission

Figure 16.1 shows the two ways to obtain tickets that are emitted by a Ticketing Server application. The first one uses an automatic selling terminal. We can consider this as a secure terminal resembling a cash dispenser, and intended to be located in bus stations or bank offices.

The automatic selling terminal communicates with the Ticketing Server through a secure and private connection within the company's network. As an additional design requirement, both terminals and client applications will not contain any sensitive data that could constitute a target for malicious attackers.

When using the remote ticket emission in public networks such as the Internet, mobility constitutes the main security problem. As mentioned before, the remote client application may be executed within a distrusted environment. Therefore, the best way to achieve practical security consists of using the client application as a gateway between the ticketing server and the smart card. Only unimportant or public information will be stored in the user disk in order to reduce transaction times.

Figure 16.1 also shows that remote client applications never gain access directly to the company's private network. Usually, critical servers are running behind some network protection facility such as a firewall.

Ticket Validation

This task is performed on the bus, and it must be accomplished off-line, without any connection to the central database. This requirement avoids having to install additional equipment other than a device with a smart card reader. However, off-line validation introduces more difficulties to assure rights contained within tickets. When a client inserts his card into the validation device on board, the software application will check the subscription to the service along with the authenticity of the card. Tickets are checked only after validating users, obtaining the rights that were granted (and signed) by the service provider. Every accepted ticket is stored in the vehicle's local database and is later committed to the company's central database.

Queries and Logs

Every transaction generates a log that is stored inside both the smart card and in the database of the vehicle. Users who want to see logs may do so using the automatic selling terminal or the remote client application. The smart card only contains an extract of the last transactions performed by the client. Therefore, if a complete log is requested, the query must be performed on-line.

16.3.1.2 System Agents

They are responsible for giving users the services defined above, and will be implemented as software applications.

The Smart Card Issuer

This agent is responsible for personalizing the cards to be used by clients. The process implies the creation of a file system within the card, along with the symmetric key set used in remote communications, in order to safely store tickets. Note that these keys are different from the asymmetric pair supplied when a secure connection is going to be performed. Therefore, this key set included with the card is intended to be used to sign commands that make read or write operations over files. The card serial identifier is also stored in a proper central database, for authentication purposes.

The issuer agent will only be installed in offices belonging to the transport company, and it will usually work in a trusted environment with no network access, because the key generation task is considered critical, especially when provider's keys or bus service instance keys are created.

Automatic Selling Terminal

As mentioned before, this terminal includes some characteristics that make it easy to use for obtaining tickets. It will have a touch-sensitive screen and a smart card reader. The software that governs the terminal will allow users to manage their PIN (Personal Identification Number), to purchase or cancel tickets, or to see logs. The existence of these devices in the system is justified for two reasons: (a) clients will be attended at any moment, and (b) box offices will be freed, avoiding client queues.

The Bus

The service offered in the ticket is provided within the vehicle. It performs the verification of users and electronic tickets. This verification process is carried out with the aid of another smart card that belongs to the bus driver. This card contains certificates of providers, along with service keys.

Remote Client Application

The remote client application provides the services offered by the automatic selling terminal to any computer with an available network connection and smart card reader. Security risks include impersonation of users or eavesdropping and manipulation of tickets before they are safely stored within the smart card. Therefore, communications will be ciphered, and additional mechanisms will be provided in order to avoid the direct management of keys by client applications.

The Ticketing Server

Client applications connect to this server in order to obtain tickets. Before getting them, both server and smart card must be authenticated to each other in order to assure subscription rights and providers' identity. The server performs the operations directly with the card, using the client software as a simple gateway between them (see Figure 16.1).

Services Database

This is the critical resource for the company. The central database stores keys and client information, including its smart card identification. It also registers a complete log of transactions, for payment purposes. The database is connected only to a private network.

16.3.2 The Electronic Ticket

A ticket is a certificate that guarantees to its owner the right to demand services written on it. We can distinguish three different parts in either traditional or electronic tickets: the issuer, the promise, and the owner. The issuer guarantees that the ticket owner has the right to claim the services written on the ticket. It is also convenient to detect any change in information (e.g. tampering). In this way, the issuer knows that the owner will not be able to alter the promise he has paid for. The customer will make sure that no one will appropriate the services written on the ticket.

Electronic tickets allow parameterization of their properties, so they greatly simplify the management of a public transport infrastructure, target of our design. This allows for many interesting possibilities such as, for instance, rights transferability among customers and a fulfilled services log. In addition, as they are coded as digital information, they can be transferred by any electronic means.

The use of a smart card as the ideal safe store for these tickets is motivated by their characteristics: e-tickets are sensitive and dynamic data, which may contain a large amount of information.

Figure 16.2 shows the electronic ticket designed for the transport system. The version field controls the number of fields and their semantic. Tickets have a provider, a card serial number, the client, and the ticket identifier. This is the information needed on board by the validation software to authenticate customers. The rights have a period validity, from the emitted date to the service date. Other important data are the bus line, origin and destination, seat, and price. The field type introduces additional information about discounts or promotions. The last 1024 bits constitute the digital signature of the ticket. There is a trade-off between key lengths and security. Using big lengths for keys

version	provider	CSN	userID	ticketN	ems date	serv date	line	org	dst	seat	price	type	signature
8 bits	16 bits	64 bits	24 bits	64 bits	32 bits	32 bits	16 bits	32 bits	32 bits	16 bits	32 bits	16 bits	1024 bits

Figure 16.2 The electronic ticket designed for the public transport system

is not a trivial consideration when using public key cryptosystems, because a lot of time will be wasted in performing cryptographic operations. There is also the problem related to the small size of the smart cards to store various tickets.

16.3.3 Choosing a Smart Card for the System

Smart cards provide very useful capabilities for storing this kind of data. Their *look and feel* is similar to traditional magnetic strip cards. In fact, print methods used on these are applicable to this technology. They can also use encryption and authentication protocols in transfers, disallowing nondesired third-party listeners to intercept or alter transactions. For more details on smart card basics, see Chapter 15.

The system presented here does not need to run any user proprietary code inside the card, although the design could be easily extended to make use of this security improvement.

Several methods can be used to communicate with the card's embedded chip. On one hand, some smart cards require a physical insertion in the reader. On the other hand, contactless cards allow talking to a reader without being physically inserted on it. The power needed for the card operation is extracted from an electromagnetic field originated in the reader. There are also cards, called *combo cards*, which provide two interfaces: a contact one and a contactless one. The latter is very suitable for our purposes. The contact interface should be used when obtaining a ticket or querying transaction logs, inserting the card in the reader. The contactless interface is more appropriate for using the ticket on board, avoiding queues and being the method preferred by users. Therefore, a main requirement for our system considers that data stored in the combo card can be accessed from both interfaces. Finally, the GemCombi/MPCOS combo card, from Gemplus, has been selected for our purposes. The GemCombi/MPCOS contactless interface is MIFARE (ISO 14443a) compatible and is not based on a file structure. Data is spread through 40 sectors and each sector is built on a variable number of 16-byte blocks (4 for sectors 0 to 31 and 16 for sectors 32 to 39). Security requirements are set up in the last block of each sector (trailer block), specifying up to two 48-bit keys, Key A and Key B, and the different operations allowed with each one.

Since the physical level of the MIFARE interface is a shared and distrusted medium (air), data interchange will be ciphered using those keys. The same keys grant an off-card application different access possibilities specified by the trailer block. Moreover, mutual authentication between entities is needed. Further authentication is not needed if subsequent accesses to the card are produced within the bounds of the same sector and if the access key is not changed.

Some data available at the contactless interface can be accessed through the contact interface via special mapping files on the file system. Concretely, sectors 32 to 39 are shared on this application. This allows great flexibility, as the contact interface can apply to those files different security policies that the contactless interface could not implement, such as PIN authentication. Obviously, these mapping files have been selected to store tickets and providers' information, as shown in Figure 16.3.

Data on the contact interface is distributed along files on a two-level hierarchical file system, corresponding to a standard ISO 7816 structure [9] as depicted in Figure 16.4. All

<table>
<tr><th>Sector</th><th>Block</th><th colspan="2">Description</th><th>Contactless access</th><th>Contact access</th></tr>
<tr><td rowspan="16">Sector 32</td><td>0</td><td colspan="2" rowspan="9">Subscription Info 0</td><td rowspan="15">Read: Public (A0A1A2A3A4A5)
Write: Forbidden</td><td rowspan="15">Read: PIN
Write: PIN + Admin key</td></tr>
<tr><td>1</td></tr>
<tr><td>2</td></tr>
<tr><td>3</td></tr>
<tr><td>4</td></tr>
<tr><td>5</td></tr>
<tr><td>6</td></tr>
<tr><td>7</td></tr>
<tr><td>8</td></tr>
<tr><td>9</td><td colspan="2" rowspan="3">User Info</td></tr>
<tr><td>10</td></tr>
<tr><td>11</td></tr>
<tr><td>12</td><td>Ticket Info 0</td><td>Ticket Info 1</td></tr>
<tr><td>13</td><td>Ticket Info 2</td><td>Ticket Info 3</td></tr>
<tr><td>14</td><td>Ticket Info 4</td><td>Ticket Info 5</td></tr>
<tr><td>15</td><td colspan="2">SECURITY(ACCESS CONDITIONS)</td><td></td><td></td></tr>
<tr><td rowspan="16">Sector 33</td><td>0</td><td colspan="2" rowspan="9">Subscription Info 1</td><td rowspan="15">Read: Public (A0A1A2A3A4A5)
Write: Forbidden</td><td rowspan="15">Read: PIN
Write: PIN + Admin key</td></tr>
<tr><td>1</td></tr>
<tr><td>2</td></tr>
<tr><td>3</td></tr>
<tr><td>4</td></tr>
<tr><td>5</td></tr>
<tr><td>6</td></tr>
<tr><td>7</td></tr>
<tr><td>8</td></tr>
<tr><td>9</td><td colspan="2" rowspan="6">(RFU)</td></tr>
<tr><td>10</td></tr>
<tr><td>11</td></tr>
<tr><td>12</td></tr>
<tr><td>13</td></tr>
<tr><td>14</td></tr>
<tr><td>15</td><td colspan="2">SECURITY(ACCESS CONDITIONS)</td><td></td><td></td></tr>
<tr><td rowspan="16">Sectors 34/35/36/37/38/39</td><td>0</td><td colspan="2" rowspan="11">Ticket 0/1/2/3/4/5</td><td rowspan="11">Read: Restricted (Issuer Key)
Write: Forbidden</td><td rowspan="15">Read: PIN
Write: PIN + Admin key</td></tr>
<tr><td>1</td></tr>
<tr><td>2</td></tr>
<tr><td>3</td></tr>
<tr><td>4</td></tr>
<tr><td>5</td></tr>
<tr><td>6</td></tr>
<tr><td>7</td></tr>
<tr><td>8</td></tr>
<tr><td>9</td></tr>
<tr><td>10</td></tr>
<tr><td>11</td><td colspan="2" rowspan="2">Service Log 0/2/4/6/8/10</td><td rowspan="4">Read: Forbidden
Write: Restricted (Issuer Key)</td></tr>
<tr><td>12</td></tr>
<tr><td>13</td><td colspan="2" rowspan="2">Service Log 1/3/5/7/9/11</td></tr>
<tr><td>14</td></tr>
<tr><td>15</td><td colspan="2">SECURITY(ACCESS CONDITIONS)</td><td></td><td></td></tr>
</table>

Figure 16.3 Memory map used to store tickets

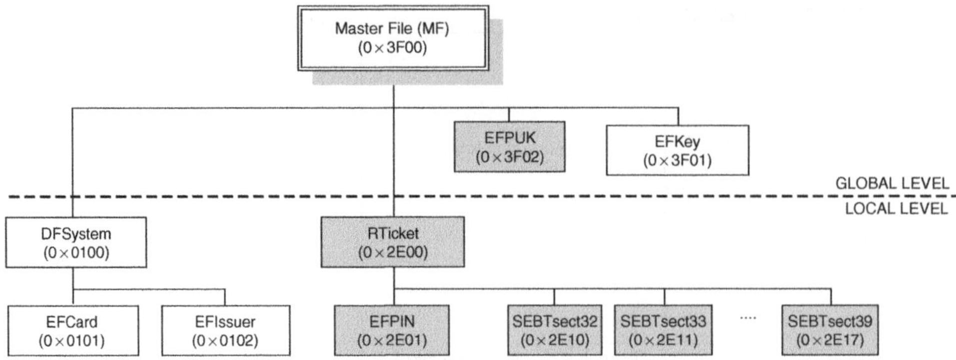

Figure 16.4 The customer card's ISO file system

the information that concerns the ticketing system is stored under a directory (dedicated file). That is, the card may support additional applications. All PIN-related files are conducted by the EFPIN file and its corresponding EFPUK file: the former stores the PIN, while the latter contains the key that lets the user unblock and replace a locked PIN. Finally, the Shared memory Exported Binary Transparent (SEBT) files act as a bridge to the corresponding sectors in the contactless memory. The PIN for read access protects them. To obtain write permissions, an additional key only known by the card and the provider is needed, so that any alteration can take place only when both user and provider agree.

16.4 Advanced Ticketing Management Using Middleware

By definition, smart card–based schemes do not allow mobility, because applications depend on being executed in the same computer that contains both a reader and a card inserted into it. This limitation may be solved using a middleware platform where a known service, that runs in the machine in which the reader and the card are connected, is available through an object request broker, and may invoke operations on the card (see Figure 16.5). Therefore, remote applications could be executed in a controlled number of machines, needing only a single reader. This proposal introduces some clear benefits, reducing costs and obtaining a better and centralized management of the environment in which cards and reader execute (drivers and software updates, virus checking, and so on). When using a middleware platform, there are strong security requirements that must be taken into account in order to avoid unrestricted access to objects. We have selected CORBA as the middleware platform in which this remote system management issue will be implemented. Application-oriented middleware with access to smart cards and readers has not been specified in any CORBA recommendation, and we have designed a new C++ API that servant implementations may use in order to homogeneously access these devices. Thus, readers and cards are exposed to the network through an IDL module. This section is focused on the specific security details of the middleware platform selected along with the distributed service and the value-added API introduced for working remotely with smart cards.

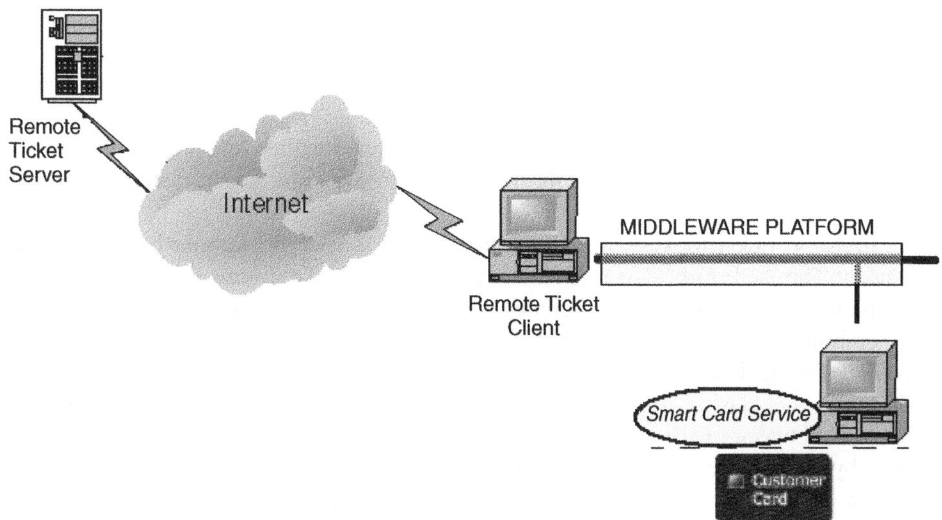

Figure 16.5 Clients are gateways between the ticketing server and the customer card

16.4.1 Middleware Platform Security

We have mentioned that e-commerce applications share, in general, some common requirements. First of all, they have to demonstrate the identity of parts involved in a transaction. Software programs that run over a computer network now substitute traditional vendors and clients of the real world. Therefore, performing authentication must be seen as a key goal of these applications. Moreover, sensitive information transmitted along the network is susceptible to be eavesdropped and manipulated (the so-called *man-in-the-middle* attacks). This means that e-commerce applications also have to protect that information, for confidentiality purposes, together with ensuring the integrity of data involved.

Public key cryptosystems are used to accomplish these stringent requirements. Using this approach, every entity that takes part in a transaction has a public/private key pair. The public key is usually stored in a directory server, and everyone may request it. However, the private key is stored in a trusted device, the smart card. Operations performed with these keys generate encrypted or signed data, respectively.

When translating e-commerce applications to a middleware platform, we have to know previously how many security elements it provides in order to separate the application business logic from the security requirements that communications need.

The first CORBA specifications did not include security. Fortunately, the OMG Security Special Interest Group (SecSIG) has defined a security reference model and an implementation architecture, which collectively establishes security policies, that is, conditions for accessing objects, in tandem with a definition of the primary actors in a secure system. Interoperation among secure system implementations is also defined.

The architecture classifies security features, depending on the needs of end users, developers, or administrators. When considering the issue of security application integration, SecSIG has defined the scope of participation of an application. Thus, CORBA Security defines three levels of participation: (a) security unaware: the middleware platform performs all the security; (b) security policy controlling: the application directs the middleware to use specific policies, but relies on the middleware for security operations; and (c) security policy enforcing: the application actively examines security context information during middleware enforcement.

CORBA provides an API for achieving security functionality levels, security replaceability, or interoperability. Most of the functionality levels are encapsulated within IDL modules.

The transport ticketing system management part that uses middleware has been developed using The ACE Orb (TAO) [21], a high performance, real-time CORBA-compliant ORB implementation that also includes support for some security functionality and interoperability features. TAO supports confidential communication through its Internet Inter-ORB Protocol (IIOP) over Secure Socket Layer (SSL) pluggable protocol, called SSLIOP [14, 17]. Using this protocol, it is possible to ensure that all remote method invocations between ORBs that implement IIOP over SSL are confidential. This is made possible by the confidentiality provided by the Secure Socket Layer [5]. Some extensions also allow certificate-based access control using X.509 certificates [8].

In TAO, all cryptographic profiles supported by the popular OpenSSL [16] library are implemented. ORBs that also support those profiles should be able to interoperate with TAO. Moreover, when authentication is needed, TAO's SSLIOP pluggable protocol allows an application to gain access to the SSL session state for the current request. It allows the application to obtain the SSL peer certificate chain associated with the current request so that the application can decide whether to reject the request. The ticket management system always makes use of SSL sessions to connect clients to the Smart Card Service that performs the operations with the smart card, as shown in Figure 16.6.

16.4.2 The Smart Card Service

Remote client applications may use this service to communicate with readers and cards. The IDL definition in Figure 16.7 follows the implementation of the object library for

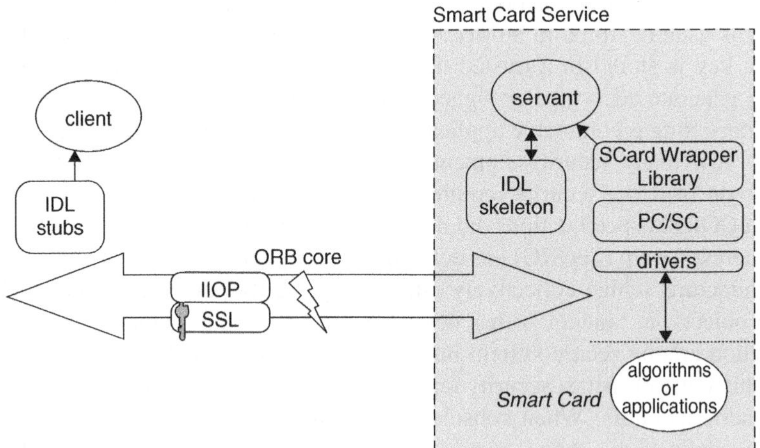

Figure 16.6 ORB security is achieved using TAO's SSLIOP protocol

```
// File: SCardServCommon.idl

module SCardService {

typedef sequence<octet> ByteArray;

interface SmartCardFamily_Factory{

SmartCardFamily getFamily (in string type)
      raises (EInvalidFamily);

};

typedef sequence<SmartCardReader> ReaderList;

interface SmartCardFamily{

readonly attribute ReaderList rlist;

SmartCardReader getReader(in string common_name)
      raises (EReaderError);

};

interface SmartCardReader{

readonly string name;

string getName();

//Iterator pattern
boolean scanForCards();

SmartCard nextCard()
      raises(ECardNotFound);
//end Iterator pattern

SmartCard findCard(in ByteArray cardSerialID)
      raises(ECardNotFound);

};

interface SmartCard{

readonly attribute string name;
readonly attribute ByteArray cardSerialID;

unsigned int sendAPDU(
      in ByteArray command,
      in unsigned int command_length;
      out ByteArray response,
      out unsigned int response_length
      ) raises(
       ECardNotPresent, EReaderError, EInvalidParam );
```

Figure 16.7 Part of the IDL for distributed smart card management

```
unsigned int sendAPDU(
      in ByteArray command,
      in unsigned int command_length;
      ) raises(

       ECardNotPresent, EReaderError, EInvalidParam );

boolean sendMessage(
      inout Message msg;
      ) raises(
       ECardNotPresent, EReaderError, EInvalidParam );

boolean performTransaction (
      inout Transaction
      ) raises(
       ECardNotPresent, EReaderError, EInvalidParam );

FileSystemMap getCardFileMap()
      raises(
       ECardNotPresent, EReaderError);

}; ...

}; // end SCardService
```

Figure 16.7 (*continued*)

homogeneous access to cards. In fact, the IDL part of Figure 16.7 corresponds to the basic functionality provided with base classes. Using these common interfaces, more complex data types such as digital bus tickets or ticketing transactions may be defined. The Smart Card object-oriented library now supports both PC/SC contact readers and the GemCombi/MPCOS reader. The MIFARE smart card is also implemented, although this contactless card will not be available for middleware applications. As part of the basic functionality provided in the IDL, there is a useful support for a BasicTicket interface together with a FileSystemMap interface that may encapsulate, for example, the blocks and sectors of the GemCombi/MPCOS card, as depicted in Figures 16.4 and 16.5.

Thus, servant implementations may benefit from the C++ object-oriented library developed. Flexibility of the library is achieved by the use of several well-known design patterns [6].

16.4.3 The Smart Card Object-Oriented Library

The IDL servant implementation code makes use of the object hierarchy depicted as a class diagram in Figure 16.8. As the first access point to the Smart Card Service, we may obtain a concrete family. We have separated functionality of contactless cards from the normal contact ones by using the Factory design pattern. This classification helps to better manage the incorporation of new specific readers and cards to the library. Object references to readers supported in a particular family may be obtained through an Iterator pattern. With a reader instance, it is possible to obtain a proper reference to a SmartCard object. The SmartCard base class introduces methods for manipulating APDUs and, when

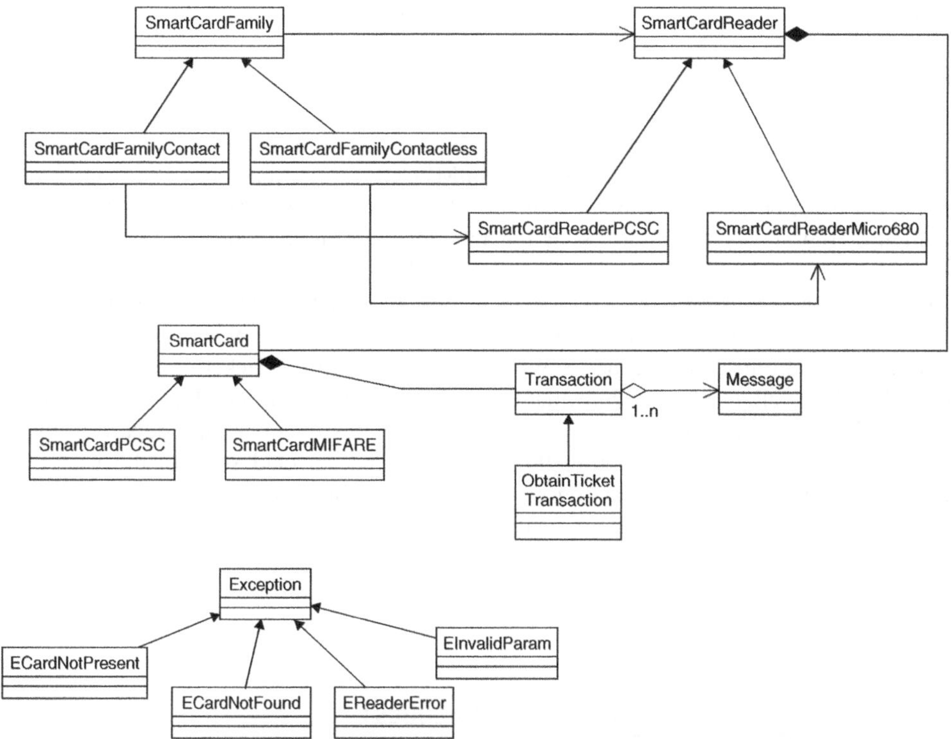

Figure 16.8 UML class diagram for the smart card API

available, file systems. We have also introduced two classes for wrapping APDUs: the Message class and the Transaction class. A Message object internally stores a byte array that constitutes the full APDU that is sent to the card. Subclasses of this object have been implemented to arrange useful APDUs for secure messaging, as they are used in the GemCombi card.

A Transaction object encapsulates an ordered sequence of Messages. When passed to the SmartCard object, the transaction may be selected to be interactive or not. In noninteractive mode, if the execution of a Message obtains an expected return code, the SmartCard tries to execute the next one in the sequence. Figure 16.8 also shows a Transaction subclass, called ObtainTicketTransaction. This class encapsulates the messages used for example when the management application for clients wants to show the interested user the tickets available on the card.

We can also consider this C++ object-oriented library an alternative to the OpenCard Framework [23] written in Java. Main contributions of the library are the extension to support contactless cards (not present in PC/SC) and the middleware bindings.

16.5 The Application Prototype

The public transport ticketing system has been implemented as a prototype, including all the infrastructure agents except the automatic selling terminal. We have used the C++

language in order to improve performance. The PC/SC library for smart cards [13, 18] has been wrapped into the Smart Card Service. Internet communications are secured using OpenSSL, and this popular library has been also used in the CORBA extensions for client mobility, through the SSLIOP protocol. An open source version of the Interbase database called it Firebird [22] is responsible for storing keys and information about clients and cards. As mentioned before, the GemCombi/MPCOS combo card has been selected as the client card, because it supports standard MIFARE and ISO 7816 specifications.

Figure 16.9 shows a screenshot of the remote client application. Its Graphical User Interface allows users to communicate with both the Remote Ticket Server and the smart card, in order to request new services or manage the existing tickets, respectively. Left menu in the application contains the program options. Figure 16.9 shows the form used within a typical transaction scenario. When a connection to the Ticketing Server has been established, the client receives information about available providers. Later queries to the Central Database provide details about services: origins, destinations, departure times, dates and prices. After completing the form with the selected service, the user must click the "Buy" button. This action sends the information to the Server, which issues the corresponding ticket. At this moment, the client application becomes a mere gateway

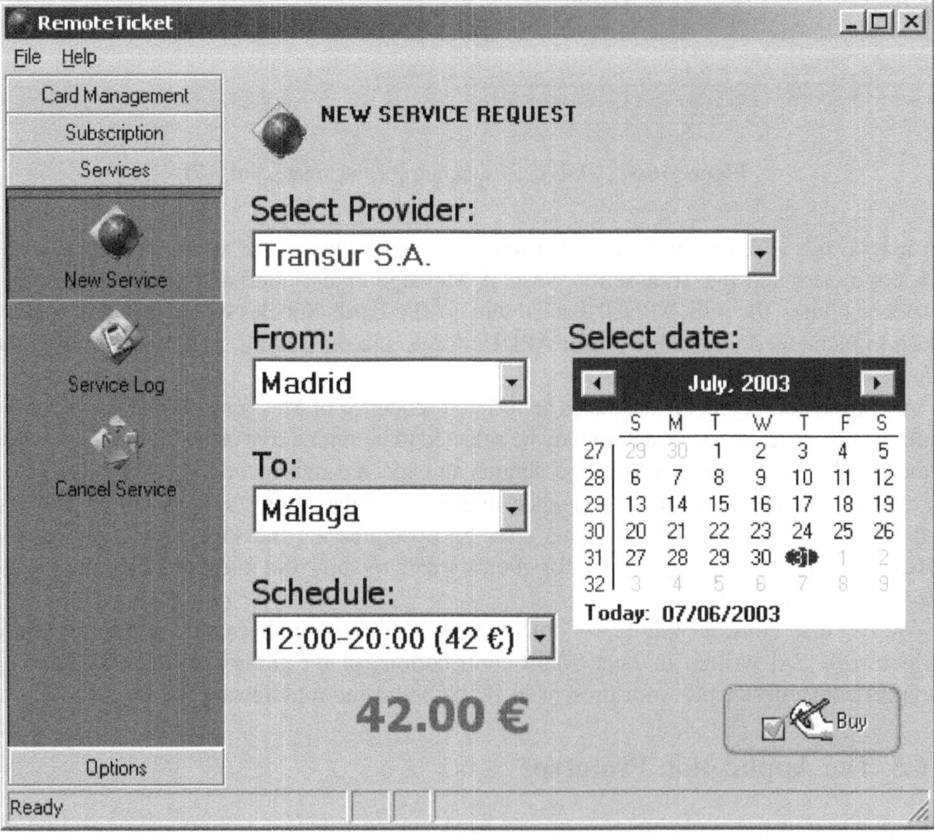

Figure 16.9 A remote client application's snapshot

between the Ticketing Server and the Smart Card Service. Therefore, the Ticketing Server will be responsible for preparing the APDU messages (using ISO 7816 secure messaging) that select the proper file to store the ticket in the card, conforming the structure shown in Figure 16.4.

The client application makes use of the middleware stubs and functionality provided by the Smart Card Service. Figure 16.10 depicts a basic code skeleton that has constituted the basis to develop the middleware part of the program. When the ORB has been initialized, we locate the Smart Card Service making use of the CORBA Naming Service. This service is a standard mechanism by which application objects can be registered by the server

```
//Included in File ''clientMgmt.cpp''
...
// Initializing the ORB
CORBA::ORB_var orb =
  CORBA::ORB_init (argc, argv);

// Using the naming service to locate SCardService
CORBA::Object_var naming_context_object =
  orb->resolve_initial_references ("NameService");
CosNaming::NamingContext_var naming_context =
   CosNaming::NamingContext::_narrow (naming_context_object.in ());

CosNaming::Name name (1);
name.length (1);
name[0].id = CORBA::string_dup ("SmartCardService");

CORBA::Object_var factory_object =
  naming_context->resolve (name);


// Downcasting the object reference to the appropriate type
SCardService::SmartCardFamily_Factory_var factory =
  SCardService::SmartCardFamily_Factory::_narrow(factory_object.in());

//Getting the contact family
SCardService::SmartCardFamily_var family =
   factory->getFamily("Contact");

//Locating the PCSC reader
SCardService::SmartCardReader_var reader =
   family->getReader("PCSC");

if(!reader->scanForCards()){
   //Display Error and close
      ...

}

SCardService::SmartCard_var card_object =
   reader->nextCard();
```

Figure 16.10 Part of the middleware skeleton code for a client

```
//Downcasting the parent object reference to the subclass type
SCardServicePCSC::SmartCardPCSC_var card =
   SCardServicePCSC::SmartCardPCSC::_narrow(card_object.in());

//Request PIN to user
retry = card->verifyPIN(pinCode);

//Establishing connection with the Ticketing Server using SSL
      ...

//From now on, the client acts as a gateway between the RemoteTicket
// Server and the CORBA SmartCard Service

//First, a (ciphered) command is received from the Remote Server
length = SSL_read(sslHandle, bufferServer, MAX_LENGTH);

//Then, the command is sent to the card Service
responseCode = card->sendAPDU( (unsigned char*)bufferServer,
                               (unsigned int)length,
                               (unsigned char*)bufferCard,
                               \&rLength);

...

//The response is returned to the Ticketing Server
SSL_write(sslHandle, responseCodeArray, 2);
if(rLength > 0)
   SSL_write(sslHandle, bufferCard, rLength);
    ...

...
orb->destroy ();
```

Figure 16.10 (*continued*)

process and accessed by distributed client applications. The client obtains the object reference by resolving the specified name to a CORBA object reference. As shown in the IDL definition (see Figure 16.7), the access point to the service is the `SmartCardFamily_Factory`. Next, we have to locate the contact family that manages contact readers and related cards. The appropriate family will give access to the PC/SC reader, which implements the Iterator pattern to find inserted cards. At last, a SmartCardPCSC instance will be returned, allowing the verification of its PIN code.

The next step consists of connecting to the Ticketing Server using the OpenSSL functions. From now on, the Server will manage the card, and the client will use `sendAPDU` methods with the data received from the network. The client does not have to understand this data, and only may check the response code from the card in order to update its graphical interface.

First tests with this prototype provide promising results in both performance and flexibility, confirming that CORBA middleware performance is largely an implementation issue as stated in [17].

16.6 Summary and Conclusions

This chapter has introduced a new public transport ticketing service that uses smart cards and a middleware platform as part of its management architecture. Smart cards are the key element within e-commerce applications because they constitute trustworthy execution environments and repositories of sensitive data. However, a usual characteristic of middleware applications, like mobility, could not be performed when using smart cards, because the application had to be executed in the same computer containing both a reader and an inserted card. This limitation can be solved using a Smart Card Service that acts as a server and gateway among network clients and the smart card. Using middleware allows clients written in different languages access to those services available within the card, through the Object Request Broker approach. Unfortunately, this poses new problems regarding security. Therefore, the main contributions of our work are: (a) the C++ class hierarchy for smart card management on servant implementations and the corresponding IDL module; (b) the design of the public transport ticketing management system, and (c) the conjunction of security available in both the TAO middleware platform and the combo card selected.

The resulting management system allows client applications to be physically decoupled from the reader and the card. This is a valuable scheme when cards are used for more complex applications than the ones for access control, and may stand inserted in the reader for a period of time. Therefore, our system not only issues personal transport cards also but departmental or familiar cards as well, intended to be used in a centralized and trusted machine containing the smart card reader. Security has been considered critical in the design, and some facilities allow agents to perform mutual authentication and ciphering transactions, such as the SSL protocol and the use of identity certificates.

As a future work, the introduced middleware API could be improved for being integrated with a CORBA Resource Access Decision Service [15]. This recommendation establishes a framework to support fine-grain access controls required by application-level security. That is, services provided by the smart card should be classified in order to enforce resource-oriented access control policies.

16.7 Acknowledgments

This work has been partially supported by the European Union and the Spanish Ministry of Science projects 1FD97-1269-C02-02 and TIC-2002-04309-C02-02.

Bibliography

[1] Chan, A. T. S. (1999) Web-Enabled Smart Card for Ubiquitous Access of Patient's Medical Record. *Proceedings of 8th International World Wide Conference, W3 Consortium*, pp. 1591–1598.

[2] Chan, A. T. S., Tse, F., Cao, J., and Va Leong, H. (2002) Enabling Distributed CORBA Access to Smart Card Applications. *IEEE Internet Computing*, **May/June**, 27–36.

[3] Diffie, W. and Hellman, M. (1976) New Directions in Cryptography. *IEEE Transactions on Information Theory*, **IT-22**(6), 644–654.

[4] Digital Cellular Telecommunications System (Phase 2+) (2000) *Specification of the SIM Application Toolkit for the Subscriber Identity Module –Mobile Equipment (SIM - ME) interface*. GSM 11.14 version 8.3.0.

[5] Freier, A., Karlton, P., and Kocher, P. C. (1996) *The SSL Protocol version 1.0*, Transport Layer Security Working Group, Internet Draft.

[6] Gamma, E., Helm, H., Johnson, R., and Vlissides, J. (1995) *Design Patterns*, Addison-Wesley Pub Co.

[7] Hansman, U. (2000) *Smart Card Application Development Using Java*, Springer-Verlag.

[8] Housley, R., Ford, W., Polk, W., and Solo, D. (1999) *Internet X.509 Public Key Infrastructure Certificate and CRL Profile*, Network Working Group, Request for Comments 2459.

[9] International Organization for Standardization (ISO) (1989) *ISO 7816 Integrated Circuit Cards with Electrical Contacts*. Part 1: Physical Characteristics. Part 2: Dimensions and Location of Contacts. Part 3: Electronic Signals and Transmission Protocols. Part 4: Interindustry Commands for Interchange.

[10] Kehr, R., Possega, J., and Vogt, H. (1999) PCA: Jini-based Personal Card Assistant. *Proceedings of CQRE '99, Lecture Notes in Computer Science 1740*, Springer-Verlag, Berlin, Germany, pp. 64–75.

[11] Lopez, J., Maña, A., Martinez, J., and Matamoros, S. (2001) Secure and Generic GSM-Ticketing. *Proceedings of the SI+T, 2001*, A Coruña, Spain, September 2001.

[12] Mana, A. and Matamoros, S. (2002) Practical Mobile Digital Signatures. *3rd International Conference on Electronic Commerce and Web Technologies*, EC-Web 2002, Aix-en-province, France.

[13] MUSCLE Project, Available at: http://www.linuxnet.com.

[14] Object Management Group (1997) *Secure Socket Layer/CORBA Security*, February 1997.

[15] Object Management Group (2001) *Resource Access Decision Facility Specification* version 1.0, April 2001.

[16] OpenSSL library, Available at: http://www.openssl.org, 2003.

[17] O'Ryan, C., Kuhns, F., Schmidt, D.C., Othman, O., and Parsons, J. (2000) The Design and Performance of a Pluggable Protocols Framework for Real-time Distributed Object Computing Middleware. *IFIP/ACM Middleware 2000 Conference*, Pallisades, New York, April 3-7.

[18] PC/SC Workgroup, Available at: http://www.pcscworkgroup.com/, 2003.

[19] Rivest, R. L., Shamir, A., Adleman, L. M. (1978) A Method for Obtaining Digital Signatures and Public-Key Cryptosystems. *Journal of the ACM*, **21**(2), 120–126.

[20] Sun Microsystems (2001) *Java Card 2.1.1 API Specification*, available at: http://java.sun.com/products/javacard/javacard21.html.

[21] The ACE ORB. Available at: http://www.cs.wustl.edu/~schmidt/TAO.html.

[22] The Firebird database project (2000) Available at: http://firebird.sourceforge.net.

[23] The OpenCard Framework, Available at: http://www.opencard.org.

17

Real-time CORBA Middleware

Arvind S. Krishna[1], Douglas C. Schmidt[1], Raymond Klefstad[2], Angelo Corsaro[3]
[1]*Vanderbilt University*
[2]*University of California, Irvine*
[3]*Washington University*

17.1 Introduction

Middleware trends.

Over the past decade, distributed computing middleware, such as CORBA [37], COM+ [32], Java RMI [63], and SOAP/.NET [56], has emerged to reduce the complexity of developing distributed systems. This type of middleware simplifies the development of distributed systems by off-loading the tedious and error-prone aspects of distributed computing from application developers to middleware developers. Distributed computing middleware has been used successfully in desktop and enterprise systems [64] where scalability, evolvability, and interoperability are essential for success. In this context, middleware offers the considerable benefits of hardware-, language-, and OS-independence, as well as open-source availability in some cases.

The benefits of middleware are also desirable for development of distributed, real-time, and embedded (DRE) systems. Because of their multiple constraints across different dimensions of performance, DRE systems are harder to develop, maintain, and evolve than mainstream desktop and enterprise software. In addition to exhibiting many of the same needs as desktop and enterprise systems, DRE systems impose stringent quality of service (QoS) constraints. For example, real-time performance imposes strict constraints upon bandwidth, latency, and dependability. Moreover, many embedded devices must operate under memory, processor, and power limitations.

As DRE systems become more pervasive, they are also becoming more diverse in their needs and priorities. Examples of DRE systems include telecommunication networks (e.g.,

Middleware for Communications. Edited by Qusay H. Mahmoud
 ISBN 0-470-86206-8

wireless phone services), tele-medicine (e.g., robotic surgery), process automation (e.g., hot rolling mills), and defense applications (e.g., total ship computing environments). The additional difficulties faced in developing these systems intensify the need for middleware to off-load time-consuming and error-prone aspects of real-time and embedded computing, as well as to eliminate the need for continual reinvention of custom solutions.

The Real-time CORBA specification [36] was standardized by the Object Management Group (OMG) to support the QoS needs of DRE systems. Real-time CORBA is a rapidly maturing middleware technology designed for applications with hard real-time requirements, such as avionics mission computing [52], as well as those with softer real-time requirements, such as telecommunication call processing and streaming video [50]. When combined with a quality real-time operating system foundation, well-tuned implementations of Real-time CORBA can meet the end-to-end QoS needs of DRE systems, while also offering the significant development benefits of reusable middleware [8].

Contributions of this chapter.

Although Real-time CORBA offers substantial benefits – and the Real-time CORBA 1.0 specification was integrated into the OMG standard several years ago – it has not been universally adopted by DRE application developers, partly due to the following limitations:

- **Lack of customization** (or the difficulty of customization), where customization is needed to enable Real-time CORBA Object Request Brokers (ORBs) to be used in diverse domains,
- **Memory footprint overhead**, stemming largely from monolithic ORB implementations that include all the code supporting the various core ORB services, such as connection and data transfer protocols, concurrency and synchronization management, request and operation demultiplexing, (de)marshaling, and error-handling, and
- **Steep learning curve**, caused largely by the complexity of the CORBA C++ mapping.

This chapter therefore presents the following contributions to the design and use of Real-time CORBA middleware implementations that address the challenges outlined above:

1. It describes how highly optimized ORB designs can improve the performance and predictability of DRE systems.
2. It shows how well-designed Real-time CORBA middleware architectures can simultaneously minimize footprint and facilitate customization of ORBs to support various classes of applications.
3. It shows that Java and Real-time Java [3] features can be applied to an ORB to simultaneously increase ease of use and predictability for Java-based DRE applications.

The material in this chapter is based on our experience in developing The ACE ORB (TAO) [52] and ZEN [25]. TAO is an open-source[1] Real-time CORBA ORB for use

[1] TAO can be downloaded from http://deuce.doc.wustl.edu/Download.html.

with C++, with enhancements designed to ensure efficient, predictable, and scalable QoS behavior for high-performance and real-time applications. ZEN is an open-source[2] Real-time CORBA ORB for use with Java and Real-time Java, designed to minimize footprint and maximize ease of use. ZEN is inspired by many of the patterns, techniques, and lessons learned from developing TAO. This chapter presents TAO and ZEN as case studies of Real-time CORBA middleware to illustrate how ORBs can enable developers to control the trade-offs between efficiency, predictability, and flexibility needed by DRE systems.

Chapter organization.
The remainder of this chapter is organized as follows: Section 17.2 provides an overview of DRE systems, focusing on the challenges involved in developing these systems; Section 17.3 discusses how Real-time CORBA can be used to ensure end-to-end predictability required for DRE systems; Section 17.4 illustrates the motivation, design, successes, and limitation of TAO; Section 17.5 details the goals, technologies, design, and successes of ZEN; Section 17.6 summarizes how our work relates to other research efforts on real-time middleware; and Section 17.7 presents concluding remarks.

17.2 DRE System Technology Challenges

DRE systems are generally harder to develop, maintain, and evolve than mainstream desktop and enterprise software since DRE systems have stringent constraints on weight, power consumption, memory footprint, and performance. This section describes the requirements and challenges present in this domain, both for contemporary DRE systems and for future DRE systems.

17.2.1 Challenges of Today's DRE Systems

Some of the most challenging problems faced by software developers are those associated with producing software for real-time and embedded systems in which computer processors control physical, chemical, or biological processes or devices. Examples of such systems include airplanes, automobiles, CD players, cellular phones, nuclear reactors, oil refineries, and patient monitors. In most of these real-time and embedded systems, *the right answer delivered too late becomes the wrong answer*, that is, achieving real-time performance end-to-end is essential. In addition, some embedded devices have limited memory (e.g., 64-512 KB) available for the platform and applications.

The three primary characteristics of DRE systems pose the following requirements for their development:

- As *distributed systems*, DRE systems require capabilities to manage connections and message transfer between separate machines.
- As *real-time systems*, DRE systems require predictable and efficient end-to-end control over system resources.

[2] ZEN can be downloaded from http://www.zen.uci.edu.

- As *embedded systems*, DRE systems have weight, cost, and power constraints that limit their computing and memory resources. For example, embedded systems often cannot use conventional virtual memory, since software must fit on low-capacity storage media, such as electrically erasable programmable read-only memory (EEPROM) or nonvolatile random access memory (NVRAM).

17.2.2 Challenges of Future DRE Systems

As hard as today's DRE systems are to develop, DRE systems of the future will be even more challenging. Many of today's real-time and embedded systems are relatively small-scale, but the trend is toward significantly increased functionality and complexity. In particular, real-time and embedded systems are increasingly being connected via wireless and wireline networks to create *distributed* real-time and embedded systems, such as total ship computing environments, tele-immersion environments, fly-by-wire air vehicles, and area/theater ballistic missile defense. These DRE systems include many interdependent levels, such as network/bus interconnects, many coordinated local and remote endsystems, and multiple layers of software.

Some of the key attributes of future DRE systems can be characterized as follows:

- Multiple quality of service (QoS) properties, such as predictable latency/jitter, throughput guarantees, scalability, dependability, and security, must be satisfied simultaneously and often in real time.
- Different levels of service will occur under different system configurations, environmental conditions, and costs, and must be handled judiciously by the system infrastructure and applications.
- The levels of service in one dimension must be coordinated with and/or traded off against the levels of service in other dimensions to achieve the intended application and overall mission results.
- The need for autonomous and time-critical application behavior requires flexible system infrastructure components that can adapt robustly to dynamic changes in mission requirements and environmental conditions.

All of these attributes are interwoven and highly volatile in DRE systems, due to the dynamic interplay among the many interconnected parts.

DRE applications are increasingly combined to form large-scale distributed systems that are joined together by the Internet and intranets. These systems can further be combined with other distributed systems to create "systems of systems." Examples of these large-scale DRE systems of systems include *Just-in-time manufacturing inventory control systems* that schedule the delivery of supplies to improve efficiency. Figure 17.1 illustrates how the combination of individual manufacturing information systems is fundamental to achieve the efficiencies of modern "just-in-time" manufacturing supply chains. Information from engineering systems is used to design parts, assemblies, and complete products. Parts manufacturing suppliers must keep pace with (1) engineering requirements upstream in the supply chain and (2) distribution constraints and assembly requirements

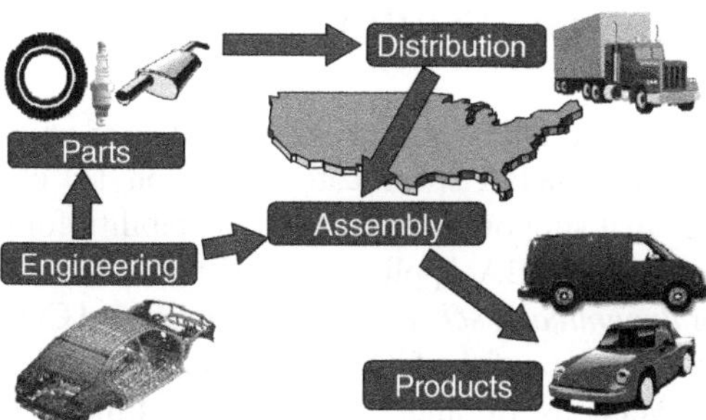

Figure 17.1 Characteristics of manufacturing system of systems

downstream. Distribution must be managed precisely to avoid parts shortages while keeping local inventories low. Assembly factories must achieve high throughput, while making sure the output matches product demand at the sales and service end of the supply chain. Throughout this process, information gathered at each stage of the supply chain must be integrated seamlessly into the control processes of other stages in the chain.

17.2.3 Limitations with Conventional DRE System Development

Designing DRE systems that implement all their required capabilities, are efficient, predictable, and reliable, and use limited computing resources is hard; building them on time and within budget is even harder. In particular, DRE applications developers face the following challenges:

- **Tedious and error-prone development** — Accidental complexity proliferates, because DRE applications are often still developed using low-level languages, such as C and assembly language.
- **Limited debugging tools** — Although debugging tools are improving, real-time and embedded systems are still hard to debug, owing to inherent complexities such as concurrency, asynchrony, and remote debugging.
- **Validation and tuning complexities** — It is hard to validate and tune key QoS properties, such as (1) pooling concurrency resources, (2) synchronizing concurrent operations, (3) enforcing sensor input and actuator output timing constraints, (4) allocating, scheduling, and assigning priorities to computing and communication resources end-to-end, and (5) managing memory.

Because of these challenges, developers of DRE systems have historically tended to rediscover core concepts and reinvent custom solutions that were tightly coupled to particular hardware and software platforms. The continual rediscovery and reinvention associated with this software development process has kept the costs of engineering and evolving DRE systems too high for too long. Improving the quality and quantity of systematically reusable software via middleware is essential to resolving this problem [8].

17.3 Overview of Real-time CORBA

To address the challenges for DRE systems described in Section 17.2, the OMG has standardized the Real-time CORBA specification. Version 1.0 of this specification [36] defines standard features that support end-to-end predictability for operations in *statically scheduled and provisioned* CORBA applications. Version 2.0 of this specification [34] defines mechanisms for *dynamically scheduled and provisioned* CORBA applications. This section first presents an overview of the CORBA reference model and its key components and then describes how versions 1.0 and 2.0 of the Real-time CORBA specification add QoS capabilities to CORBA.

17.3.1 Overview of CORBA

CORBA Object Request Brokers (ORBs) allow clients to invoke operations on distributed objects without concern for object location, programming language, OS platform, communication protocols and interconnects, and hardware [20]. Figure 17.2 illustrates the following key components in the CORBA reference model [37] that collaborate to provide this degree of portability, interoperability, and transparency:

Client.
A client is a *role* that obtains references to objects and invokes operations on them to perform application tasks. A client has no knowledge of the implementation of the object but does know the operations defined via its interface.

ORB Core.
An ORB core is the layer of a CORBA ORB implementation that is responsible for connection and memory management, data transfer, endpoint demultiplexing, and concurrency control. When a client invokes an operation on an object, the ORB Core is

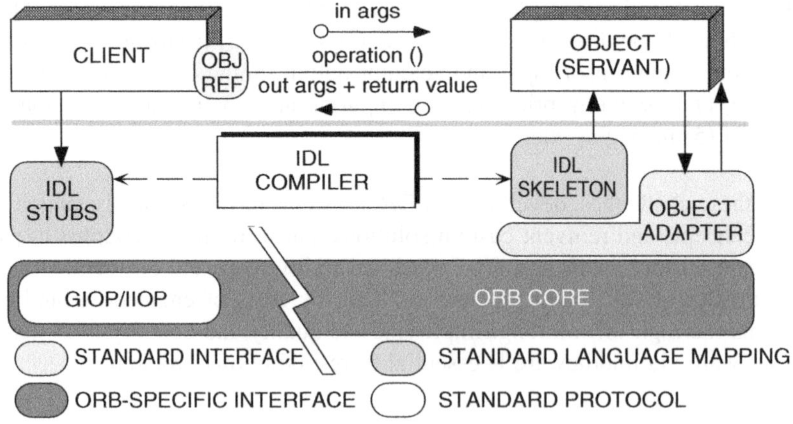

Figure 17.2 Key components in the CORBA reference model

responsible for delivering the request to the server and returning the response, if any, to the client. For remote objects, the ORB Core transfers requests using the General Internet Inter-ORB Protocol (GIOP) that runs atop many transport protocols, including TCP/IP and many embedded systems interconnects.

Object.
In CORBA, an object is an instance of an OMG Interface Definition Language (IDL) interface. Each object is identified by an *object reference*, which associates one or more paths through which a client can access an object on a server. Over its lifetime, an object can be associated with one or more servants that implement its interface.

Servant.
This component implements the operations defined by an OMG IDL interface. In object-oriented (OO) languages, such as C++ and Java, servants are implemented using one or more class instances. In non-OO languages, such as C, servants are typically implemented using functions and `structs`. A client never interacts with servants directly, but always through objects identified via object references.

OMG IDL Stubs and Skeletons.
IDL stubs and skeletons serve as a "glue" between the client and servants, respectively, and the ORB. Stubs implement the *Proxy* pattern [4] and marshal application parameters into a common message-level representation. Conversely, skeletons implement the *Adapter* pattern [12] and demarshal the message-level representation back into typed parameters that are meaningful to an application.

Object Adapter.
An Object Adapter is a composite component that associates servants with objects, creates object references, demultiplexes incoming requests to servants, and collaborates with the IDL skeleton to dispatch the appropriate operation upcall on a servant. Object Adapters enable ORBs to support various types of servants that possess similar requirements. Even though different types of Object Adapters may be used by an ORB, the only Object Adapter defined in the CORBA specification is the Portable Object Adapter (POA) [43].

17.3.2 Overview of Real-time CORBA 1.0

The Real-time CORBA 1.0 specification is targeted for *statically scheduled and provisioned* DRE systems, where the knowledge of applications that will run on the system and/or the priorities at which they execute are known *a priori*. This specification leverages features from the CORBA standard (such as the GIOP protocol) and the Messaging specification [38] (such as the QoS policy framework) to add QoS control capabilities to regular CORBA. These QoS capabilities help improve DRE application predictability by bounding priority inversions and managing system resources end-to-end. Figure 17.3 illustrates the standard features that Real-time CORBA provides to DRE applications to enable them to configure and control the following resources:

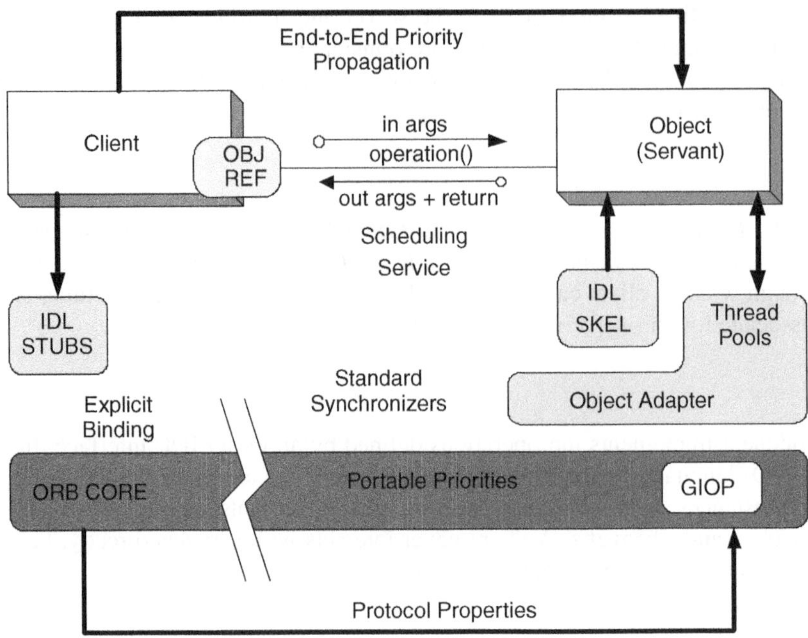

Figure 17.3 Real-time CORBA 1.0 features

- **Processor resources** via thread pools, priority mechanisms, intra-process mutexes, and a global scheduling service for real-time applications with fixed priorities. To enforce strict control over scheduling and execution of processor resources, Real-time CORBA 1.0 specification enables client and server applications to (1) determine the priority at which CORBA invocations will be processed, (2) allow servers to predefine pools of threads, (3) bound the priority of ORB threads, and (4) ensure that intra-process thread synchronizers have consistent semantics in order to minimize priority inversion.
- **Communication resources** via protocol properties and explicit bindings to server objects using priority bands and private connections. A Real-time CORBA endsystem must leverage policies and mechanisms in the underlying communication infrastructure that support resource guarantees. This support can range from (1) managing the choice of the connection used for a particular invocation to (2) exploiting advanced QoS features, such as controlling the ATM virtual circuit cell rate.
- **Memory resources** via buffering requests in queues and bounding the size of thread pools. Many DRE systems use multithreading to (1) distinguish between different types of service, such as high-priority versus low-priority tasks [18] and (2) support thread preemption to prevent unbounded priority inversion. Real-time CORBA specification defines a standard *thread pool* model [53] to preallocate pools of threads and to set certain thread attributes, such as default priority levels. Thread pools are useful for real-time ORB endsystems and applications that want to leverage the benefits of multithreading, while bounding the amount of memory resources, such as stack space, they

consume. Moreover, thread pools can be optionally configured to buffer or not buffer requests, which provides further control over memory usage.

17.3.3 Overview of Real-time CORBA 2.0

The Real-time CORBA 2.0 specification is targeted for *dynamically scheduled and provisioned* DRE systems, where the knowledge of applications that will run on the system and/or the priorities at which they execute is *not* known *a priori*. Real-time CORBA 2.0 extends version 1.0 by providing interfaces and mechanisms to plug in dynamic schedulers and interact with them. It allows applications to specify and use the scheduling disciplines and parameters that most accurately define and describe their execution and resource requirements, for example, this specification supports the following capabilities:

- It allows application developers to associate any scheduling discipline, for example, Earliest Deadline First or Maximum Urgency First, with the scheduler.
- It enables the scheduling parameters to be associated with the chosen discipline at any time.
- It provides a set of ORB/Scheduler interfaces that support the development of portable (i.e., ORB-independent) schedulers.

The Real-time CORBA 2.0 specification defines the following two key capabilities that provide significant enhancements compared to version 1.0:

- **Distributable thread.** Dynamic DRE systems require more information than just priority, that is, fixed-priority propagation is not sufficient to ensure end-to-end timeliness requirements. Instead, an abstraction is required that identifies a schedulable entity and associates with it the appropriate scheduling parameters. The Real-time CORBA 2.0 specification defines a *distributable thread* abstraction that enables a thread to execute operations on objects without regard for physical node boundaries. This abstraction is the loci of execution that spans multiple nodes and scheduling segments. Figure 17.4 shows a distributable thread *DT*1 that spans two nodes Host 1 and Host 2 as a part of a two-way invocation on a remote object.
- **Scheduling service architecture.** To facilitate the association of various scheduling disciplines with a dynamic scheduler, the Real-time CORBA 2.0 specification defines a Scheduling Service interface that provides mechanisms for plugging in different schedulers. An application passes its scheduling requirements to the scheduler via these interfaces. Similarly, the ORB also interacts with the scheduler at specific scheduling points to dispatch and share scheduling information across nodes.

As discussed above, Real-time CORBA versions 1.0 and 2.0 provides mechanisms to meet the needs of a wide variety of DRE applications that possess different constraints and priorities. These differing needs, however, must be met through appropriate designs and implementations of ORBs. In particular, DRE system designers need configurable ORBs to provide a flexible range of choices to meet the needs of their particular system's functional and QoS requirements. Research and implementation experience gained while developing

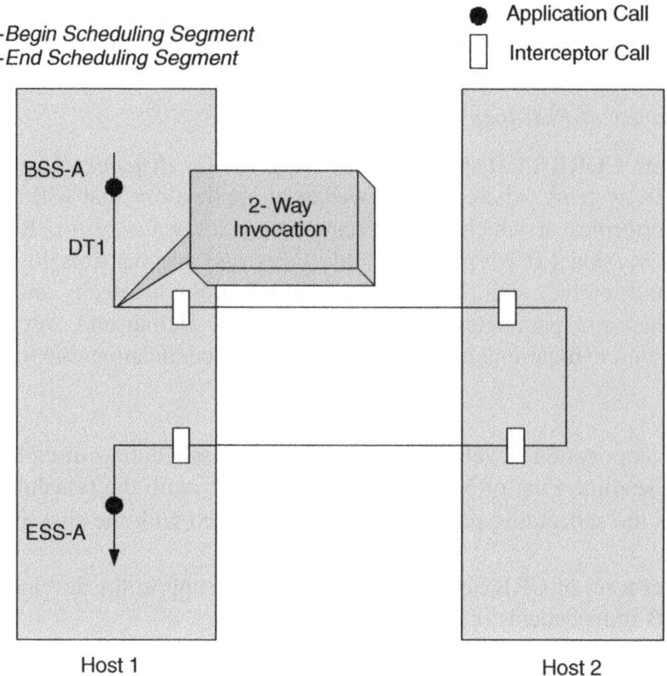

Figure 17.4 Distributable thread abstraction

Real-time CORBA ORBs have yielded insights, which in turn have led to the evolution of a palette of ORB design alternatives to meet those specialized needs. Sections 17.4 and 17.5 illustrate various ways in which ORBs can offer differing advantages, providing the flexibility of different alternatives for developers of DRE applications.

17.4 TAO: C++-based Real-time CORBA Middleware

17.4.1 Motivation

Traditional tools and techniques used to develop DRE software are often so specialized that they cannot adapt readily to meet new functional or QoS requirements, hardware/software technology innovations, or emerging market opportunities. Standard commercial-off-the-shelf (COTS) middleware could therefore be implemented and deployed in an efficient and dependable manner, to provide advantages of COTS middleware to DRE applications. Until recently, however, it was not feasible to develop mission-critical DRE systems using standard COTS middleware because of its inability to control key QoS properties, such as predictable latency, jitter, and throughput; scalability; dependability; and security.

Over the past decade, researchers at Washington University, St Louis, University of California, Irvine, and Vanderbilt University have worked with hundreds of developers and companies from around the world to overcome the problems outlined above. The result has been an open-source standard CORBA Object Request Broker (ORB) called

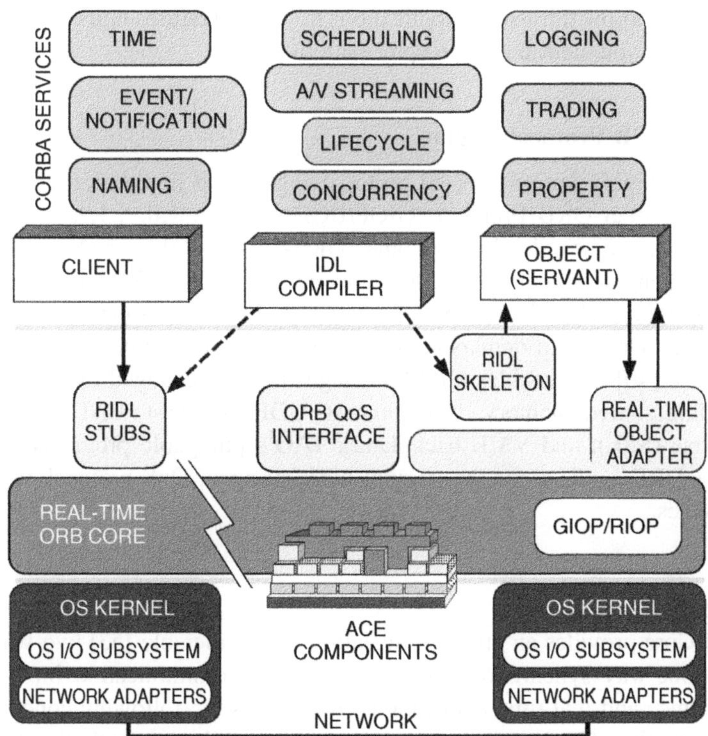

Figure 17.5 TAO ORB end system architecture

The ACE ORB (TAO) [52]. The remainder of this section describes TAO and summarizes its current status and impact.

17.4.2 TAO Architecture and Capabilities

TAO is a C++ ORB that is compliant with most of the features and services defined in the CORBA 3.0 specification [37], as well as the Real-time CORBA specification [36]. The latest release of TAO contains the following components shown in Figure 17.5 and outlined below:

17.4.2.1 IDL Compiler

TAO's IDL compiler [16] is based on the enhanced version of the freely available SunSoft IDL compiler. The latest CORBA 3.0 IDL-to-C++ mapping has been implemented, including the latest CORBA Component Model and POA features. The IDL compiler also supports the Object-by-Value specification and the CORBA Messaging specification, as well as asynchronous method invocations. In addition, TAO's IDL compiler can generate stubs/skeletons that can support either native C++ exceptions or the more portable `CORBA::Environment` approach. Finally, TAO's IDL compiler generates codes for smart proxies that allow third-party applications to "plug" features into clients and portable interceptors that implement the Interceptor pattern [54].

17.4.2.2 Inter-ORB Protocol Engine

TAO contains a highly optimized [15] protocol engine that implements the CORBA 3.0 General/Internet Inter-ORB Protocol (GIOP/IIOP), version 1.0, 1.1, and 1.2. TAO can therefore interoperate seamlessly with other ORBs that use the standard IIOP protocol. TAO's protocol engine supports both the static and dynamic CORBA programming models, that is, the SII/SSI and DII/DSI, respectively. TAO also supports Dynamic Anys, which facilitate incremental demarshaling. In addition, TAO supports a pluggable protocols framework [40] that enables GIOP messages to be exchanged over non-TCP transports, including shared memory, UDP unicast, UDP multicast, UNIX-domain sockets, Secure Sockets (SSL), and VME backplanes. TAO's pluggable protocols framework is important for DRE applications that require more stringent QoS protocol properties than TCP/IP provides.

17.4.2.3 ORB Core

TAO's ORB Core provides an efficient, scalable, and predictable [53] two-way, one-way, and reliable one-way synchronous and asynchronous communication infrastructure for high-performance and real-time applications. It provides the following concurrency models [51]: (1) reactive, (2) thread-per-connection, (3) thread pool (including the Real-time CORBA thread pool API), and (4) reactor-per-thread-priority (which is optimized for deterministic real-time systems). TAO's ORB Core is based on patterns and frameworks in ACE [46, 47], which is a widely used object-oriented toolkit containing frameworks and components that implement key patterns [54] for high-performance and real-time networked systems. The key patterns and ACE frameworks used in TAO include the Acceptor and Connector, Reactor, Half-Sync/Half-Async, and Component/Service Configurator.

17.4.2.4 Portable Object Adapter

TAO's implementation of the CORBA Portable Object Adapter (POA) [43] is designed using patterns that provide an extensible and highly optimized set of request demultiplexing strategies [44], such as perfect hashing and active demultiplexing, for objects identified with either persistent or transient object references. These strategies allow TAO's POA to provide constant-time lookup of servants based on object keys and operation names contained in CORBA requests.

17.4.2.5 Implementation and Interface Repositories

TAO's Implementation Repository automatically launches servers in response to client requests. TAO also includes an Interface Repository that provides clients and servers with runtime information about IDL interfaces and CORBA requests.

In addition, TAO provides many of the standard CORBA services, including Audio/Video Streaming Service [33], Concurrency Service, Telecom Logging Service, Naming Service, Notification Service, Property Service, Time Service, and Trading Service. TAO also provides nonstandard services that are targeted for various types of DRE application domains, including a Load Balancing service [41], a Real-time Event Service [48], and a Real-time Scheduling Service [13].

17.4.3 TAO Successes

The TAO project has been active since 1997. The ACE project, which provides the reusable frameworks and components upon which TAO is based, has been active since 1991. During this time, results from the ACE and TAO projects have had a significant impact on middleware researchers and practitioners, as described below:

17.4.3.1 Research Innovations

For the past decade, research on TAO has focused on optimizing the efficiency and predictability of the ORB to meet end-to-end application QoS requirements by vertically integrating middleware with OS I/O subsystems, communication protocols, and network interfaces. TAO is designed carefully using architectural, design, and optimization patterns [54] that substantially improve the efficiency, predictability, and scalability of DRE systems. The optimization-related research contribution of the TAO project illustrated in Figure 17.6 includes the following:

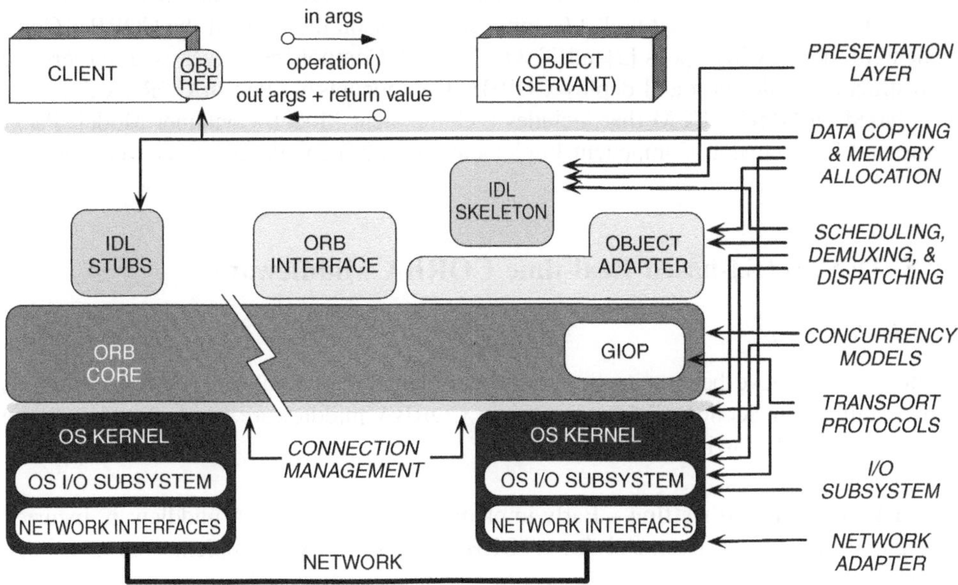

Figure 17.6 Optimizations in TAO ORB

- An ORB Core that supports deterministic real-time concurrency and dispatching strategies. TAO's ORB Core concurrency models are designed to minimize context switching, synchronization, dynamic memory allocation, and data movement.
- Active demultiplexing and perfect hashing optimizations that associate client requests with target objects in constant time, regardless of the number of objects, operations, or nested POAs.
- A highly optimized CORBA IIOP protocol engine and an IDL compiler that generates compiled stubs and skeletons. TAO's IDL compiler also implements many of the optimizations pioneered by the Flick IDL compiler [10].
- TAO can be configured to use a nonmultiplexed connection model, which avoids priority inversion and behaves predictably when used with multirate real-time applications.
- TAO's pluggable protocols allow the support of real-time I/O subsystems [28] designed to minimize priority inversion interrupt overhead over high-speed ATM networks and real-time interconnects, such as VME.
- TAO's Real-time Event Service and static and dynamic Scheduling Services integrate the capabilities of TAO ORB described above. These services form the basis for next-generation real-time applications for many research and commercial projects, including ones at Boeing, Cisco, Lockheed Martin, Raytheon, Siemens, and SAIC.

17.4.3.2 Technology Transitions

Now that TAO has matured, thousands of companies around the world have used it in a wide range of domains, including aerospace, defense, telecom and datacom, medical engineering, financial services, and distributed interactive simulations. In addition, a number of companies began to support it commercially. Open-source commercial support, documentation, training, and consulting for TAO is available from PrismTech and Object Computing Inc. (OCI). OCI also maintains the TAO FAQ and anonymous CVS server. iCMG has developed its K2 Component Server on the basis of the CORBA Component Model (CCM) specs [2]and TAO. The K2 Component Server is a server-side infrastructure to develop and deploy CORBA Components written in CORBA 3.0 IDL. It is based on OMG's CCM that includes a Component Model, Container Model, Packaging and Deployment, Component Implementation Framework, and Inter-working with EJB 1.1.

17.5 ZEN: RTSJ-based Real-time CORBA Middleware

17.5.1 Motivation

Although there have been many successful deployments of Real-time CORBA (such as those outlined in Section 17.4.3), Real-time CORBA middleware has suffered to some extent from the following limitations:

- **Lack of feature subsetting** – Early implementations of CORBA middleware incurred significant footprint overhead due to ORB designs that were implemented as a large body of monolithic code, which stymies feature subsetting and makes it hard to minimize middleware footprint.

- **Inadequate support for extensibility** – Distributed systems not only require a full range of CORBA services but also need middleware to be adaptable, that is, meet the needs of a wide variety of applications developers. Current Real-time CORBA middleware designs are not designed with the aim of applicability in various domains.
- **Increased complexity** – A key barrier to the adoption of Real-time CORBA middleware arises from steep learning curve caused by the complexity of the CORBA C++ mapping [49, 65]. Real-time CORBA middleware should therefore be designed using programming languages that shield application developers from type errors, memory management, real-time scheduling enforcement, and steep learning curves.

Custom software development and evolution is labor-intensive and error-prone for complex DRE applications. Middleware design should therefore be simultaneously extensible, provide feature subsetting, and be easy to use, thereby minimizing the total system acquisition and maintenance costs.

In recent years, the Java programming language has emerged as an attractive alternative for developing middleware. Java is easier to learn and program, with less inherent and accidental complexity than C++. There is also a large and growing community of Java programmers, since many schools have adopted Java as a teaching language. Java also has other desirable language features, such as strong typing, dynamic class loading, introspection, and language-level support for concurrency and synchronization. Implementation in Java could therefore provide an easy-to-use Real-time CORBA middleware tool, shielding application developers from the steep learning curve due to type errors, memory management, and real-time scheduling enforcement.

Conventional Java runtime systems and middleware have historically been unsuitable for DRE systems, however, due to

1. The under-specified scheduling semantics of Java threads, which can lead to the most eligible thread not always being run.
2. The ability of the Java GC to preempt any other Java thread, which can yield unpredictably long preemption latencies.

To address these problems, the Real-time Java Experts Group has defined the RTSJ [3], which extends Java in several ways, including (1) new memory management models that allow access to physical memory and can be used *in lieu* of garbage collection and (2) stronger guarantees on thread semantics than in conventional Java. Real-time Java offers middleware developers a viable means of producing middleware with a simpler programming model that still offers control over memory and threading necessary for acceptable predictability.

17.5.2 ZEN Architecture and Capabilities

The ZEN ORB developed at the University of California, Irvine, leverages the lessons learned from our earlier efforts on TAO's design, implementation, optimization, and benchmarking. ZEN is a Real-time CORBA ORB implemented using Java and Real-time

Java [3], which simplifies the programming model for DRE applications. To address the challenges specific to DRE systems, the ZEN project has the following research goals:

- Provide an ORB that increases ease of use by leveraging the advantages of Java.
- Reduce middleware footprint to facilitate memory-constrained embedded systems development, yet provide a full range of CORBA services for distributed systems.
- Demonstrate the extent to which COTS languages, runtime systems, and hardware can meet the following QoS performance requirements: (1) achieve low and bounded jitter for ORB operations; (2) eliminate sources of priority inversion; and (3) allow applications to control Real-time Java features.

Our experience in developing TAO taught us that achieving a small memory footprint is only possible if the architecture is designed to support this goal initially. Implementing a full-service, flexible, specification-compliant ORB with a monolithic ORB design can yield a large memory footprint, as shown in Figure 17.7. ZEN has therefore been designed using a micro-ORB architecture. Sidebar 17.5.1 discusses the advantages and disadvantages of the monolithic- and micro-ORB architectures shown in Figures 17.7 and 17.8.

Monolithic versus Micro-ORB Architectures

In a *monolithic ORB architecture*, the ORB is a single component that includes all the code including those of configuration variations. This component is loaded as one executable. The advantage of a monolithic design is that its runtime performance is often efficient, it is relatively easy to code, and it can support all CORBA services. The main disadvantage, however, is that a monolithic ORB implementation can incur excessive memory footprint and therefore must rely on OS virtual memory, even if only a small subset of its features is used.

Basing an ORB architecture on patterns [54, 12] can help resolve common design forces and separate concerns effectively. For instance, the pluggable design framework can lead to a micro-ORB architecture that substantially reduces middleware footprint and increases flexibility. In a *Micro-ORB architecture*, only a small ORB kernel is loaded in memory, with various components linked and loaded dynamically on demand. The advantage of this design is the significant reduction in footprint and the increase in extensibility. In particular, independent ORB components can be configured dynamically to meet the needs of different applications. The disadvantage is that dynamic linking on-demand incurs jitter, which may be undesirable for many DRE systems.

In ZEN, we generalized TAO's pluggable protocol framework to other modular services within the ORB so that they need not be loaded until they are used. ZEN's flexible and extensible *micro-ORB design* (rather than monolithic-ORB design) is used for all CORBA services. In particular, we applied the following design process systematically:

1. Identify each core ORB service whose behavior may vary. Variation can depend upon (1) a user's optional choice for certain behavior and (2) which standard CORBA features are actually used.

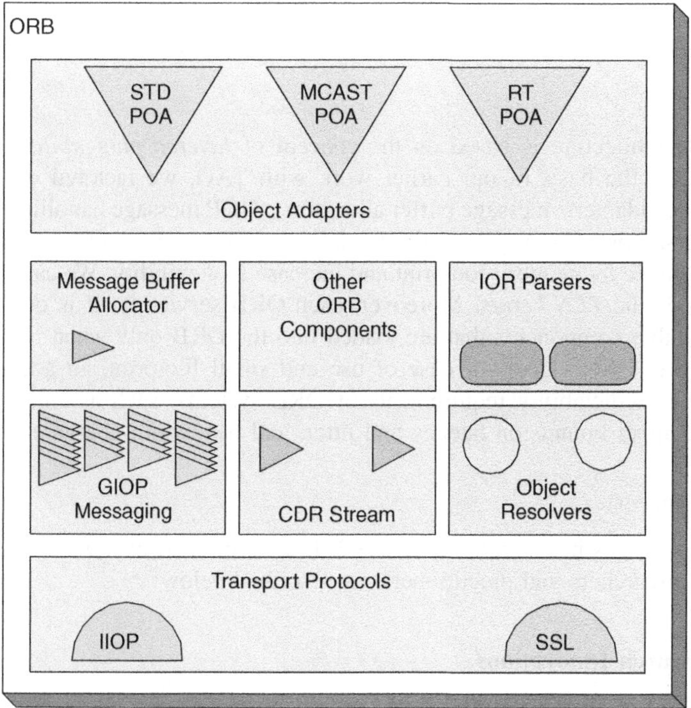

Figure 17.7 Monolithic ORB architecture

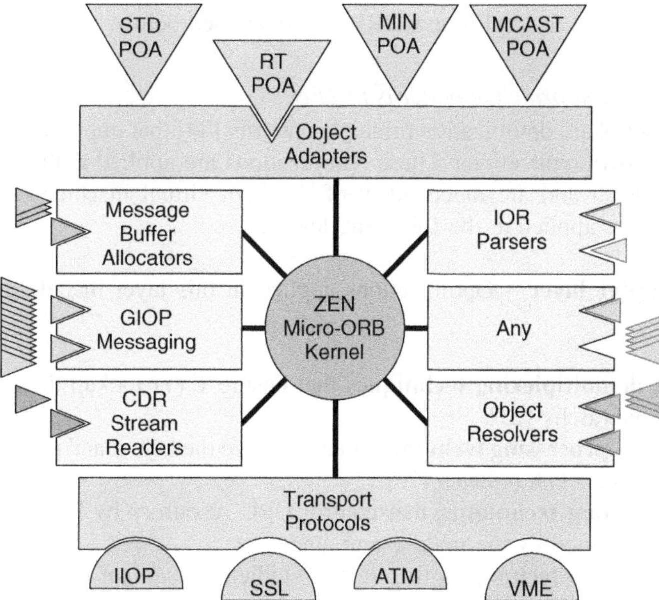

Figure 17.8 Micro-ORB architecture

2. Move each core ORB service out of the ORB and apply the Virtual Component pattern [6] to make each service pluggable dynamically.
3. Write concrete implementations of each abstract class and factories that create instances of them.

ZEN's ORB architecture is based on the concept of *layered pluggability*, as shown in Figure 17.8. On the basis of our earlier work with TAO, we factored eight core ORB services (object adapters, message buffer allocators, GIOP message handling, CDR Stream readers/writers, protocol transports, object resolvers, IOR parsers, and Any handlers) out of the ORB to reduce its memory footprint and increase its flexibility. We call the remaining portion of code the *ZEN kernel*. Moreover, each ORB service itself is decomposed into smaller pluggable components that are loaded into the ORB only when needed.

In addition to ZEN's goals of ease of use and small footprint, an additional goal is to support the predictability requirements of DRE systems, such as end-to-end priority preservation, upper bounds on latency and jitter, and bandwidth guarantees.

17.5.3 ZEN Successes

The ZEN project has been active since 1999. In a short span of time, ZEN is impacting middleware researchers and practitioners, as described below:

17.5.3.1 Research Innovations

Similar to TAO, research on ZEN has focused on optimizing the efficiency and predictability of the ORB to meet end-to-end application QoS requirements. In particular, research on ZEN focuses on achieving predictability by applying optimization principle patterns and the RTSJ to Real-time CORBA as discussed below.

Applying optimization principle patterns to ZEN.
At the heart of ZEN are optimization principle patterns [44] that improve the predictability as required by DRE applications. These optimizations are applied at the algorithmic and data structural level and are independent of the Java virtual machine (JVM). In ZEN, these strategies are applied at the following levels:

- **Object Adapter layer** – Optimizations applied in this layer include predictable and scalable

    1. **Request demultiplexing techniques** that ensure $O(1)$ lookup time irrespective of the POA hierarchy [27],
    2. **Object key–processing techniques** that optimize the layout and processing of object keys based on POA policies [26],
    3. **Thread-pooling techniques** that control CPU resources by bounding the number of threads created by the middleware, and
    4. **Servant lookup techniques** that ensure predictable-servant to-object-id operations in the POA.

- **ORB Core layer** – Optimizations applied in this layer include the following:
  1. **Collocation optimizations**, that minimize the marshaling/demarshaling overhead based on object/ORB location,
  2. **Buffer-allocation strategies**, that optimize the allocation and caching of buffers to minimize garbage collection, and
  3. **Reactive I/O** using Java's nio [22] package that allows asynchronous communication.

Applying RTSJ features to ZEN.
The OMG Real-time CORBA specification was adopted several years before the RTSJ was standardized. Real-time CORBA's Java mapping therefore does not use any RTSJ features. To have a predictable Java-based Real-time CORBA ORB like ZEN, however, it is necessary to take advantage of RTSJ features to reduce interference with the GC and improve predictability. In these optimizations, RTSJ features are directly associated within key ORB core components to enhance predictability accrued from the optimization principle patterns.

Our goal for apply RTSJ features to ZEN is to (1) comply with the Real-time CORBA specification and (2) be transparent to developers of DRE applications. Our predictability-enhancing improvements of ZEN begin by identifying the participants associated with processing a request at both the client and server sides. For each participant identified, we associate the component with nonheap regions and resolve challenges arising from this association. RTSJ features can be utilized [26] to implement mechanisms in a Real-time CORBA ORB, such as priority-banded thread pool lanes, to improve ORB predictability. These mechanisms can be implemented in each of the following design layers within ZEN:

1. **I/O layer**, for example, Acceptor/Connector and Reactor,
2. **ORB Core layer**, for example, CDR streams and Buffer Allocators, and
3. **Object Adapter layer**, for example, Thread Pools and the POA.

After the key ORB core components are allocated within scoped and immortal memory and RealtimeThreads are used for request/response processing, predictability will improve. Currently, ZEN has been ported [27] to both ahead-of-time compiled RTSJ platforms such as jRate [5] and interpreted platforms such as OVM [42].

17.5.3.2 Technology Transitions

The ZEN project is part of the DARPA PCES [39] program, which provides language technology to safely and productively program and evolve crosscutting aspects to support DRE middleware and "plug & play" avionics systems. Although ZEN's development is still in its early stages, ZEN is being used to support several other research projects. The distributed automated target recognition (ATR) project [9] developed at MIT uses ZEN to transmit images of identified targets to another vehicle over wireless ethernet. ZEN supports the FACET [21] real-time event channel (RTEC) implemented in RTSJ, whose design is based on that of the TAO RTEC. ZEN is also being extended to support the CORBA component model (CCM), which is designed to support DRE system development by composing well-defined components. The Cadena project [19], provides an integrated environment for building and modeling CCM system. ZEN is currently being integrated with Cadena to model CCM implementations using a combination of the ZEN ORB, ZEN CCM, and FACET.

17.6 Related Work

In recent years, a considerable amount of research has focused on enhancing the predictability of real-time middleware for DRE applications. In this section, we summarize key efforts related to our work on TAO and ZEN.

QoS middleware R&D.
An increasing number of efforts are focusing on end-to-end QoS properties by integrating QoS management into standards-based middleware.

URI [62] is a Real-time CORBA system developed at the US Navy Research and Development Laboratories (NRaD) and the University of Rhode Island (URI). The system supports expression and enforcement of dynamic end-to-end timing constraints through timed distributed method invocations (`TDMIs`) [11].

ROFES [45] is a Real-time CORBA implementation for embedded systems. ROFES uses a microkernel-like architecture [45]. ROFES has been adapted to work with several different hard real-time networks, including SCI [29], CAN, ATM, and an ethernet-based time-triggered protocol [30].

The *Quality Objects* (QuO) distributed object middleware, developed at BBN Technologies [66], is based on CORBA and provides the following support for agile wide area–based applications: the runtime performance tuning and configuration through the specification of *QoS regions*, and reconfiguration strategies that allow the QuO runtime to trigger reconfiguration adaptively as system conditions change.

The *Time-triggered Message-triggered Objects* (TMO) project [24] at the University of California, Irvine, supports the integrated design of distributed OO systems and real-time simulators of their operating environments. The TMO model provides structured timing semantics for distributed real-time object-oriented applications by extending conventional invocation semantics for object methods, that is, CORBA operations, to include (1) invocation of time-triggered operations based on system times and (2) invocation and time-bounded execution of conventional message-triggered operations.

The *Kokyu* project, at Washington University, St Louis, provides a multiparadigm strategized scheduling framework. Kokyu has been implemented within TAO's Real-Time Event Service. Kokyu enables the configuration and empirical evaluation of multiple scheduling paradigms, including static (e.g., Rate Monotonic (RMS) [31]), dynamic (e.g., earliest deadline first (EDF) [31]) and hybrid (e.g., maximum urgency first (MUF) [57]) scheduling strategies.

The *Component Integrated ACE ORB* [60] project is an implementation of the CORBA Component Model (CCM) [35] specification. CIAO provides a component-oriented paradigm to DRE system developers by abstracting DRE-critical systemic aspects, such as

QoS requirements and real-time policies, as installable/configurable units supported by the CIAO component framework [61].

The *Model Driven Architecture* (MDA) [17] adopted by the OMG is a software development paradigm that applies domain-specific modeling languages systematically to engineer DRE systems. MDA tools are being combined with QoS-enabled component middleware to address multiple QoS requirements of DRE systems in real time [59].

RTSJ middleware research.
RTSJ middleware is an emerging field of study. Researchers are focusing at RTSJ implementations, benchmarking efforts, and program compositional techniques.

The TimeSys corporation has developed the official RTSJ Reference Implementation (RI) [58], which is a fully compliant implementation of Java that implements all the mandatory features in the RTSJ. TimeSys has also released the commercial version, JTime, which is an integrated real-time JVM for embedded systems. In addition to supporting a real-time JVM, JTime also provides an ahead-of-time compilation model that can enhance RTSJ performance considerably.

The *jRate [5]* project is an open-source RTSJ-based real-time Java implementation developed at Washington University, St Louis. jRate extends the open-source GCJ runtime system [14] to provide an ahead-of-time compiled platform for RTSJ.

The *Real-Time Java for Embedded Systems* (RTJES) program [23] is working to mature and demonstrate real-time Java technology. A key objective of the RTJES program is to assess important real-time capabilities of real-time Java technology via a comprehensive benchmarking effort. This effort is examining the applicability of Real-time Java within the context of real-time embedded system requirements derived from Boeing's Bold Stroke avionics mission computing architecture [55].

Researchers at Washington University in St Louis are investigating automatic mechanisms [7] that enable existing Java programs to become storage-aware RTSJ programs. Their work centers on validating RTSJ storage rules using program traces and introducing storage mechanisms automatically and reversibly into Java code.

17.7 Concluding Remarks

DRE systems are growing in number and importance as software is increasingly used to automate and integrate information systems with physical systems. Over 99% of all microprocessors are now used for DRE systems [1] to control physical, chemical, or biological processes and devices in real time. In general, real-time middleware (1) off-loads the tedious and error-prone aspects of distributed computing from application developers to middleware developers, (2) defines standards that help reduce total ownership costs of complex software systems, and (3) enhances extensibility for future application needs. In particular, Real-time CORBA has been used successfully in a broad spectrum of DRE systems, conveying the benefits of middleware to the challenging requirements of applications in domains ranging from telecommunications to aerospace, defense, and process control.

As Real-time CORBA middleware has evolved to meet the needs of DRE systems, middleware developers are designing ORB implementations that offer unique advantages and allow DRE systems developers to control the trade-offs between different development concerns. Universal adoption of early versions of Real-time CORBA middleware has been hampered to some extent, however, by the steep learning curve of the CORBA C++ mapping and the difficulty of obtaining skilled C++ developers. Java and Real-time Java have emerged as an attractive alternative, since they are simpler to learn and have less inherent and accidental complexity.

The next generation of Real-time CORBA middleware, such as the ZEN ORB described in this chapter, is designed to reduce the difficulties of earlier middleware by combining Real-time Java with Real-time CORBA. The result is an easy-to-use, extensible, flexible, and standards-based middleware with an appropriate footprint and QoS to meet the needs of many DRE systems. As illustrated by the TAO and ZEN case studies described in this chapter, future generations of middleware will continue to evolve to meet the changing and demanding needs of DRE systems.

Bibliography

[1] Alan, B. and Andy, W. (2001) *Real-Time Systems and Programming Languages*, 3rd Edition, Addison-Wesley Longman.

[2] BEA Systems, et al. (1999) *CORBA Component Model Joint Revised Submission*, OMG Document orbos/99-07-01 edn, Object Management Group.

[3] Bollella, G., Gosling, J., Brosgol, B., Dibble, P., Furr, S., Hardin, D., and Turnbull, M. (2000) *The Real-Time Specification for Java*, Addison-Wesley.

[4] Buschmann, F., Meunier, R., Rohnert, H., Sommerlad, P., and Stal, M. (1996) *Pattern-Oriented Software Architecture—A System of Patterns*, Wiley & Sons, New York.

[5] Corsaro, A. and Schmidt, D. C. (2002) The Design and Performance of the jRate Real-Time Java Implementation, in *On the Move to Meaningful Internet Systems 2002: CoopIS, DOA, and ODBASE* (eds R. Meersman, and Z. Tari), Lecture Notes in Computer Science 2519, Springer-Verlag, Berlin, pp. 900-921.

[6] Corsaro, A., Schmidt, D. C., Klefstad, R., and O'Ryan, C. (2002) Virtual Component: A Design Pattern for Memory-Constrained Embedded Applications. *Proceedings of the* 9th *Annual Conference on the Pattern Languages of Programs*, Monticello, IL.

[7] Deters, M., Leidenfrost, N., and Cytron, R. K. (2001) Translation of Java to Real-Time Java Using Aspects. *Proceedings of the International Workshop on Aspect-Oriented Programming and Separation of Concerns*, Lancaster, UK, pp. 25–30. Proceedings published as Tech. Rep. CSEG/03/01 by the Computing Department, Lancaster University.

[8] Douglas, C. S. and Buschmann, F. (2003) Patterns, Frameworks, and Middleware: Their Synergistic Relationships, *International Conference on Software Engineering (ICSE)*, IEEE/ACM, Portland, OR.

[9] Dudgen, D. E. and Lacoss, R. T. (1993) An Overview of Automatic Target Recognition. *MIT Lincon Laboratory Journal*, **6**, 3–10.

[10] Eide, E., Frei, K., Ford, B., Lepreau, J., and Lindstrom, G. (1997) Flick: A Flexible, Optimizing IDL Compiler, *Proceedings of ACM SIGPLAN '97 Conference on Programming Language Design and Implementation (PLDI)*, ACM, Las Vegas, NV.

[11] Fay-Wolfe, V., Black, J. K., Thuraisingham, B., and Krupp, P. (1995) Real-time method invocations in distributed environments. Technical Report 95-244, University of Rhode Island, Department of Computer Science and Statistics.

[12] Gamma, E., Helm, R., Johnson, R., and Vlissides, J. (1995) *Design Patterns: Elements of Reusable Object-Oriented Software*, Addison-Wesley, Reading, MA.

[13] Gill, C. D., Levine, D. L., and Schmidt, D. C. (2001) The Design and Performance of a Real-Time CORBA Scheduling Service. Real-Time Systems. *The International Journal of Time-Critical Computing Systems*, special issue on Real-Time Middleware, **20**(2).

[14] GNU is Not Unix (2002) GCJ: The GNU Compiler for Java, http://gcc.gnu.org/java.

[15] Gokhale, A. and Schmidt, D. C. (1998) Principles for Optimizing CORBA Internet Inter-ORB Protocol Performance. *Hawaiian International Conference on System Sciences*, Hawaii, USA.

[16] Gokhale, A. and Schmidt, D. C. (1999) Optimizing a CORBA IIOP Protocol Engine for Minimal Footprint Multimedia Systems. *Journal on Selected Areas in Communications*, special issue on Service Enabling Platforms for Networked Multimedia Systems, **17**(9).

[17] Gokhale, A., Schmidt, D. C., Natarajan, B., Gray, J., and Wang, N. (2003) Model Driven Middleware, in *Middleware for Communications* (ed. Q. Mahmoud), Wiley & Sons, New York.

[18] Harrison, T. H., Levine, D. L., and Schmidt, D. C. (1997) The Design and Performance of a Real-time CORBA Event Service, *Proceedings of OOPSLA '97*, ACM, Atlanta, GA, pp. 184–199.

[19] Hatcliff, J., Deng, W., Dwyer, M., Jung, G., and Prasad, V. (2003) Cadena: An Integrated Development, Analysis, and Verification Environment for Component-Based Systems. *Proceedings of the International Conference on Software Engineering*, Portland, OR.

[20] Henning, M. and Vinoski, S. (1999) *Advanced CORBA Programming with C++*, Addison-Wesley, Reading, MA.

[21] Hunleth, F. and Cytron, R. K. (2002) Footprint and Feature Management Using Aspect-Oriented Programming Techniques, *Proceedings of the Joint Conference on Languages, Compilers and Tools for Embedded Systems*, ACM Press, pp. 38–45.

[22] Hutchins, R. (2002) *Java NIO*, O'Reilly & Associates.

[23] Jason, L. (2001) Real-Time Java for Embedded Systems (RTJES), http://www.opengroup.org/rtforum/jan2002/slides/java/lawson.pdf.

[24] Kim, K. H. K. (1997) Object Structures for Real-Time Systems and Simulators. *IEEE Computer*, **30**(7), 62–70.

[25] Klefstad, R., Schmidt, D. C., and O'Ryan, C. (2002) The Design of a Real-time CORBA ORB Using Real-time Java, *Proceedings of the International Symposium on Object-Oriented Real-time Distributed Computing*, IEEE.

[26] Krishna, A. S., Schmidt, D. C., and Klefstad, R. (2004) Enhancing Real-Time CORBA via Real-Time Java, Submitted to the *24th International Conference on Distributed Computing Systems (ICDCS)*, IEEE, Tokyo, Japan.

[27] Krishna, A. S., Schmidt, D. C., Klefstad, R., and Corsaro, A. (2003) Towards Predictable Real-Time JAva Object Request Brokers, *Proceedings of the 9th Real-time/Embedded Technology and Applications Symposium (RTAS)*, IEEE, Washington, DC.

[28] Kuhns, F., Schmidt, D. C., and Levine, D. L. (1999) The Design and Performance of a Real-time I/O Subsystem, *Proceedings of the* 5th *IEEE Real-Time Technology and Applications Symposium*, IEEE, Vancouver, British Columbia, Canada, pp. 154–163.

[29] Lankes, S. and Bemmerl, T. (2001) Design and Implementation of a SCI-Based Real-Time CORBA, *4th IEEE International Symposium on Object-Oriented Real-Time Distributed Computing (ISORC 2001)*, IEEE, Magdeburg, Germany.

[30] Lankes, S. and Jabs, A. (2002) A Time-Triggered Ethernet Protocol for Real-Time CORBA, *5th IEEE International Symposium on Object-Oriented Real-Time Distributed Computing (ISORC 2002)*, IEEE, Washington, DC.

[31] Liu, C. and Layland, J. (1973) Scheduling Algorithms for Multiprogramming in a Hard-Real-Time Environment. *Journal of the ACM*, **20**(1), 46–61.

[32] Morgenthal, J. P. (1999) Microsoft COM+ Will Challenge Application Server Market, www.microsoft.com/com/wpaper/complus-appserv.asp.

[33] Mungee, S., Surendran, N., and Schmidt, D. C. (1999) The Design and Performance of a CORBA Audio/Video Streaming Service. *Proceedings of the Hawaiian International Conference on System Sciences*, Hawaii.

[34] Obj (2001) *Dynamic Scheduling Real-Time CORBA 2.0 Joint Final Submission*, OMG Document orbos/2001-06-09 edn.

[35] Obj (2002a) *CORBA Components*, OMG Document formal/2002-06-65 edn.

[36] Obj (2002b) *Real-time CORBA Specification*, OMG Document formal/02-08-02 edn.

[37] Obj (2002c) *The Common Object Request Broker: Architecture and Specification*, 3.0.2 edn.

[38] Object Management Group (1998) *CORBA Messaging Specification*, OMG Document orbos/98-05-05 edn, Object Management Group.

[39] Office DIE (2001) Program Composition for Embedded Systems (PCES), www.darpa.mil/ixo/.

[40] O'Ryan, C., Kuhns, F., Schmidt, D. C., Othman, O., and Parsons, J. (2000) The Design and Performance of a Pluggable Protocols Framework for Real-time Distributed Object Computing Middleware, *Proceedings of the Middleware 2000 Conference*, ACM/IFIP.

[41] Othman, O., O'Ryan, C., and Schmidt, D. C. (2001) Designing an Adaptive CORBA Load Balancing Service Using TAO. *IEEE Distributed Systems Online*, **2**(4), computer.org/dsonline/.

[42] OVM/Consortium (2002) OVM An Open RTSJ Compliant JVM, http://www.ovmj.org/.

[43] Pyarali, I. and Schmidt, D. C. (1998) An Overview of the CORBA Portable Object Adapter. *ACM StandardView*.

[44] Pyarali, I., O'Ryan, C., Schmidt, D. C., Wang, N., Kachroo, V., and Gokhale, A. (1999) Applying Optimization Patterns to the Design of Real-time ORBs, *Proceedings of the* 5th *Conference on Object-Oriented Technologies and Systems*, USENIX, San Diego, CA, pp. 145–159.

[45] RWTH Aachen (2002) ROFES, http://www.rofes.de.

[46] Schmidt, D. C. and Huston, S. D. (2002a) *C++ Network Programming, Volume 1: Mastering Complexity with ACE and Patterns*, Addison-Wesley, Boston, MA.

[47] Schmidt, D. C. and Huston, S. D. (2002b) *C++ Network Programming, Volume 2: Systematic Reuse with ACE and Frameworks*, Addison-Wesley, Reading, MA.

[48] Schmidt, D. C. and O'Ryan, C. (2002) Patterns and Performance of Real-time Publisher/Subscriber Architectures. *Journal of Systems and Software*, special issue on Software Architecture – Engineering Quality Attributes.

[49] Schmidt, D. C. and Vinoski, S. (2000) The History of the OMG C++ Mapping. *C/C++ Users Journal.*

[50] Schmidt, D. C., Kachroo, V., Krishnamurthy, Y., and Kuhns, F., 2000a Applying QoS-Enabled Distributed Object Computing Middleware to Next-generation Distributed Applications. *IEEE Communications Magazine*, **38**(10), 112–123.

[51] Schmidt, D. C., Kuhns, F., Bector, R., and Levine, D. L. 1998a The Design and Performance of an I/O Subsystem for Real-time ORB Endsystem Middleware. submitted to the *International Journal of Time-Critical Computing Systems*, special issue on Real-Time Middleware.

[52] Schmidt, D. C., Levine, D. L., and Mungee, S., (1998b) The Design and Performance of Real-Time Object Request Brokers. *Computer Communications*, **21**(4), 294–324.

[53] Schmidt, D. C., Mungee, S., Flores-Gaitan, S., and Gokhale, A. (2001) Software Architectures for Reducing Priority Inversion and Non-determinism in Real-time Object Request Brokers. *Journal of Real-time Systems*, special issue on Real-time Computing in the Age of the Web and the Internet.

[54] Schmidt, D. C., Stal, M., Rohnert, H., and Buschmann, F., (2000b) *Pattern-Oriented Software Architecture: Patterns for Concurrent and Networked Objects*, Volume 2, Wiley & Sons, New York.

[55] Sharp, D. C. (1998) Reducing Avionics Software Cost Through Component Based Product Line Development. *Proceedings of the 10th Annual Software Technology Conference.*

[56] Snell, J. and MacLeod, K. (2001) *Programming Web Applications with SOAP*, O'Reilly & Associates.

[57] Stewart, D. B. and Khosla, P. K. (1992) Real-Time Scheduling of Sensor-Based Control Systems, in *Real-Time Programming* (eds W. Halang, and K. Ramamritham), Pergamon Press, Tarrytown, New York.

[58] TimeSys (2001) Real-Time Specification for Java Reference Implementation, www.timesys.com/rtj.

[59] Trask, B. (2000) A Case Study on the Application of CORBA Products and Concepts to an Actual Real-Time Embedded System, *OMG's First Workshop On Real-Time & Embedded Distributed Object Computing*, Object Management Group, Washington, DC.

[60] Wang, N., Schmidt, D. C., Gokhale, A., Gill, C. D., Natarajan, B., Rodrigues, C., Loyall, J. P., and Schantz, R. E. (2003a) Total Quality of Service Provisioning in Middleware and Applications. *The Journal of Microprocessors and Microsystems*, **27**(2), 45–54.

[61] Wang, N., Schmidt, D. C., Gokhale, A., Rodrigues, C., Natarajan, B., Loyall, J. P., Schantz, R. E., and Gill, C. D. (2003b) QoS-Enabled Middleware, in *Middleware for Communications* (ed. Q. Mahmoud), Wiley & Sons, New York.

[62] Wolfe, V. F., DiPippo, L. C., Ginis, R., Squadrito, M., Wohlever, S., Zykh, I., and Johnston, R. (1997) Real-Time CORBA. *Proceedings of the Third IEEE Real-Time Technology and Applications Symposium*, Montréal, Canada.

[63] Wollrath, A., Riggs, R., and Waldo, J. (1996) A Distributed Object Model for the Java System. *USENIX Computing Systems*.

[64] Zahavi, R. and Linthicum, D. S. (1999) *Enterprise Application Integration with CORBA Component & Web-Based Solutions*, John Wiley & Sons, New York.

[65] ZeroC, I. (2003) The Internet Communications Engine^{TM}, www.zeroc.com/ice.html.

[66] Zinky, J. A., Bakken, D. E., and Schantz, R. (1997) Architectural Support for Quality of Service for CORBA Objects. *Theory and Practice of Object Systems*, **3**(1), 1–20.

18

Middleware Support for Fault Tolerance

Diana Szentiványi, Simin Nadjm-Tehrani
Linköping University

18.1 Introduction

Our society is increasingly dependent on service-oriented applications built from complex pieces of software. Significant questions for achieving high availability in multitier service architectures are (1) do we gain from separation of code that implements functional and non-functional requirements in a service? In particular, what are the benefits of doing the separation by implementing support for fault tolerance (FT) in the middleware? There are two extreme answers to this question:

- A lot! all FT-related code should be implemented in middleware.
- Nothing! FT-related code can be tailor-made in each application.

The truth could of course be somewhere in between. Some applications will benefit from FT support via middleware, and the implementation of the support could benefit from some application knowledge. The follow up question, assuming that performance is a major requirement in most applications, is (2) how is performance affected by increased availability? Specifically, will different techniques for achieving fault tolerance have different benefits and penalties, and how are these factors affected if the support is implemented in middleware? Our research agenda is to provide quantifiable evidence in answering these important questions.

In telecommunication systems, CORBA is a widespread middleware due to these systems' long lifetime and interoperability requirements [18]. Moreover, a review of the related work for other middleware shows that CORBA has the most elaborate standardized

Middleware for Communications. Edited by Qusay H. Mahmoud
 ISBN 0-470-86206-8

schemes. Telecom management systems are large and complex pieces of software that are costly to maintain. One factor affecting this complexity is that system availability is typically dealt with at the application level by ad hoc fault tolerance mechanisms. Thus, multiple applications reproduce almost the same recovery routines over and over again. Building fault tolerance capabilities in middleware has the benefit that reliable code (for implementing fault tolerance) is produced once, there is less application code to maintain, and the failover procedures are transparent. The application writer has to write very little FT-specific code, and the client need not even notice that a server goes down while the service continues. The remaining question is therefore the adequacy of average response times for serving a client request.

Extending an Object Request Broker (ORB) for dealing with fault tolerance properties affects timing behavior of an application. Compared to the use of a standard non-FT ORB, we would expect the client to experience longer response times due to continuous saving of information that is recaptured after failures. To measure the response time, one could sum up the chunks of time spent by a CORBA request on different segments while traveling from the client to the server object, getting processed and returning as a reply. In an ORB that does not support fault tolerance, the time intervals consist of the time spent in the ORB layers and the time spent on the network transport. We will refer to their sum as *roundtrip time*. In an extended fault-tolerant ORB, the time intervals spent on segment i of the named path will increase by a certain amount (δ_i). The sum of δ_i on all segments gives the so-called *overhead*. This overhead is encountered constantly while the system is running under no-failure conditions.

Similarly, when a failure occurs, all requests sent around the moment of server crash will potentially experience a larger overhead. The reason for the increase is the time spent for reconfiguration of the server, to designate a new request processing replica, and to set its state up-to-date. This time is called *failover time*.

When quantifying the performance-availability trade-offs, one can either consider only application objects' failures or consider eliminating all single points of failure. We have studied both approaches by building and evaluating two extensions of an open source ORB, Exolab OpenORB [4]. One extension closely follows the FT-CORBA standard and has no infrastructure replication. FT-CORBA standard creates new infrastructure objects that, unless replicated, can act as central units that become single points of failure. The other, called the fully available platform (FA), uses distributed infrastructure units for dealing with recovery. Failover support is implemented by a combination of leader election and consensus primitives. In our evaluation of the trade-offs, we have experimented with a generic service from Ericsson Radio Systems on top of both extensions.

In our experiments, we used roundtrip time overhead (in absence of failures) and failover times as performance metrics. Having these quantified, the telecom engineer, depending on the availability requirements of the particular service, can make the relevant decisions for his/her replicated application configuration. The decision will be highly influenced by how critical is the delivery of that service; hence, how fast are the desired reactions to failures. The decision is also dependent on how much latency can the client tolerate when receiving a reply to a request in a no-failure scenario.

18.2 Terminology

In this section, we review some important terms used in the rest of the chapter. These terms are widely used in the fault tolerance literature but we repeat them here to make the text self-contained.

18.2.1 Replication Styles

The most straightforward way to obtain fault tolerance is to replicate processing units, that is, to distribute the same application functionality on several nodes. Depending on whether only one or all of the replicas process every arriving client request, we can distinguish two main replication styles:

- Primary-backup, also known as passive replication
- Active replication

In the first case, only the primary replica processes incoming requests, while its state is read (checkpointed) and stored in a log from time to time [8]. When the primary fails, one of the backups has to take over its role, using a failover protocol. The last state of the primary will be transferred to the backup and the requests that arrived since the last checkpoint are replayed. In the second case, all replicas receive and process the requests from clients. To ensure that the same result is sent to the client no matter which replica responds, replica computations need to be deterministic [21]. In both cases, the client perceives the replica set as a single server, and all state saving and failover operations are transparent to the client.

18.2.2 Consensus

In some applications, distributed servers have to agree on some information, for example, the outcome to be sent to the client. The algorithm that achieves agreement is called a *consensus algorithm*. There are two main steps to perform: putting forward the own proposal, that is, the value to agree upon, and deciding on a common value. There are intermediate steps performed in a number of rounds to achieve this.

A consensus algorithm must guarantee that all replicas decide the same value (agreement), the decided value was proposed by at least one of the involved parties (validity) and that all of the replicas manage to decide within a finite interval of time (termination). Essentially, the properties involve replicas that have not failed.

When failures are expected to happen, a major assumption affects the solution:

- The distributed system is synchronous: upper bounds on message delivery delays are known; also, the different computation nodes have a processing rate with a known relation (when a process performs a computation step, every other process performs n computation steps for some $n \geq 1$).
- The distributed system is asynchronous: no knowledge about message delivery delays, or computation rates exists.

Fischer et al. presented a major result in this research area in 1985 [7]. It states that there are no general algorithms to implement consensus in the asynchronous setting even if a single node can be expected to crash. Intuitively, the difficulty arises because there is no way to know whether a message from a node is only late or will never arrive (since the node crashed). The theory of unreliable failure detectors was developed to circumvent this impossibility result by making further assumptions (see next section).

18.2.3 Unreliable Failure Detectors

Formalized by Chandra and Toueg in 1996 [2], an unreliable failure detector is a distributed entity used as an element in consensus algorithms for asynchronous systems. The distributed failure detector provides information about which nodes are "correct" (have not crashed) and can participate in agreement protocols. The information about failed units comes in the form of so-called suspicions. Since the failure detectors are unreliable, from time to time, the information may be wrong in one of the following two senses: either some nodes that failed are not detected (lack of *completeness*) or correct nodes are suspected to have failed (lack of *accuracy*). By allowing such mistakes to a certain extent, consensus can be solved even in a failure-prone asynchronous system. Chandra and Toueg classified failure detectors in terms of properties that characterize correctness of their detections (in the above two senses). Then they showed that consensus can indeed be solved in an asynchronous setting for given types of failure detectors. In particular, even using the weakest failure detector consensus can be solved if a majority of the nodes are always correct.

18.2.4 Broadcast

Consensus algorithms typically use the primitive operation *broadcast* in order to send the same message to a set of multiple receivers [9]. The abstraction encapsulates the requirement that all destination nodes should receive (and use) the same message. In failure-prone systems, this is not trivial to obtain. It is possible that the sender fails in the process of sending the message to a bunch of receivers. To ensure that in such a case either all or none of the receivers use a given message, we need to build the broadcast mechanism on top of other primitives: we distinguish between receiving and *delivering* a message. A broadcast message is only delivered (used in an application) if all the nodes in the destination set have received the message. This notion is then used to define a useful broadcast, namely, an algorithm that implements a *reliable broadcast*. This primitive basically guarantees that *all* correct receivers deliver the same messages, messages that are delivered have been sent by some sender, and all messages sent by the correct senders are eventually delivered.

18.3 Background

Bakken defines middleware as "a class of software technologies designed to help manage the complexity and heterogeneity inherent in distributed systems" [1]. One of the complexities in distributed systems arises from potential failures. Thus, there have been several attempts to build support for fault tolerance in middleware. In this section, we first briefly review the combination of FT support and middleware in general, and then go on to present related approaches for supporting FT in CORBA.

18.3.1 Middleware Fault Tolerance

To replace fault tolerance code at application level, one would expect an FT-supporting middleware to provide at least client transparency to failures, and automatic replica management with state saving and restoring features. Middleware exist in several flavors: those incorporating component/object as well as communication management (e.g., CORBA, COM/DCOM), those dealing with communication management (e.g., Java RMI), those that support resource allocation in a dynamic way (e.g., Jini), and those that offer containers for objects that can communicate using several offered technologies (e.g., EJB).

Microsoft's COM/DCOM allows customizing different elements of the remoting architecture[1] in order to accommodate communications with multiple server copies and transparent failover. The COMERA extension by Wang and Lee [24] was needed in order to provide efficient ways especially for replica migration where connections have to be restored in a transparent way and logged messages to be replayed.

Java RMI, as a basic technology, does not include fault tolerance features. On the other hand, built on top of this communication infrastructure, middleware such as the Jgroup toolkit [13] or the AROMA [14] system provide transparent fault tolerance to applications. In Jgroup, clients transparently access replica groups as one entity. Replica group members, when processing a query, exchange messages ensuring proper state logging. Failure of a replica is transparently handled by the group. AROMA intercepts RMI messages, and sends them in the same order to all server replicas, thereby assuring consistent states at recovery time. Replication style can be chosen by the application writer on the basis of the fault tolerance needs and failover time allowed.

Jini addresses a different approach to fault tolerance. As a technology inherently used for distributed, potentially mobile, networked applications, it does not advocate explicit replication for fault tolerance. Therefore, no automatic application server or client replication is supported; neither is application state saving and restoring. On the other hand, by offering the Lookup Service (possibly replicated) and the lease-based resource allocation approach, failures of different service providers can be made transparent to the client [12, 17].

EJB technology uses clustering and transactions to provide transparent fault tolerance to client applications [10]. Still, this is not sufficient for the servers to be available in case of failures. Enhanced with group communication [19], the architecture can provide transparent failover to clients as well as proper state logging and recovery.

18.3.2 CORBA and Fault Tolerance

CORBA provides by far the most developed support for fault tolerance in middleware. Prior to the specification of the CORBA extension for FT (April 2000), there are few reported works on alternative augmentations of CORBA with a process (object) group module. Some of these have provided a basis for the development of the standard.

[1] Infrastructure connecting clients to server objects

Felber et al. present three different approaches for introducing object groups in CORBA, dependent on the position of the group communication module relative to the ORB [6]. These are the interception approach, the integration approach, and the service approach. The work covers the third alternative. The object group service (OGS) uses different CORBA objects to implement fault tolerance related services.

Narasimhan et al. developed and implemented the idea of operating system level interceptors to provide fault tolerance to CORBA [15]. The result of research efforts in this direction is the Eternal System. With this approach, an ORB's functionality can be enhanced for fault tolerance without changes in the ORB, or in the application.

DOORS is a fault-tolerant infrastructure built using the service approach [3, 16]. Application objects register to DOORS in order to be made fault-tolerant. Fault tolerance services are realized with two components: ReplicaManager and WatchDog.

A framework for fault-tolerant CORBA services by using aspect-oriented programming is presented by Polze et al. [20]. Their goal is to provide the application writer with the possibility to build a fault-tolerant application by choosing the types of faults to be tolerated: crash faults, timing faults, design faults. The toolkit then chooses the appropriate fault tolerance strategy.

Killijan and Fabre [11] describe a framework for building fault-tolerant CORBA applications by using reflection. They define a meta-object protocol, and use an open compiler, to be able to extend CORBA objects with wrappers of modifier methods. The wrappers always save object state when the wrapped modifier method is called. Therefore, application writers do not have to write state reading or applying methods.

Most of the above works differ from the work we present in this chapter in that they do not provide trade-off studies or study artificially created applications on top of the platform. None of them follows the actual FT-CORBA standard.

18.4 Standard Support for FT in CORBA

The CORBA standard was extended to support basic mechanisms for fault tolerance in December 2001 [18]. At that time, no major studies documenting the performance/FT trade-offs in an FT-CORBA-based implementation existed. To provide some evidence for conditions under which the suggested primitives would be useful in a real application, an experimental platform in compliance with FT-CORBA was built [23]. In Section 18.7, we present the results of these studies. But before that, we provide a short overview of the FT-CORBA standard in this section and explain the architecture of the implemented platform. This can be seen as an infrastructure over which FT applications can be built with little effort.

18.4.1 The FT-CORBA Standard

Applications are supported for fault tolerance through replicating objects. Temporal replication is supported by request retry, or request redirection. Replicated objects are monitored for failure detection. The failover procedure depends on the replication strategy used.

Support is provided for use of active, warm and cold passive, and stateless replication. In cold passive replication, backups are totally passive and are neither updated with primary state nor with information about processed method calls. The state of the primary is checkpointed periodically. Also, information about method calls that change the state of the replica object is stored in a log. These are called *update* method calls, as opposed to *read-only* ones that do not modify the state. At failover, all this information is transferred to the backup that has to take over the role of the primary. For warm passive replication, on the other hand, state checkpointing at the primary coincides with transfer of that information to all the backups. Also, information about update method calls executed at the primary are broadcast to the backups and stored in a log, without being executed. At failover, all necessary information is present at the backup that will be promoted to the primary role. Stateless replication is used only when the server does not have state. For active replication, the standard strongly recommends the use of a gateway for accessing the active replica group. This gateway plays the role of a relay for method calls – it broadcasts the method calls to all replicas that have to execute them in the same order.

The basic building blocks in the infrastructure are the Replication Manager, the Fault Notifier, and the Fault Monitor. The Replication Manager implements interfaces that provide operations for setting group properties, managing replicated groups by the application, as well as creating groups[2] on demand, and handling failover.

The Fault Notifier interface contains methods for registering fault consumers, for announcing faults, as well as for creating event filters. Fault consumers are those nodes in the application or the infrastructure that have to know about the occurrence of faults. An event filter is related to a fault consumer and is used to send a fault notification exactly to those parties that are interested in it.

To be able to manage large applications, the notion of *fault tolerance domains* is introduced. Each fault tolerance domain contains several hosts and object groups and a separate pair of Replication Manager and Fault Notifier is associated with it. On every host, there has to be one Object Factory and one Fault Monitor. The Object Factory creates group replicas on demand. The Fault Monitor has to detect failures of group members running on its host.

Every application object has to implement interfaces for failure detection and checkpointing. Logging and recovery mechanisms, needed for passive stateful replication, are automatically used when infrastructure-controlled consistency and membership is chosen by the application. In active and stateless replication styles, request replies can be logged in order to avoid repeated execution of a method call.

18.4.2 Architecture Units

Figure 18.1 shows the overall architecture of our prototype FT-CORBA platform. In particular, it depicts the service objects used as building blocks.

The building blocks of the infrastructure are as follows:

- A collection of service objects (in the named boxes and ovals in Figure 18.1)
- CORBA portable interceptors (explained later)
- ORB class extensions (more details in [22])

We now go on to describe the basic building blocks. Note, however, that the elements of the infrastructure are not replicated. The assumption is that the most vulnerable units

[2] This functionality appears in the Object Factory in the form of replica creation.

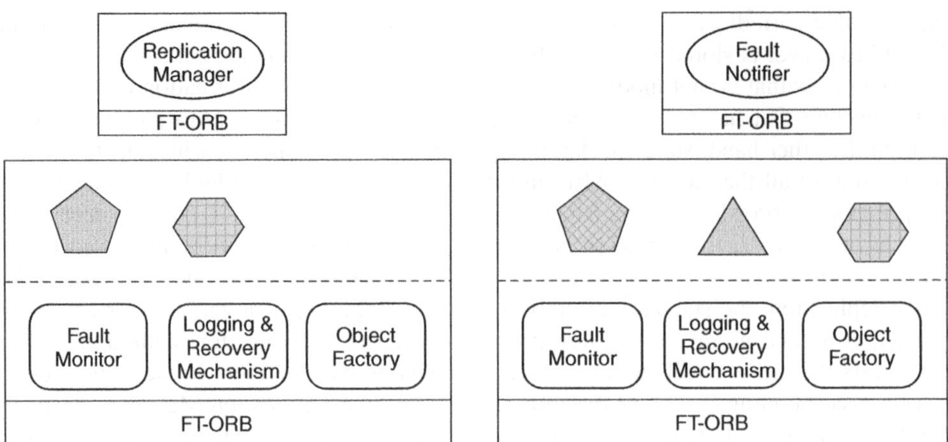

Figure 18.1 The overall architecture of the FT-CORBA platform, with one replicated server

are application objects (the different shapes in Figure 18.1), while infrastructure units run in processes or machines that are not failure prone.

18.4.2.1 Service Objects

The service objects consist of a set of CORBA objects for which the roles were clearly defined by the standard; but the standard leaves some implementation choices open. So for the implemented platform, each had to be considered separately:

- The Object Factory, Replication Manager, and Fault Notifier are described as interfaces in the standard. Therefore, they were implemented as CORBA objects.
- Fault Monitors were implemented as CORBA objects.
- The implementation of the Logging and Recovery Mechanism needed some careful considerations.
- For the active replication style, the standard leaves the choice of a mediator gateway object open. The creation of such an object as part of the infrastructure was chosen as a solution.

Next, a more detailed view of the above object categories will be provided.

Standard Service Objects

In a fault tolerance domain, there is one Replication Manager, CORBA object in charge of creating object groups, whenever the application demands it. A more dynamic and important role of the Replication Manager is to coordinate the failover procedure after it receives a failure report from the Fault Notifier. For passive replication, the procedure includes the promotion of a backup node to the position of primary. In either case (active and passive), the "failed" replica must be destructured. Note that this applies even if the suspicion is inaccurate. Even if the object was temporarily unavailable, it is permanently removed from the set of replicas. The FT-CORBA platform, thus, does not use unreliable failure detectors.

Object Factories are distributed over every host in the fault tolerance domain in which application object replicas can be created and run. Upon request from the Replication Manager, the Object Factory creates a new instance of an application object and starts it in a different process on the machine. Replicas of an object group are created on different hosts. The list of hosts that can be used is given in the application's request to the Replication Manager.

Replica failures are detected by Fault Monitors running on each replica's host. For each object to be monitored, the Fault Monitor CORBA object creates a separate thread. A pull-based monitoring style is used, that is, the Fault Monitor calls the **`is_alive`** method of the application CORBA object periodically to find out whether the replica is still able to process client requests. As soon as this method call times out, or it signals a failure, the Fault Monitor reports this to the Fault Notifier. The latter sends fault notifications to its registered consumers, one of them being the Replication Manager.

Gateway for Active Replication

Instead of the client broadcasting its request directly to all active server replicas, our platform uses a gateway CORBA object. This object is created as soon as an actively replicated group is deployed, and started on one of the server machines. The gateway fulfills a double role in the context of processing requests by the active group.

First and foremost, the gateway broadcasts incoming client requests as method calls to the replicas. Secondly, when the replicas, in their turn, become clients for a third-tier server, the gateway acts as a duplicate suppressor. Thus, outgoing method calls from replicas are routed through the gateway, whose reference is known by the replicas' ORBs. The reply received from the server is sent back to all replicas via the same path. Finally, the client will be sent the reply provided by the fastest active server replica.

The gateway participates in the group reconfiguration procedure when an active replica fails. To begin with, on its creation, the gateway registers as another consumer (besides the Replication Manager) of the Fault Notifier. After a failure, the gateway finds out the new configuration of the group and directs the future broadcast requests to the current set of active replicas. When a new replica is created and joins the group, the gateway manages the state transfer procedure. It stops requests that arrive after the start of the state reading, and broadcasts them to the group only after the new replica is successfully set up.

Logging and Checkpointing Object

The logging and recovery mechanism is used only for cold and warm passive replication (see Figure 18.2). The mechanism is implemented on each host using a separate CORBA object. This object's interface offers operations for logging method call and reply information. Other operations help in retrieving the logged information, when needed at failover. Only information about update method calls that arrived since the last checkpoint must be logged, because only those are needed in the failover procedure.

We will come back to the logging and recovery mechanism in Section 18.4.2.3.

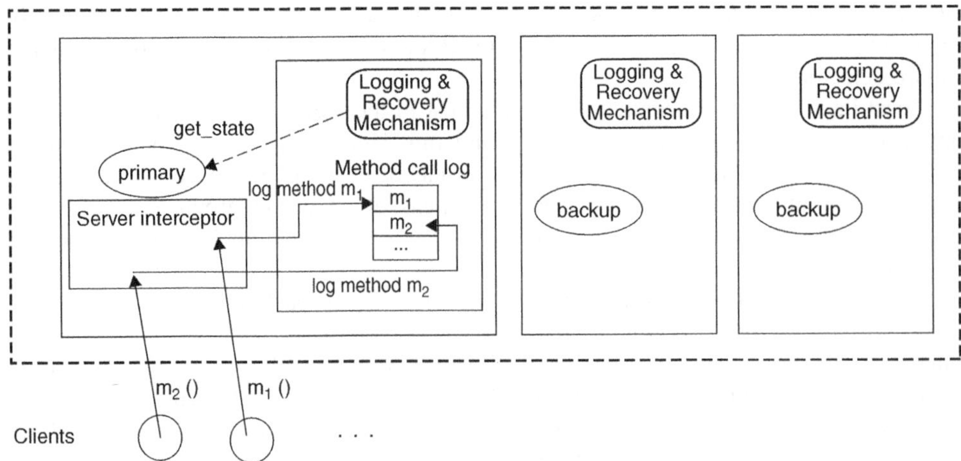

Figure 18.2 Checkpointing and logging in primary-backup replication

18.4.2.2 CORBA Portable Interceptors

The CORBA standard offers an elegant solution, portable interceptors, for performing operations around a CORBA request at the client side, as well as before the start or after the finishing of the computation performed at the server application object. In the presented platform, CORBA interceptors were used to perform the following operations on the request:

- Adding extra information to the request: the server group reference version known to the client, as well as a request identifier that remains the same even after resending the request, and a request expiration time. This information is placed in a service context at the level of the client portable request interceptor.
- Recording information about a call and its reply in a log. These operations take place in the server request interceptor.
- Tracing of timing information when performance measurements are needed. This operation is done both in the client and the server interceptors.

For different replication styles, application server objects are equipped with different portable server interceptors. For example, in active replication, no logging of method call information is needed. In cold passive replication, method calls are logged only at the primary (see Figure 18.2).

18.4.2.3 Logging and Recovery Mechanism

This section provides more details about how the logging and checkpointing mechanism works and some important implementation decisions.

Checkpointing

As shown in Figure 18.2, the logging and recovery mechanism calls `get_state` on the primary replica to perform state checkpointing. Execution of `get_state` has to wait until all update methods that arrived earlier finish their execution. Analogously, execution of update method calls that arrive while the state is being read has to wait until `get_state` finishes execution. In this context, some further questions arise: should checkpointing be done periodically? If yes, how often? When the state does not change so often, it would be better to save it as soon as a change occurs. This would lead to *event-triggered* checkpointing.

When the primary state is obtained, it must be stored for reuse on recovery from failures. Another issue is related to storing only state change records instead of entire state records. This way of checkpointing (`get_update`, as opposed to `get_state`) is efficient especially when the state can become large, but through small increments. However, there is a problem with this approach: when the fields of the object are complex structures, a change to the state might be difficult to express. As the `get_state` and `get_update` methods have to be written by the application developer, it is important that the demands are feasible. Note that if `get_update` methods are used, *all* state update records have to be kept in the log, as opposed to entire state records that are removed as soon as a new checkpoint is performed (in fact, in the latter case, the log always contains one single record). So, there are also issues related to the extensions to the footprint of the middleware that have to be considered.

Logging Method Call and Reply Information

As well as logging the state information, one needs to log the arriving requests for later replay purposes. Also, one needs to record the evidence that a request was executed once on the server object, at least in the case of update method calls. Executing an update method twice on the server can lead to inconsistent updates of the state, whereas a read-only method reply can potentially be sent twice.

Update method call information has to be logged in a certain order. In our platform, this is given by the order of execution of those methods on the primary replica. This is due to the need for later replaying the methods on the backup, in the same order that they were executed on the primary. This order is only available from the log. Further, since logging is done separately from execution, at the level of the server-side portable interceptor, the middleware has to enforce sequential method execution, independent of the threading policy in the ORB. The evidence that an update method was executed may consist of a simple note, or, when there is a return value, of the result structure.

The type of a method call (update or read-only) is known in the server-side interceptor. When creating a replica of the server group, the Object Factory (written by the application writer) provides this information together with method signature information to the ORB in which the replica is started.

This concludes our overview of the implementation of a platform in accordance with the FT-CORBA standard.

To summarize the characteristics of the platform, it supports three replication strategies, but for active replication, it builds on the replicas being deterministic. It does not support failures in the infrastructure units, and it works in a multitier application architecture provided that all the tiers provide support for FT with the same platform. We now go on to describe an alternative platform, also based on CORBA, and set the stage for a comparative evaluation of these in Section 18.6.

18.5 Adding Support for Full Availability

This section describes the architecture of a fully available (FA) middleware. In this approach, the infrastructure units are distributed and may only fail together with application objects, and therefore an algorithm that takes care of application replica failures is sufficient for assuring the availability of the whole system.

We build our fully available middleware on top of CORBA because of its attractive properties, for example, scalability and interoperability. Also, this gives us an opportunity to evaluate FT-CORBA in comparison with another CORBA-based FT support. As an ingredient in the FA infrastructure, we employ an improved version of an algorithm (called *universal construction*) originally developed by Dutta et al. [5]. The emphasis in this section is put on how the algorithm was embedded in a CORBA infrastructure that should provide fault tolerance and transparency to applications. Thus, architecture elements are presented together with CORBA-specific implementation issues. The improvement made to the algorithm does not affect its availability aspects (these were already present in the original algorithm), but affects its performance aspects on recovery from a failure. The last part of this section will describe this improvement for the interested reader.

18.5.1 Architecture Units

The conceptual building blocks of the fully available infrastructure are two distributed components: (1) a consensus service used to ensure the agreement among the replicas on the result to be returned to a client request and the state change enforced on the replicas, and (2) a leader election service used to designate the replica that is supposed to deal with the request.

The algorithm incorporates unreliable failure detectors, and hence the detection of failures is in terms of suspicions. The leader election service might elect different leaders when wrong failure suspicions arise. In this case, the consensus service helps resolve the conflict. The leader election service at each node relies on information from its (unreliable) failure detector oracle (that corresponds to the weakest failure detector in [2]). The failure detector is implemented as distributed modules that exchange heartbeat messages. A server replica that is not leader is simply called witness in the sense that it helps its actual leader resolve a conflict with other possible leaders, and stores the result of the real leader's actions.

Figure 18.3 illustrates the deployment of the FA infrastructure units in a CORBA environment. Each server replica uses the following units when executing the steps of the consensus algorithm: *the leader election unit* (LEU) and *the consensus object* (CO). Another element involved in the processing of a request is the local copy of the application server object (ASO) within the server interceptor. All operations on the application object are executed at the level of the server-side portable interceptor (as explained shortly). The failure detector is set up as a separate server thread, attached to the application CORBA object. As part of the leader election unit, the failure detector server is supposed to receive *I-am-alive* messages from other LEUs at other application replicas.

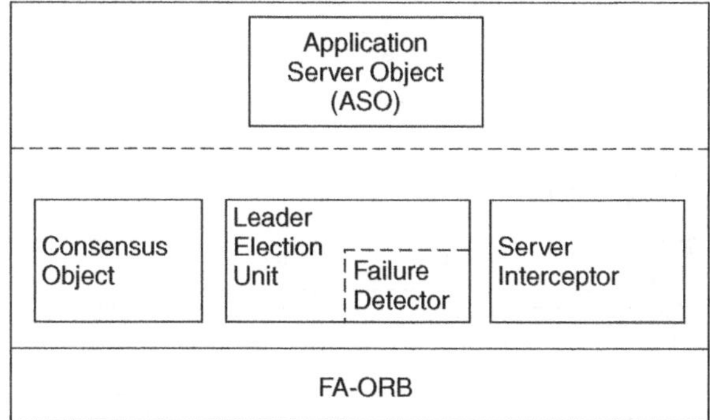

Figure 18.3 Server replica architecture

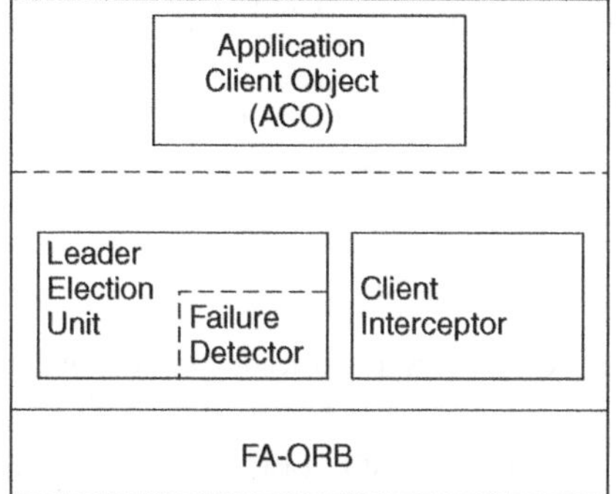

Figure 18.4 Client-side architecture

Note that it can be assumed that the leader election unit (and thus failure detection unit) sends "correct" I-am-alive messages on behalf of the application server replica. In other words, the fault tolerance approach properly deals with crashes of the whole node. Thus, the absence of an I-am-alive message leads to the suspicion of the non-availability of the application server, although the message comes from a separate unit. When trying to introduce the outcome of a request in the total order of outcomes recorded by all replicas, the leader election unit is queried for the present leader. The unit will return the correct process with the lowest index.

Figure 18.4 depicts the client side of the architecture. The client does not have a consensus object, but it has a leader election unit that identifies the replica that is currently in charge of servicing the client queries, similar to the case of server replicas.

18.5.2 Infrastructure Interactions

A view of major interactions inside the fully available infrastructure is shown in Figure 18.5. The picture visualizes the message flows between similar units within each server replica, as well as between client and server.

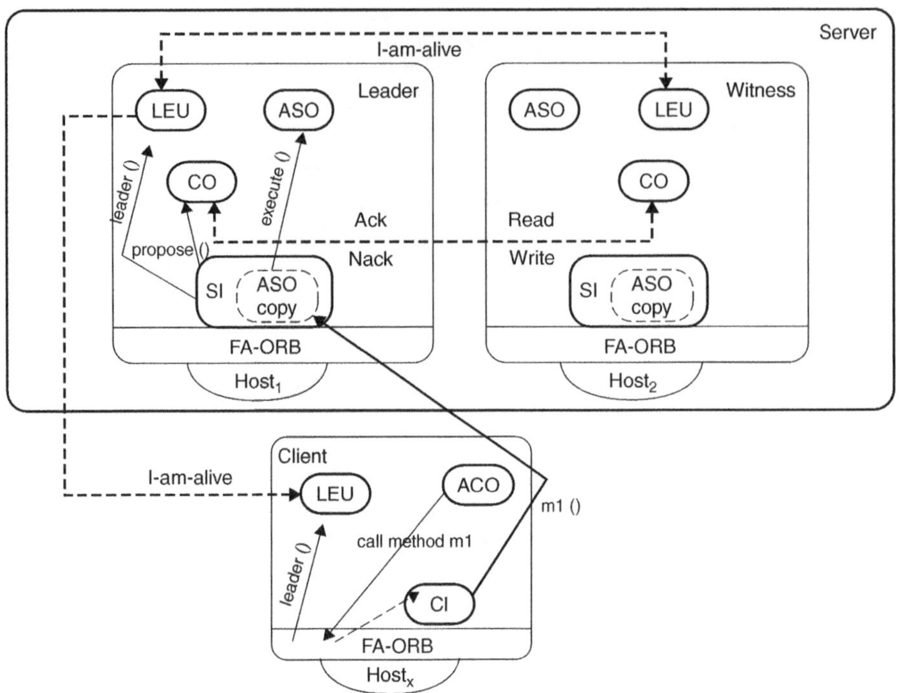

Figure 18.5 Client-server and leader-witness interaction view

The dashed line boxes containing ASO copy suggest that the server interceptors call methods on the application object, from the interceptor level. The dashed arrows that go between two consensus object boxes, or two leader election unit boxes, represent message flow by using simple socket connections, not involving CORBA calls. The dashed arrows that appear inside the solid "host" box, suggest internal operations within a process (e.g. in the client box, the dashed arrow represents a portion of the flow of a method call from the client to the server, which spans from the ORB to the client interceptor CI). The thick solid arrows represent flow of CORBA calls. The thin ones show method calls that are not performed via CORBA (e.g. `leader()` or `propose()`), but are done via socket connections.

I-am-alive messages received from replicas are used to determine who the leader is. The client, in order to enable its leader election unit to receive I-am-alive messages from server replicas, has to register with the server group. Registration is achieved by sending an *I-am-in* message to all members of the replica group. This message has piggy-backed the address in the client where the future I-am-alive messages have to be sent, that is, the address of the client's leader election unit.

Let us now concentrate on the solid arrows (thin and thick) in Figure 18.5. The working of the algorithm in a nonfailure scenario can be visualized as follows:

1. arrow from ACO to FA-ORB: client initiates method call `m1;`
2. arrow from FA-ORB to LEU: the LEU is queried about the identity of the current leader, as seen by the client;

3. arrow from client CI to server interceptor SI on $Host_1$: client sends request corresponding to method call `m1`;
4. arrow from SI (ASO copy) to ASO: call to `execute()` that means tentative execution of the request on the application object. Tentative means that no durable change of the state of the object is made so far;
5. arrow from SI to LEU: the LEU is queried about the identity of the current leader;
6. arrow from SI to CO: if the LEU on $Host_1$ indicates leadership, then `propose()` is called with the outcome (meaning state update and result) of `execute()`.

While performing (6) above, the leader tries to write its outcome in a unique position of a total order of state updates maintained at all replicas[3]. It achieves its goal only if a majority of the replicas acknowledge its leadership. The dashed arrow between CO on $Host_1$ and CO on $Host_2$ "implements" this by a series of READ, WRITE, and ACK/NACK messages.

18.5.3 Platform Implementation

This section provides a couple of details in the implemented platform.

18.5.3.1 Portable Interceptors

The main CORBA-specific elements employed to implement the algorithm and the infrastructure are portable interceptors. Client-side portable interceptors are used to add unique request identifying information to the client request. Portable server-side interceptors are used mainly for intercepting I-am-in registration messages from clients, and for intercepting client requests. Also, all the operations devised in the algorithm are performed in these interceptors. Thus:

- When receiving an I-am-in message in the interceptor (sent by dynamic CORBA method invocation), the call is stopped from reaching the application replica by throwing an exception, handled in a well-defined part of the ORB. Before the exception is thrown, however, the leader election unit is informed about the address where the I-am-alive messages have to be sent later.
- The algorithm operations executed in the server interceptor are those described in Section 18.5.2. The tentative execution of the request on the server replica is simulated by three steps: (a) reading the replica state using **`get_state`**; (b) executing the method call on the server replica, so that the result, if any, is returned and the state of the replica is changed; and (c) reading of the state update. After this execution, if the current process is a leader, but its proposed outcome is not the decided one, or if it is not the leader, the state read before the method execution is reinstated.

Note that if the request is read-only, the server replica does not execute the consensus algorithm. After the request execution, the replica sends the reply to the client and stops, returning to the point where it waits for a new request to arrive.

---

[3] From the leader's point of view, the other replicas are witnesses.

It is also important to note that the only published references (addresses) are those of the application server replicas. Therefore, the I-am-in messages are sent via CORBA dynamic request invocations to these addresses.

I-am-alive and other infrastructure-related messages are sent using simple reliable socket connections. Thus, CORBA-specific overhead is not encountered for these messages.

18.5.3.2 State Transfer

To clarify our improvement of the original algorithm, let us detail the activities performed by a witness including the moment when it becomes a leader.

During normal processing periods, that is, when no replica fails or is wrongly suspected to have failed, the leader transfers state updates to the witnesses as soon as a request is processed. These updates are stored at witnesses, possibly in a structure residing in memory. With no further intervention (as in the published algorithm [5]), the storage can be filled up with update structures. Therefore, one limitation of the algorithm was the possible infinite growth of memory needs.

Another limitation is related to the moment when a witness becomes a leader. This can happen either because the previous leader really failed or because the leader suddenly became very slow and the client as well as the other replicas do not recognize its leadership anymore. After this, when the witness-leader replica receives the first request from a client, it has to bring its state to the value found in the previous leader at the time of its abdication. Before the improvement we made, this state was built in the new leader by incrementing the initial replica state, using the updates stored at the other replicas. Obviously, the time for this procedure grows in an unbounded manner the further we get from the start-up time.

The original paper already proposed a way to solve the problem in the future. When a witness receives a state update from the leader replica, it will immediately apply it on the application object, instead of only storing it. Thus, storage needs are reduced almost to zero. Also, the new leader does not need to reconstruct its state by using information from witnesses anymore, since it has the current state itself.

Our platform implements a different improvement. In this setting, from time to time, the leader sends entire state information to witnesses. When receiving it, witnesses prune their storages, that is, replace the set of existing records with only one state record. Thus, the storage is dimensioned so that it accommodates a limited (quite small) number of records. Consequently, the failover process in the new leader will take a limited and relatively short time to execute. It is possible for the leader to trigger the pruning periodically (say every 5 or 10 s). A second alternative is to send the state after executing a certain number of requests: say after every 20 executed requests, the leader sends its current state to the witnesses.

The second alternative offers the possibility to dimension the storages to accommodate a known number of records. In the present implementation, the periodic alternative was used, since it offered a direct way to compare with the passive replication case in the FT-CORBA implementation, where checkpointing of state was done periodically.

18.6 Experiments with a Telecom Application

This section will briefly describe the application that was run on top of the two extended ORBs. Also, the set up for the experiments is presented.

18.6.1 The Service

The platforms were evaluated using a real application from Ericsson Radio Systems. The server is a generic service in the operations and management (O&M) part of a future generation radio network and is called Activity Manager. The role of this server is to create jobs and activities to be scheduled immediately or at later times. Also, the server receives activity status and progress reports. There are 15 operations (methods) offered by the server to fulfill the functionality it promises. The Manager has internal structures where the activity and job objects are stored. These structures make up the state of the server. They are managed according to what the clients (other parts of the O&M network) ask for. This service was chosen by our industrial partners as it has some generic characteristics that are representative for other applications and some methods that implement a multitier call structure.

18.6.2 Experiment Setup

The experiments were performed in a host park connected via an IP network with no isolation from other tasks on the hosts. The reason for not performing measurements in a controlled host environment was to mimic the realistic setting in which the service will eventually run. For both platforms and in all experiments, the clients were unchanged.

There were two goals to performing experiments: measuring roundtrip time overheads, and measuring failover times. As a baseline for comparison, a non-replicated server was employed where no fault tolerance support in the ORB was included. Note that such a comparison preserves neutrality to different replication styles (compared to ad hoc application-level replication that may resemble one of the styles). To avoid the probe effect, there were null probes inserted in the non-replicated case in the same position for each interceptor unit.

For the measurements on failover times, the baseline was chosen as the time taken from the crash of a (non-replicated) server to its restart (possibly manually as it is the case in the real setting). In the different platforms, the failover time is affected by the replication style and is computed as follows, for the FT-CORBA and FA-CORBA platforms respectively:

- In cold passive replication, it is the time taken to set the state on the new primary, plus the time needed to replay the requests arrived since the last checkpoint.
- In warm passive replication, it is the time spent in replaying requests.
- In active replication, it is the time spent to reconfigure the server group.
- For the FA-CORBA platform, to measure failover times, the current leader is forced to fail after a certain number of processed requests. Then, a new leader is elected and forced to bring its state up-to-date as soon as a next query arrives.

To obtain relevant results, the numbers of requests processed before the failure were chosen to be the same for both warm and cold passive replication in the FT-CORBA platform and for the experiments with the FA-CORBA platform.

For measuring overhead, the clients used in the tests were simple: they called six of the update methods of the server in a loop. The clients called the methods in loops of 100 and 200 iterations. Some of the methods were called once per iteration (Method 1, 2, 3, and 5), others two (Method 6), and four (Method 4) times, respectively. The results are presented as averages of the measurements for every method call computed over runs in both types (100 and 200 iterations) of loops (hereafter called *one experiment*). The averaging was done in order to take the different network and processor loads at the time of the experiments into consideration.

In both platforms, group sizes of 2, 3, 4 and 8 replicas were used. In the FA-CORBA platform, the group size was of special relevance, since it reflects different majority rules. In the FT-CORBA platform, three different checkpointing intervals were used: 1, 5, and 10 s. In the FA-CORBA platform, the pruning interval was chosen in a similar way to make valid comparisons.

All in all, 12 experiments were performed for each of the following styles: warm passive, cold passive, fully available replicas (corresponding to 4 replica configurations and 3 checkpointing intervals). For the active replication, 4 experiments were performed (corresponding to 4 replica configurations).

18.6.3 Measuring Overheads

The slices in the roundtrip time were used to identify dominant parts of the overhead. The slices can be described as follows: the time spent in the client interceptor (when adding information to the request), the time spent on traveling from the client-side node to the server-side node and until the request is taken up for processing, the time spent in the server-side interceptor, when a request arrives at the server and when it returns to the client as reply.

For both platforms, the server-side time includes time spent waiting for all earlier arrived update methods to finish execution on the server. In the FT-CORBA platform, the time for logging the method call information is included, as well as the time spent recording reply information. The approximate computation time is counted from the moment of leaving the interceptor at receiving the request, until it is entered again when sending the reply to the client. In the FA-CORBA platform, the situation is different: the time spent in the server interceptor includes the method execution time, as well as the consensus execution time and state reading and updating times. The time taken by the reply to travel back from the server to the client node is traced in the same way in both platforms.

18.7 Trade-off Studies

This section presents experimental results for both the FT-CORBA and the FA-CORBA platforms. First, overheads are presented, followed by the time taken for failover.

18.7.1 Overheads

Table 18.1 summarizes overheads in the FT-CORBA platform (columns three to five) and the FA-CORBA platform (column six). Each row of the table corresponds to one of the six method calls. Column two contains average roundtrip times for the different method calls when the client was calling them on the non-replicated server. The results in the next columns are presented as average percentages. Each average is computed as follows. Consider the highlighted cell in Table 18.1 that contains the value 62%. Here, the average roundtrip time value for method 3 on the non-replicated server was 65 ms. The % means that the average roundtrip time in the replicated experiments was $65 + 0.62 \times 65 = 105$ ms. This was the lowest roundtrip time measured within the 12 experiments. The highlighted cell shows that the average overhead when calling Method 3 on the server, while using cold passive replication style, ranged between of 62 and 163% for group sizes of two, three, four, or eight replicas and checkpointing intervals 1 s, 5 s, or 10 s. In particular, 62% corresponds to a group size of three and checkpointing interval of 10 s, while 163% corresponds to a group size of eight and checkpointing interval of 1 s.

We observed during our experiments that the group size slightly influenced the overheads in warm passive replication. In cold passive replication, the group size had almost no effect on the overheads. The variation in overhead for this replication style is mainly given by the different checkpointing intervals. For lower values of the checkpointing interval, the overhead was slightly higher, due to higher degree of interference between method calls and state checkpointing. Another interesting observation is that the overhead percentages are not much different between cold and warm. For active replication, on the other hand, the overheads are large, and highly affected by group size. The large variations present in Table 18.1 are due to variation over the number of replicas. The least value in the range corresponds to a group size of two replicas, while the largest to a group size of eight replicas.

The FA-CORBA overheads were also influenced by the group size, but also by the chosen pruning interval. Both parameters influence the average time taken for execution of the consensus primitive which is part of the overhead, in the following ways:

Table 18.1 Summary of overhead percentages in the FT-CORBA and FA-CORBA platforms (FT-CORBA figures from Szentiványi, D. and Nadjm-Tehrani, S. (2002) Building and Evaluating a Fault-Tolerant CORBA Infrastructure. *Proceedings of the DSN Workshop on Dependable Middleware-Based Systems*, Washington, DC, June 23-26. © 2003 IEEE)

	Non repl.	Cold passive	Warm passive	Active	FA-CORBA
Method 1	130 ms	55%-117%	53%-134%	1745%-5000%	141%-472%
Method 2	61 ms	77%-221%	110%-285%	770%-3360%	265%-950%
Method 3	65 ms	62%-163%	74%-240%	270%-360%	213%-861%
Method 4	80 ms	76%-447%	100%-550%	1790%-5100%	308%-1020%
Method 5	133 ms	44%-333%	59%-363%	905%-4300%	223%-726%
Method 6	106 ms	37%-419%	68%-383%	800%-3100%	280%-844%

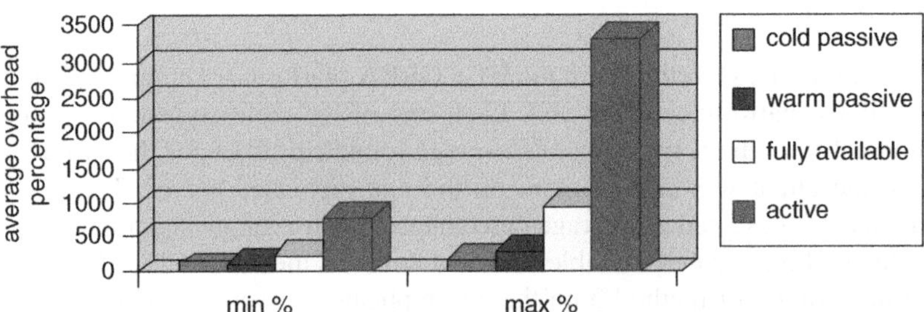

Figure 18.6 Bird's eye view of roundtrip time overhead percentages in the two platforms

- The group size gives the number of nodes to which messages have to be sent, and from which (their majority) answers must be received, as part of the consensus execution.
- The pruning interval affects the time spent in execution of consensus, due to the pruning action on the registers. Thus, the shorter the pruning interval, the more often the time spent in consensus is larger than in a no pruning situation. So, the average roundtrip time grows, and so does the overhead.

Figure 18.6 gives a bird's eye view of the overhead percentages for the different methods. We have chosen to show the figures for Method 2 that are representative for other methods too. The chart represents the lower end (min %) of the overhead ranges described in Table 18.1 as well as the higher end (max %). It shows that

- the overheads in the FT-CORBA platform for passive replication are lower than those in the FA platform;
- the overheads for active replication are much higher than the overheads in the FA platform;
- the overheads for cold passive replication style are slightly lower than those for warm passive.

The slices of the roundtrip times for the replicated scenarios can be further investigated to find out where the overhead is coming from. It turns out that

- in the FT-CORBA passive styles, as well as in the FA-CORBA platform, the main part of the overhead comes from the wait time when executing methods in a serialized way. This wait time is the sum of times t_r spent by all requests arrived before the one currently waiting, where t_r denotes the time from the arrival of the request in the server interceptor until the result of its processing is sent back to the client. In the FA-CORBA platform, this wait time, an inherent feature of the algorithm, is even larger because individual t_rs are increased due to the execution of the consensus primitive (see Figure 18.7 for the values of the consensus time);
- in the case of active replication, the average overhead values are generally large. Since the larger overheads appear only in case of method calls that themselves made outgoing calls, it was deduced that it was because of the gateway's duplicate suppression mechanism. Method 3 did not call other methods, and thus experienced a lower overhead.

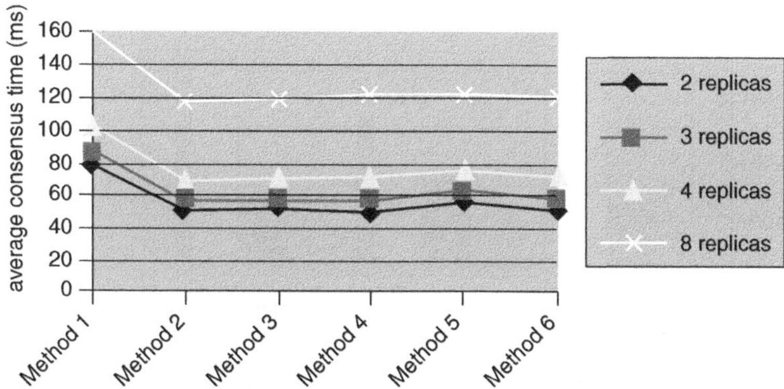

Figure 18.7 Dependence of the average time spent in consensus on the number of replicas

The chart in Figure 18.7 visualizes the variation with the group size of the average time spent performing the consensus for different methods. Here, absolute timing values were used to depict the results. The tendency is that the higher the number of replicas, the higher the time spent in consensus. A clear difference can be seen for eight replicas where the majority was five nodes, as compared with two nodes in case of two or three replicas, and three nodes in case of four replicas, respectively. Also, it can be noticed that for a fixed group size the time spent in consensus stayed fairly constant independently of the called method.

18.7.2 Failover Times

As mentioned in the introduction, support by middleware requires writing little extra code by the application writer, so in that sense it is directly comparable to a non-replicated scenario. In this section, we quantify the benefit to the application writer in terms of shorter failover times (compared to manual restarts). We summarize failover time performance in both platforms and relate to the parameters influencing it.

In the FT-CORBA platform, there is a significant difference between failover times using passive or active replication. In active replication, the failover time was extremely short: 70–75 ms. In passive replication, when using the warm style, the faster failover was due to the fact that no state transfer was needed at the moment of the reconfiguration, as opposed to cold style. In both passive styles, the failover time was dependent on the number of method calls logged since the last checkpoint that had to be replayed on the backup. This dependence was stronger in the warm passive case.

In the FA-CORBA platform, the failover time was dependent on the number of state changes that happened in the application object since the last witness register pruning. These state changes (updates) have to be applied at failover on the application object inside the new leader. Let us consider the operations associated with applying an update on the new leader at failover as similar operations to replaying a method in the FT-CORBA platform. These operations consist of reading the update to be applied while performing the consensus primitive as in no-failure cases, followed by applying the read update on the application object.

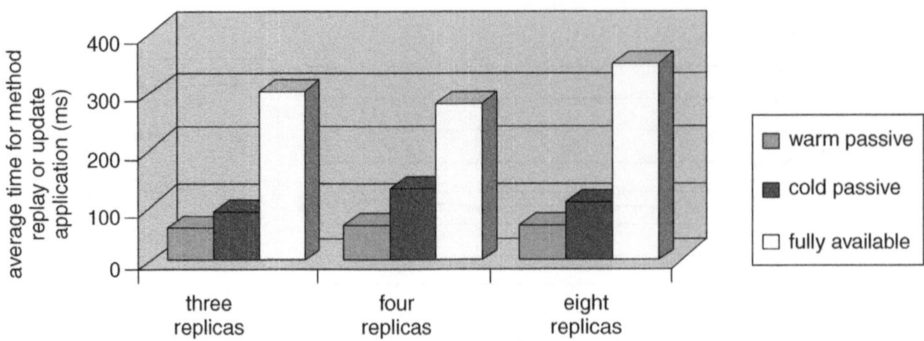

Figure 18.8 Comparison of the average method replay or update application times (in ms)

Figure 18.8 shows how the average time per method replay (FT-CORBA) or update application (FA-CORBA) depends on the group size in a scenario where the server failed after approximately 600 requests, and the checkpointing (respectively pruning) interval was 5 s. The general tendency is that replay time is lowest in warm passive replication, followed by the cold passive replication. The highest average time per individual replay (update application) was experienced in the FA-CORBA platform. The explanations are as follows: first, cold passive recovery includes, besides the time for replaying the methods, the time to set the last checkpointed state on the new replica. Second, the average time for applying the stored state changes in the recovering leader (in FA-CORBA) is higher because each such operation implies execution of the consensus algorithm.

The length of the checkpointing/pruning interval makes a difference in the number of requests to replay (updates to apply). In our experiments, the numbers become larger as the intervals grow.

Figure 18.9 shows the general tendency of the average time to execute one update application in the FA-CORBA platform. This average time grows with the size of the replica group (due to the consensus execution). Here, we show the experiments with a pruning interval of 10s.

Table 18.2 presents examples of absolute failover times for different group sizes (3, 4 or 8) and checkpointing/pruning intervals (5 or 10 s), when the server failed after approximately 400 requests. All values are given in seconds.

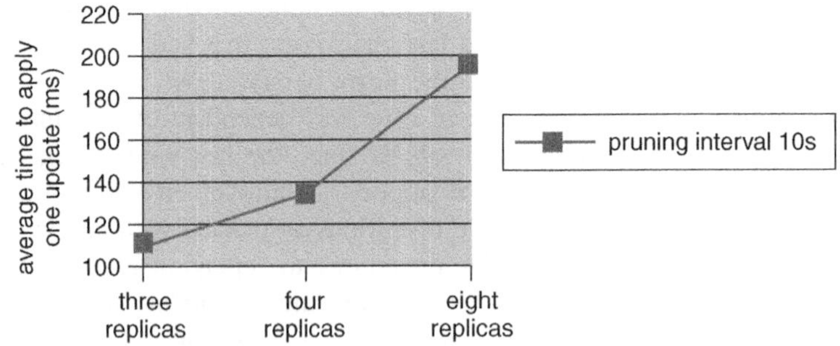

Figure 18.9 Dependence of average time to apply one update on the number of replicas

Table 18.2 Example failover times in different platforms and mechanisms (in s)

	3 replicas		4 replicas		8 replicas	
	5 s	10 s	5 s	10 s	5 s	10 s
Cold passive	4	6	6	10	4	8
Warm passive	4	6	3	4	6	9
Fully available	3	11	1	2	2	4

The above figures are selected from a series of experiments, the results of which are summarized below. The scenarios were generated by forcing failures at different points in the request execution sequence. The measured failover times ranged over intervals

- between 2 and 10 s in cold passive style;
- between 2 and 11 s in warm passive style;
- between 1 and 16 s in the fully available platform.

These values should be compared with the ~70 ms failover time of the active replication. Considering the lower overheads, the higher failover times may well be acceptable in a given application. The next question is whether to pay the larger overhead penalty in the FA-CORBA platform for obtaining the full availability. The answer could be yes, since there is no need to worry about extra failure-prone infrastructure units.

18.8 Conclusions

This chapter presented experiences with building two different platforms supporting fault tolerance in CORBA applications. One of them (called FT-CORBA) deals only with failures of the application. The other (called FA-CORBA) uniformly handles failures of the application as well as the infrastructure, hence labeled fully available. Both platforms provide transparency for the application writer so that only very little change has to be applied in the application run on top of the platforms. The main purpose of the study was to find out whether the prospects of the support from middleware would be appealing to the application writer, considering factors such as performance overheads and gains in failover time. At the time of implementation, there were no ORBs available in full compliance with the CORBA standard or with fully available property. The performance-related insights are obviously affected by the ORB implementation over which the platforms were built. However, we believe that the wisdom from relative comparison of the different replication styles carries over to other ORBs. A different ORB is expected to increase/decrease the overheads in all styles by similar amounts.

Our findings are supported by measurements on a telecom application using both platforms. Again, one must be careful about drawing general conclusions on the basis of this one application. However, all the earlier published studies on the support for fault tolerance based on CORBA had used synthetic experiments for evaluation. We believe that this work is at least a step in the right direction. Given that the FT technology finds its way into new (commercial) ORBs, the work presented should guide the application engineer in methods for profiling new development environments.

Varying the number of replicas has generated conclusions that are likely to carry over to other middleware. The number of replicas made a significant difference only for active replication and the FA styles. For the FA-CORBA platform, the group size made a difference in both overheads and failover times. This was due to the need for executing the consensus primitive in both scenarios. The corresponding notion for state transfer, pruning interval, also influenced the two metrics.

To sum up, providing fault tolerance embedded in middleware and keeping transparency for application writers involves a price. This price is the extra delay experienced by a client while receiving a reply from a replicated server. The price becomes higher in an infrastructure in which failures of nonapplication units are handled together with those of application objects (employing a distributed algorithm).

The experience with building the FT-CORBA platform shows that it is feasible to implement the FT-CORBA standard by extending an existing ORB. On the other hand, with no other failure-handling mechanisms, except the application object related ones, the infrastructure units have to run as processes that do not fail. Ad hoc replication might be employed to cope with the problem. Still, the rigid failure detection mechanism is inherent to this platform: timers are explicit and visible at a very high level.

In contrast, the FA-CORBA platform ensures that failures of infrastructure units do not impair the availability of the application. Besides the higher price paid in terms of overheads, there is another drawback in this approach: a majority of replicated units must be up at all times. This requires some initial hardware investment beyond a redundant pair. Note that if the group has two replicas, none of them may fail! In the FT-CORBA platform, no such restrictions exist. At the extreme (although not really desirable), the system could run with only one replica. Thus, for example, seven out of eight replicas may fail, and the application will still not block, as opposed to the case in FA-CORBA.

The failover comparisons show that active replication in the FT-CORBA platform is unlikely to be worthy of attention because of its high overheads. In our experiments, the extremely short failover time ($\sim$70 ms) cannot be considered essential in a high-assurance system if the no-failure roundtrip time is around 6 s (after adding the overheads). Another limiting factor might be the need to ensure replica determinism. In such a competition, the FA-CORBA platform will win, since nondeterminism of replicas is tolerated. In both cases, the available resources further influences the decision, since the number of replicas in the group gives the maximum number of failures tolerated.

When looking at overheads, the two passive replication styles in FT-CORBA score well: no need for determinism, and failover is in range of seconds. Warm passive replication does not exhibit larger overheads than the cold passive style, but the failover time is lower in this case, especially if the checkpointing interval is low.

The bottom line is that the quest for higher availability lies in the likelihood of failure in the infrastructure units. As long as the costs for hardening such units is not prohibitive, the FT-CORBA platform may be considered as adequate since the overheads are smaller and the failover times have comparable values.

18.9 Acknowledgments

This work is supported by the European Commission IST initiative and is included in the cluster of projects EUTIST-AMI on Agents and Middleware Technologies applied in real industrial environments (www.eutist-ami.org). It was also supported by the European project Safeguard (IST-2001-32685). The FA-CORBA platform was implemented by Isabelle Ravot on a visit at Linköping university from Lausanne. The authors wish to thank Prof. Rachid Guerraoui for valuable discussions during preparations of a poster presentation for Middleware'03 on an earlier version of the FA-CORBA platform that has influenced the presentation in this chapter. Also, our thanks go to Torbjörn Örtengren and Johan Moe from Ericsson Radio Systems for priceless cooperation.

Bibliography

[1] Bakken, D. E. (2002) Middleware, http://www.eecs.wsu.edu/~bakken/middleware.pdf, April 2004.

[2] Chandra, T. D. and Toueg, S. (1996) Unreliable Failure Detectors for Reliable Distributed Systems. *Journal of the ACM*, **43**(2), 225–267.

[3] Chung, P. E., Hung, Y., Yajnik, S., Liang, D., and Shih, J. (1998) DOORS: Providing Fault Tolerance for CORBA Applications. *Poster Session at IFIP International Conference on Distributed Systems Platforms and Open Distributed Processing (Middleware'98)*, The Lake District, England, September 15–18.

[4] Daniel, J., Daniel, M., Modica O., and Wood, C. *Exolab OpenORB webpage*, http://www.exolab.org, August 2002.

[5] Dutta, P., Frølund, S., Guerraoui, R., and Pochon, B. (2002) An Efficient Universal Construction for Message-Passing Systems. *Proceedings of the 16th International Symposium on Distributed Computing*, Toulouse, France, October 28–30.

[6] Felber, P., Guerraoui, R., and Schiper, A. (2000) *Replication of CORBA Objects, Lecture Notes in Computer Science, 1752*, Springer-Verlag, Berlin, pp. 254–276.

[7] Fischer, M. J., Lynch, N. A., and Paterson, M. S. (1985) Impossibility of Distributed Consensus with One Faulty Process. *Journal of the ACM*, **32**(2), 374–382.

[8] Guerraoui, R. and Schiper, A. (1997) Software-Based Replication for Fault Tolerance. *IEEE Computer*, **30**(4), 68–74.

[9] Hadzilacos, V. and Toueg, S. (1993) Fault-Tolerant Broadcasts and Related Problems, in *Distributed Systems*, 2nd edition (ed. S. Mullender), ACM Press Addison-Wesley, New York, pp. 97–147.

[10] IONA. *EJB Level Clustering*, Available at http://www.iona.com/support/docs/e2a/asp/5.1/j2ee/cluster/Intro-cluster9.html, April 2004.

[11] Killijan, M. -O. and Fabre, J. -C. (2000) Implementing a Reflective Fault-Tolerant CORBA System. *Proceedings of the 19th IEEE Symposium on Reliable Distributed Systems (SRDS'00)*, Nürnberg, Germany, October 16–18.

[12] Larsen, J. E. and Spring, J. H. (1999) *GLOBE (Global Object Exchange) A Dynamically Fault-Tolerant and Dynamically Scalable Distributed Tuplespace for Heterogeneous, Loosely Coupled Networks.* Master of Science Dissertation, University of Copenhagen, Denmark.

[13] Montresor, A. and Meling, H. (2002) Jgroup Tutorial and Programmer's Manual, Available as Technical report at http://www.cs.unibo.it/projects/jgroup/papers/2000-13.pdf, April 2004.

[14] Narasimhan, N., Moser, L. E., and Melliar-Smith, P. M. (2000) Transparent Consistent Replication of Java RMI Objects. *Proceedings of the International Symposium on Distributed Objects and Applications (DOA)*, Antwerp, Belgium, September 21–23.

[15] Narasimhan, P., Moser, L. E., and Melliar-Smith, P. M. (1999) Using Interceptors to Enhance CORBA. *IEEE Computer*, **32**(7), 62–68.

[16] Natarajan, B., Gokhale, A., Yajnik, S., and Schmidt, D. C. (2000) DOORS: Towards High-Performance Fault Tolerant CORBA. *Proceedings of the International Symposium on Distributed Objects and Applications (DOA)*, Antwerp, Belgium, September 21-23, pp. 39–48.

[17] Nikander, P. (2000) Fault Tolerance in Decentralized and Loosely Coupled Systems. *Proceedings of Ericsson Conference on Software Engineering*, Stockholm, Sweden, September 13–14.

[18] OMG. *Fault Tolerant CORBA*. Chapter 23 in Common Object Request Broker Architecture: Core Specification, pp. 23-1–23-106, Available at http://www.omg.org/docs/formal/02-12-06.pdf, April 20.

[19] Pasin, M., Riveill, M., and Silva Weber, T. (2001) High-Available Enterprise JavaBeans Using Group Communications System Support. *Proceedings of the European Research Seminar on Advances in Distributed Systems (ERSADS)*, University of Bologna, Italy, May 14-18.

[20] Polze, A., Schwarz, J., and Malek, M. Automatic Generation of Fault-Tolerant CORBA Services. *Proceedings of Technology of Object Oriented Languages and Systems (TOOLS)*, Santa Barbara, CA, July 30–August 3; IEEE Computer Society Press.

[21] Schneider, F. B. (1990) Implementing Fault-Tolerant Services Using the State Machine Approach: A Tutorial. *ACM Computing Surveys*, **22**(4), 299–319.

[22] Szentiványi, D. (2002) *Performance and Availability Trade-Offs in Fault-Tolerant Middleware*. Licentiate Dissertation no. 982, ISBN 91-7373.467-5, Linköping University, Sweden.

[23] Szentiványi, D. and Nadjm-Tehrani, S. (2002) Building and Evaluating a Fault-Tolerant CORBA Infrastructure. *Proceedings of the DSN Workshop on Dependable Middleware-Based Systems*, Washington, DC, June 23-26.

[24] Wang, Y. -M. and Lee, W. -J. (1998) COMERA: COM Extensible Remoting Architecture. *Proceedings of the 4th USENIX Conference on Object-Oriented Technologies and Systems (COOTS)*, Santa Fe, Mexico, April 27–30.

Index

Middleware for Communications. Edited by Qusay H. Mahmoud
 ISBN 0-470-86206-8

Printed and bound by CPI Group (UK) Ltd, Croydon, CR0 4YY
16/04/2025
14658475-0003